The Marine Fauna of the British Isles and North-West Europe

Volume 2

Molluscs to Chordates

The Marine Fauna of the British Isles and North-West Europe

VOLUME 2

Molluscs to Chordates

Edited by

P. J. HAYWARD AND J. S. RYLAND

Marine, Environmental and Evolutionary Research Group,
School of Biological Sciences,
University College of Swansea

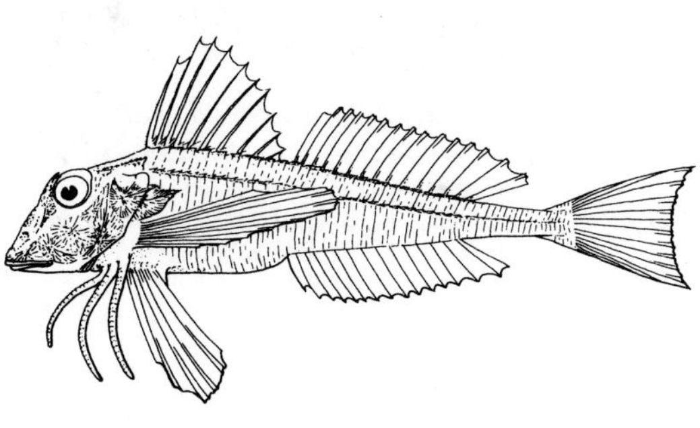

CLARENDON PRESS · OXFORD

1990

Oxford University Press, Walton Street, Oxford OX2 6DP
Oxford New York Toronto
Delhi Bombay Calcutta Madras Karachi
Petaling Jaya Singapore Hong Kong Tokyo
Nairobi Dar es Salaam Cape Town
Melbourne Auckland
and associated companies in
Berlin Ibadan

Oxford is a trade mark of Oxford University Press

Published in the United States
by Oxford University Press, New York

British Library Cataloguing in Publication Data
The Marine fauna of the British Isles and North-West Europe.
Vol. 2 Molluscs to chordates.
1. Europe. North-western Europe. Coastal waters. Marine animals
I. Hayward, P. J. II. Ryland, J. S.
591.916336
ISBN 0–19–857515–7

Library of Congress Cataloging in Publication Data
The Marine fauna of the British Isles and North-West Europe / edited by P. J. Hayward and J. S. Ryland.
p. cm.
Includes bibliographical references and index.
Contents: v. 1. Introduction and protozoans to arthropods—v. 2. Molluscs to chordates.
1. Marine fauna—Great Britain. 2. Marine fauna—Europe.
I. Hayward, P. J. (Peter Joseph) II. Ryland, J. S. (John Stanley)
QL255.M29 1990 591.92′13—dc20 90–7629
ISBN 0–19–857356–1 (v. 1)
ISBN 0–19–857515–7 (v. 2)

Printed in Great Britain by
Butler & Tanner Ltd, Frome, Somerset

PREFACE

THIS work was conceived in 1969 by members of the Zoology Department of the University College of Swansea, during the annual marine field course, then held at Galway, Ireland, over the period of the autumn equinoctial tides. It was an exciting time of expansion for marine biology, in which field studies seemed to have an increasingly important role to play. The traditional marine field class had long been an integral part of zoology teaching at most British universities, but by 1969 marine ecology was a fast-growing discipline attracting more students and researchers than ever before. This was matched by a renascent amateur interest, partly stimulated by the increasing popularity of SCUBA diving, and partly by a new awareness of the need for marine resources management and conservation programmes. Correct identification of species is a necessary prerequisite for any research in the field, or based on specimens collected in the field. Our Galway courses were served by the small but excellent *Guide to the sea shore*, by John Barrett and C. M. Yonge, while the more scholarly, but older and poorly illustrated, *Littoral fauna of the British Isles*, by N. B. Eales, provided dichotomous keys and an introduction to the specialist literature increasingly needed to identify the catch. As more ecologists and students visited an ever widening spectrum of shores, and investigated more habitats, the shortcomings of available identification keys became increasingly obvious. Yet, for many groups the specialist literature, dating from the mid- to late-nineteenth century, employed obsolete and unscientific terminology, and was frequently difficult to use. In the past two decades, stimulated partly by need and partly by the availability of new methods of study, there has been a great deal of systematic revision, and many new insights into the biology of marine animals. The fruits of this work can be seen, for example, in the new series of *Synopses of the British fauna*, published at first by the Linnean Society of London, alone, and later jointly with the Estuarine and Brackish Water Sciences Association. However, there remains an urgent need for an illustrated identification manual which bridges the considerable gap between the semi-popular field guides, a number of which are now available, and the specialist treatises or series of synopses. Moreover, whereas the works of Eales and of Barrett and Yonge dealt solely with intertidal faunas, their successor single volume needs also to cover the common species of the sublittoral. These volumes are designed to fill the gap, providing dichotomous keys, illustrations, brief descriptions, and notes on distributions for the most frequently occurring intertidal and subtidal marine organisms.

The Introduction deals essentially with the marine environment of the British Isles and adjacent areas of the North Sea and north-east Atlantic. Most of the animal species occurring here are found also on the coast of mainland Europe. North of Bergen the marine fauna becomes increasingly Arctic in character, while many north-west European species reach their southern limits on the coast of Brittany. These two points mark the limits of coverage of this book. In some groups a few species known only from either the extreme northern or southern parts of this region have been excluded, at least partly because of continuing uncertainty over their taxonomic identity. For most groups, however, the keys provided will suffice for the identification of all but the rarest coastal species occurring between Bergen and Brittany.

The years, over which preparation of this book have been spread, have seen many changes.

The Zoology Department of the University College of Swansea has now been amalgamated with others into its School of Biological Sciences. The Marine Biology honours BSc degree, instituted in 1975, flourishes, and the new Marine, Environmental and Evolutionary Research Group sustains most of the marine research interests of the School. In Britain as a whole, despite constraints imposed by financial cutback, field studies are regarded as an important component in secondary and tertiary level biology teaching, as well as of numerous higher education programmes, and amateur interest continues to grow. Conservation is perceived as an urgent priority around the coasts and in the shallow seas of western Europe, and it seems that marine environmental and management studies will continue to grow in importance.

With just three exceptions, all of the contributors to this work are, or have been, associated with the former Departments of Zoology and Oceanography, at Swansea. The roll expanded steadily as the book developed, while one or two of the original participants regrettably felt obliged to withdraw their commitment during its long gestation. One of the original few, David N. Huxtable, deserves special mention. A gifted and devoted student of marine biology, David early developed considerable expertise with the Anthozoa, and acquired an impressive knowledge of the species found in south Wales, north Devon, and on the Galway coast. As a graduate student he worked tirelessly on the behaviour of burrowing sea anemones, but made time also to produce a draft key to the British marine hydroids, which he intended to develop into the first modern account of this group. David Huxtable's promising career was tragically ended by a fatal accident in 1974. It is fitting that we remember him now, as our joint venture finally achieves publication.

Swansea P. J. H.
1989 J. S. R.

ACKNOWLEDGEMENTS

The editors wish to express their appreciation of the many, patient undergraduate students who have participated in testing these identification keys through many drafts. We wish to thank the illustrators for their long hours of detailed and painstaking work, and also Mr. M. R. Tootill, course leader in Scientific Illustration at Blackpool and the Fylde College, for his much valued help in encouraging graduate illustrators to take on this task. We are grateful to the Natural Environment Research Council for the generous award of a grant towards illustration costs. Mrs. Margaret Thomas and Mrs. Nancy Taylor gave indispensable support with their swift and efficient typing, and we deeply appreciate their efforts.

In several instances the text has benefited from the critical attention of specialist colleagues from other institutions. The section on Stauromedusae owes much to the freely given expertise of Mr. P. G. Corbin (MBA, Plymouth), who also advised on sandeels. Professor A. A. Myers (University College, Cork) critically reviewed the text for Amphipoda, and Dr. D. M. Holdich (University of Nottingham) suggested improvements to parts of Chapter 9.

Finally, we should like to thank all of the contributors for their work over many years, and special thanks are reserved for the Oxford University Press for its remarkable patience during the long gestation of this book.

The editors are pleased to acknowledge the cited sources from which the following figures have been redrawn:

Fig. 2.1
Prorocentrum lima, Sinophysis ebriola, Thecadinium petasatum, Amphidinium scissum, A. britannicum, A. herdmanae, Gyrodinium lebourae, Oxyrrhis marina, Herdmania litoralis, Katodinium asymmetricum, Glenodinium monensis, Gonyaulax polyedra, Polykrikos lebourae, Adenoides eludens, Amphidiniopsis swedmarkii: after Dodge (1982). *Gymnodinium arenicolus, Pontosphaera roscoffensis, Trachelmonas abrupta*: after Dragesco (1965). *Cyathomonas truncata*: after Hollande (1952). *Tetraselmis convolutae*: after Parke and Manton (1967). *Urceolus cyclostomus, Petalomonas mediocanellata*: after Huber-Pestalozzi (1955). *Sphenomonas elongata, Pentamonas spinifera*: after Lackey (1962).

Fig. 2.6
Stephanopogon colpoda: after Dragesco (1963a). *Tracheloraphis phoenicopterus, Geleia fossata, Remanella rugosa, Prorodon marinus*: after Dragesco (1960). *Kentrophoros latum, Pseudoprorodon arenicola, Conchostoma longissima*: after Fenchel (1969). *Metacystis truncata, Coleps tesselatus, Lachrymaria coronata, Spathidium procerum, Mesodinium pulex, Loxophyllum verrucosum, Litonotus fasciola*: after Kahl (1930–35). *Coelosomides marina*: after Faure-Fremiet (1950).

Fig. 2.7
Scaphidiodon navicula, Discotricha papillifera: after Dragesco (1965). *Paranassula microstoma, Chlamydodon triquetrus, Dysteria aculeata*: after Kahl (1930–35). *Hypocoma acinetarum*: after Deroux (1975). *Heliochona sessilis*: after Fenchel (1965*b*). *Ancistrocoma pelseneeri*: after Kozloff (1946). *Conidophrys pilisuctor, Foettingeria actiniarum, Gymnodiniodes inkystans, Chromidina elegans*: after Chatton and Lwoff (1935). *Ephelota gemmipara, Acineta tuberosa, Discophyra lyngbyei*: after Kahl (1933–34).

Fig. 2.8
Frontonia marina, Uronema marinum, Platynematum denticulatum, Cyclidium glaucoma, Vorticella nebulifera, Zoothamnium alternans, Scyphidia patellae, Pyxicola ligiae, Leiotrocha patellae, Trichodina ophiotrichis: after Kahl (1930–35). *Pleuronema coronatum*: after Dragesco (1960). *Peniculostoma mytili, Ancistrum mytili, Hemispeira antedonis*: after Fenchel (1965*b*). *Anoplophryopsis ovata*: after de Puytorac (1954). *Epistylis gammari*: after Fenchel (1965*b*).

Fig. 2.9
Blepharisma clarissimum, Metopus contortus, Condylostoma arenarium, Peritromus faurei, Caenomorpha levanderi, Licnophora auerbachi, Saprodinium halophilum, Urostrongylum caudatum, Holosticha arenicola, Trachelostyla pediculiformis, Aspidisca pulcherrima: after Kahl (1930–35). *Lagotia viridis*: after Das (1949). *Strombidium sauerbreyae*: after Faure-Fremiet (1960). *Tintinnopsis beroidea*: after Marshall (1969). *Diophrys scutum*: after Dragesco (1963*b*).

Fig. 4.3
Diagrams mainly after Millard (1975).

Fig. 4.5
Ectopleura dumortieri colony: after Lagardère and Tardy (1980, *Cah. Biol. mar.*, **21**). *Hybocodon prolifer*: after Agassiz (1862). *Sarsia loveni, Sarsia tubulosa*: after C. Edwards (1978, *JMBA*, **58**).

Fig. 4.6
Podocoryne borealis, P. carnea: after C. Edwards (1977, *JMBA*, **52**).

Fig. 4.7
Rhizorhagium roseum: after Russell (1907) and W. J. Rees (1938, *JMBA*, **23**). *Garveia nutans, Dicoryne conferta*: after Allman (1871). *Dicoryne conybeari*: after Ashworth and Ritchie (1917). *Leuckartiara octona*: after W. J. Rees (1938, *JMBA*, **23**).

Fig. 4.8
Bougainvillia ramosa: after Allman (1871). *B. pyramidata, B. britannica*: after C. Edwards (1964, *JMBA*, **44**). *B. muscoides*: after W. J. Rees (1938, *JMBA*, **23**).

Fig. 4.9
Proboscidactyla stellata: after Hincks (1872). *Eudendrium capillare*: after Allman (1871).

Fig. 4.22
Monotheca obliqua: ♀ gonotheca after Lister (1838); ♂ gonotheca after J.-M. Gili i Sarda (1982, *Tret. Inst. Catalana Hist. Nat.*, **10**).

Fig. 4.23
Polyplumaria flabellata: gonotheca after Billard (1906). *Schizotricha frutescens*: gonotheca after Vervoort (1946, *Fauna Ned.*).

Fig. 4.24
Aglaophenia acacia: colony after Allman (1883).

Fig. 4.26
Alcyonium glomeratum: after Manuel (1981).

Fig. 4.27
Swiftia pallida: polyp after Manuel (1981).

Fig. 6.30
Branchellion borealis, Calliobdella lophii, Calliobdella nodulifera, Ganymedebdella cratere, Janusion scorpii: after Leigh-Sharpe (1914, 1915, 1917, 1933). *Brumptiana lineata*: after Llewellyn and Knight-Jones (1984). *Branchellion torpedinis*: after Harding (1910). *Hemibdella soleae*: after Llewellyn (1965). *Heptacyclus myoxocephali, Malmiana bubali*: after Srivastra (1966).

Fig. 6.31
Malmiana yorki: after Srivastra (1966). *Pontobdella muricata*: after Llewellyn (1966). *Pontobdella vosmaeri*: after Llewellyn (1966) and Harding (1910). *Platybdella anarrhichae*: after Leigh-Sharpe (1916). *Pterobdellina jenseni*: after Bennicke and Bruun (1939). *Sanguinothus pinnarum*: after de Silva and Burdon-Jones (1961).

Fig. 7.1
Echiurus echiurus, Bonellia viridis, Maxmuelleria lankesteri: after Stephen and Edmonds (1972). *Halicryptus spinulosus, Priapulus caudatus, Thalassema thalassemum*: after Shipley (1910). *Amalosoma eddystonense*: after Saxena (1986). *Siboglinum* sp.: after A. J. Southward *et al.* (1986). *Siboglinum fjordicum*: after Webb (1963). *Siboglinum holmei*: after E. C. Southward (1963). *Siboglinum norvegicum*: after Ivanov (1963).

Fig. 7.2
Aspidosiphon muelleri, Golfingia elongata, Golfingia vulgaris, Nephasoma minuta, Nephasoma rimicola, Onchnesoma squamatum Phascolion strombi, Phascolosoma granulatum, Sipunculus nudus. Thysanocardia procera: after Gibbs (1977).

Fig. 7.3
Loxosomella nitschei: after Ryland (1961, *Sarsia*, **1**). *L. antedonis*: after Ryland and A. P. Austin (1960, *Proc. Zool. Soc. Lond.*, **133**). *Barentsia mutabilis*: after Ryland (1961, *Ann. Mag. nat. Hist.*, **13** (3)). Other *Barentsia* spp. drawn from photographs in P. Emschermann (1972), unpublished, 'Key for the determination of the Barentsiidae of the World'. *Pedicellina cernua*: colony after Brien (1959, *Traite de Zoologie* **5**(1)). Other *Pedicellina species*: after Ryland (1965, *Trans. Roy. Soc. N.Z.*, **6**).

Fig. 8.4
All figures after Sars (1903, 1918).

Fig. 8.5
All figures after Sars (1911).

Fig. 8.6
All figures after Sars (1911).

Fig. 8.7
All figures after Sars (1911).

Fig. 8.8
All figures after Sars (1911).

Fig. 8.9
Botrillophilus brevipes, Notodelphys allmani, Doropygus pulex, Enterocola bilamellata, Ascidicola rosea, Notopterophorus papilio, Modiolicola insignis, Lichomolgus agilis, Sabelliphilus elongatus, Eunicicola clausi: after Sars (1921). *Thespesiopsyllus paradoxus*: after J. Bresciani and J. Lützen (1962, *Vidensk. Medd. dansk. naturh. Foren.*, **124**). *Splanchnotrophus brevipes*: after A. Hancock and A. M. Norman (1863, *Trans. Linn. Soc.*, **24**). *Leptinogaster histrio*: after C. Bocquet and J. H. Stock (1958, *Arch. zool. exp. gen.*, **96**). *Ergasilus gibbus, Holobomolochus confusus, Acanthochondria cornuta*: after Kabata (1979).

Fig. 8.10
Artotrogus orbicularis, Cancerilla tubulata: after Sars (1921). *Lernaeocera branchialis*: after T. Scott and A. Scott (1913, *British Parasitic Copepoda*, London). *Hatschekia cluthae, Caligus elongatus, Lepeophtheirus pectoralis, Clavella adunca, Haemobaphes cyclopterina*: after Kabata (1979). *Nicothoe astaci, Sarsilenium crassirostris, Eurysilenium truncatum*: after W. H. Leigh-Sharpe (1928, *Parasitology*, **17**). *Lamippe rubra*: after J. Bresciani and J. Lützen (1962, *Vidensk. Medd. dansk, naturh. Foren.*, **124**). *Aspidoecia normani*: after H. J. Hansen (1897), *Choniostomatidae*, Copenhagen). *Phyllocola petiti*: after D.-Deboutteville and Laubier (1960). *Thaumaleus rigidus*: after Malaquin (1901). *Thaumaleus malaquini*: after M. Caullery and F. Mesnil (1914), *Bull. Scient. Fr. Belg.*, **48**). *Monstrilla*

helgolandica: after P. Pelseneer (1914, *Bull. Scient. Fr. Belg.*, 48).

Fig. 8.11
Lepas hilli, L. anserifera: after Broch (1960).

Fig. 8.12
Chthalamus montagui, C. stellata: partly after Southward (1976).

Fig. 8.14
All figures after Hoeg and Lutzen (1985).

Fig. 8.15
All figures after Hoeg and Lutzen (1985).

Fig. 8.16
All figures after Hoeg and Lutzen (1985).

Fig. 9.8
All figures after Tattersall and Tattersall (1951).

Fig. 9.9
All figures after Tattersall and Tattersall (1951).

Fig. 9.17
Acidostoma obesum, Hippomedon denticulatus, Lysianassa ceratina, Lysianassa plumosa, Nannonyx goesi, Lepidepecreum longicorne: after Lincoln (1979).

Fig. 9.18
Perriella audouiniana, Socarnes erythrophthalmus, Scopelocheirus hopei, Tryphosella sarsi, Orchomene nana, Orchomene humilis: after Lincoln (1979).

Fig. 9.19
Tmetonyx cicada, Tryphosites longipes, Ampelisca brevicornis, Ampelisca gibba, Ampelisca macrocephala: after Lincoln (1979).

Fig. 9.20
Ampelisca spinipes, Ampelisca typica, Ampelisca tenuicornis, Haploops tubicola: after Lincoln (1979). *Iphimedia obesa, Iphimedia minuta*: after Myers, McGrath and Costello (1987).

Fig. 9.21
Iphimedia perplexa, Iphimedia spatula: after Myers, McGrath and Costello (1987). *Pereionotus testudo, Amphilochus manudens, Amphilochus neapolitanus, Amphilochus spencebatei, Colomastix pusilla*: after Lincoln (1979).

Fig. 9.22
Gitana sarsi, Peltocoxa damnoniensis, Paramphilochoides odontonyx, Leucothoe spinicarpa: after Lincoln (1979). *Leucothoe incisa, Leucothoe lilljeborgi*: after Myers and Costello (1986).

Fig. 9.23
Metopa alderi, Metopa pusilla, Stenothoe monoculoides, Stenothoe marina, Stenothoe valida, Stenula rubrovittata, Parametopa kervillei: after Lincoln (1979).

Fig. 9.24
Orchestia gammarellus, Orchestia mediterranea, Talorchestia deshayesii, Talorchestia brito: after Lincoln (1979).

Fig. 9.25
Hyale nilssoni, Hyale perieri, Hyale pontica, Chaetogammarus marinus, Chaetogammarus pirloti, Chaetogammarus stoerensis: after Lincoln (1979).

Fig. 9.26
Eulimnogammarus obtusatus, Gammarus chevreuxi, Gammarus crinicornis, Gammarus duebeni, Gammarus finmarchicus, Gammarus insensibilis: after Lincoln (1979).

Fig. 9.27
Gammarus locusta, Gammarus oceanicus, Gammarus salinus, Gammarus tigrinus, Gammarus zaddachi, Pectenogammarus planicrurus: after Lincoln (1979).

Fig. 9.28
Abludomelita obtusata, Allomelita pellucida, Ceradocus semiserratus, Cheirocratus assimilis, Cheirocratus intermedius, Cheirocratus sundevallii: after Lincoln (1979).

Fig. 9.29
Elasmopus rapax, Gammarella fucicola, Maera grossimana, Maera othonis, Maerella tenuimana, Melita gladiosa: after Lincoln (1979).

Fig. 9.30
Melita palmata, Melita hergensis, Bathyporeia elegans, Bathyporeia guilliamsoniana, Bathyporeia nana, Bathyporeia pelagica: after Lincoln (1979).

Fig. 9.31
Bathyporeia pilosa, Bathyporeia sarsi, Bathyporeia tenuipes, Urothoe elegans, Urothoe marina, Urothoe poseidonis, Urothoe brevicornis: after Lincoln (1979).

Fig. 9.32
Argissa hamatipes, Monoculodes carinatus, Perioculodes longimanus, Pontocrates altamarinus, Pontocrates arenarius, Synchelidium haplochelės, Westwoodilla caecula: after Lincoln (1979).

Fig. 9.33
Harpinia antennaria, Harpinia pectinata, Harpinia crenulata, Metaphoxus fultoni, Metaphoxus pectinatus, Phoxocephalus holbolli: after Lincoln (1979).

Fig. 9.34
Megaluropus agilis, Melphidippella macra, Liljeborgia kinahani, Liljeborgia pallida, Eusirus longipes, Apherusa bispinosa, Apherusa cirrus: after Lincoln (1979).

Fig. 9.35
Apherusa clevei, Apherusa jurinei, Apherusa henneguyi, Gammarellus homari, Gammarellus angulosus, Calliopius laeviusculus: after Lincoln (1979).

Fig. 9.36
Parapleustes bicuspis, Stenopleustes nodifer, Atylus falcatus, Atylus swammerdami, Atylus guttatus, Atylus vedlomensis: after Lincoln (1979).

Fig. 9.37
Dexamine spinosa, Dexamine thea, Guernea coalita, Tritaeta gibbosa, Ampithoe rubricata, Ampithoe ramondi: after Lincoln (1979).

Fig. 9.38
Ampithoe gammaroides, Ampithoe helleri, Amphitholina cuniculus, Sunamphitoe pelagica, Lembos longipes, Lembos websteri: after Lincoln (1979). *Aora gracilis*: after Myers and Costello (1984).

Fig. 9.39
Leptocheirus pilosus, Leptocheirus hirsutimanus, Microdeutopus anomalus, Microdeutopus chelifer,

Microdeutopus gryllotalpa, *Microdeutopus versiculatus*: after Lincoln (1979).

Fig. 9.40
Gammaropsis maculata, *Gammaropsis nitida*, *Gammaropsis palmata*, *Gammaropsis sophiae*, *Isaea elmhirsti*, *Isaea montagui*: after Lincoln (1979).

Fig. 9.41
Photis longicaudata, *Megamphopus cornutus*, *Microprotopus maculatus*, *Protomedeia fasciata*: after Lincoln (1979). *Photis pollex*, *Photis reinhardi*: after Myers and McGrath (1981).

Fig. 9.42
Corophium acherusicum, *Corophium acutum*, *Corophium affine*, *Corophium arenarium*, *Corophium bonelli*, *Corophium crassicorne*: after Lincoln (1979).

Fig. 9.43
Corophium insidiosum, *Corophium lacustre*, *Corophium sextonae*, *Corophium multisetosum*, *Corophium volutator*: after Lincoln (1979). *Siphonoecetes kroyeranus*, *Siphonoecetes striatus*: after Myers and McGrath (1979).

Fig. 9.44
Unciola crenatipalma, *Unciola planipes*, *Chelura terebrans*: after Lincoln (1979). *Ericthonius fasciatus*, *Ericthonius rubricornis*, *Ericthonius punctatus*, *Ericthonius difformis*: after Myers and McGrath (1984).

Fig. 9.45
Ischyrocerus anguipes, *Jassa falcata*, *Jassa ocia*, *Jassa marmorata*, *Jassa pusilla*, *Parajassa pelagica*, *Microjassa cumbrensis*: after Lincoln (1979).

Fig. 9.46
Dyopedos monacanthus, *Dyopedos porrectus*, *Dulichia fulcata*, *Podocerus variegatus*: after Lincoln (1979). *Hyperia galba*: after Sars (1894).

Fig. 10.11
All figures after Selbie (1915).

Fig. 10.22
All figures after Ingle (1980).

Fig. 10.24
Cancer bellianus: after Christiansen (1969).

Fig. 10.27
Pilumnus spinifer: after Christiansen (1969).

Fig. 10.28
Geryon affinis: after Christiansen (1969).

Fig. 11.22
Polyxenus lagurus, *Thalassisobates littoralis*, *Ommatoiulus sabulosus*, *Brachyiulus pusillus*: after Blower (1985). *Necrophloeophagus longicornis*, *Hydroschendyla submarina*, *Geophilus fucorum seurati*, *Strigamia maritima*: after Eason (1964).

Fig. 12.3
Partly after Fretter and Graham (1962).

Fig. 12.20
Aplysia punctata, *Berthella plumula*, *Alderia modesta*, *Elysia viridis*, *Limapontia capitata*, *Tritonia hombergi*, *Cuthona caerulea*, *Polycera quadrilineata*: after Thompson and Brown (1976).

Fig. 12.21
Colpodaspis pusilla, *Retusa obtusa*, *R. truncatula*, *R. umbilicata*, *Roxania utriculus*, *Cylichna cylindracea*, *Scaphander lignarius*: whole animals after Thompson and Brown (1976).

Fig. 12.22
Philine aperta, *P, scabra*, *P, catena*, *P, punctata*, *Berthella plumula*: whole animals after Thompson and Brown (1976).

Fig. 12.24
Alderia modesta, *Hermaea dendritica*, *H. bifida*, *Limapontia capitata*, *L, senestra*, *L, depressa*: after Thompson and Brown (1976).

Fig. 12.26
Goniodoris castanea, *G. nodosa*: after Thompson and Brown (1976).

Fig. 12.27
Palio dubia: after Thompson and Brown (1976).

Fig. 12.28
Rostanga rubra, *Jorunna tomentosa*: after Thompson and Brown (1976).

Fig. 12.29
Embletonia pulchra: after Thompson and Brown (1976).

Fig. 12.32
Tergipes tergipes, *Tenellia adspersa*: after Thompson and Brown (1976).

Fig. 13.18
Sepiola aurantiaca, *Sepietta oweniana*, *S. neglecta*: after Jaeckel (1968).

Fig. 13.20
Todaropsis eblanae, *Todarodes sagittatus*: after B. J. Muus (1963, *Fisches d'identification du zooplancton*, **96**).

Fig. 14.1
Phoronis hippocrepia (top left) and *P. muelleri*: after Cori (1932, 1939). *P. hippocrepia*: whole animal and LS lophophore base (top right) after C. Dawydoff (1959, *Traite de Zool.*, **5**(1)). *P, psammophila*: after Selys-Longchamps (1907). *P. ovalis* (left): after Harmer (1917, *Q. J. micros. Sci.*, **62**). *P. ovalis* lophophore: after Cori (1932, 1939). *P. pallida*: after Silen (1952, *Ark. Zool.*, **4**(4)). All nephridia after C. C. Emig (1974, *Z. zool. Syst. Evolut.-forsch.*, **12**).

Fig. 14.5
All figures after Davidson (1886–88).

Fig. 15.1
Partly after Dales (1969 (ed) *Practical Invertebrate Zoology*. London)

Fig. 15.2
Partly after Dales (1969 (ed) *Practical Invertebrate Zoology*. London)

Fig. 15.10
Globiferous pedicellariae ossicles: after Mortensen (1927).

Fig. 16.1
Figure A: after Stebbing (1970A); Figures B–F, and L: after Horst (1927–39); Figure G: after Brambell and

Cole (1939); Figure M: after Burdon-Jones and McIntyre (1960).

Fig. 16.2
Open body diagrams after P. Mather (1984, 'A Coral reef handbook', Brisbane).

Fig. 16.3
Diagrams after P. Mather (1984, 'A Coral reef handbook', Brisbane).

Fig. 16.4
Diazona colony: after Thompson (1933, *Fish, Scotl. Sci. Invest.*). *Polycitor*: after Kott (1952). All others: after Berrill (1950).

Fig. 16.6
All after Berrill (1950), except *A. densum, A. 'nordmanni'*: after Medioni (1970b); *A. pallidum*: after Van Name (1945, *Bull. Am. Mus. nat. Hist.*, **84**).

Fig. 16.7
Perophora: after Lister (1834). *D. spongiforme*: zooid clusters after Kott (1952). *A. obliqua*: after Alder and Hancock (1905). *A. prunum*: after Van Name (1945, *Bull. Am. Mus. nat. Hist.*, **84**).

Fig. 16.8
Didemnid zooids all after Lafargue (1968, 1975*a*, 1976), except *D. maculosum*: after Medioni (1970*a*).

Fig. 16.9
Didemnid zooids all after Lafargue (1968, 1975*b*), except *Polysyncraton bilobatum* (right): after Medioni (1970*a*); *Leptoclinides faeroensis*: after Millar (1966).

Fig. 16.11
Whole ascidians *S. coriacea, C. mollis, P. fibrosa* and *P. gracilis*: after Alder and Hancock (1907); body walls: after Van Name (1945, *Bull. Am. Mus. nat. Hist.*, **84**), except *P. pomaria, P. rustica* and *P. gracilis*: after Berrill (1950).

Fig. 16.12
Body walls of *D. variolosus* and *S. socialis*: after Lacaze-Duthiers and Delage. Body wall of *D. grossularia*: after Van Name (1945, *Bull. Am. Mus. nat. Hist.*, **84**).

Fig. 16.13
Whole ascidians *M. claudicans, P. microcosmos* (upper), *P. tessellata, P. squamulosa* and *B. echinata* after Alder and Hancock (1907). *P. microcosmos* (lower) and *E. arenosa* after Thomson (1932, *Fish. Scotl. Sci. Invest.*). Body walls of *M. claudicans* and *P. microcosmos* after Berrill (1950). *P. squamulosa* body wall: after Lacaze-Duthiers and Delage. Remainder after Lacaze-Duthiers.

Fig. 17.15
Callionymus reticulatus: after H. W. Chang (1951, *JMBA.*, **30**).

ILLUSTRATORS

GRAHAM AUSTIN
Blackpool and the Fylde College
Figs. 8.4–8.8, 9.26–9.46.

F. T. BANNER
Dept. of Geological Sciences, University College, London
Figs. 2.2–2.5.

SIMON CHEW
Blackpool and the Fylde College
Figs. 6.2–6.18, 16.6, 16.7, 16.10–16.12.

DANIEL COLE
Blackpool and the Fylde College
Figs. 6.2–6.18, 16.5.

NIGEL GERKE
Carmarthen School of Art
Figs. 12.4–12.19.

R. GIBSON
Dept. of Biology, Liverpool Polytechnic
Figs. 5.1–5.4.

TONI HARGREAVES
Blackpool and the Fylde College
Figs. 15.1–15.5, 15.10, 15.11.

P. J. HAYWARD
School of Biological Sciences, University College of Swansea
Figs. 1.1–1.5, 3.1–3.11, 4.2, 6.1, 6.28, 6.29, 8.1, 8.9, 9.10–9.15, 11.1, 11.22, 11.29, 12.1–12.3, 12.20–12.32, 13.1, 13.18–13.20, 14.2–14.4, 14.13, 14.14, 16.2–16.4 16.8, 16.9.

PHYLLIS KNIGHT-JONES
School of Biological Sciences, University College of Swansea
Figs. 6.19–6.27.

PAUL J. LLEWELLYN
School of Biological Sciences, University College of Swansea
Figs. 4.1, 4.3–4.31, 6.2, 6.3, 6.9, 6.11, 6.30, 6.31, 7.1, 7.2, 8.2, 8.3, 8.10–8.16, 9.1–9.7, 9.16–9.25, 9.27–9.48, 10.1–10.28, 11.2–11.20, 11.23–11.25, 11.27, 11.28, 11.30–11.32, 13.2–13.17, 15.1, 15.2, 15.6–15.9, 15.12–15.14, 16.1, 16.5, 16.10, 16.11, 16.13, 17.1–17.18.

JULIET POWELL
Carmarthen School of Art
Figs. 2.1, 2.6–2.9.

P. J. A. PUGH
School of Biological Sciences, University College of Swansea
Figs. 11.2–11.16.

J. S. RYLAND
School of Biological Sciences, University College of Swansea
Figs. 7.3, 14.1, 14.5–14.12, 14.15.

NATHALIE YONOW
School of Biological Sciences, University College of Swansea
Figs. 9.8, 9.9, 11.17–11.21.

CONTENTS

CONTRIBUTORS

F. T. BANNER
Dept. of Geological Sciences, University College London
Chapter 2, Sarcodina.

P. F. S. CORNELIUS
Dept. of Zoology, British Museum (Natural History), London
Chapter 4, Hydrozoa (with J. S. Ryland).

P. E. J. DYRYNDA AND ELIZABETH A. DYRYNDA
School of Biological Sciences, University College of Swansea
Chapter 3.

P. ELLIOTT
School of Biological Sciences, University College of Swansea
Chapter 11, Coleoptera (with P. E. King).

R. GIBSON
Dept. of Biology, Liverpool Polytechnic
Chapter 5, Nemertea (with E. W. Knight-Jones).

P. J. HAYWARD
School of Biological Sciences, University College of Swansea
Chapter 1; Chapter 12, Polyplacophora and Scaphopoda; Chapter 13, Bivalvia.

M. J. ISAAC
Dept. of Extra Mural Studies, University College of Swansea
Chapter 8, Ostracoda and Copepoda; Chapter 9, Leptostraca, Cumacea, and Tanaidacea.

P. E. KING
School of Biological Sciences, University College of Swansea
Chapter 11, Pycnogonida, Insecta (with P. Elliott, Alison Smith and Sarah L. A. Wheeler), Tardigrada (with C. I. Morgan).

E. W. KNIGHT-JONES
School of Biological Sciences, University College of Swansea
Chapter 2, Mastigophora, Ciliophora (with J. M. Wright);
Chapter 5, Nemertea (with R. Gibson); Chapter 6, Hirudinea;
Chapter 7, Priapulida, Sipuncula, Echiura, and Pogonophora;
Chapter 16, Hemichordata.

PHYLLIS KNIGHT-JONES
School of Biological Sciences, University College of Swansea
Chapter 6, Sabellariidae, Sabellidae, Serpulidae, and Spirorbidae.

P. MAKINGS
School of Biological Sciences, University College of Swansea
Chapter 9, Mysidacea.

R. L. MANUEL
Dept. of Zoology, University of Oxford
Chapter 4, Anthozoa.

C. I. MORGAN
Field Studies Council Research Centre, Popton Fort
Chapter 11, Tardigrada (with P. E. King).

J. MOYSE
School of Biological Sciences, University College of Swansea
Chapter 8, Cirripedia;
Chapter 10, Reptantia; Chapter 15, Echinoidea, Asteroidea.

E. NAYLOR
School of Ocean Sciences, University College of North Wales
Chapter 9, Isopoda.

A. NELSON-SMITH
School of Biological Sciences, University College of Swansea
Chapter 6, Polychaeta (except for Sabellariidae, Sabellidae, Serpulidae, and Spirorbidae), Oligochaeta.

P. J. A. PUGH
School of Biological Sciences, University College of Swansea
Chapter 11, Arachnida.

J. S. RYLAND
School of Biological Sciences, University College of Swansea
Chapter 4, Scyphozoa, Hydrozoa (with P. F. S. Cornelius);
Chapter 7, Entoprocta; Chapter 14; Chapter 16, Urochordata;
Chapter 17.

G. SMALDON
formerly University College of Swansea
Chapter 9, Caprellidea; Chapter 10, Natantia.

ALISON SMITH
School of Biological Sciences, University College of Swansea
Chapter 11, Diptera (with P. E. King).

P. A. TYLER
Dept. of Oceanography, University College of Swansea
[*present address: Dept. of Oceanography, University of Southampton*]
Chapter 15, Ophiuroidea, Holothurioidea.

SARAH L. A. WHEELER
School of Biological Sciences, University College of Swansea
Chapter 11, Collembola (with P. E. King).

G. D. WIGHAM
Department of Biological Sciences, Polytechnic South-West
Chapter 12, Prosobranchia.

R. G. WITHERS
formerly University College of Swansea
Chapter 9, Gammaridea.

J. M. WRIGHT
formerly University College of Swansea
Chapter 2, Ciliophora (with E. W. Knight Jones).

NATHALIE YONOW
School of Biological Sciences, University College of Swansea
Chapter 12, Opisthobranchia, Pulmonata.

MOLLUSCS
TO
CHORDATES

12 MOLLUSCA I
POLYPLACOPHORA, SCAPHOPODA, AND GASTROPODA

PHYLUM MOLLUSCA

THE molluscs constitute one of the largest groups of marine organisms and, with the exception of the Monoplacophora, representatives of all classes may be found in British waters. The basic molluscan body plan has responded adaptively to environmental pressures and opportunities, perhaps more so than any other phylum of marine organisms, and has been profoundly modified in many different ways. Living molluscs occur in terrestrial, freshwater, and marine ecosystems, and in the latter case have become adapted to almost every kind of habitat from abyssal oozes to oceanic surface currents. Nevertheless, the distinctive molluscan morphology is recognizable in each of the many disparate living orders.

Molluscs are unsegmented, bilaterally symmetrical animals. The body comprises typically a muscular *foot*, a variously developed *head*, and a soft, non-muscular *visceral mass*. The foot may be adapted in different ways for grasping the substratum, for locomotion, burrowing, or feeding, and is frequently closely associated with the head. In all Gastropoda the head is usually well-developed, with paired sensory organs and often specialized feeding apparatus. Within the buccal cavity lies a complex, eversible feeding structure, the *odontophore*, associated with a chitinous, toothed ribbon, the *radula*. Adaptive radiation in the Gastropoda is reflected in feeding methods, and consequent modifications of the radula. In the Bivalvia the head has regressed, and feeding and sensory functions are performed by other parts of the body. In cephalopods the head and foot are not distinguishable separately; the two parts of the body are integrated in the anterior, tentacle-bearing end of the animal, which displays marked cephalization and advanced sensory development.

The main concentration of body organs is contained within a soft, non-muscular visceral mass, primitively situated dorsally towards the posterior end of the animal. Around the periphery of the visceral mass a wide fold of body-wall constitutes the *mantle*, or *pallium*, another major characteristic of the Mollusca. The mantle edge is responsible for secreting the calcareous external shell, characteristic of the molluscs, although secondarily absent in many families and orders. The shell is three-layered, with an uncalcified, organic outer layer, the *periostracum*, a middle layer of columnar calcite, and an inner laminated layer, often nacreous. The mantle skirt covers the whole body of the animal and typically incorporates a cavity, the *mantle* or *pallial cavity*, considered to be primitively posterior in position. The mantle cavity has particular functional and evolutionary significance to the Mollusca. In primitive gastropods it contains the anus and the paired openings of the gonads and kidneys. It is also the site of respiration, containing, primitively, paired gills, or *ctenidia*, together with sensory organs, *osphradia*, and a mucus-secreting *hypobranchial* gland. Circulation and sanitation within the pallial cavity are achieved through the action of cilia and mucus-secreting cells. Modifications of mantle cavity functions and morphology are adaptive features of many molluscan groups. In the sedentary bivalves, in which the entire animal comes to lie within an enlarged mantle cavity, the ciliary/mucus system of gill cleansing has been adapted to allow microphagous feeding; the head has largely disappeared as a consequence, the mouth being equipped with pronounced *labial palps* for the reception and ingestion of detritus-laden mucus.

In all molluscs the coelom is reduced to a cavity surrounding the heart and the gonads. The blood circulatory system is open, and both blood and coelomic fluid circulate freely through extensive haemocoelic spaces in the body of the animal, which serve as an efficient hydrostatic skeleton. In primitive molluscs sexes are separate, with external fertilization; embryos develop as planktonic *trochophore* larvae. Although most marine molluscs are still dioecious, hermaphroditism is the rule in terrestrial groups; simultaneous or consecutive hermaphrodites are known among certain families of marine molluscs, while the subclass Opisthobranchia is entirely monoecious. Fertilization and developmental processes vary widely throughout the phylum.

CLASSIFICATION OF THE PHYLUM MOLLUSCA

Class **Monoplacophora**

Limpet-like animals with a conical, dorsal shell and a broad, ventral foot. Head conspicuous, although lacking eyes and tentacles. Anus posterior, median. Gills in paired, lateral series within a peripheral pallial groove. A single genus, *Neopilina*, with a number of species known from abyssal zones in the Pacific, Atlantic, and Indian Oceans.

Class **Polyplacophora**

Chitons, or coat-of-mail shells. A single living order, Neoloricata, contains about 600 species worldwide. Animal oval, with a dorsal shell of eight overlapping plates embedded peripherally in a toughened girdle of mantle; foot broad, ventral. Head lacking eyes and tentacles. Gills variable in number, paired, lateral, within a peripheral pallial groove. Anus posterior, median. Entirely marine; mostly limited to shallow coastal waters.

Class **Aplacophora**

Elongate, worm-like molluscs lacking a shell, the body usually completely enclosed by the mantle. From coastal waters to abyssal depths, typically infaunal in soft substrata. Approximately 250 species worldwide.

Class **Scaphopoda**

Body elongate, enclosed within a cylindrical shell, open at both ends. Head highly modified, with numerous sensory and captorial tentacles. Mantle cavity extending along whole of ventral surface. Gills lacking. Infaunal, from coastal waters to abyss; 350 species worldwide.

Class **Gastropoda**

The largest class of living molluscs, including more than 60 000 species in marine, freshwater, and terrestrial habitats. Body asymmetrical. Head well-developed, with eyes and sensory tentacles. Visceral mass rotated through 180° in early embryogeny, a process known as torsion, giving the looped gut and nervous system, and anteriorly situated mantle cavity characteristic of the class. Secondary detorsion occurs in some groups. Shell of one piece, primatively conical, typically spiralled. Ventral foot well-developed, used for crawling, burrowing, and swimming. Feeding modes diverse.

Subclass **Prosobranchia**

Gastropods displaying torsion, with marked subsequent modification of the anterior mantle cavity. Respiration conducted via single (primitively paired) gill within the mantle cavity. Head with pronounced snout, tentacles, and eyes. Shell typically coiled, often massively developed. Feeding and reproductive modes highly varied. Mostly marine, occurring abundantly from the supralittoral to the abyss.

Subclass **Opisthobranchia**

Gastropods with varying degree of detorsion. Shell rarely conspicuous, typically reduced, and lacking entirely in a majority of families. Respiration achieved via a single gill in a posteriorly located mantle cavity, or via newly developed structures such as pallial gills, pseudobranchs, or cerata. Mostly hermaphroditic. Herbivores and carnivores occur, often highly specialized. Entirely marine.

Subclass **Pulmonata**

Gastropods with partly or wholly detorted body; shelled or shell-less. Gills lacking; respiration carried out by the mantle cavity, which functions as a lung. Entirely hermaphroditic. Mostly freshwater or terrestrial, with a few marine species occurring in shallow coastal waters.

Class **Bivalvia**

About 7500 species in numerous distinctive families; classifications still subject to dispute. Bilaterally symmetrical molluscs, the body laterally compressed and entirely enclosed by a two-valved shell. Head lacking. Feeding almost entirely microphagous, through filtration and sorting of suspended detritus, or resuspended deposits, by cilial tracts of the enlarged gills. Reproduction largely employing external fertilization. Sedentary, rarely free-moving; infaunal, epifaunal, or sessile, displaying diverse adaptations to habitat. Freshwater and marine.

Class **Cephalopoda**

Highly specialized marine molluscs divided into two distinctive subclasses, the shelled Nautiloidea and the Coeloidea, in which the shell is reduced and internal, or lacking entirely. About 600 species worldwide, typically nektonic or bentho-pelagic, fast-moving, predatory. Foot modified to form grasping arms, intimately associated with the head, which has greatly developed eyes, a marked development of the cerebral ganglia, and a powerful beak.

REFERENCES

Hochachka, P.W. (1983*a*). *The Mollusca*, **1**. *Metabolic biochemistry and molecular biomechanics*. Academic Press, London and New York.

Hochachka, P.W. (1983*b*). *The Mollusca*, **2**. *Enviromental biochemistry and physiology*. Academic Press, London and New York.

Hyman, L.H. (1967). *The invertebrates*. **6**. *Mollusca* **1**. McGraw Hill, New York and London.

Moore, R.C. (1964) (ed.). *Treatise on invertebrate paleontology*, **1**. *Mollusca* **1**. General features. Scaphopoda. Amphineura. Monoplacophora. Gastropoda—general features. Archaeogastropods and some (mainly Paleozoic) Caenogastropods and Opisthobranchia. Geological Society of America, and University of Kansas Press.

Moore, R.C. (1964). *Treatise on invertebrate paleontology*, **K**. *Mollusca* **3**. Cephalopoda—general features. Endoceratoidea. Actino-

ceratoidea. Nautiloidea. Bactritoidea. Geological Society of America, and University of Kansas Press.
Moore, R. C. (1968) (ed.). *Treatise on invertebrate paleontology*, L. *Mollusca* 4. Cephalopoda. Ammonoidea. Geological Society of America, and University of Kansas Press.
Moore, R. C. (1969) (ed.). *Treatise on invertebrate paleontology*, N. *Mollusca* 6. Bivalvia. Geological Society of America, and University of Kansas Press.
Morton, J. E. (1967). *Molluscs*, 4th edn. Hutchinson, London.
Purchon, R. D. (1977). *The biology of the Mollusca*, (2nd edn). Pergamon, Oxford.
Russell-Hunter, W. D. (1983). *The Mollusca*, 6. *Ecology*. Academic Press, London and New York.
Saleuddin, A. S. M. and Wilbur, K. M. (1983*a*). *The Mollusca*, 4. *Physiology* Part 1. Academic Press, London and New York.
Saleuddin, A. S. M. and Wilbur, K. M. (1983*b*). *The Mollusca*, 5. *Physiology* Part 2. Academic Press, London and New York.
Tompa, A. S., Verdonk, N. H., and van den Biggelaar, J. A. M. (1984). *The Mollusca*, 7. *Reproduction*. Academic Press, London and New York.
Verdonk, N. H., van den Biggelaar, J. A. M., and Tompa, A. S. (1983). *The Mollusca*, 3. *Development*. Academic Press, London and New York.
Yonge, C. M. and Thompson, T. E. (1976). *Living marine molluscs*. Collins, London.

A. Class **Polyplacophora**

Chitons have an elongate, oval, limpet-like body. Dorsally, there is a shell of eight, arched transverse plates, interlocked along their anterior-posterior margins and anchored laterally to the mantle. The mantle skirt extends around the whole periphery of the animal as a toughened girdle, of variable width, often fringed with spines and bearing calcified granules, spines, or small plates on its upper surface. The head is small, anteriorly situated, and hidden from dorsal view by the girdle. The larger part of the ventral surface is occupied by the muscular foot. The mantle cavity is essentially posterior in position and contains the anus, renal pores, genital openings, osphradium and, in primitive species, the gills. In all chitons, however, the mantle cavity is continuous with deep channels which extend anteriorly on each side of the animal between the girdle and the foot. In most families the gills have increased from the primitive pair to paired series which are distributed part or all of the way along the lateral pallial grooves.

All polyplacophorans are herbivorous grazers adapted for living on hard substrata; like limpets they have considerable powers of adhesion, are able to withstand wave shock, and probably lead fairly sedentary lives. Some genera, such as *Tonicella* may be particularly associated with certain crustose algae. Sexes are separate, embryos develop as short-swimming lecithotrophic larvae.

The inner layer of the chiton shell is usually produced anteriorly in each of the six intermediate plates as an articulating flange, the *articulamentum*. Additionally, it may be produced laterally as notched *insertion plates*, which serve to attach the plates to the body of the animal. On the head and tail valve this layer may be similarly produced as a series of projecting insertion plates around the convex anterior (or posterior) border. In some cases it is necessary to remove and clean shell plates to be certain of the species identity. Other features of taxonomic importance are valve sculpture, and the granulation or spinulation of the girdle.

REFERENCES

Dell'Angelo, B. and Tursi, A. (1978). Guida bibliografica ai chitoni (Polyplacophora): 1970–1978. *Oebalia*, **4**, 79–151.
Dell'Angelo, B. and Tursi, A. (1981). Guida bibliografica ai chitoni (Polyplacophora): 1970–1979—primo supplemento. *Oebalia*, **7**, 43–133.
Jones, A. M. and Baxter, J. M. (1987). Mollusca: Caudofoveata, Solenogastres, Polyplacophora and Scaphopoda. *Synopses of the British Fauna (New Series)*, No. 37. Leiden, E. J. Brill, for the Linnean Society of London, and the Estuarine and Brackish-water Sciences Association.
Matthews, G. (1952). A key for use in the identification of British chitons. *Proceedings of the Malacological Society of London*, **29**, 241–8.

KEY TO FAMILIES

1. Girdle with prominent, erect, dense tufts of bristles; up to 10 pairs **4. Acanthochitonidae**
   Girdle smooth, granular or scaly; with or without spines, but not with erect tufts of bristles **2**

2. Girdle covered with small, flat scales, sometimes interspersed with fine, recumbent spines, and fringed with spines **3**
   Girdle with fine or coarse granulation, or appearing smooth; with or without recumbent spines, usually fringed with spines **3. Ischnochitonidae**

3. Intermediate valves rounded, without a keel; or, if keeled then without spines on the girdle. No insertion plates **1. Lepidopleuridae**

   Intermediate valves keeled, girdle with randomly distributed long spines on its surface, and a fringe of equally long spines. Head valve with insertion plates **2. Hanleyidae**

1 LEPIDOPLEURIDAE

Small chitons with relatively narrow girdle bearing spines or scales; shell valves enclosing most of animal, typically with granular ornamentation. Articulamentum only slightly produced; insertion plates poorly developed or absent. Gills posterior.

1. Shell valves markedly keeled, with prominent posterior beaks, sculpture fine. Girdle with rectangular, longitudinally striated scales ***Leptochiton asellus***

   Shell valves arched, without marked keel, beaks scarcely developed **2**

2. Shell valves smoothly arched, with uneven granular sculpture. Girdle with rectangular, longitudinally striated scales. Foot red ventrally in live specimens. South-west only ***Leptochiton scabridus***

   Shell valves smoothly and strongly arched; regular, fine granulation giving chequered effect. Girdle with fine flat scales with curled tips ***Leptochiton cancellatus***

Leptochiton asellus (Gmelin) FIG. 12.1

Up to 18 mm, rarely larger; broadly oval, about one and a half times long as wide. Shell valves off-white or yellow, often with black or brownish organic incrustations; with distinct keel and small posterior beaks. Sculpture of fine granules, becoming coarser towards margins and accentuating growth lines. Edges of plates finely denticulate; articulamentum semi-elliptical, with smooth edge. Girdle relatively narrow, covered with overlapping, elongate rectangular scales, each with distinct longitudinal striations; fringed with delicate marginal spines.

Lower shore and sublittoral. All coasts, widespread and common.

Leptochiton cancellatus (Sowerby) FIG. 12.1

Up to 9 mm long, elongate oval, about twice as long as broad. Shell valves yellowish white to fawn, smoothly and strongly arched, rounded dorsally, without a keel. Sculpture of fine, regularly ordered granules giving chequered (or cancellate) effect, more pronounced towards margins where growth lines may be enhanced. Edges of plates finely denticulate; articulamentum rounded triangular, with smooth edge. Girdle narrow, covered with overlapping thin, flat, scales, rectangular and varying in size, with free edge often curled; fringed with delicate marginal spines.

Lower shore and shallow sublittoral. Predominantly southern and western, extending as far north as Shetland, perhaps absent from north-east coasts.

Leptochiton scabridus (Jeffreys) FIG. 12.1

Up to 8 mm long, elongate oval, about two and a half times long as broad. Shell valves yellowish white, smoothly arched, rather low, without keel or beaks. Sculpture of uneven granules, indistinct dorsally, coarse and well-marked towards margins; particularly coarse, and more or less cancellate on head shield and anal shield. Girdle relatively narrow, covered with overlapping, elongate rectangular scales, each with fine longitudinal striations; fringed with fine spines. Ventral surface of foot red in living specimens.

Lower shore and sublittoral. Channel Isles and extreme south-western shores only; rare.

2 HANLEYIDAE

Small to medium-sized chitons; with relatively narrow girdle bearing spicules and granules, often interspersed with spines, with a dense fringe of slender spines around the edge. Valves highly arched, often keeled; with variable, often coarse, granulation. Head valve with an insertion plate, lacking notches, but intermediate valves without insertion plates. Articulamentum well developed.

Hanleya hanleyi (Bean in Thorpe) FIG. 12.1

Up to 22 mm long, elongate oval, about two and a half times long as broad. Shell valves off-white to yellowish, highly arched, with distinct sharp keel and pronounced beaks. Sculpture of coarse granules, in longitudinal series, particularly marked laterally, and on head and anal valves, where they are developed as protruding tubercles. Girdle narrow, covered with small, flat scales interspersed with fine, recumbent spines; fringed with similar spines.

Lower shore and sublittoral. Northern coasts only; extending south to Northumberland in the east, and to the Isle of Man and western Ireland in the west.

3 ISCHNOCHITONIDAE

Small to medium-sized chitons, with relatively broad girdle bearing granules or small, recumbent spines. Shell valves smooth or finely granular, typically with diagonal ridges on intermediate valves; head valve usually longer than tail valve. Articulamentum well-developed, with up to twelve slits in margin on head and tail valves; lateral insertion plates with one to three notches.

1. Girdle covered with densely packed, recumbent acicular spines. Head valve with about 15 notches on anterior ventral rim. Intermediate plates with three pairs of notches in edge of insertion plates ***Callochiton achatinus***

Girdle with dense or sparse covering of granules, spines limited to fine marginal fringe only. Head valve with up to ten notches on anterior ventral rim. Intermediate plates with no more than two notches on each side **2**

2. Up to 15 pairs of gills, restricted to posterior parts of lateral pallial grooves. Girdle with densely packed, ovoid granules. Valves appear almost smooth ***Tonicella rubra***

   More than 15 pairs of gills, extending anteriorly to occupy greater part of pallial groove on each side **3**

3. Shell valves smooth and glossy. Girdle broad, appearing smooth and leathery, with minute widely scattered granules ***Tonicella marmorea***

   Shell valves finely granular. Girdle relatively narrow, with conspicuous, rounded granules, of variable size, fringed with blunt, fusiform spines ***Lepidochitona cinereus***

   Shell valves very finely granular, growth lines not pronounced. Girdle narrow, with large, oval or rectangular granules, fringed with fine spines ***Ischnochiton albus***

Lepidochitona cinereus (Linnaeus) FIG. 12.1

Up to 24 mm long, broadly oval, about one and a half times long as wide. Colour variable: red, brown, yellow to green, in patches and bands. Shell valves smoothly arched, rather low, with distinct, low keel and pronounced beaks. Sculpture of very regular, fine granulation, becoming slightly coarser towards margins where growth lines may be enhanced. Head valve with eight or nine anterior ventral notches; intermediate plates typically with one indistinct notch laterally (rarely two), articulamentum developed on each side as a gently convex, semi-elliptical plate. Girdle relatively narrow, densely covered with fine rounded granules; fringed with fusiform spines.

Middle to lower shore, and sublittoral. All coasts, widespread and common.

Tonicella marmorea (Fabricius) FIG. 12.1

Up to 40 mm long, broadly oval, about one and a half times long as wide. Shell valves smooth and glossy, chestnut-brown with variable white or fawn marbling; not highly arched, keel scarcely developed but beaks pronounced. Sculpture of distinct growth lines and some very fine granulation. Head valve with eight major notches on anterior ventral edge; intermediate valves with one major notch on each side, articulamentum comprising a single process with smooth, evenly concave margin. Girdle rather broad, thin, appearing leathery, reddish brown, with minute, sparsely distributed granules, fringed with tiny, somewhat flattened marginal spines.

Lower shore and sublittoral. Predominantly northern distribution; extending to Yorkshire coast in the east, to west Scotland and Irish Sea on west coast.

Tonicella rubra (Linnaeus) FIG. 12.1

Up to 20 mm long, broadly oval, about one and a half times long as wide. Shell valves smooth and glossy, pale buff to deep chestnut-brown, richly marbled with white; gently arched, keel scarcely developed but beaks well-marked. Sculpture more distinct than in *T. marmorea*: very fine granulation dorsally, becoming rather pronounced marginally; growth lines conspicuous. Girdle marbled brown and white, densely covered with spherical granules, fringed with slightly flattened spines.

Lower shore and sublittoral. All coasts, except south-east.

Callochiton achatinus (Brown) FIG. 12.1

Up to 30 mm long, oval, about twice as long as broad. Shell smooth, greenish or reddish-brown with white or greenish-white marbling; gently arched, keel distinct, beaks well-marked. Sculpture of distinct growth lines, and fine dorsal granulation, becoming coarser marginally. Head valve with about 15 notches in anterior ventral edge; intermediate valves with three deep notches on each side, articulamentum developed as a single plate along whole of anterior margin. Girdle relatively broad, covered with densely packed, overlapping acicular spines, and with a fringe of similar spines.

Lower shore and sublittoral. Widespread, all coasts.

Ischnochiton albus (Linnaeus) FIG. 12.2

Up to 9 mm long, elongate oval, two to two and a half times long as wide. Shell valves unicolorous, off-white to yellowish, only moderately arched, with rounded keel and small beaks. Sculpture of delicate growth lines and fine granulation. Head valve with up to 15 notches in anterior ventral margin; intermediate plates with a single notch on each side, articulamentum projecting along whole of anterior border of valve. Girdle relatively narrow, covered with densely packed, overlapping, rectangular granules, fringed with fine spines.

Predominantly northern distribution, south to the Isle of Man on the west, and to north Kent on the east coast. Not common.

4 ACANTHOCHITONIDAE

Small to large chitons, with broad girdle thickly covered with coarse spines, and typically bearing conspicuous tufts of large bristles. Shell valves partially overlapped by girdle; coarsely granular, with well-marked keel. Articulamentum well-developed; about five slits in margin of head valve, lateral insertion plates with a single notch.

---

1. Valve sculpture of large pyriform tubercles unevenly distributed ***Acanthochitona crinitus***

   Valve sculpture of fine rounded granules, densely and regularly distributed ***Acanthochitona fascicularis***

Acanthochitona crinitus (Pennant) FIG. 12.2

Up to 34 mm long, elongate oval, about two and a half times long as broad. Colour very variable: off-white, grey, yellowish, chestnut, green, olive, brown, and shades of blue; in patches

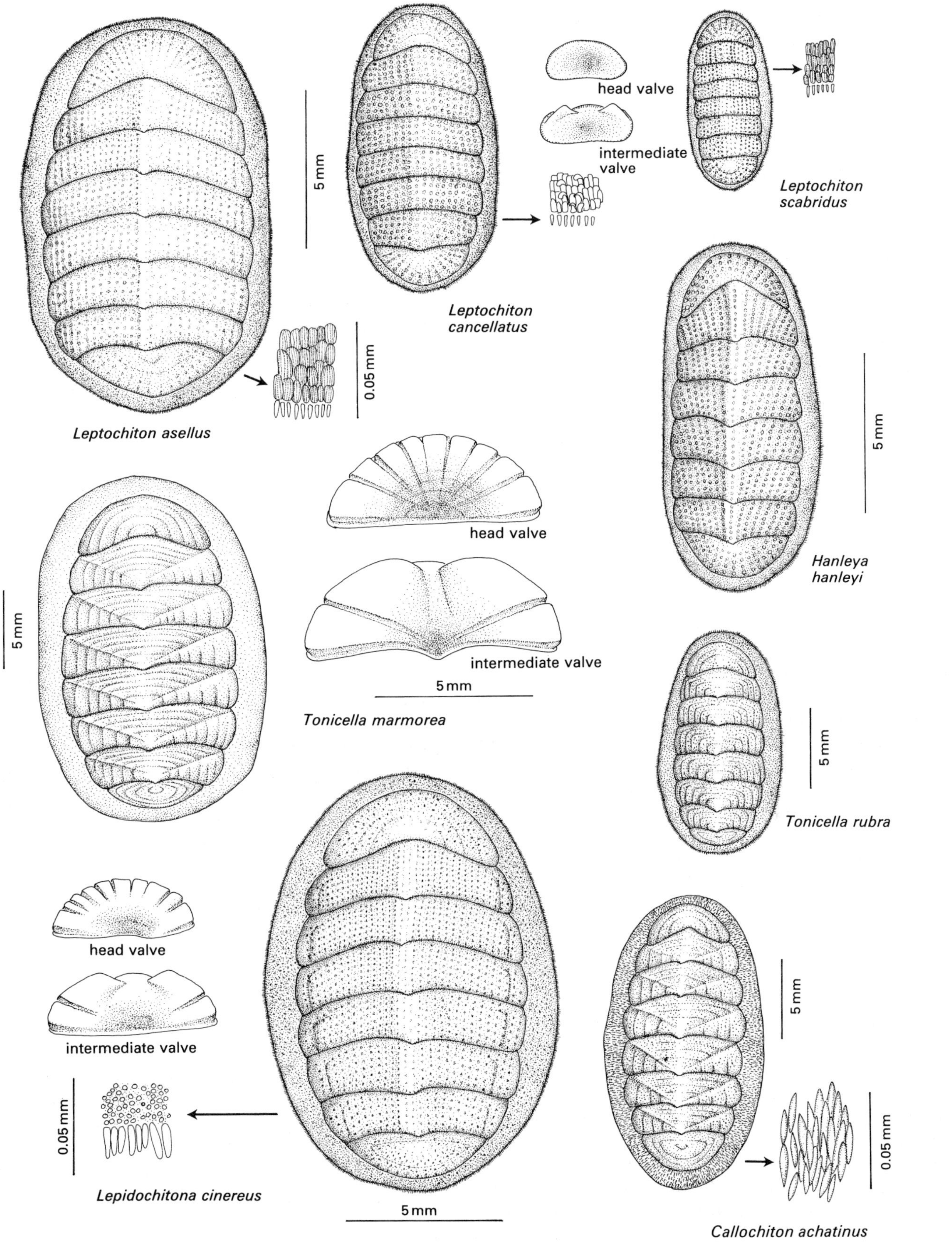

5 mm
0.05 mm
Leptochiton asellus
Leptochiton cancellatus
head valve
intermediate valve
Leptochiton scabridus
5 mm
Hanleya hanleyi
5 mm
head valve
intermediate valve
5 mm
Tonicella marmorea
5 mm
Tonicella rubra
head valve
intermediate valve
0.05 mm
Lepidochitona cinereus
5 mm
5 mm
0.05 mm
Callochiton achatinus

Fig. 12.1

and streaks, or marbled, rarely unicolorous. Shell valves strongly arched, with moderate rounded keel and pronounced beaks. Sculpture of longitudinal ridges on keel, laterally with large, flat-topped, pyriform granules, varying in size and unevenly spaced. Girdle with 18 dense tufts of bristles (up to 1 mm long), one situated, on each side, close to the junction between each pair of plates, and four in an arc anterior to the head plate; surface with fine, irregular granules interspersed with recumbent spines, fringed with dense series of long acicular spines (up to 0.5 mm).

Lower shore and sublittoral. All coasts, common.

Acanthochitona fascicularis (Linnaeus) FIG. 12.2

Up to 60 mm long, elongate oval, twice as long as broad. Colour variable, marbled with off-white, grey, yellowish, or brown. Shell valves strongly arched, with rounded keel and prominent beaks. Sculpture of longitudinal ridges on keel, laterally with fine, densely packed, oval or rounded granules. Girdles with 18 dense tufts of bristles (up to 1.5 mm long), one on each side at posterior plate boundaries, and four in an arc anterior to head plate; surface covered with densely packed, recumbent spines (feeling velvety to touch) and fringed with dense series of longer spines (up to 1 mm).

Lower shore and sublittoral. Recorded distribution from north Kent, south and west to Channel Isles, northwards to Anglesey, and on west coast of Ireland.

Acanthochitona discrepans (Brown) FIG. 12.2

Up to 34 mm, elongate oval, twice as long as wide. Shell morphology, colour and sculpture as for *A. crinitus*. Girdle with 19 or 20 tufts of bristles: 18 distributed as in *A. crinitus*, one or two additional tufts situated posterior to anal plate; surface covered with minute, recumbent, acicular spines, not visible to unaided eye.

Recorded from the Channel Isles, Pembrokeshire, Lulworth Cove, and Strangfjord Lough. The distinction between *A. crinitus* and *A. discrepans* is unclear. Jones and Baxter (1987) state that the characteristics of the two species seem to intergrade, and consider that only one variable species may exist.

B. Class **Scaphopoda**

The Scaphopoda is a small class of few genera. Only five species are likely to be encountered frequently in British waters, although none is common.

The scaphopod body is anterio-posteriorly elongate; the pallium is largely fused along the whole of the ventral side and secretes a tubular shell, open at each end. The head, closely associated with a short, cylindrical, burrowing foot is protruded from the broad, anterior end of the shell. The mantle cavity extends the whole length of the shell, and both inhalant and exhalant respiratory currents pass through the narrower, posterior end. The mouth is surrounded by short, ciliated tentacles or *captacula*; scaphopods are microphagous deposit feeders. All scaphopods live sedentary lives in offshore, soft-bottom deposits. The broad anterior end of the shell is buried vertically in the substratum, with the posterior end projecting above it.

REFERENCE

Jones, A. M. and Baxter, J. M. (1987). Molluscs: Caudofoveata, Solenogastres, Polyplacophora and Scaphopoda. *Synopses of the British Fauna (New Series)*, No. 37. Leiden, E. J. Brill, for the Linnean Society of London, and the Estuarine and Brackish-water Sciences Association.

1. Shell a solid, thickly calcified tube, slightly tapered and gently curved. Narrow (posterior) end with distinct longitudinal striations ... ***Antalis vulgare***
   Not as described ... **2**
2. Shell fairly thin, smooth, large (up to 35 mm); markedly tapered and distinctly curved. Narrow (posterior) end occluded by a septum with a central spout ... ***Antalis entalis***
   Shell scarcely tapered; not, or scarcely, curved. Less than 10 mm long ... **3**
3. Shell almost straight, gently tapered at each end, more or less bottle-shaped ... ***Cadulus jeffreysii***
   Shell almost straight, gently tapered at posterior end. Rather thick, with marked concentric growth lines ... ***Pulsellum lofotense***

1 DENTALIIDAE

Shell smooth or sculptured, curved, regularly tapering. Foot conical, with laterally expanded sheath.

Antalis entalis Linnaeus FIG. 12.2

Up to 40 mm long, maximum width of 5 mm at anterior end. Shell glossy white, translucent, smooth, with no growth rings or ridges; markedly tapered to posterior end and with a pronounced curve. Anterior aperture circular; posterior end obliquely truncate, aperture oval with a notch on convex side, occluded by a calcified septum with a central tube (*pipe*), having a pyriform orifice.

Offshore in sand. All coasts, but uncommon in south; not present around Channel Isles.

Antalis vulgare (da Costa) FIG. 12.2

Up to 60 mm long, maximum width of 6 mm at anterior end. Shell thick; dull, opaque white, with a few marked, oblique, concentric growth lines or ridges: gently tapered towards posterior end, less curved than *A. entalis*. Posterior portion of shell with fine, closely spaced, longitudinal striations. Anterior aperture circular; posterior end obliquely truncate, aperture circular with smooth rim, occluded by septum, with central pipe bearing a circular orifice.

Offshore, in sand. Distributed off south-western coasts only, in the Channel and off western Ireland. Not common.

2 SIPHONODENTALIIDAE

Shell small, smooth, typically tapered at both ends. Foot simple, vermiform, or distally expanded to form a disc.

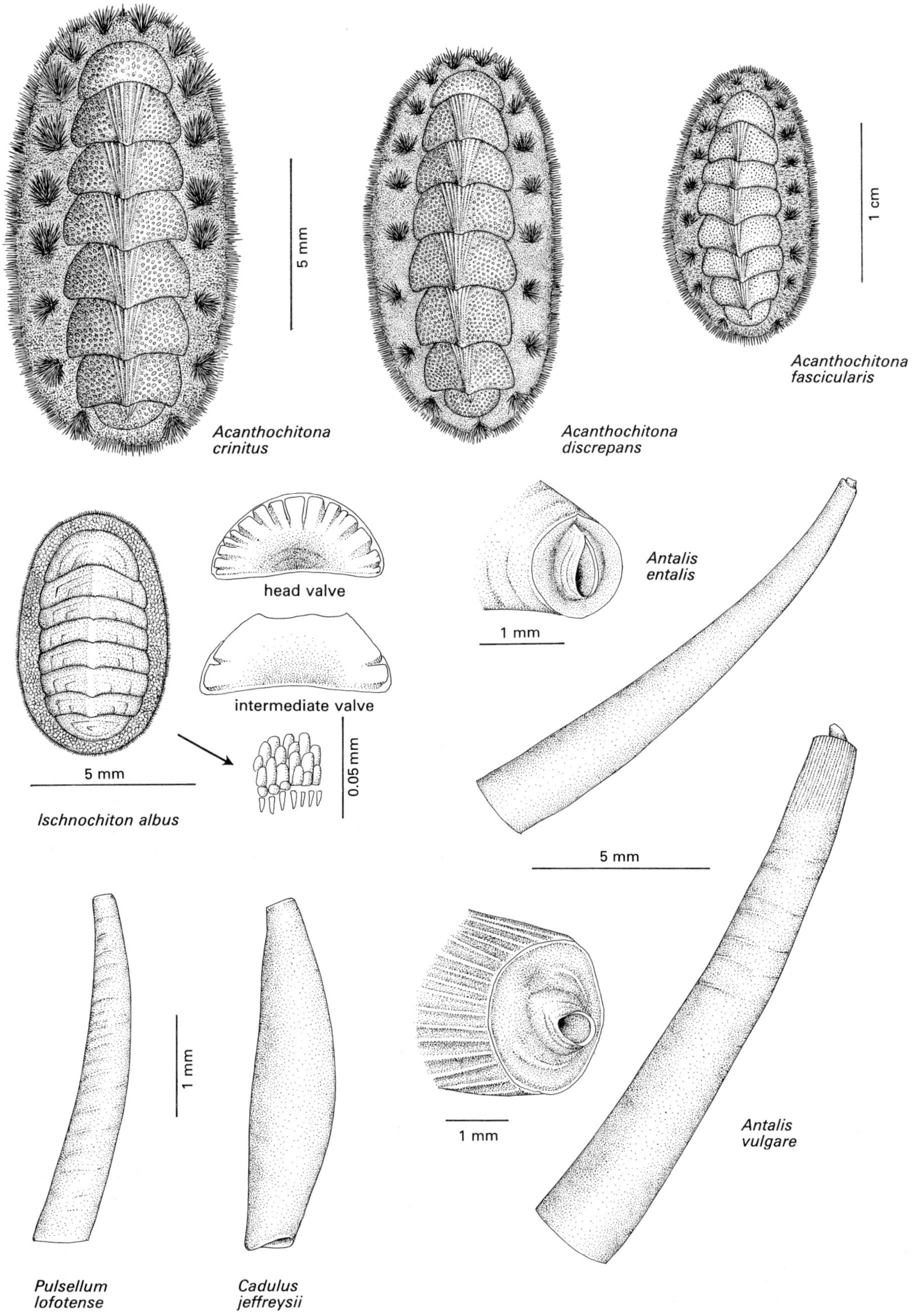
5 mm
Acanthochitona crinitus
Acanthochitona discrepans
1 cm
Acanthochitona fascicularis
head valve
intermediate valve
5 mm
0.05 mm
Ischnochiton albus
Antalis entalis
1 mm
5 mm
1 mm
Antalis vulgare
1 mm
Pulsellum lofotense
Cadulus jeffreysii

Fig. 12.2

Pulsellum lofotense (M. Sars) FIG. 12.2

Up to 6 mm long, with maximum width of 0.5 mm at anterior end. Shell thin, transparent, white, with well-marked oblique, concentric growth lines; tapered gently to posterior, but scarcely curved. Anterior and posterior apertures circular; no septum or pipe at posterior end, rim of aperture with four indistinct notches.

Offshore. Rarely reported; known from western coasts of Ireland and Scotland, and from Shetland. *P. affine* (M. Sars) is very similar but has a smooth, shiny shell up to 4.5 mm long, lacking growth lines. It has been reported from Shetland and south-west Ireland.

Cadulus jeffreysii (Monterosata) FIG. 12.2

Up to 4 mm long. Shell thin, transparent; white, smooth and glossy. Fusiform in shape: ventral side straight, dorsal arched; tapering towards both ends, anterior aperture scarcely wider than posterior.

Offshore. Rarely reported; known from Shetland, Hebrides, and south-west Ireland.

Class **Gastropoda**

C. Subclass **Prosobranchia**

There is no single character by which all members of the subclass Prosobranchia can be distinguished from other gastropod groups; the commonly used operculum is not only absent in some prosobranchs (e.g. in limpets and cone shells), but is present in some opisthobranch gastropods. Similarly the coiled shell, typical of most gastropods, is not a feature of many limpets, while in other prosobranchs it may be temporarily (e.g. in cowries) or permanently (e.g. in *Lamellaria* spp.) hidden below fleshy extensions of the body. Thus, for a limited number of species, affinities with major groups may not be apparent at first inspection: small coiled shells may belong to pulmonate or opisthobranch gastropods, while prosobranchs may take on the appearance of apparently shell-less opisthobranchs or non-operculate pulmonates.

Typically, the prosobranch gastropod body is enclosed in a *dextrally* coiled shell, that is, a shell which, when viewed from above the tip or *apex*, spirals in a clockwise direction (Fig. 12.3); an anticlockwise spiral is referred to as *sinistral*. The spiral coils, or *whorls*, meet at *sutures* and vary in profile from flat-sided to strongly convex, or tumid; whorls are widest at their periphery. The initial apical whorls represent the larval shell, or *protoconch*; this may be eroded in some shells, while in others it remains as a prominent and often diagnostic feature. The lower whorls of the spire comprise the adult shell, or *teleoconch*; the last whorl terminates in the *aperture*, from which the body emerges. In dextral shells the aperture lies to the right of the axis of coiling, or *columella*. It is bounded to the right by the *outer lip* and to the left by the *inner lip*, which may meet to form a continuous boundary, or *peristome* (in some instances the completion of the peristome may take the form of a glaze on the *parietal* region of the shell); either lip may be thickened internally and bear ridges, or *teeth*. A complete peristome may not always be found and in some shells it may be interrupted *adapically* (towards the apex) by an *anal sinus*, or *abapically* (away from the apex) by a *siphonal canal*, or both. The columella may be hollow, and open basally as an *umbilicus*.

The shell surface may be enhanced by various sculptural elements, the most common being the successive growth lines generated at the margin of the outer lip as the shell grows; these may be microscopic in some species, while in others they are represented by prominent raised elements on the shell. Temporal variations in shell secretion at the outer lip may be superimposed upon the pattern of growth lines, which are often irregular, as a series of regular, raised, longitudinal ridges, or *costae*. Where the costae are out-turned or flared they produce a *varix*, which when situated at the outer lip is termed a *labial varix*. Where costae and growth lines are in line with the vertical axis of the shell they are described as being *orthocline* (Fig. 12.3). If they incline to the right adapically, and hence to the left abapically, they are *opisthocline*, while the reverse inclination is *prosocline*. The inclination of longitudinal sculptural elements reflects the angle at which the aperture is inclined to the vertical shell axis. Spatial variations in the secretion of calcareous material along the outer lip produces a spiral element of ridges, or *striae*, and grooves of variable dimension. Where both costae and striae are present, their interaction may form a *reticulate* or *decussate* pattern of squared or rectangulate blocks (see *Hinia*, Fig. 12.3). The outer organic covering, or *periostracum*, while eroded from most shells, may present a prominent horny layer on others (see *Trichotropis*, p. 674).

According to the spatial relationship between successive whorls the shell may be a disc-like *planospiral* or present a more conical shape, with a variable degree of spire development. The profile of the spire, as drawn through the periphery of the whorls, may be straight-sided or slightly convex (*cyrtoconoid*) or concave (*coeloconoid*) (Fig. 12.3). In some shells (e.g. cowries) the last whorl envelops the spire to form a convolute shell which apparently lacks a spire. The reduction of the spire in limpets produces a cap-shaped shell dominated by the enlarged last whorl.

The body of most prosobranchs can be withdrawn, head first, into the shell. Once the body is withdrawn the aperture is then sealed by a horny plug, or *operculum* carried on the posterio-dorsal surface of the muscular foot. The operculum may be a spiral of many turns (*polygyrous*), or few (*oligogyrous*), originating from a central or marginal nucleus, or it may comprise a series of concentric rings (Fig. 12.3). The operculum is calcified in the family Tricoliidae. In addition to the operculum the foot may bear *epipodial tentacles* (Fig. 12.3), which arise from a lateral *epipodial fold*; in some families one or two *metapodial tentacles* (Fig. 12.3) may arise from the opercular fold or posterior tip of the foot. A groove housing the mucus-secreting anterior *pedal gland* separates the anterior *propodium* from the *mesopodium* on which the animal creeps. Other mucus glands, the *ventral pedal gland* and the *posterior pedal gland* are variously represented on the pedal sole of prosobranchs, while in the shell-boring muricids an *accessory boring organ* is also present.

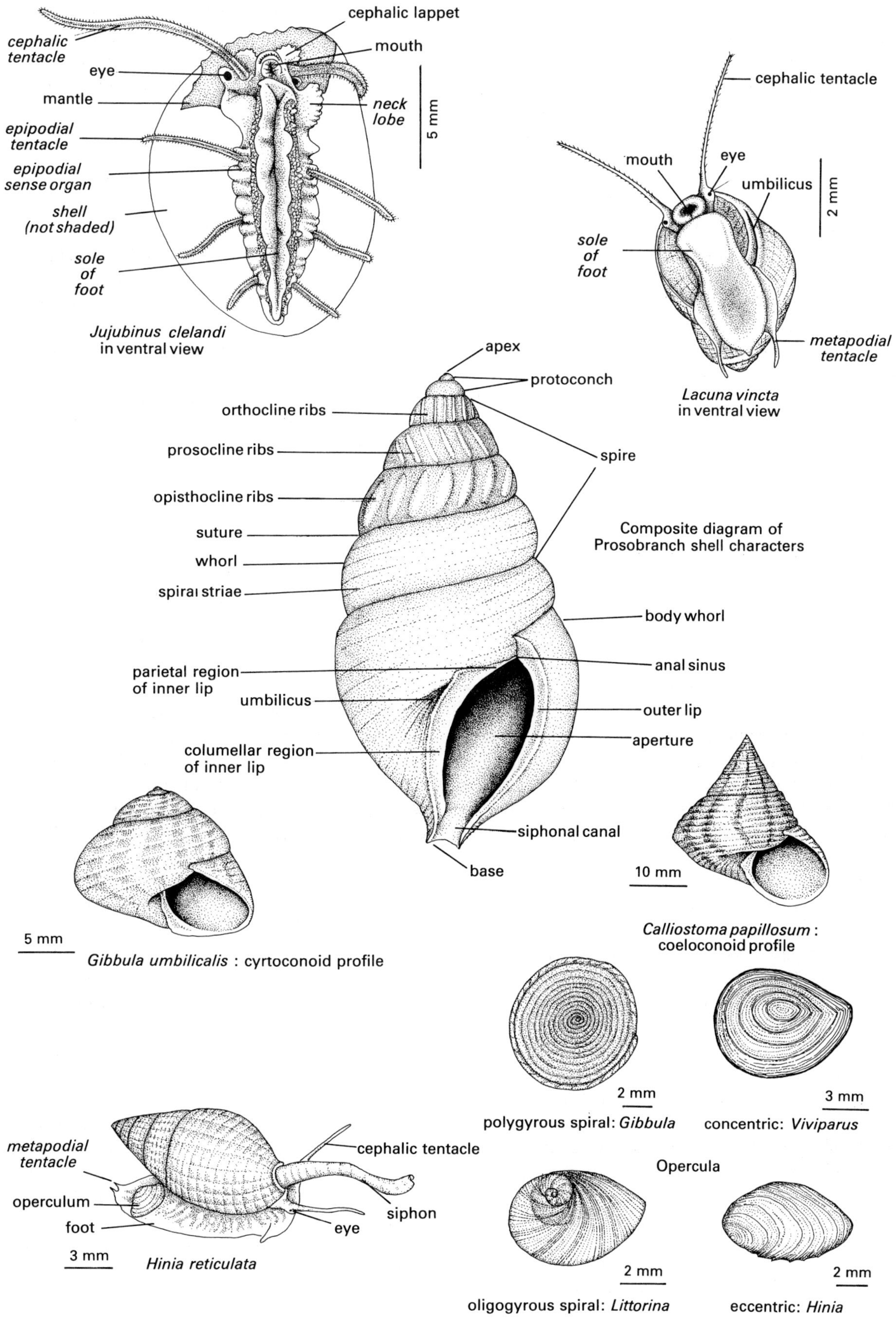
cephalic lappet
cephalic tentacle
mouth
eye
mantle
neck lobe
5 mm
epipodial tentacle
epipodial sense organ
shell (not shaded)
sole of foot
Jujubinus clelandi in ventral view
cephalic tentacle
mouth
eye
umbilicus
2 mm
sole of foot
metapodial tentacle
Lacuna vincta in ventral view
apex
protoconch
orthocline ribs
prosocline ribs
spire
opisthocline ribs
suture
whorl
spiral striae
Composite diagram of Prosobranch shell characters
body whorl
parietal region of inner lip
anal sinus
umbilicus
outer lip
aperture
columellar region of inner lip
siphonal canal
base
10 mm
5 mm
Gibbula umbilicalis : cyrtoconoid profile
Calliostoma papillosum : coeloconoid profile
2 mm
3 mm
polygyrous spiral: Gibbula
concentric: Viviparus
Opercula
metapodial tentacle
cephalic tentacle
operculum
siphon
foot
eye
3 mm
Hinia reticulata
2 mm
2 mm
oligogyrous spiral: Littorina
eccentric: Hinia

Fig. 12.3

The head bears a pair of *cephalic tentacles*; eyes are set on lateral eyestalks which extend for a variable distance alongside, and often fuse with, the tentacles (Fig. 12.3). The mouth is borne below the head on the end of a snout of variable dimension; in higher mesogastropods and in neogastropods (e.g. the whelks) this opening is not that of the true mouth but of the *proboscis sac*, the mouth being at the end of an eversible *proboscis*. Behind, and dorsal to, the head is the *mantle cavity* in which the gills, or *ctenidia*, and their associated sense organ (the *osphradium*) and mucus gland (the *hypobranchial gland*) are housed, together with the combined genital/renal and the anal papillae. Males possess a penis, which is often a conspicuous feature on the right of the mantle cavity; in some species, usually those with tall and needle-like shells, the male is aphallic. In limpets the mantle cavity extends as a shallow *pallial groove* around the base of the foot, the deeper part of the mantle cavity behind the head being the *nuchal cavity*; in *Patella* the mantle cavity contains numerous, plate-like *pallial gills*, while a single ctenidium is present in the nuchal cavity of *Acmaea*; in *Lepeta* and *Propilidium* both pallial gills and ctenidia are absent. The mantle cavity has as its roof the *mantle skirt*, the free edge of which may be seen to lie along the outer lip of the shell and, in some species bears *pallial tentacles*; most rissoaceans have one or two, turritellids many. In higher mesogastropods and in neogastropods an extension of the mantle edge, the *siphon*, forms an inhalant channel housed within, and often projecting beyond, the siphonal canal at the base of the shell (Fig. 12.3).

While shell characters alone may suffice in the identification of many prosobranch species, it is important in the separation of some families, as for instance in the rissoacean superfamily, to have living animals; with species of Littorinidae it is necessary to view the mantle anatomy of fresh specimens. In general the identification of all prosobranch species is greatly facilitated by access to clean shells, living specimens, a good binocular microscope, and directional lighting to enhance shell topography.

Specimens retained for later examination are best narcotized, fixed, and preserved. All molluscan groups display varying responses to standard narcotization procedures, and trial and error may be necessary. Slow-moving animals may respond to magnesium chloride, at a concentration of 75 g/l sea-water. Others may be relaxed in sea-water deoxygenated with carbon dioxide; a rechargeable siphon is filled with sea-water, which is then gently and slowly added to the dish of water containing the animals. Some species will relax overnight if menthol crystals are scattered thinly over the surface of a dish of sea-water containing them, others may respond to 70% ethanol slowly added to the dish to give a concentration of 10% by volume. Finally, propylene phenoxetol, added as small globules or as a shaken emulsion with sea-water, is a good, slow-acting narcotic. Specimens may be fixed in 5% sea-water formalin and stored in 70% methanol or 1% aqueous propylene phenoxetol.

REFERENCES

Fretter, V. and Graham, A. (1962). *British Prosobranch molluscs*. Ray Society, London.

Fretter, V. and Graham, A. (1976). The Prosobranch molluscs of Britain and Denmark. Part 1. Pleurotomariacea, Fissurellacea and Patellacea. *Journal of Molluscan Studies*, (Supplement **1**), 1–38.

Fretter, V. and Graham, A. (1977). The Prosobranch molluscs of Britain and Denmark, Part 2. Trochacea. *Journal of Molluscan Studies*, (Supplement **3**), 39–100.

Fretter, V. and Graham, A. (1978*a*). The Prosobranch molluscs of Britain and Denmark, Part 3. Neritacea, Viviparacea, Valvatacea, terrestrial and freshwater Littorinacea and Rissoacea. *Journal of Molluscan Studies*, (Supplement **5**), 101–52.

Fretter, V. and Graham, A. (1978*b*). The Prosobranch molluscs of Britain and Denmark, Part 4. Marine Rissoacea. *Journal of Molluscan Studies*, (Supplement **6**), 153–241.

Fretter, V. and Graham, A. (1980). The Prosobranch molluscs of Britain and Denmark, Part 5, Marine Littorinacea. *Journal of Molluscan Studies*, (Supplement **7**), 243–84.

Fretter, V. and Graham, A. (1981). The Prosobranch molluscs of Britain and Denmark, Part 6. Cerithiacea, Strombacea, Hipponicacea, Calyptraeacea, Lamellariacea, Cypraeacea, Naticacea, Tonnacea, Heteropoda. *Journal of Molluscan Studies*, (Supplement **9**), 285–363.

Fretter, V. and Graham, A. (1982). The Prosobranch molluscs of Britain and Denmark, Part 7. Heterogastropoda (Cerithiopsacea, Triforacea, Epitoniacea, Eulimacea). *Journal of Molluscan Studies*, (Supplement **11**), 363–434.

Fretter, V. and Graham, A. (1985). The Prosobranch molluscs of Britain and Denmark, Part 8. Neogastropoda. *Journal of Molluscan Studies*, (Supplement **15**), 435–556.

Graham, A. (1989). British Prosobranch and other operculate gastropod molluscs. *Linnean Society Synopses of the British Fauna* (n.s.), **2**. 2nd. edition. Leiden: E. J. Brill, for the Linnean Society of London, and the Estuarine and Brackish Water Sciences Association.

Moyse, J., Thorpe, J. P., and Al-Hamadani, E. (1982). The status of *Littorina aestuarii* Jeffreys. An approach using morphology and biochemical genetics. *Journal of Conchology*, **31**, 7–15.

Smith, D. A. S. (1976). Disruptive selection and morph-ratio clines in the polymorphic snail *Littorina obtusata* (L.). (Gastropoda: Prosobranchia). *Journal of Molluscan Studies*, **42** (1), 114–35.

Williamson, P. and Kendall, M. A. (1981). Population, age structure and growth of the trochid *Monodouta lineata* determined from shell rings. *Journal of the Marine Biological Association of the U. K.*, **6**, 1011–26.

KEY TO FAMILIES

1. Shell external, possibly covered by mantle folds in active animals which withdraw when touched **2**

   Shell internal, permanently hidden beneath papillate mantle; animal resembles dorid nudibranch but has smooth tentacles and lacks a dorsal gill circlet **31. Lamellariidae**

2. Shell a short curved tube, open basally, sealed apically with a conical or rounded calcareous plug **23. Caecidae**

   Shell not this shape **3**

3. Shell with marginal slit, or with one or several holes in addition to aperture **4**

   Shell imperforate apart from aperture **7**

4. Shell with marginal slit **5**

   Shell with one or more holes **6**

5. Shell a small, globose spiral of two or three reticulate whorls **2. Scissurellidae**

   Shell an open limpet- or cap-shaped cone, with recurved beak at apex **3. Fissurellidae**

6. Shell an open limpet-shaped cone with single hole near apex **3. Fissurellidae**

   Shell a flattened, ear-shaped spiral with several holes **1. Haliotidae**

7. Shell not coiled; or if coiled, convolute so that coiling is not evident at apex **8**

   Shell coiled, not convolute; apical whorls may be raised above body whorl to form a spire, or coiled in one plane so that spire is absent **13**

8. Shell convolute, ovoid or spindle-shaped with long, narrow aperture; mantle covers shell in active animals **9**

   Shell an open, limpet-shaped cone; apex central or toward anterior margin **10**

9. Shell ovoid, rounded at each end, with transverse ridges **32. Eratoidae**

   Shell elongate, drawn out into apical and basal canals; smooth and polished **33. Ovulidae**

10. Shell with tongue-like shelf below posterior end of foot, aperture nearly circular; apex pointed, with small beak; body with expanded lateral lobes on neck; ctenidium large **30. Calyptraeidae**

    Shell without internal shelf; without neck lobes; ctenidium small or absent **11**

11. Shell smooth, depressed, with apex toward anterior margin; pale with branching rays of pink or brown; single ctenidium in mantle cavity above head **4. Acmaeidae**

    Shell smooth or with radiating ridges; not markedly depressed, often much raised; no ctenidium; with or without accessory (pallial) gills on mantle skirt around foot **12**

12. Pallial gills present on mantle skirt, edge of which also bears numerous pallial tentacles; pigmented eyes at base of cephalic tentacles; shell smooth, with blue rays near apex (animal on weed), or with radiating ridges (animal on rocks and stones); littoral **5. Patellidae**

Without pallial gills or pallial tentacles; no pigmented eyes; shell small (rarely > 10 mm in length), finely reticulate; never littoral **6. Lepetidae**

---

13. Shell an elongate cone of up to five whorls, with strong spiral striae over which a thick periostracum is raised into bristles; aperture triangular, tapering basally to a shallow spout; body with short, non-retractile proboscis; sublittoral **28. Trichotropidae**

Shell and body not like this **14**

---

14. Shell glossy, porcellaneous, with short spire and large, rather inflated last whorl; aperture long, narrow, with or without teeth on inner and outer lips; colourful mantle covers shell in active animals; no operculum **32. Eratoidae**

Shell and body not like this **15**

---

15. Shell without siphonal canal **16**

Shell with siphonal canal **56**

---

16. Shell with elevated spire **29**

Spire very depressed (not rising much above body whorl), or absent **17**

---

17. Shell a minute (< 1 mm diameter), biconcave disc, spire absent **20. Omalogyridae**

Shell not like this **18**

---

18. Shell with umbilicus **19**

Shell without umbilicus **25**

---

19. Umbilicus wide, funnel-shaped, exposing all older whorls; shell fragile, minute (< 2.5 mm diameter), almost disc-like, with three or four whorls and low spire **20**

Umbilicus deep, with narrow round or oval mouth, or chink-like with $\wedge$-shaped umbilical groove **21**

---

20. Shell smooth with irregular growth lines; whorls almost circular in section, aperture with circular peristome; up to 1.5 mm diameter; deep chestnut-brown **18. Skeneopsidae**

Shell with spiral striae confined to base of body whorl; whorls almost circular in section, aperture with circular peristome; up to 1.2 mm diameter; colourless or white; foot with three pairs of epipodial tentacles **6. Skeneidae**

Shell with complex sculpture of keels, costae, and growth lines; aperture without peristome, with prominent anal canal; up to 2.5 mm diameter; whitish **19. Tornidae**

---

21. Umbilicus chink-like, with $\wedge$-shaped umbilical groove; aperture height almost equal to shell height; foot with two pairs of metapodial tentacles **10. Lacunidae**

Umbilicus deep, with narrow, round or oval mouth; foot without metapodial tentacles **22**

---

22. Shell smooth, globular, polished; umbilicus deep, partly occluded by callus-like extension of inner lip; foot with greatly enlarged propodium forming plough-like shield over head and front of shell; creeps through and over sand **34. Naticidae**

Shell and animal not like this **23**

23. Shell minute (<2 mm with three or four whorls), globose (height = breadth); aperture with complete peristome; foot with three pairs of epipodial tentacles or none **24**

Shell small (>4 mm with four or five whorls), depressed globose (height $<$ breadth); aperture without complete peristome; foot with five or six pairs of epipodial tentacles **7. Trochidae**

24. Shell with spiral striae around umbilicus or over entire surface, umbilicus oval with spiral ridge; foot with three pairs of epipodial tentacles; operculum transparent, polygyrous with central nucleus **8. Skeneidae**

Shell smooth, without spiral striae, umbilicus round, without spiral ridge; foot without epipodial tentacles; operculum with nucleus on columella margin, with peg-like process and three radiating ridges on underside **21. Rissoellidae**

25. Shell cap-shaped, almost hemi-ellipsoid, with slightly coiled subterminal apex; internally a large shell-like partition lies at posterior end below apex; body with expanded neck lobes and large ctenidium; often found in stacks of ascending size **30. Calyptraeidae**

Shell and body not like this **26**

26. Shell minute (2 mm or less), ear-shaped with up to two and a half rapidly expanding whorls; without periostracum; aperture very large, though animal cannot fully withdraw into shell; no operculum; in crevices and empty barnacle tests on upper shore **Pulmonata: *Otina***

Shell and body not like this **27**

27. Shell thinly calcified, with thick periostracum; no operculum; usually sublittoral **28**

Shell solid, without periostracum; with operculum; common intertidally **11. Littorinidae**

28. Shell cap-shaped (or, more strictly, involute, i.e. coiled in one plane) with large, almost circular aperture applied to substratum so that apex coils posteriorly; periostracum around aperture forms fringe, with earlier fringes evident on expanded last whorl; shell not covered by mantle **29. Capulidae**

Shell ear-shaped, not involute; periostracum not drawn out into frills; mantle edge swollen, partly covering shell in active animals **31. Lamellariidae**

29. Spire tall, shell awl-shaped with at least ten whorls in adult **30**

Spire short, less than ten whorls; shell more broadly conical **37**

30. Shell smooth or with microscopic growth lines only; often highly polished **31**

Shell with obvious ornament of costae or spiral striae or both **32**

31. Whorls flat-sided, sutures extremely shallow; shell highly polished **38. Eulimidae**

Whorls tumid, sutures deep; shell <6 mm in height **37. Aclididae**

32. Sculpture of spiral striae only **33**
Sculpture of costae, with or without spiral striae **34**

---

33. Shell large (up to 20 whorls, 55 mm high); mantle edge with pinnate tentacles, operculum with fringed margin; no umbilicus **22. Turritellidae**
Shell small (up to ten whorls, 6 mm high); mantle edge smooth, operculum plain; with umbilicus **37. Aclididae (*Aclis*)**

---

34. Sculpture of costae only **35**
Sculpture of costae and spiral striae producing reticulate or tuberculate surface **36**

---

35. Shell large (up to 40 mm high), whorls very timid, linked at sutures by thin, widely spaced costae; aperture circular, with complete peristome **35. Epitoniidae**
Shell small (up to 8 mm high), whorls almost flat-sided, costae numerous and flexuous; aperture ear-shaped; apex of shell sinistral **Opisthobranchia: Pyramidellidae, *Turbonilla***

---

36. Shell minute (2–3 mm high), almost transparent, delicately reticulate with nine or ten narrow, tumid whorls and bulbous apex; slight umbilical groove **37. Aclididae (*Graphis*)**
Shell solid, opaque, commonly 8–15 mm high with 10–15 slightly tumid whorls; reticulate ornament of strong tubercles, base of last whorl with spiral striae only; no umbilical groove **24. Cerithiidae**

---

37. Shell globose-conic with the initial two or three whorls forming a styliform apex; mantle partly covers shell in life; parasitic on echinoids **39. Stiliferidae**
Shell not like this **38**

---

38. Shell delicate, purple or pale violet, broadly conical or globose; pelagic, usually washed ashore on south-western beaches **36. Janthinidae**
Shell not like this **39**

---

39. Shell smooth, globular and polished; umbilicus deep, partly occluded by a callus-like extension of the inner lip; foot with greatly enlarged propodium forming plough-like shield over head and front of shell; creeps through and over sand **34. Naticidae**
Shell and body not like this **40**

---

40. Shell oval-conic, smooth, with four or six tumid whorls patterned with zigzag streaks of variable colour; operculum calcareous, white, convex lens-shaped; foot with three pairs of epipodial tentacles **9. Tricoliidae**
Shell not like this; operculum horny, foot with or without epipodial tentacles **41**

---

41. Adult shell decollate (apical whorls lost), columnar with three or four whorls and blunt apex, usually finely ribbed; immature shell (before decollation) of six or seven tapering whorls, the apical two smooth; snout very long, used with foot in caterpillar-like 'looping' locomotion; high on shore in muddy areas with Sea Blite; south coast **13. Truncatellidae**
Shell and body not like this **42**

---

42. With at least three pairs of epipodial tentacles; operculum polygyrous, circular with central nucleus; shell depressed-globular or pyramidal (breadth usually greater than height); aperture in markedly prosocline plane (about 45° or more to vertical) **43**

Without epipodial tentacles; operculum not circular, with few turns; shell globose-conic or a more oval or elongate cone (higher than broad) **44**

43. Shell minute (<2 mm high), globular or depressed globular; aperture with complete peristome; protoconch without pointed tip **8. Skeneidae**

Shell not minute (>2 mm high), depressed globular or pyramidal; aperture without complete peristome, usually with nacreous lining; protoconch with broad point on tip of initial whorl **7. Trochidae**

44. Shell smooth, polished, oval or barrel-shaped with incised sutures; aperture long, narrow with tooth on columella; pinkish or brown with white bands; shallow burrower in sand **Opisthobranchia: Acteonidae**

Shell not like this **45**

45. Shell with chink-like umbilicus and $\wedge$-shaped groove on columella; foot with two prominent metapodial tentacles (Fig. 12.3) **10. Lacunidae**

Shell with or without umbilicus, but if present without $\wedge$-shaped groove leading to it; foot with or without metapodial tentacles **46**

46. Shell turreted, with fine striae and prosocline costae below thick periostracum; foot with two prominent metapodial tentacles **10. Lacunidae**

Shell and body not like this **47**

47. (*In this dichotomy it is important that shell size be related to the number of whorls*) Shell rarely up to 9 mm, commonly <6 mm in height at five or six whorl stage, 2 mm or less at three to four whorls; oval-conic or globular; smooth, or with costae, or spiral striae, or both; aperture with complete peristome; foot with or without metapodial tentacle, mantle with one or two pallial tentacles or none **48**

Shell minimum 9 mm, frequently larger, at five or six whorl stage, 4 mm or more at three to four whorls; drop-shaped with short, slightly coeloconoid spire of flat-sided whorls, or more globular with tumid whorls; smooth or with spiral striae, never costate; aperture without distinct peristome, which is completed by no more than a thin glaze in the parietal region; without metapodial or pallial tentacles **11. Littorinidae**

48. Apex of shell sinistral, spire otherwise dextral; smooth or with fine decussation, with or without tooth on columella; cephalic tentacles grooved on outer sides, with eyes set between them; a shelf-like projection (*mentum*) lies between the underside of the head and the propodium **Opisthobranchia: Pyramidellidae**

Shell and body not like this **49**

49. Shell smooth, oval or spindle-shaped with flat-sided whorls and shallow sutures; last whorl large (>60% of shell height), aperture narrow, scimitar-shaped with two or three prominent teeth on columella; no operculum, cephalic tentacles short, retractile, with eyes at base; in crevices, usually on upper shore **Pulmonata: Ellobiidae**

Shell and body not like this **50**

50. Cephalic tentacles short, semicircular lobes with large eyes at tip; semi-terrestrial in salt-marshes between Kent and Humber (shell oval-conic) or high on shore in caves and crevices on south coasts (shell globose-conical, rare) **14. Assimineidae**

Cephalic tentacles cylindrical, single or bifid, with eyes at or behind their base **51**

51. Shell smooth, with five or six slightly tumid whorls; usually cream with single spiral band of red (may be uniform red or white); operculum crimson with internal peg-like process; foot with red opercular lobes and slightly bifid posterior tip; no metapodial or pallial tentacles **16. Barleeidae**

Shell and body not like this **52**

52. Mantle edge with pallial tentacle on right (protrudes from adapical angle of aperture), with or without similar tentacle on left **53**

Mantle edge without pallial tentacle on left or right **54**

53. Metapodial tentacle present, though it may not extend to posterior tip of foot; shell smooth or with costae or spiral striae or both; marine **15. Rissoidae**

Metapodial tentacle absent, shell smooth; in estuarine and brackish waters **12. Hydrobiidae**

54. Adult shell 5–6 mm with six whorls; thin periostracum, occasionally drawn out into peripheral keel of bristles, often with black encrustations on spire; umbilicus at most a minute chink; in fresh and brackish waters only **12. Hydrobiidae (*Potamopyrgus*)**

Adult shell minute (<2 mm with three to four whorls), without periostracum; umbilicus usually obvious, chink-like or deep and round; foot with prominent opening of mucus gland medially on sole; marine **55**

55. Shell whitish, glossy, and slightly iridescent, its transparency revealing distinctive dark spots on mantle; umbilicus a narrow chink or deep and round; snout deeply bifid, each half tentacle-like; cephalic tentacles single or bifid; operculum concentric with nucleus at middle of columella edge, a peg-like process and three radiating ridges on its inner surface **21. Rissoellidae**

Shell cream or light horn colour with three red-brown bands on body whorl (young shells may be uniform brown); with or without umbilical chink; snout bifid but not tentacle-like, cephalic tentacles single, foot with short triangular metapodial tentacle, operculum spiral, without ridges **17. Cingulopsidae**

56. Shell dextral **57**

Shell sinistral **26. Triphoridae**

57. Shell with tall spire of tumid whorls bearing crescentic costae and fine spiral striae; mature shells with palmate extension to outer lip, juveniles (up to eight whorls) with sharply pointed siphonal canal but without palmate outer lip **27. Aporrhaidae**

Shell not like this **58**

58. Shell elongate, needle-like, of up to fourteen whorls, each with three rows of tubercles; aperture small with short siphonal notch at base **25. Cerithiopsidae**
Shell not like this **59**

59. Shell smooth, or with growth lines only **41. Buccinidae**
Shell with costae or spiral striae or both **60**

60. Shell costate, with or without spiral striae **61**
Shell with spiral striae only **71**

61. Spiral striae absent **40. Muricidae (*Boreotrophon*)**
Spiral striae present, though may be fine **62**

62. Shell with twelve or more costae on penultimate whorl **63**
Penultimate whorl with less than twelve costae **68**

63. Shell breadth equals 50% or more of total height **64**
Shell breadth less than 50% of height **65**

64. Shell breadth >60% of height, whorls tumid, costae strongly crescentic, fading below periphery of last whorl; aperture broadly oval, without teeth on outer lip; shell large (up to 110 mm high) **41. Buccinidae**
Shell less than 30 mm in height, costae not strongly crescentic, outer lip thickened and ridged internally; foot with two metapodial tentacles **42. Nassariidae**

65. Siphonal canal long (equal to aperture length, or nearly so), narrow **66**
Siphonal canal not as described **67**

66. Shell of seven or eight turreted whorls, with numerous narrow costae and spiral striae producing deep reticulation; spiral element absent from subsutural band on shoulder of each whorl; aperture oval or triangular, without anal sinus, siphonal canal straight, narrow **40. Muricidae (*Trophonopsis*)**
Shell of up to eleven tumid whorls, with numerous costae terminating apically at a broad, concave subsutural band; aperture elongate with deep anal sinus; no operculum **43. Turridae (*Comarmondia*)**

67. Shell up to 30 mm, spire tall, almost flat-sided, with reticulate pattern; aperture may have teeth on inner and outer lips, siphonal canal separated from base of shell by deep spiral groove; foot with two metapodial tentacles, operculum present **42. Nassariidae**
Shell not as above, <20 mm in height; foot without metapodial tentacles, with or without operculum **43. Turridae**

68. Shell breadth equal to or greater than half the height **69**
Shell breadth less than half the height **70**

69. Shell <12 mm with seven to nine whorls; no operculum	**43. Turridae**
Shell >15 mm (often twice this) with seven or eight whorls; with operculum	**40. Muricidae**
70. Adult shell up to 5 mm with five or six whorls; penultimate whorl with four flat spiral striae and nine or ten costae; aperture without anal sinus, outer lip with varix; shell chestnut-brown; with operculum	**41. Buccinidae (*Chauvetia*)**
Shell not as above; adult >5 mm, commonly up to 10 mm, aperture with anal sinus; no operculum	**43. Turridae**
71. Spiral ridges strong and cord-like, five or six on penultimate whorl; aperture narrow with deep anal sinus, outer lip thin, crenulate; protoconch with diamond-shaped reticulation; no operculum	**43. Turridae (*Teretia*)**
Shell not like this; aperture broad, anal sinus slight or absent, protoconch simple, animal with operculum	**72**
72. Spire short, last whorl $>80\%$ of shell height; outer lip thick, with internal teeth in mature shells	**40. Muricidae (*Nucella*)**
Spire long, last whorl $<80\%$ of shell height	**41. Buccinidae**

1 HALIOTIDAE

Only one British representative of this family.

Haliotis tuberculata Linnaeus FIG. 12.4

Shell a flattened, ear-shaped spiral with a series of funnel-shaped holes, only the most recently formed (usually five to seven) perforate, surmounting a flattened peripheral band. Blunt keel and shell sculpture enhanced by narrow spiral ridges, traversed by growth lines accentuated at intervals to form ridges, probably corresponding to winter growth checks. Outer shell variously coloured, mottled and, in older shells, often supporting epizoites. Inner surface of shell iridescent mother-of-pearl. Size up to 90×60 mm. Foot large, notched anteriorly, with numerous, extensile epipodial tentacles. Mantle split below respiratory openings through which it extends to form short exhalant siphons. Operculum absent.

Ranges from Mediterranean to the Channel Islands where it is confined to rocky shores from ELWST to 13 m. Absent from British and Irish mainlands.

2 SCISSURELLIDAE

Only one British representative of this family.

Scissurella crispata Fleming FIG. 12.4

Small, delicate, white shell forming a globose spiral of two to five finely reticulate whorls, with prominent peripheral groove or 'slit-band'. For a short distance back from outer lip this groove penetrates body whorl as a narrow slit through which setose pallial tentacles extend in living specimens. Size up to 3×4 mm. Cephalic tentacles extensile, with pinnately branching setae. Foot with seven epipodial tentacles on each side, some setose; posteriorly a thin, concentrically marked operculum.

Found in dredgings (8–2000 m) on stony, shelly, or gravelly bottoms, off north and west coasts, and western approaches to Channel.

3 FISSURELLIDAE

Shell conical or cap-shaped with reticulate sculpture of radiating ribs and concentric growth lines; often retaining one- or two-whorled protoconch. Shell perforated apically or sub-apically by a single hole or marginally by a narrow slit; no operculum; epipodial tentacles well-developed.

1. Shell with marginal slit	***Emarginula fissura***
Shell with apical or sub-apical hole	**2**
2. Shell with posteriorly directed, recurved apex; slit-like hole anterior to apex	***Puncturella noachina***
Shell conical with single apical hole	***Diodora graeca***

Emarginula fissura (Linnaeus) FIG. 12.4

Shell solid, strongly reticulate, cap-shaped with posteriorly directed, recurved apex (beak) of two spirally coiled whorls. A sunken slit band arising anterior to beak, perforate distally, forming a marginal exhalant slit above mantle cavity. Size up to $10 \times 8 \times 6$ mm. Foot shield-shaped, broader anteriorly, with ten epipodial tentacles on each side. An additional epipodial tentacle arises from back of head on right side. No operculum.

From lower shore (LWST) to 265 m, on rocks and boulders bearing sponges, upon which it feeds. On all suitable west European coasts from Norway and Faroe Islands to Canary Islands.

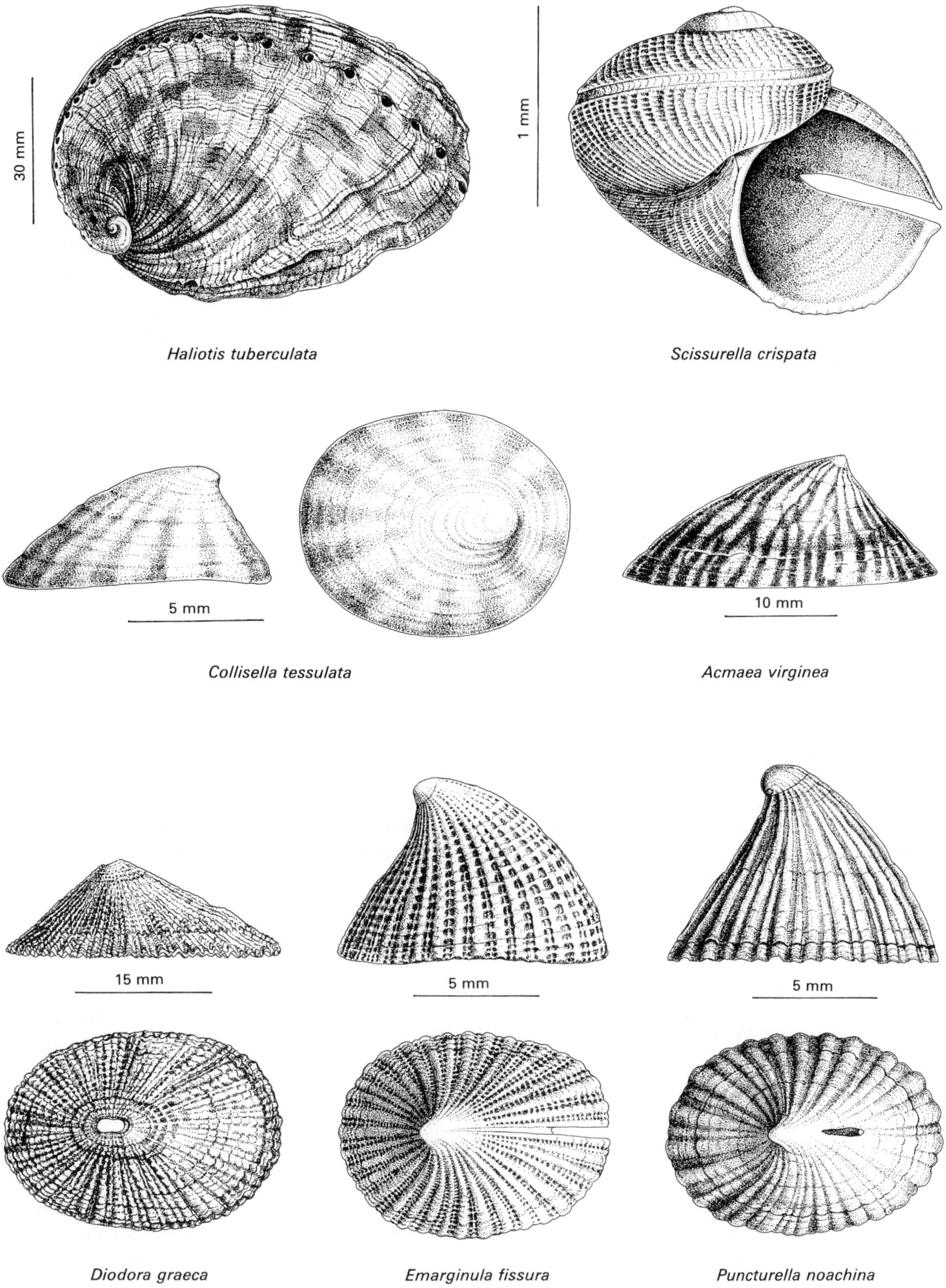
30 mm
1 mm
Haliotis tuberculata
Scissurella crispata
5 mm
10 mm
Collisella tessulata
Acmaea virginea
15 mm
5 mm
5 mm
Diodora graeca
Emarginula fissura
Puncturella noachina

Fig. 12.4

Emarginula conica Lamarck and *E. crassa* Sowerby are sometimes encountered in offshore dredgings. *E. conica* (up to 6 × 4 × 4 mm) has a thinner shell than *E. fissura* with more prominently recurved beak, and 24–28 epipodial tentacles. It occurs off northern France, the Channel Isles, Scilly Isles, and south-west England; single specimens recorded from Arran and the Clyde Sea. The shell of *E. crassa* (up to 33 × 25 × 17 mm) is finely reticulate with about 50 radiating costae (*fissura* has about 30); foot with about 30 epipodial tentacles. A northern species limited in Britain to west Scotland, Orkney and Shetland. Also recorded from south-west Ireland and Antrim.

Puncturella noachina (Linnaeus) FIG. 12.4

Shell cap-shaped with recurved beak terminating in a small spire of one or one and one-half whorls; slit-like hole opens on anterior slope just below beak, in dead shells an internal hemi-conical septum is seen below this opening. Size up to 7 × 5 × 4 mm. Body whitish; foot oval, truncated at front end with 20–30 epipodial tentacles around its base. A bifid tentacle behind eyestalk at base of right cephalic tentacle.

A circumpolar species which extends south to Portugal and the Mediterranean. In British waters confined to sublittoral rocky and stony substrata (especially those rich in sponges) between 20–165 m, and does not occur south of a line between Liverpool Bay in the west and Dogger Bank in the east.

Diodora graeca (Linnaeus) FIG. 12.4

Shell strongly reticulate, conical, oval in plan view. Apex about one-third of shell length behind anterior margin, perforated by a single oval opening, with somewhat flared posterior lip. Size up to 25 × 16 × 10 mm. Foot oval, straight-sided, with 30–35 epipodial tentacles at base; inner lamella of mantle forming thick papillose skirt which hangs below margin of shell. Body cream coloured, deep orange or red.

Under boulders and overhangs on rocky shores from LWST to 250 m. Distributed from Faroes to the Canary Islands. Absent from west coast of Norway, North Sea coasts of England, and eastern parts of Channel.

4 ACMAEIDAE

Shell a shallow, smooth cone with apex towards anterior margin. Foot oval or slightly rectangular, no epipodial tentacles or operculum. Nuchal cavity behind head with single prominent ctenidium, protruding beyond shell margin in active animals; pallial gills absent. Animals firmly attached to hard surfaces.

1. Shell white or pinkish with pink or brownish rays, head scar white; lips round mouth smooth; edge of mantle smooth, white with pink bands	***Acmaea virginea***
Shell white or greenish with chestnut-brown rays, head scar chocolate-brown; lips round mouth fringed; edge of mantle cuprous green, bearing short tentacles	***Collisella tessulata***

Acmaea virginea (Müller) FIG. 12.4

Shell conical, apex toward anterior margin in intertidal specimens, more centrally placed in shells from sublittoral habitats. Anterior profile slightly concave, lateral and posterior profiles convex. Sculpture of fine radiating ridges and concentric growth lines, the latter mostly faint though three or four form conspicuous ledges (annual rings?) in the shell profile. Shell pale pink, yellowish, or white, overlain by broad, often branching rays of pink or brown; rays occasionally broken to form a chequered pattern or, more rarely, absent; pure white shells may occur. Inside of shell white, pale green, or pinkish, often with red V-shaped mark near apex. Size up to 10 × 9 × 4 mm. On the snout, smooth outer lips form a broadly lobed frill, incomplete ventrally, drawn out postero-laterally into extensile lobes. Mantle edge lacking tentacles, with minute, possibly sensory, papillae; light coloured, with pink banding corresponding to shell rays, studded with chalk white repugnatorial glands.

On boulders and small, smooth stones, particularly those encrusted with red algae such as *Lithothamnion*; from LWST to 100 m on all British coasts. Distributed from Iceland to the Azores but absent from the west coast of Jutland and the Baltic.

Collisella tessulata (Müller) FIG. 12.4

Shell conical, apex toward anterior margin with anteriorly projecting beak (more discernible in young shells). Profile more variable than in *A. virginea*; often showing a marked apical inflection in both posterior profile (mainly convex) and anterior profile (mainly straight). Sculpture as in *A. virginea*. Shell with light background overlain by series of irregularly branched brown bands radiating from apex to produce a tessellate pattern. Inside of shell white with outside pattern visible marginally; apical region chocolate brown. Size up to 30 × 24 × 10 mm, commonly 14 × 10 × 4 mm in littoral habitats. Outer lips on snout with narrow, somewhat stiff, outer fringe; incomplete ventrally where a broad, lobed region bounds mouth. Mantle edge cuprous green with two staggered rows of short tentacles.

On boulders and small, smooth stones, particularly those encrusted with red algae such as *Lithothamnion*; from LWNT to 50 m. A circumpolar species extending south to Dublin, Liverpool and the Humber; absent from west coast of Ireland.

5 PATELLIDAE

Shell conical or cap-shaped with no spiral coiling; operculum absent. Foot large and oval, with horseshoe-shaped muscle attachment to shell. Mantle cavity a shallow groove around base of foot, deepening to nuchal cavity behind the head. Mantle edge bears pallial gills; no ctenidium in nuchal cavity. Animals firmly attached to rock (*Patella* spp.) or weed (*Helcion*).

1. Shell with radial ribs; pallial gills along entire length of mantle skirt	**2**
Shell usually smooth or with marked concentric ridges; pallial gills do not extend around anterior margin of nuchal cavity	**4**

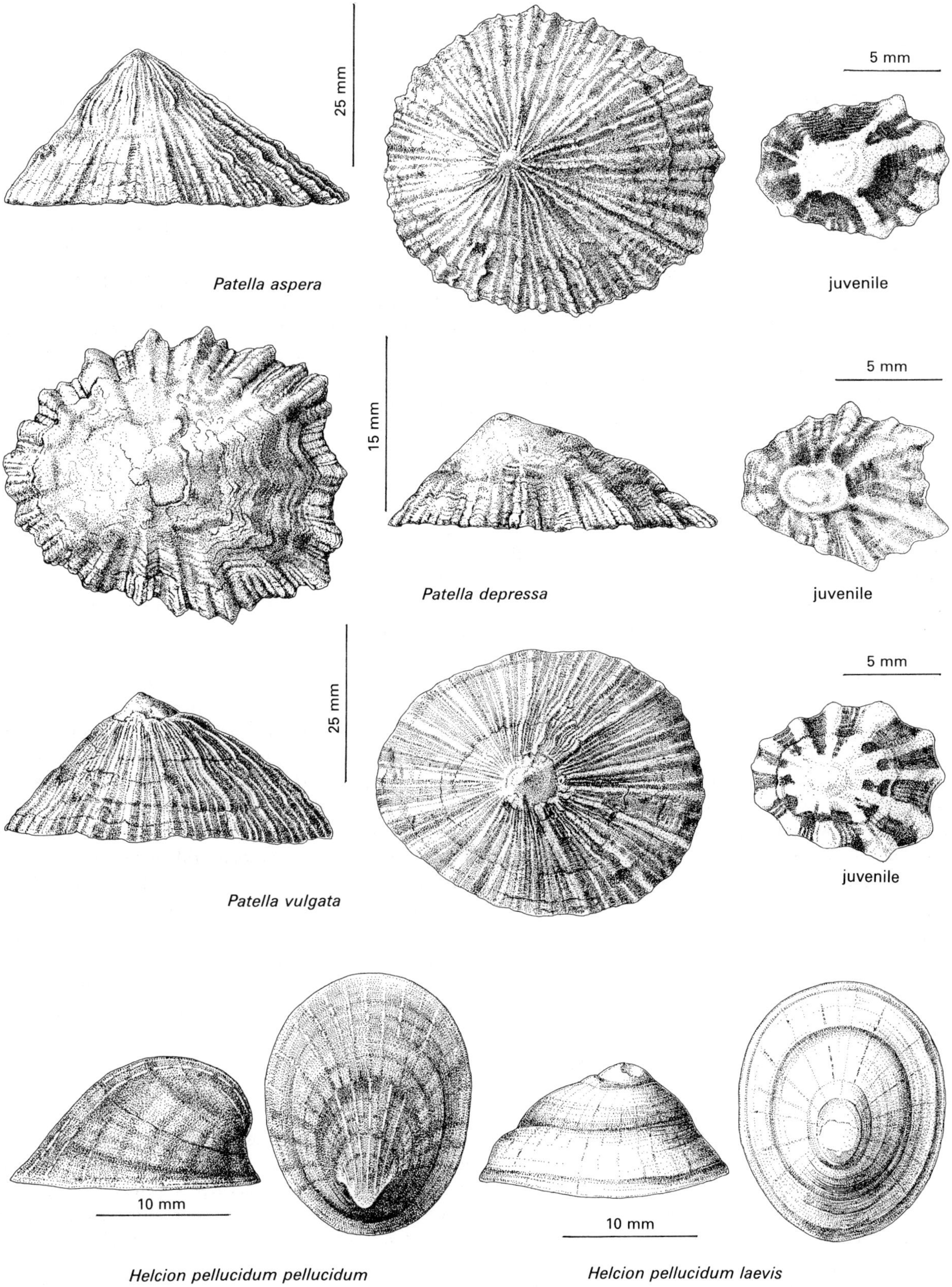
25 mm
5 mm
Patella aspera
juvenile
15 mm
5 mm
Patella depressa
juvenile
25 mm
5 mm
juvenile
Patella vulgata
10 mm
10 mm
Helcion pellucidum pellucidum
Helcion pellucidum laevis

Fig. 12.5

2. Pallial tentacles white **3**

   Pallial tentacles translucent or grey; foot olive, grey, or khaki; inside of shell grey/green, with bluish iridescence ***Patella vulgata***

3. Pallial tentacles transcluscent white; foot apricot-yellow or cream; inner surface of shell porcellaneous, often with cream or orange head scar ***Patella aspera***

   Pallial tentacles chalky white; foot dark olive-black; inner surface of shell cream-orange with chocolate-brown marginal rays ***Patella depressa***

4. Shell small, translucent, depressed, with bright blue rays radiating from apex; on fronds of *Laminaria* ***Helcion pellucidum pellucidum***

   Shell solid, conical, with ledged profile; blue rays on apex of shell only; from *Laminaria* holdfasts ***Helcion pellucidum laevis***

Patella vulgata Linnaeus FIG. 12.5

Shell conical with irregular radiating costae. Apex central or slightly anterior; apical angle variable, smaller (70–80°) in high shore specimens, larger (105–115°) in low shore and young animals. Inner surfaces grey-green. Size up to 60 × 50 × 30 mm. Mantle skirt fringed with translucent pallial tentacles arranged in three series of different lengths, internal to which lies a complete circlet of pallial gills.

From between MHWST–MHWNT downward to ELWST on all suitable rocky shores; upper limit of distribution increased by shade and exposure. Tolerant of low salinity but generally confined to habitats with salinity >25‰. Distributed from Mediterranean to Lofoten Islands, Norway.

Patella aspera Röding FIG. 12.5

Shell conical, commonly with lower profile (apical angle 110–120°) than *P. vulgata* and with apex noticeably anterior to mid-line. Radiating ridges finer than in *P. vulgata*, tending to show characteristic pattern of alternating single and triple ridges, particularly at somewhat straightened posterior margin. Porcellaneous interior, often tinged orange apically, is characteristic, as is combination of cream-orange foot with white pallial tentacles arranged in two series of different sizes; in *P. vulgata* there are three series of pallial tentacles. Size up to 50 × 40 × 20 mm.

On rocky shores at MLWNT and below, favouring areas such as gullies and rock pools; most common on exposed shores avoiding extreme shelter and low salinities. A southern species distributed from the Mediterranean to the British Isles where it achieves its northern limit. Absent from British shores between the Isle of Wight and the Humber, and from continental shores east of Barfleur (near Cherbourg).

Patella depressa Pennant FIG. 12.5

Shell a flattened cone (apical angle approx. 120°) with fine radiating ribs; distinctive orange-brown marginal rays on inner surface. Foot dark olive or black, pallial tentacles chalky white; body otherwise similar to *P. vulgata*. Size up to 30 × 25 × 12 mm.

On exposed rocky shores only, from MHWN to MLWN, prefers vertical surfaces. A Mediterranean species reaching its northern limit on shores of south-west England and Wales; rare on west coast of Ireland.

Helcion pellucidum (Linnaeus) FIG. 12.5

This limpet is almost confined to *Laminaria* plants where it occurs in two forms, *H.p.pellucidum* on the fronds (less frequently on fronds of *Fucus serratus*) and *H.p.laevis* within the holdfast.

The shell of *H.p.pellucidum* is a thin, smooth, depressed cone, with apex close to, or at, anterior margin; posterior profile markedly convex. Translucent, horn-coloured, patterned with bright blue rays. Size up to 20 × 15 × 7 mm.

The taller and more robust shell of *H.p.laevis* has prominent concentric growth lines, giving it a ledged profile, invariably surmounted by a distinct apical region exhibiting features of the *pellucidum* shell. Elsewhere, shell dull and devoid of blue rays, though it may have red rays. Size up to 20 × 8 × 10 mm.

Occurs intertidally on *Laminaria*, and occasionally on low level (LWST) *Fucus serratus*; extends sublittorally to 27 m. On all British and Irish coasts. Distributed from Iceland to Portugal; absent from Belgium, Holland, east coast of Denmark, and Baltic Sea.

6 LEPETIDAE

Shell small, cap-shaped or conical, with or without spirally whorled apical protoconch; sculpture finely reticulate. No operculum. Foot oval or shield-shaped, tapering towards rear. Mantle cavity a shallow groove around base of foot, deepening to nuchal cavity behind head. Cephalic tentacles without pigmented eyes at base, ctenidia and pallial gills absent. Animals sublittoral, attached to stones.

---

1. Shell with recurved apex bearing spiral protoconch; internally a small septum below apex ***Propilidium exiguum***

   Shell apex without spiral protoconch; no internal septum **2**

2. Shell white, conical, with apex central or slightly anterior ***Lepeta caeca***

   Shell orange-yellow (rarely white), conical, with apex near anterior margin ***Iothia fulva***

Lepeta caeca (Müller) FIG. 12.6

Shell small, conical, white, opaque; apex central or slightly anterior, often eroded to reveal nacreous layer below. Sculpture of fine reticulation; radial ridges may be papillate, especially towards shell margins. No internal apical septum, anterior limit of horseshoe-shaped muscle scar half-way between apex and anterior margin. Size up to 17 × 13 × 7 mm (commonly half this

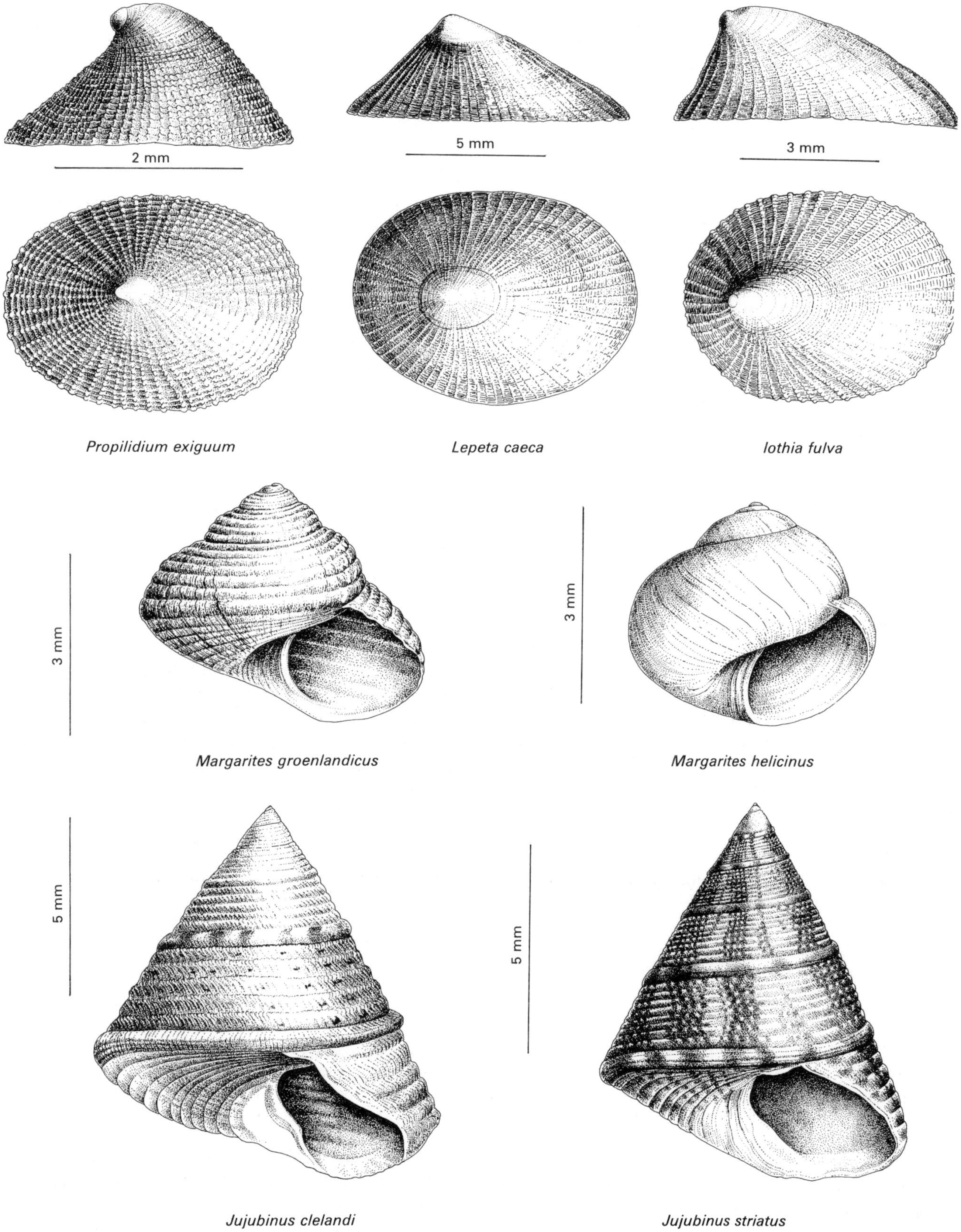
2 mm
5 mm
3 mm
Propilidium exiguum
Lepeta caeca
Iothia fulva
3 mm
3 mm
Margarites groenlandicus
Margarites helicinus
5 mm
5 mm
Jujubinus clelandi
Jujubinus striatus

Fig. 12.6

size). Body patelliform, though lacking pallial gills and tentacles; cephalic tentacles without pigmented eyes. Foot tapering posteriorly; mantle edge simple; mouth transversely elongated.

Sublittoral on hard substrata, gravels and clay. Circumpolar distribution, extending south at increasing depths to the Azores. Rare in the British Isles, records of living animals are confined to dredgings off the Shetlands and the west coast of Scotland.

Iothia fulva (Müller) FIG. 12.6

Shell small, conical, orange-yellow (occasionally white), opaque; apex near anterior margin, with blister-like protoconch. Sculpture of fine reticulation, radial ridges may be papillate. No internal apical septum, anterior limit of horseshoe-shaped muscle scar lies at same level as apex. Size up to $7 \times 5 \times 2.5$ mm. Body patelliform, similar to *Lepeta*.

Sublittoral, on stony ground from 5–200 m. Distributed from Iceland southwards, at increasing depth (to 600 m), to the Azores. More common than *L. caeca* in the British Isles, *I. fulva* has been recorded off the west coast of Scotland, Irish Sea, west coast of Ireland and British (not continental) North Sea coasts.

Propilidium exiguum (Thompson) FIG. 12.6

Shell small, cap-shaped, white, semi-transparent; apex central, recurved posteriorly, with spiral protoconch. Radiating ridges and growth lines produce finely reticulate surface, often more discernible on younger parts of shell. Internally, a small septum projects from posterior face of shell below apex. Size up to $4 \times 3 \times 2.5$ mm. Body patelliform, similar to *Lepeta* but with mouth elongated in anterio-lateral direction and mantle edge slightly crinkled.

Sublittoral only, on hard substrata, from 7 to 170 m, deepest distribution in southern parts of range. Off Shetlands, west coast of Scotland and Ireland. Distributed from Norway south to Azores and Mediterranean; absent from Baltic, North Sea, and English Channel.

7 TROCHIDAE

Shell spiral, conical (often pyramidal); frequently umbilicate, mother-of-pearl inside often revealed at eroded apex. Plane of aperture markedly prosocline. Foot wth extensile epipodial tentacles; operculum polygyrous, circular. Detrital sweepers, with the exception of *Calliostoma* spp. which are probably grazing carnivores.

1. Shell a sharply pointed cone; whorls flat **2**
   Shell bluntly pointed, rather cyrtoconoid; whorls tumid **5**

2. Shell height approximately equal to breadth **3**
   Shell height approximately twice breadth; body whorl with eight or nine smooth spiral ridges ***Jujubinus striatus***

3. Nodular spiral ridges and growth lines interact to produce spiral rows of lozenge-shaped dimples; oral lips without tubular extension; sublittoral ***Jujubinus clelandi***
   Sculpture not as described; oral lips drawn out mid-ventrally as a spout-like projection to right **4**

4. Spiral ridges smooth; spire regularly conical; four or five pairs of epipodial tentacles ***Calliostoma zizyphinum***
   Spiral ridges papillate; spire coeloconoid; three pairs of epipodial tentacles; rarely intertidal ***Calliostoma granularis***

5. Columella with tooth or bulge; outside of shell not nacreous, or only so at eroded apex; animal with three pairs of epipodial tentacles **6**
   Columella a smooth simple curve, without bulge; shell externally nacreous; animal with five to seven pairs of epipodial tentacles **11**

6. Umbilicus a small depression; finely patterned dark shell often eroded to underlying nacre at apex; height > breadth; columella tooth prominent ***Monodonta lineata***
   Umbilicus open or closed; shell height ⩽ breadth; bulge on columella does not form prominent tooth **7**

7. Upper part of each whorl tuberculated; shell profile stepped; umbilicus wide and deep, its margins limited by a spiral ridge arising from the base of the columella ***Gibbula magus***
   Upper part of each whorl not tuberculated **8**

8. Shell profile slightly turreted; body whorl with 16–18 spiral ridges above and 14–18 below periphery; umbilicus deep but occluded to varying degree by inner lip; mainly sublittoral ***G. tumida***
   Shell not turreted, spiral ridges less numerous; umbilicus variable or absent; mainly littoral **9**

9. Shell light grey/yellowish in colour with fine (0.5 mm or less) striped pattern of darker colour; umbilicus narrow (occasionally absent in older shells) ***G. cineraria***
   Shell light greenish/horn coloured with broad stripes (1 mm or more) of red or purple; more depressed, cyrtoconoid profile **10**

10. Umbilicus round and deep (occluded in some older specimens); reddish purple stripes are usually entire on base of shell ***G. umbilicalis***
    Umbilicus in young specimens only; coloured stripes are deeper purple and often give rise to a chequered pattern of rectangular, pigmented spots on the base of the shell. Channel Isles only ***G. pennanti***

11. Shell smooth, nacreous, glossy; spiral ridges confined to region around umbilicus; animal with six pairs of epipodial tentacles ***Margarites helicinus***
    Shell sculpture of weak or strong spiral ridges on all whorls; seven pairs of epipodial tentacles ***M. groenlandicus***

Margarites helicinus (Fabricius) FIG. 12.6

Shell depressed, smooth, conical, somewhat globular; with four to five tumid whorls dipping abapically into sutures. Close examination reveals sculpture of fine, sinuous growth lines, with spiral ridges and grooves on basal part of last whorl. Umbilicus oval, conspicuous, partly occluded by inner lip. Orange-red or horn-coloured with green or purple refringency; often with brown band on last whorl. Size up to 9 × 11 mm, smaller in south. Cephalic and epipodial tentacles, and dorsal surface of snout, with numerous sensory papillae. Foot narrow, with anterior cleft (pedal gland) and irregularly ridged sides. The six (sometimes five) pairs of epipodial tentacles have, in addition to their sensory papillae, a pigmented sense-organ below base. Operculum polygyrous (10–12 turns), slightly concave.

Under stone, in pools and on weed, on lower parts of rocky shores, and sublittoral. A circumboreal species, extending south to shores of Yorkshire on the east coast and the Isle of Man and North Wales on the west. Locally abundant.

Margarites groenlandicus (Gmelin) FIG. 12.6

Shell depressed, conical, with five to six tumid whorls dipping sharply and abapically into sutures. Sculpture variable; usually spiral ridges and distinct growth lines; occasional shells smooth or with only one or two furrows at periphery and fine sculpturing near umbilicus. Latter form common in Greenland, spiral ridging common in European and American waters. Umbilicus wide, partly occluded by inner lip. Cream, pale tan, or pale orange, with blue-green refringence internally. Size up to 7 × 8 mm. Foot with broad, double-edged anterior and smooth sides; seven pairs of epipodial tentacles each with sense-organ below base. Operculum polygyrous (10–12 turns), markedly concave.

Under stones, on weeds and on *Laminaria* holdfasts from ELWST to 75 m. A northern species extending south from Arctic waters to north-west coasts of British Isles and west coast of Norway. Not recorded from North Sea; rare on British coasts.

Margarites argentatus (Gould) is a circumboreal species which occurs rarely off western Scotland. It has a globose shell of four tumid whorls, with fine spiral sculpture and growth lines; olive, grey, cream, or white with a green or purple refringence. The foot has only five pairs of epipodial tentacles.

Gibbula cineraria (Linnaeus) FIG. 12.7

Shell solid, conical, slightly cyrtoconoid, with five or six slightly tumid whorls; last whorl angulate, base flat. Sculpture of fine spiral ridges and grooves, 10–17 (commonly 13–14) on base of last whorl, traversed by prosocline, S-shaped, growth lines. Umbilicus small, oval or egg-shaped, deep. Umbilical cavity revealed as a small pinhole at apex of eroded shells. Greyish or straw coloured with many thin opisthocline stripes of reddish-brown or purple resulting from alignment, or near alignment of pigmented spots on successive spiral straie. Size up to 16 × 15 mm. Sensory papillae on snout, cephalic tentacles and three pairs of epipodial tentacles. Foot oval, fringed with papillae, without anterior pedal gland.

On rocky shores, on weed and under stones, below LWNT (may occur higher in rock pools, especially in north), extending sublittorally to about 130 m. Distributed from Gibraltar to north Norway and Iceland. Abundant on all British coasts.

In populations of *G. cineraria* from the Channel coasts of Devon and Cornwall many individuals have been found in which the umbilicus is either wholly or partially occluded by the inner lip. This tendency towards umbilical occlusion appears to increase with age.

Gibbula umbilicalis (da Costa) FIG. 12.7

Shell similar to *G. cineraria* but distinguished by a markedly cyrtoconoid shape (especially in older shells), a more angular last whorl (especially in younger shells), fewer (8–11) spiral ridges on base of last whorl, a larger and more circular umbilicus and a cream or greenish background colour with broad stripes or bands of red or reddish-purple. Size up to 16 × 22 mm. Body similar to *G. cineraria*; papillae fringing foot more pronounced. Non-umbilicate forms have been recorded from south-east Cornwall and south-west Devon.

Habitat similar to that of *G. cineraria*, with which it is often found; *G. umbilicalis* is more tolerant of emersion and therefore exhibits higher vertical zonation (MHWS–MLWS). A more southern species than *G. cineraria*, extending along Atlantic coast of France and west coast of Britain, Ireland, and northwards to Orkney. Absent from North Sea and British coasts in eastern basin of Channel.

Gibbula pennanti (Philippi) FIG. 12.7

Shell similar to *G. umbilicalis* but with upper whorls tending to be more tumid, with sutures more prominent, and with slightly turreted profile, reminiscent of *G. tumida* and *G. magus*. Umbilicus open in young shells, closing later to form a depression against inner lip; occasionally a small umbilical chink may persist. Shell colour similar to *G. umbilicalis*, but with reticulate rather than striped pattern on base of last whorl; some shells homogeneous coppery red, green, or black, with or without red or white spiral bands at periphery. Size up to 16 × 15 mm (large shells may be taller than broad).

Habitat similar to *G. cineraria* and *G. umbilicalis*; particularly common in low-shore pools. A southern species occurring on the Atlantic coasts of Spain and France and in the Channel Islands. Absent from mainland Britain and French Channel coasts east of Barfleur (near Cherbourg).

Gibbula magus (Linnaeus) FIG. 12.7

Shell solid, conical, with stepped profile and marked peripheral keel on last whorl; on upper whorls keel lies immediately above deep suture line. Up to eight whorls. Sculpture of short, nodular costae on apical part of each whorl, below which are a variable number of spiral ridges and grooves. In intertidal specimens ridges are often smooth, in specimens from deeper water they may be markedly imbricate. Umbilicus large and deep, partly occluded by inner lip, edged by thickened spiral ridge. White, grey, or yellowish in colour with irregular brown, red, or purple streaks and blotches. Size up to 30 × 35 mm. Snout densely papillated, cephalic tentacles setose; foot with papillate margin and three pairs of epipodial tentacles. Operculum polygyrous (13–15 turns), slightly concave.

On weed, under stones, and on shingle beaches at ELWST. Sublittoral, on muddy gravel, to 70 m. Distributed from Mediterranean and Azores to western basin of English Channel, west coast of Britain, and north to Shetland.

Gibbula tumida (Montagu) FIG. 12.7

Shell solid, cyrtoconoid, with stepped profile of six to seven slightly tumid, flat-shouldered whorls meeting at well-defined sutures. Sculpture of fine prosocline growth lines traversing spiral grooves and ridges. Umbilicus open and deep in young specimens, occluded by inner lip with age, and in larger shells appearing as an oval opening or small chink. Light fawn with broken brown bands across whorls; usually with greenish-blue iridescence. Size up to 9 × 10 mm. Body similar to *G. magus*; snout not so papillate, sides of foot more so.

Sublittoral only; on *Laminaria*, and down to 1200 m on stones on sandy-muddy bottoms. A northern species extending south from Iceland and north Norway to British, Irish, French, and Iberian coasts.

Monodonta lineata (da Costa) FIG. 12.7

Shell solid, conical, with five to six tumid, rapidly expanding whorls. Sculpture of prosocline lines and occasional clear growth disturbances which may reflect annual increments (Williamson and Kendall 1981); young specimens with thin spiral ridges and grooves on first three whorls, in older shells often abraded to reveal underlying nacre, subsequent whorls being smooth. Columella thick, with prominent tooth near base; inner lip reflected over umbilicus which remains as slight depression or chink in adults shells, umbilicus of juveniles may be open. Cream or horn-coloured, heavily overlain with reddish-brown, purple, or green zigzag streaks; base of shell less heavily pigmented. Iridescent nacre on inner surface. Size up to 30 × 25 mm. Cephalic tentacles with lobed sides; foot with finely papillate margin and tuberculate sides; three pairs of epipodial tentacles, each with two sensory papillae at base. Operculum flat, polygyrous (around 15 turns).

On rocky or boulder shores from MHWS–MLWN, avoiding excessive exposure and deposits of shingle or sand. A southern species with a rather patchy distribution in the British Isles, occurring in the western basin of the English Channel, the Welsh coast north to Anglesey, and the south and west coasts of Ireland.

Calliostoma zizyphinum (Linnaeus) FIG. 12.8

Shell regularly conical, straight-sided, up to 12–13 whorls; sculpture of regular spiral grooves and ridges traversed by fine prosocline growth lines. Last whorl with prominent peripheral keel bearing two broad ridges; ridges above suture in preceding whorls. Base of shell rather flat, inner lips reflected over shallow umbilical groove. Colour variable, yellowish, pale pink, or violet with streaks and blotches of brown, red, or purple. Blotches on keel generally darker, more frequent and more regular than on other parts of shell. Pure white or violet specimens occasionally found. Size up to 30 × 30 mm. Head with papillate snout and finely papillate tentacles; mouth T-shaped when closed. Foot with anterior pedal gland, transversely ridged sole and papillate sides; four or five pairs of epipodial tentacles, lacking basal sense-organ. Operculum polygyrous (about 15 turns), shallowly concave.

Below LWNT on weed covered rocky shores, extending sublittorally on hard and soft bottoms to 300 m; tolerant of salinities down to 21‰. Distributed from Mediterranean, Canaries and Azores to Norway. All British and Irish shores.

Calliostoma papillosum (da Costa) FIG. 12.8

Shell with distinctly coeloconoid (concave) spire of seven to nine whorls. Sculpture of regularly papillate spiral ridges, the most peripheral of which are more prominent, forming a pronounced keel on last whorl, and clearly distinguished above the sutures on preceding whorls. Base of shell more tumid than in *C. zizyphinum*, inner lip reflected over umbilicus which may persist as a narrow chink. Cream, yellow, or orange, with brown or reddish blotches on last whorl and spire, and small spots on basal ridges. Size up to 35 × 35 mm. Body similar to *C. zizyphinum*, but with more papillate snout, mouth I-shaped or Y-shaped when closed, cephalic tentacles setose, foot with only three pairs of epipodial tentacles.

Sublittoral only, on soft bottoms from 7–300 m. Distributed from Canaries and Mediterranean to Shetland. Absent from North Sea, rare up to Scottish west coast and the eastern basin of the English Channel. More regularly encountered in the western basin though never common.

Dredgings off British North Sea coasts, north of the Humber, may yield *C. formosa* (McAndrews and Forbes). This species may be up to 12 × 11 mm, the shell is conical and nacreous with seven or more semi-transparent whorls ornamented by prominent, keel-like spiral ridges.

Jujubinus clelandi (Wood) FIG. 12.6

Shell solid, pyramidal, with seven to eight straight-sided whorls, with a slight peripheral keel. Sculpture of prosocline growth lines and spiral grooves and ridges, producing lozenge-shaped dimples. With or without narrow V-shaped umbilicus. White, greenish, or pinkish, often with pink or brown spots. Size up to 18 × 18 mm but commonly 12 × 12 mm. Snout without papillae, mouth I-shaped when closed; cephalic tentacles long, slender, with fine sensory papillae. Foot with papillate sides; three pairs of papillate epipodial tentacles, with basal sense-organs. Operculum pale yellow, polygyrous, with 11–13 turns.

Sublittoral only, 35–800 m, not uncommon on stone or gravel bottoms, less common on sand or muddy sand. Distributed from Mediterranean north to Norway. More common in north of range, absent from eastern basin of English Channel.

Jujubinus striatus (Linnaeus) FIG. 12.6

Shell solid, conical with six or eight flat-sided whorls. Sculpture of spiral ridges; growth lines fine, may give nodular appearance to spiral ridges, especially in apical and basal areas of shell. Inner lip occludes umbilicus in adults, young shells with open umbilicus. White, creamy, or greenish with brown blotches

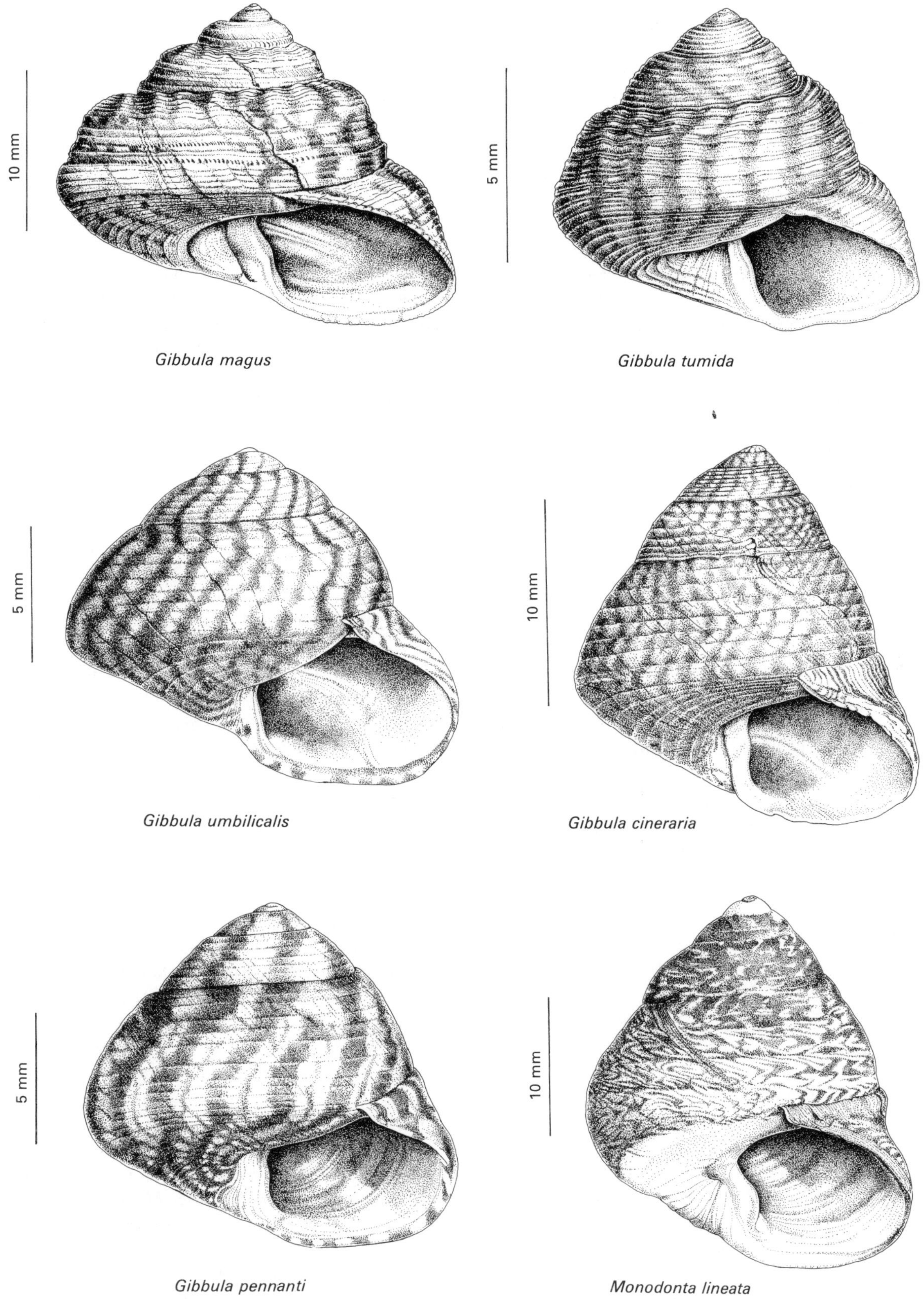
10 mm
Gibbula magus
5 mm
Gibbula tumida
5 mm
Gibbula umbilicalis
10 mm
Gibbula cineraria
5 mm
Gibbula pennanti
10 mm
Monodonta lineata

Fig. 12.7

and stripes. Apical two or three whorls often pink. Size up to 10 × 8 mm. Snout with some sensory papillae, cephalic tentacles setose, mantle edge double and slightly lobed. Foot with papillae on sides, and edging sole; anterior margin double. Three pairs of epipodial tentacles, each with basal sense-organs.

On weeds, especially *Zostera*, at LWST and down to 200 m. A southern species extending northward from the Mediterranean to south-west coasts of the British Isles; north as far as Isle of Man, and east to the Channel Isles where it is common.

J. exasperatus (Pennant) has similar distribution and habitat but is less common. It may be distinguished from *J. striatus* by more nodular ornamentation to the spiral ridges, which are fewer in number (four to five compared with eight to nine) on the upper part of the body whorl. Size up to 8 × 7 mm.

Dredgings over stony or gravel bottoms between 10–200 m off the north-east coast, over the Dogger Bank and in the western basin of the English Channel may yield *J. montagui* (Wood). The body is similar to both *J. striatus* and *J. exasperatus* but the shell is more cyrtoconoid, with six or seven slightly tumid whorls bearing rather broad, flat spiral ridges. Size up to 8 × 6 mm.

8 SKENEIDAE

Minute white or colourless shells with three to four tumid and rapidly expanding whorls; often with sculpture of fine spiral striae, though this may be confined to basal areas of last whorl. Aperture with distinct peristome; umbilicus wide. Protoconch smooth, inflated, and without pointed tip found in other trochacean families (Trochidae, Turbinidae, and Tricoliidae). Body more clearly trochacean; foot with three pairs of epipodial tentacles, operculum polygyrous. Often overlooked because of small size.

1. Umbilicus wide, revealing underside of all whorls; spiral ridges cover base of last whorl only — ***Skenea serpuloides***
   Umbilicus oval and narrow — **2**
2. Spiral ridges confined to area immediately adjacent to umbilicus — ***Skenea nitens***
   Spiral ridges on all whorls — ***Skenea cutleriana***

Skenea serpuloides (Montagu) FIG. 12.8

Shell small, white or colourless; thin and glossy with three to four tumid, rapidly expanding whorls. Last whorl constitutes most of shell; adapically smooth, as are preceding whorls, underside enhanced by sharp spiral ridges and grooves. Aperture almost circular, with complete peristome; inner lip not reflected over umbilicus, which is wide, revealing underside of spire. Size up to 1.5 × 1.2 mm. Cephalic tentacles broad and flat basally, fringed with sensory papillae; an additional tentacle, the postoptic, behind right cephalic tentacle. Papillae also on mantle edge and epipodial tentacles. Foot with anterio-lateral tentaculate extensions. Operculum polygyrous (about six turns), transparent.

On rocky shores among weeds and detrital sediments at LWST; sublittoral on gravelly and shelly sand down to 50 m. A southern species extending north to the British Isles where it has been recorded off the west coast of Ireland, south-west England, and North Sea coast as far south as Scarborough. Patchy distribution may result in part from collectors overlooking these small shells.

Skenea nitens (Philippi) FIG. 12.8

Shell small, white or pale horn-coloured; smooth, glossy with low spire of three to four tumid, rapidly expanding whorls. Sculpture confined to fine, irregular growth lines, augmented in umbilical region by a few spiral ridges. Aperture with sinus, or embayment, at periphery and base; inner lip reflected over and partly occluding oval umbilicus. Size up to 1 × 1 mm. Body as for *S. serpuloides*, but lacking postoptic tentacle and with shorter anterio-lateral processes on foot.

Habitat and distribution as for *S. serpuloides*, though recorded also from Norway; generally considered to be less common in British waters.

Skenea cutleriana (Clark) FIG. 12.8

Shell similar to *S. nitens* but with taller spire of three whorls, all enhanced by sculpture of spiral ridges and grooves. Umbilicus narrow, partly occluded by inner lip; aperture with complete but slightly irregular peristome. Size up to 1 × 1 mm. Body as for *S. serpuloides* but lacking postoptic tentacle and with broader, shorter anterio-lateral processes on foot.

Habitat similar to that of *S. serpuloides* and *S. nitens* but with more southern distribution. In the British Isles it is limited to south-west coasts, north to the Isle of Man; absent from North Sea and Scandinavia.

Skenea basistriata (Jeffreys) (Fretter and Graham 1977: 91); may be dredged rarely from deep water (90–1400 m) off the Atlantic coasts of Britain, Norway, and France. It may be distinguished from the previous species by a more turreted outline to shell, a narrow, circular umbilicus, and a groove on the apertural margin of inner lip into which the operculum fits. Spiral sculpturing may occur over all of the shell but is more common on the base of the body whorl.

9 TRICOLIIDAE

Only one British representative of this family.

Tricolia pullus (Linnaeus) FIG. 12.8

Shell conical or turban-shaped, with deep sutures; solid, smooth, and glossy. Last whorl approximately two-thirds shell length; aperture pear-shaped with shallow umbilical groove. Cream or white with irregular overlay of red or purple streaks and blotches; streaks often produce a flowing zigzag pattern. Columella, inner lip, and operculum milk-white. Specimens lacking a coloured overlay occasionally found. Size up to 9 mm. Cephalic tentacles long, slender, papillate; neck lobes comb-like with 11–14 papillate processes. Foot narrow, smooth, with double edge anteriorly and laterally; three pairs of papillate epipodial tentacles. Operculum calcareous, markedly convex in profile. *Tricolia*

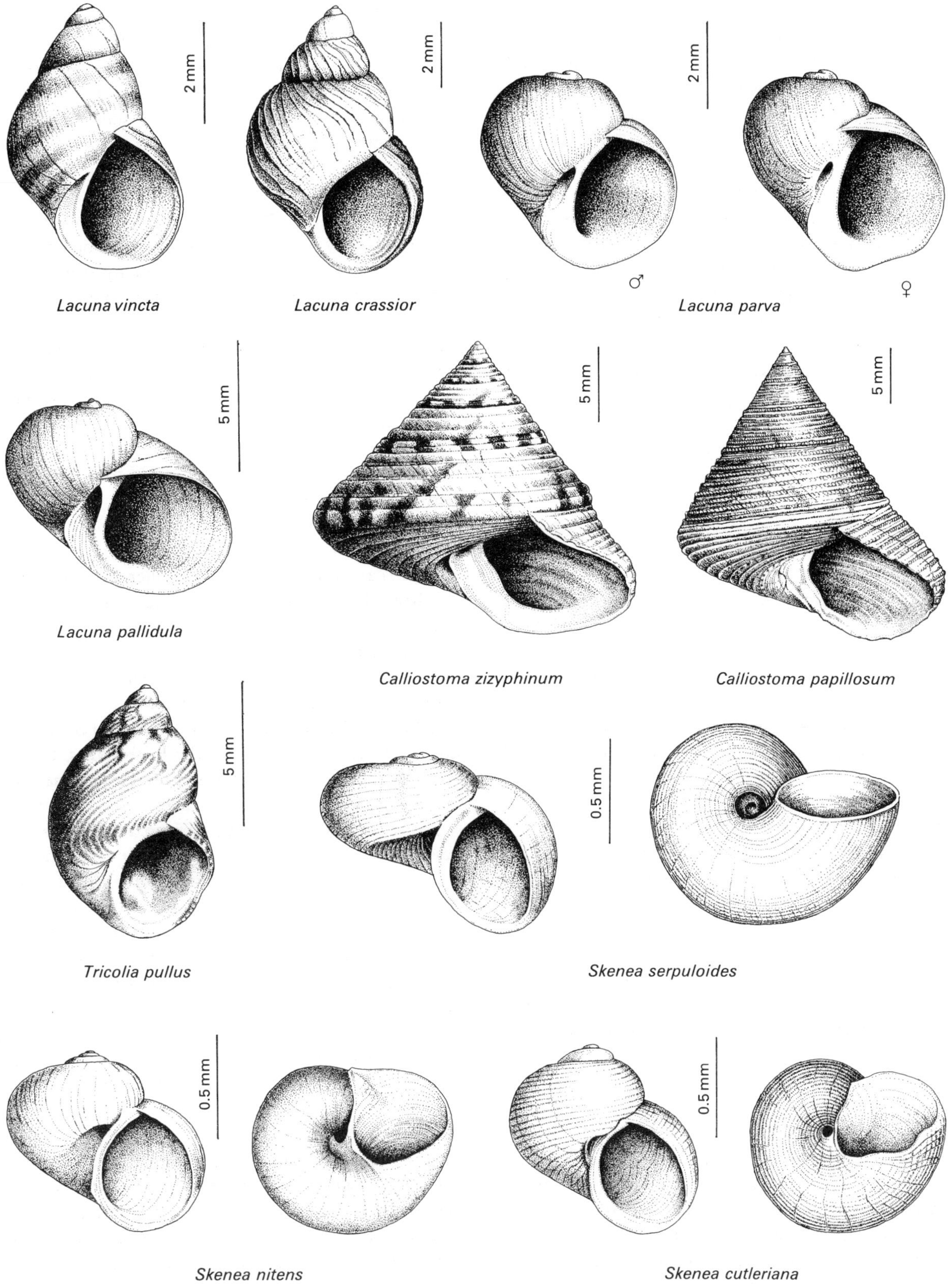
2 mm
2 mm
2 mm
Lacuna vincta
Lacuna crassior
♂
♀
Lacuna parva
5 mm
5 mm
5 mm
Lacuna pallidula
Calliostoma zizyphinum
Calliostoma papillosum
5 mm
0.5 mm
Tricolia pullus
Skenea serpuloides
0.5 mm
0.5 mm
Skenea nitens
Skenea cutleriana

Fig. 12.8

pullus is the only British prosobranch with such an operculum.

On rocky shores near LWST and sublittorally to 35 m; intertidally most abundant on tufted red algae such as *Lomentaria*, *Laurencia*, and *Chondrus*. A southern species extending from the Mediterranean to the west and south-west coasts of Britain and Ireland; rare in the North.

10 LACUNIDAE

Shell a tumid or turreted spire, or globular with last whorl prominent and spire much reduced; usually small, thin and smooth, often with an umbilical chink or groove on the columella. Foot with two prominent metapodial tentacles; operculum a rapidly expanding spiral or two or three turns, thin and horny.

---

1. Shell with five to seven tumid or turreted whorls; aperture drawn out, angulate below columella — **2**

   Shell globular, spire much reduced; aperture not drawn out at base, or only slightly so — **3**

2. Shell solid; six to seven turreted whorls with fine striae and costae below thick periostracum; no colour banding, with or without umbilicus — ***Lacuna crassior***

   Shell thin; five to six smooth, tumid whorls, usually with brown banding; umbilicus obvious — ***Lacuna vincta***

3. Aperture nearly equal to shell height; without colour bands on last whorl — ***Lacuna pallidula***

   Aperture much less than this, usually with red-brown banding on last whorl — ***Lacuna parva***

Lacuna vincta (Montagu) FIG. 12.8

Shell thin, with five to six tumid whorls; periostracum thin, though thickened patches may line sutures. Shell smooth to naked eye, but with fine spiral ridges and prosocline growth lines. Umbilicus $\wedge$-shaped, with a prominent groove or chink bounded on apertural margin by inner lip, and on abapertural side by upturned outer lip; umbilical groove smooth, aperture below drawn out and slightly angular. Pale horn colour with two to four red-brown bands on last whorl; upper whorls with two colour bands, becoming faint towards apex which is generally purplish. Banding occasionally faint or absent. Size up to 10 × 5 mm. Snout long, broad, with slender cephalic tentacles arising from base. Foot elongate, with double edged anterior margin; two short, flat metapodial tentacles project from below operculum.

Common on seaweeds, occasionally on *Zostera*, at LWST, sublittorally to 40 m. Tolerant of salinities down to 20‰. A circumboreal species extending south in Europe as far as the Channel and, less frequently, the Atlantic coasts of France. On suitable weed-covered shores on all British and Irish coasts.

Lacuna crassior (Montagu) FIG. 12.8

Shell strong, yellow, with six to seven turreted whorls. Fine spiral striae and vertical costae mostly obscured by thick periostracum, forming a series of irregular folds on last whorl. Umbilicus occasionally closed and umbilical groove reduced; aperture drawn out and angulate at base of columella. Size up to 14 × 10 mm. Body as in *L. vincta* but with metapodial tentacles white and less flattened.

Uncommon. Occurs on mud/gravel at LWST and sublittorally to 90 m. A circumboreal species, extending south into British waters where it has been recorded from Dorset, Milford Haven, and the Celtic Sea, and north-east coasts north of Scarborough. It may be widespread.

Lacuna parva (da Costa) FIG. 12.8

Shell small, globular, with three rapidly expanding whorls; sutures deep. Apparently smooth but with delicate ornamentation of fine spiral striae and growth lines. Last whorl occupies most of spire; aperture rounded in male; more ovate, with basal flare or spout, in female. Umbilicus as in *L. vincta*; umbilical groove with longitudinal ridges. Red-brown or horn-coloured, often with rufous bands on last whorl and lilac tinge to apex. Size up to 4 × 4 mm.

On seaweeds, lower shore (occasionally higher, where it may occur on *Fucus vesiculosus*) and sublittorally to 40 m. Distribution similar to that of *L. vincta* but extending further south to Spanish coasts.

Lacuna pallidula (da Costa) FIG. 12.8

Shell globular, thin and smooth, with three or four rapidly expanding whorls; sutures deep. Last whorl occupies most of spire; aperture very large, almost equal to shell height. Umbilicus deep, open, revealing much of interior surface of spire; umbilical groove wide (wider in males than in females), with longitudinally ridged surface. Yellowish green, occasionally white; unbanded. Size up to 12 × 6 mm in females, 6 × 3 mm in males. Body as in *L. vincta*.

On seaweeds at LWST, extending sublittorally to 70 m. Distribution similar to that of *L. vincta*.

11 LITTORINIDAE

Shell solid, with or without a moderately high spire, often pigmented. Sculpture of growth lines, with or without spiral striae; never costate. Foot shield-shaped, with truncated, double edged anterior; without epi- or metapodial tentacles. Operculum a spiral of two to three rapidly expanding turns.

Because of the degree of overlap in shell characteristics and internal morphology between some species it may not be possible to reach a decision as to their identity without access to a number of live specimens. This is particularly so for *L. saxatilis*, *L. rudis*, and *L. arcana*, where dissection is invaluable, though not infallible. In particular the difference in the ciliated field alongside the rectum between *Littorina arcana* and *L. rudis*/*L. saxatilis* is not by itself a reliable character.

---

1. Spire of shell low (apical angle >110°) or absent; apex blunt or completely flat — **2**

   Shell with well-developed spire; apex pointed — **3**

2. Shell with small but distinct spire; outer lip arises below level of apex; aperture wide **Littorina obtusata**

   Spire very low or absent; outer lip arises level with apex; aperture narrow, with constricted throat **Littorina mariae**

3. Spire of flattened or slightly tumid whorls and shallow sutures; outer lip meets last whorl tangentially **4**

   Spire of tumid whorls and deep sutures; outer lip meets last whorl almost at right angle **5**

4. Adult shell large (up to 38 mm high); usually with spiral ridges (these may be more obvious in young specimens); columella white; cephalic tentacles with transverse black stripes **Littorina littorea**

   Adult shell small (up to 9 mm high), smooth; columella pigmented; periostracum extends beyond outer lip; cephalic tentacles with longitudinal black stripes **Littorina neritoides**

5. Shell small (usually <5 mm), globular, with short spire; thin and superficially smooth, with brown spiral band entering aperture; floor of mantle cavity dark; penis with long tip; ovoviviparous **Littorina neglecta**

   Shell with well-developed spire; often with spiral striae, though these may be eroded or otherwise indistinct; no brown band entering aperture; floor of mantle with diagonal white band; oviparous or ovoviviparous **6**

6. Animal found on sheltered brackish shores, commonly on weeds, often sublittoral; shell superficially smooth, dark and thin **Littorina tenebrosa**

   Animals from the upper levels of fully saline shores, in crevices and under stones; shell usually with spiral ridges, solid **7**

7. Shell with strap-like spiral ridges often with a shallow central furrow; grooves between ridges often darkly pigmented; penis with very short tip and up to 12 glands; oviparous **Littorina nigrolineata**

   Shell usually with sharper spiral ridges, though strap-like ridges may occur; penis with moderately long tip; oviparious or ovoviviparous **8**

8. Oviparous; prostate does not reach level of anus and has a small ciliated field alongside; penis with tapering tip; found with *L. rudis* from which it is indistinguishble in most cases in respect of shell form **Littorina arcana**

   Ovoviviparous; prostate reaches beyond anus and has a large ciliated field alongside; penis with blunt tip **9**

9. Spiral ridges usually with crest at their adapical edge; aperture with prominent everted area at base of columella; in *Pelvetia* zone, among weeds and on stones **Littorina rudis**

   Spiral ridges with central crest; only small everted area at base of aperture; in high-shore crevices (*Pelvetia* zone and above) **Littorina saxatilis**

Littorina littorea (Linnaeus) FIG. 12.9

Shell solid, with five or six slightly tumid whorls; sutures shallow, in some specimens upper whorls may be flat-sided or even coeloconoid. Spire prominent, pointed, more so in young shells, though last whorl may occupy up to 85% of shell height. Shell may appear smooth, especially in older specimens, but has prosocline growth lines and numerous, slight spiral ridges; latter may be strap-like on last whorl. Aperture ear-shaped, outer lip arising tangential to last whorl; inner lip thick, reflected over base of columella. No umbilicus or umbilical groove. Generally black or dark grey-brown, often lighter towards apex, with heavier pigmentation between spiral ridges; occasional red, orange, or white individuals occur. Columella region white, outer lip with a darkly pigmented band internally, contrasting with lighter pigmentation of throat. Size up to 32 × 25 mm. Cephalic tentacles rather flat and broad, with many transverse black stripes and, ventrally, a single longitudinal line. Foot and operculum of typical littorinid pattern.

Common on all but the most exposed rocky shores; mainly intertidal, the upper range increasing with exposure, but extends sublittorally to 60 m in the north. In sheltered conditions it may be found undertaking feeding excursions over muddy shores; tolerant of brackish water. Distributed from northern Spain to White Sea. On all British coasts, though rare or absent in Isles of Scilly and Channel Isles.

Littorina obtusata (Linnaeus) FIG. 12.9

Shell solid, with blunt apex and very much reduced spire, last whorl tumid, comprising >90% of shell height; sutures shallow, lying well above periphery of preceding whorl so as to obscure much of it from view. Shell appears smooth but has very fine reticulation of numerous spiral striae and prosocline growth lines. Aperture broadly oval, drawn out adapically to an anal sinus, giving a tear-drop shape. Inner lip reflected over columellar region, outer lip thickens rapidly toward the throat; no umbilicus or umbilical groove. Juvenile aperture more elongate, its base drawn out into a slight spout, with a more pronounced anal sinus, throat not thickened. Colour extremely variable and a number of colour morphs have been recognized: *olivacea* (uniform light to dark green), *citrina* (uniform yellow to orange, with white throat), and *reticulata* (green, brown, or black, with reticulate pattern) are most common. Size up to 15 × 17 mm. Body similar to *L. littorea*, but colour tends to follow that of shell; cephalic tentacles with two longitudinal lines and no transverse stripes. Examination of the penis (flat and with short tip), penial glands (16–53 glands in two to three rows) and ovipositor (pigmented) may be necessary to confirm separation of this species from *L. mariae*.

On weeds, especially *Fucus vesiculosus* and *Ascophyllum nodosum*, from MHWN to MLWS; occasionally in shallow sublittoral. Tolerant of low salinity and a wide range of exposure.

Colour morphs appear to have separate, though overlapping habitat preferences (Smith 1976). Distributed from western Mediterranean to northern Norway; on all British coasts.

The brackish water *L. aestuarii* (Jeffreys) has a higher spire and narrower aperture than *L. obtusata*. Its taxonomic status is discussed by Moyse *et al.* (1982), who regard it as an ecomorph of *L. obtusata*.

Littorina mariae Sacchi and Rastelli FIG. 12.9

Shell similar to that of *L. obtusata*. Spire reduced or absent, last whorl constituting most or all of shell. Aperture lacking anal sinus, with a more restricted throat; outer lip impinges high on last whorl, almost at right angles to it. Size up to 11 × 12 mm. Body essentially as in *L. obtusata*, but differs in the penis (long tapering tip), arrangement of penial glands (single row of 5–16 glands) and lack of pigmentation in ovipositor.

Tends to live lower on the shore than *L. obtusata*, otherwise habitat and distribution similar.

Littorina neritoides (Linnaeus) FIG. 12.9

Shell solid, with five or six slightly tumid whorls; last whorl more distinctly tumid, spire prominent, and flat-sided. Surface with numerous irregular growth lines covered by well-developed periostracum; a surface bloom, rather like that on black grapes, is a feature of all but the youngest shells. Aperture somewhat narrow, tapering sharply towards adapical region. Outer lip meets body whorl tangentially, with a thin periostracal extension along margin and a low ridge internally; inner lip reflected over columella region; both lips grooved. No umbilicus. Commonly black or dark brown; last whorl occasionally with spiral banding or vertical stripes. Uniformly pale shells may also occur. Operculum and interior of aperture dark. Size up to 9 × 7 mm. Body essentially similar to *L. littorea*; cephalic tentacles lack transverse stripes and have two longitudinal dark lines on dorsal surface.

In crevices and empty barnacle cases on upper shore and in splash zone; often several metres above EHWST on exposed shores. Distributed from Black Sea to western Norway. Absent from southern North Sea; on British Channel coast it does not penetrate eastwards beyond Isle of Wight. Local distribution may be patchy.

Littorina nigrolineata Gray FIG. 12.9

Shell solid, with well-developed spire of five or six moderately tumid whorls; last whorl comprises >75% of shell height. Sculpture of irregular prosocline growth lines and strap-like spiral striae; a shallow central groove on striae of last whorl. Outer lip meets last whorl almost at right angles, margin slightly crenulate, internally thickening rapidly towards throat; inner lip reflected over columella region, extending to meet outer lip adapically as a glaze on last whorl. No umbilicus or umbilical groove. Whitish, yellow or deep orange-red, often with dark brown pigmentation in grooves between spiral striae; peristome white, throat brown. Size up to 30 × 27 mm, commonly 20 × 18 mm. Body similar to *L. littorea*; penis with very short tip, close to which are 12 large glands. Female oviparous, ovipositor conspicuous below right eye. A pale streak extends diagonally from right to left across floor of mantle and through the penis/ovipositor.

Among weeds and stones; higher on exposed shores (barnacle/*Fucus spiralis* zone), extending down to *Ascophyllum* belt in sheltered areas. Distribution imperfectly known; probably occurs on all British coasts other than those in south-east.

Littorina neglecta Bean FIG. 12.9

Shell globular, rather thin, with three or four tumid, superficially smooth whorls; spire short, blunt, last whorl comprising >85% shell height. Sculpture of very fine spiral striae and prosocline growth lines, latter often more pronounced on last whorl. Aperture very rounded; outer lip thin but lacking periostracal flap (cf. *L. neritoides*) and with no ridge on internal surface, meeting last whorl almost at right angles; inner lip reflected over columella region, extending as a glaze on last whorl towards origin of outer lip. No umbilicus or umbilical groove. Colour variable, pale with dark, often tessellated markings, or darker (brown) with light areas; frequently a dark spiral band, bordered by cream, follows periphery of last whorl and enters aperture. Columella region and throat brown, base of aperture white. Size up to 5 × 4.5 mm, often smaller. Body similar to *L. littorea*; cephalic tentacles without transverse stripes, penis with elongate tip and single row of three to seven glands; female ovoviviparous, with brood pouch. No light diagonal band across mantle floor.

Most common on exposed or moderately exposed shores, in empty barnacle shells and crevices towards lower limits of barnacles, and below (recorded from *Laminaria* holdfasts and *Corallina*). Less common on sheltered shores but may be abundant in salt-marshes. Distribution imperfectly known; on western British coasts but may be more widespread.

Littorina saxatilis (Olivi) FIG. 12.9

Shell solid, opaque, with three or four tumid whorls. Sculpture variable, either with strong V-shaped spiral striae, often symmetrical in profile, or with fine spirals giving a superficially smooth appearance. Aperture large, rounded, outer lip everted where it meets columella but not forming prominent spout. No umbilicus or umbilical groove. Colour variable, usually with some pattern (tessellation or spiral lines) and with dark throat. Size up to 18 × 14 mm. Ovoviviparous. Extensive ciliary field (red in live animals) between genital tract and edge of columella muscle; prostate extends beyond anus, penis with elongate tip and variable number (4–26) of small glands arranged in one or two rows along its length but rarely extending to its curved base.

In *Pelvetia* zone and above, in crevices and empty barnacle shells. Widely distributed, from Mediterranean to Arctic.

The taxonomy of the *L. saxatilis* complex (*nigrolineata, neglecta, arcana, saxatilis, rudis,* and *tenebrosa*) has received a great deal of attention over a number of years (for a review see Fretter and Graham 1980) and while the separation of *L. nigrolineata, L. neglecta*, and *L. arcana* from *L. saxatilis* has achieved general approval, there is still some uncertainty as to the relationship and status of the residual trio of *L. saxatilis, L. rudis*, and *L. tenebrosa*.

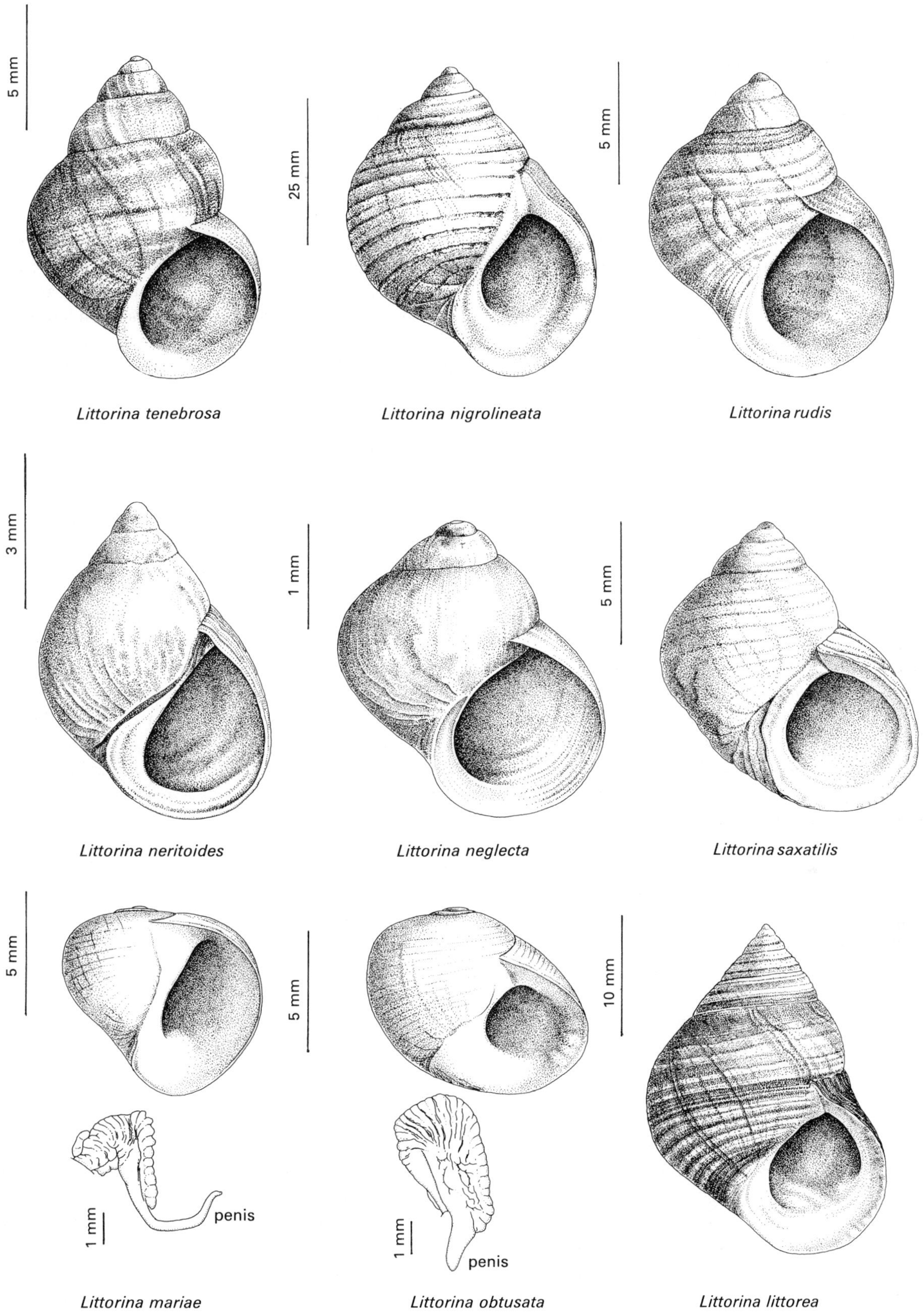
5 mm
25 mm
5 mm
Littorina tenebrosa
Littorina nigrolineata
Littorina rudis
3 mm
1 mm
5 mm
Littorina neritoides
Littorina neglecta
Littorina saxatilis
5 mm
5 mm
10 mm
1 mm
penis
1 mm
penis
Littorina mariae
Littorina obtusata
Littorina littorea

Fig. 12.9

Littorina rudis (Maton) FIG. 12.9

Shell solid, opaque, with four to five tumid whorls. Sculpture usually of strong spiral striae which may be strap-like or V-shaped in profile; V-shaped ridges frequently asymmetrical, with crest of ridge lying towards shell apex, though this is not a reliable diagnostic character. Aperture with prominent spout where outer lip meets columella. No umbilicus or umbilical groove. Shell usually monochrome, without patterning or dark throat. Size up to 18 × 14 mm. Body similar to *L. saxatilis*

In Pelvetia zone, lower than *L. saxatilis*, preferring damp places among stones rather than dry crevices. Widely distributed, from Mediterranean to Arctic.

Littorina tenebrosa (Montagu) FIG. 12.9

Shell thin, sometimes transparent, with five or six tumid whorls, deep sutures and somewhat turreted profile. Superficially smooth but with growth lines and very fine spiral striae. Aperture slightly squared, outer lip slightly everted below columella but not forming prominent spout as in *L. rudis*. No umbilicus or umbilical groove. Usually dark, often with a tessellated pattern; columella and throat brown. Body similar to *L. rudis*, of which it may be a brackish water ecomorph.

Characteristically on weed in sheltered areas with reduced salinities; often permanently submerged. Distribution uncertain, probably widespread on British and European coasts.

Littorina arcana Hannaford Ellis

Shell essentially like that of *L. rudis*. Female oviparous; presence of glandular oviduct and absence of brood pouch can be used to separate females of this species from those of *L. saxatilis* and *L. rudis*, with which it is found. In males the prostate does not extend to level of anus, penial glands frequently extend to or around curved base of penis; ciliated field adjacent to genital tract is neither extensive nor red, as in *L. saxatilis* or *L. rudis*.

In upper barnacle and *Verrucaria* zone, particularly on exposed shores. Probably widely distributed.

12 HYDROBIIDAE

Shell small, conical, smooth; aperture ear-shaped or oval, with complete peristome; operculum spiral and horny. Animal with long, bifid snout; mantle edge with or without pallial tentacle; foot folding transversely into anterior and posterior halves, latter with a median mucus gland.

---

1. Umbilicus absent or very much reduced, whorls tumid, sometimes with spiral keel and/or periostracal bristles; snout dark with light transverse band anteriorly, tentacles pale, dark at base. Mantle edge without pallial tentacle. In fresh or brackish water **_Potamopyrgus jenkinsi_**

   Shell with umbilical chink; mantle edge with pallial tentacle on right side. In brackish water **2**

2. Whorls flat-sided or only slightly tumid, last whorl angulate; tentacles with black ring near tip, snout with black transverse bar. Often on open shores **_Hydrobia ulvae_**

   Whorls more tumid, slightly transparent; pigmentation not as above. In more confined brackish situations **3**

3. Whorls distinctly tumid; tentacles with longitudinal black line near tip; snout with triangular black area between eyes; penis with flagellum **_Hydrobia ventrosa_**

   Whorls moderately tumid; spire moderately slender; tentacles with conical black marks near tip; penis without flagellum **_Hydrobia neglecta_**

Hydrobia ulvae (Pennant) FIG. 12.10

Shell an elongate cone, length/breadth ratio 2:1, slightly transparent, glossy, with six to seven flat-sided or slightly tumid whorls, apex rounded; last whorl comprising >50% shell height, slightly angulate at periphery, especially in younger shells. Superficially smooth but with microsculpture of fine prosocline growth lines and finer spiral striae. Aperture slightly pear-shaped (more angulate in young shells), outer lip arising tangentially to last whorl, forming a slight spout basally where it meets columella; inner lip reflected over last whorl, leaving a small umbilical chink. Periostracum horn-coloured, underlying shell white; viscera showing through apical whorls give live shells a darker appearance. Size up to 6 × 3 mm. Body with long bifid snout, often with transverse band of dark pigment anteriorly and a triangular or shield-shaped pigmented area between cephalic tentacles; a single, small pallial tentacle projects from mantle edge near junction of last whorl and outer lip. Cephalic tentacles long, slender, setose, left slightly thicker than the right; eyes at base. A dark pigmented band placed at a distance greater than its own length from the tentacle tips. Penis large, a smooth sickle-shaped curve, with pointed tip.

A brackish water species favouring substrata of firm mud and muddy sand; most abundant above MTL, but found sublittorally down to 20 m. Widely distributed from northern Norway to Senegal.

Hydrobia ventrosa (Montagu) FIG. 12.10

Shell transparent, glossy, with five or six markedly tumid whorls, no visible ornamentation (fine growth lines may be seen on closer inspection); last whorl comprising >50% of shell height, not angulate at periphery. Aperture egg-shaped, with thin peristome; outer lip abuts almost at right angles to last whorl, slightly flared basally, though less so than in *H. ulvae*; inner lip reflected over umbilicus, leaving a shallow groove. Shell colour as in *H. ulvae*. Size up to 5 × 2 mm. Body similar to *H. ulvae* but lacks dark transverse band on snout; no dark ring on tentacles, instead a dark dorsal line often present near tips. Penis smaller than in *H. ulvae*, with a narrow flagellate tip.

Not as common as *H. ulvae*, occurs in more brackish waters, preferring isolated lagoons, creeks, and ditches to open coasts. Patchily distributed from Norway and the Baltic to the Mediterranean and Black Sea.

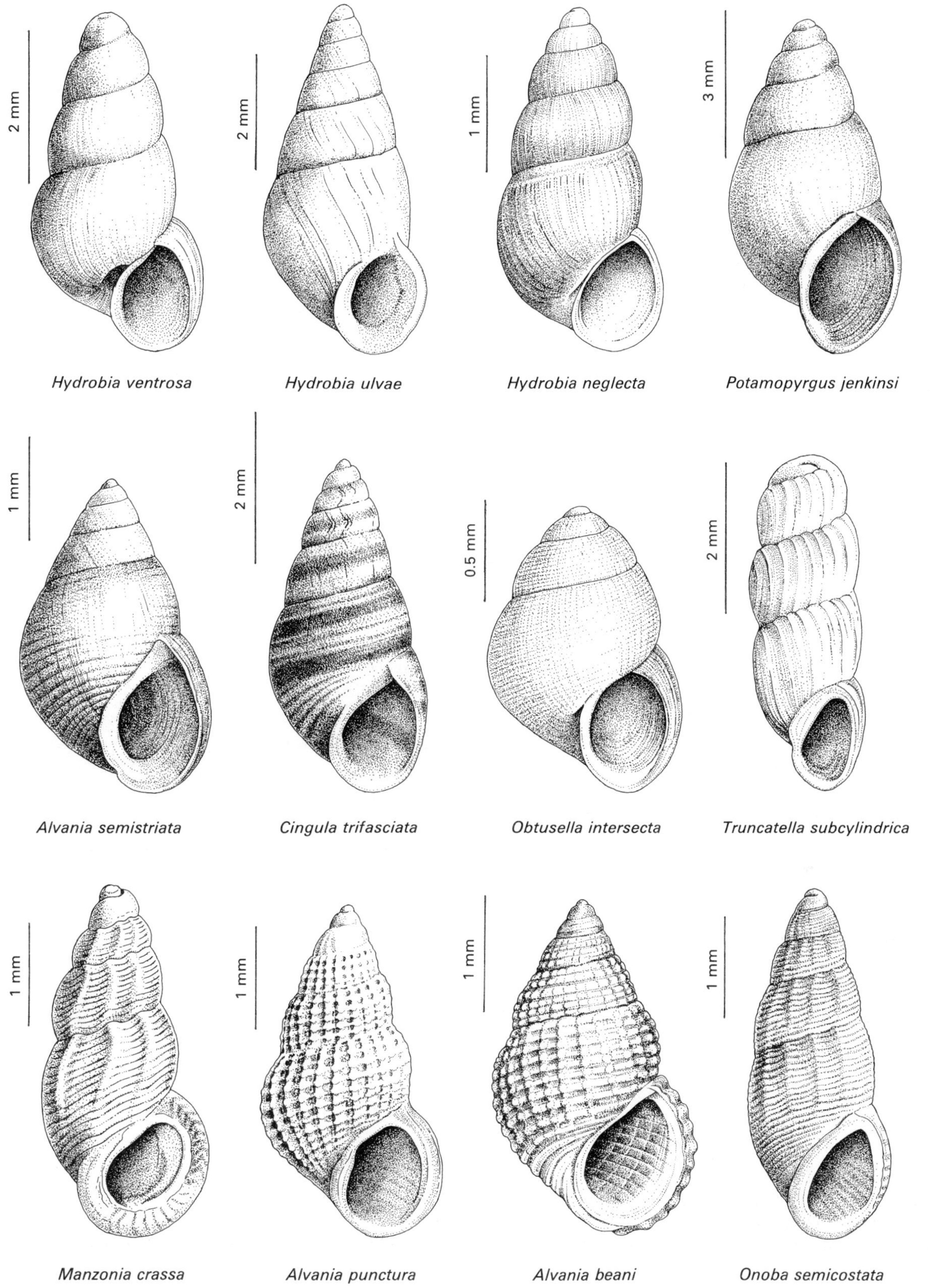
2 mm
Hydrobia ventrosa
2 mm
Hydrobia ulvae
1 mm
Hydrobia neglecta
3 mm
Potamopyrgus jenkinsi
1 mm
Alvania semistriata
2 mm
Cingula trifasciata
0.5 mm
Obtusella intersecta
2 mm
Truncatella subcylindrica
1 mm
Manzonia crassa
1 mm
Alvania punctura
1 mm
Alvania beani
1 mm
Onoba semicostata

Fig. 12.10

Hydrobia neglecta Muus FIG. 12.10

Shell more slender than in *H. ventrosa*, with apex less sharp, whorls less tumid; spire slightly cyrtoconoid in profile. Size up to 4 × 1.5 mm. Body differs from *H. ulvae* and *H. ventrosa* in tentacle pigmentation, with black pigment in cone-shaped areas set less than their own length away from tentacle tips; *H. ulvae* has rings of pigment, *H. ventrosa* has, at most, a darkish streak on dorsal surface near the tip. The structure of the penis may also be used to distinguish *H. neglecta*; it lacks the flagellum of *H. ventrosa*, and the recurved anterior end is angulate rather than smoothly sickle-shaped as in *H. ulvae*.

Habitat similar to that of *H. ventrosa*, though appears to be less tolerant of salinities below 10‰ and seems to favour the presence of *Potamogeton* and *Zostera*. First described in 1963 and subsequently recorded from North Sea coasts, Guernsey, and the western coasts of Scotland and Ireland.

Potamopyrgus jenkinsi (Smith) FIG. 12.10

Shell of six tumid whorls, often encrusted with black deposits so that little of the appearance of the spire is discernible. Last whorl large; aperture ear-shaped or oval, with peristome in mature shells; inner lip occludes umbilicus. In some shells, notably from brackish water populations, a spiral keel may be present, with or without periostracal bristles; some have a spiral row of bristles but lack keel. Beneath black encrustations shell is pale horn-colour. Size up to 6 × 3 mm.

Body with long, narrow, bifid snout, evenly and darkly pigmented, with pale transverse band anteriorly. Cephalic tentacles long and setose, pale with darker base; white spot behind each eye. Mantle edge without pallial tentacle.

Common in a variety of freshwater and brackish (< 16‰) habitats, often with *H. ventrosa* in the latter. Widely distributed throughout Europe. In the British Isles it is scarce or absent in mid-Wales and large parts of the Scottish highlands.

13 TRUNCATELLIDAE

Only one British representative of this family.

Truncatella subcylindrica Linnaeus FIG. 12.10

Adult shell of three or four slightly tumid, though flat-sided whorls; spire cylindrical or slightly tapered; apex blunt. Sculpture of fine costae and delicate spiral striae; some shells (var. *laevigata* Risso) smooth. Young shells, prior to loss of part of spire, with six or seven whorls, gently tapered spire and smooth protoconch of two whorls. Umbilicus narrow in young shells, occluded by inner lip in mature specimens. Buff coloured, digestive gland appearing dark below semi-transparent apex. Size up to 5 × 2 mm. Snout long, cylindrical, with short cephalic tentacles at base. Foot short, with double edged anterior; operculum narrow, set across foot.

In high shore, muddy habitats, among *Suaeda maritima* and *Halimione portulacoides*. Now very restricted, perhaps limited to a few localities on the Dorset coast. Moves in a distinctive 'looping' fashion, using foot and snout. Distributed from Mediterranean and Black Sea to southern shores of British Isles.

14 ASSIMINEIDAE

Shell small (less than 5 mm), conical or globose conical. Snout bilobed; cephalic tentacles short, rounded lobes with eyes at their tips. Family includes two British species of somewhat doubtful affinity.

1. Shell conical; whorls flat-sided or slightly tumid; no umbilicus. A groove on each side of head from mantle cavity to foot. ***Assiminea grayana***

   Shell globular, light horn-coloured, with three or four tumid whorls; small umbilicus. No lateral grooves on head ***Paludinella littorina***

Assiminea grayana Fleming FIG. 12.11

Shell solid, semi-transparent, conical, with six or seven flat-sided or slightly tumid whorls; apex sharp. Appears smooth but with fine irregular growth lines and faint spiral lines. Aperture small, oval or ear-shaped, with peristome. Inner lip reflected over base of last whorl; no umbilicus. Dark horn or tan coloured, often with broad rufous band on last whorl. Size up to 5 × 3 mm. Snout bilobed, drawn out into lateral keels, below which ciliated groove passes forward from mantle cavity down each side of foot. Cephalic tentacles short, rounded lobes with eyes at tips. Mantle edge without pallial tentacle. Operculum spiral.

Limited to upper regions of salt-marshes, where it is often found crawling on mud and in brackish pools among vegetation; locally common. Confined to North Sea coast between Kent and Humber, and from the Netherlands to Denmark.

Paludinella littorina (delle Chiaje) FIG. 12.11

Shell solid, semi-transparent, globular, with three or four tumid, rapidly expanding whorls; last whorl about three-quarters of shell height. Appears smooth, but with many prosocline growth lines and few, faint spiral grooves. Aperture D-shaped, with peristome. Inner lip reflected over umbilical groove; umbilicus small. Light horn coloured, with pale columella. Size up to 2 × 2 mm. Snout bilobed, expanded laterally; cephalic tentacles broad, rounded lobes with eyes near tip. Foot without lateral grooves; operculum spiral.

Rare. In crevices in splash zone and among *Pelvetia*. May also occur at HWM among stones on muddy shores. Distributed from south coast and Channel Islands to Mediterranean.

15 RISSOIDAE

Shell small (typically 4–5 mm or less), with conical spire; aperture ear-shaped or oval, usually with complete peristome; operculum spiral and horny. Shell sculpture variable: smooth, or with spiral striae, costae or reticulations; may vary within species according to environmental conditions. Body with long bifid snout; mantle edge smooth, with one or two pallial tentacles. Foot folds transversely, the posterior sole having a median mucus gland; a single metapodial tentacle projects from below the rear of the operculum.

1. Shell with reticulate pattern of pits on last whorl; with or without costae — ***Rissoa violacea***

   Shell ornament not so — **2**

2. Shell without ornament, or with spiral striae, sometimes with costae on subsutural parts of whorls only — **3**

   Shell with costae and spiral striae, the latter conspicuous or microscopic; costae extend from suture to suture on whorls of spire and reach at least to periphery of last whorl; uppermost whorls of spire may lack costae — **10**

3. Shell slender, breadth <50% of height — **4**

   Shell oval-conic, breadth ⩾50% of height — **5**

4. Shell with conspicuous, strap-like spiral striae and short, subsutural costae; last whorl >60% of shell height — ***Onoba semicostata***

   Shell thin, smooth (spiral striae microscopic), without costae; last whorl <50% of shell height; brackish habitats — ***Rissostomia membranacea***

5. Shell uniform yellowish or white, with numerous, microscopic spiral striae; globose, with blunt apex — ***Obtusella intersecta***

   Shell with spiral or transverse colour bands (or both) on last whorl; apex blunt or sharp — **6**

6. Shell yellowish-white, usually with three spiral bands of reddish-brown on the last whorl, no transverse pigmentation; spiral striae on last whorl, conspicuous at base; outer lip arises tangential to last whorl — ***Cingula trifasciata***

   Shell without this combination of characters — **7**

7. Shell with two rows of transverse, reddish-brown marks on last whorl; whorls almost flat-sided, with fine strap-like spiral striae below sutures and at base of last whorl; outer lip arises tangential to last whorl — ***Alvania semistriata***

   Shell without this combination of characters — **8**

8. Shell often with prominent, white varix behind outer lip, traversed in most mature shells by a brown comma-shaped mark (juveniles lack mark and varix, aperture with angulate periphery); colour variable, spiral bands, with or without transverse marks, apex diffuse lilac; up to 5 mm long and eight whorls — ***Rissoa parva* v. *interrupta***

   Shell without this combination of characters; transverse brown marks on penultimate and last whorls; up to 3 mm long, five or six whorls — **9**

9. [*In their non-costate form the next two species are difficult to separate on shell alone.*]

   Apex bronze; shell slightly globular, whorls tumid, sutures deep; often with labial varix; metapodial tentacle flattened — ***Rissoa sarsi***

   Apex dark purple; shell not globular, whorls slightly tumid; often without labial varix; metapodial tentacle cylindrical — ***Rissoa inconspicua***

10. Uppermost whorls of teleoconch without transverse ornament, other whorls costate to a variable degree; metapodial tentacle single and filiform; pallial tentacle on right — **11**

    All whorls of teleoconch with transverse ornament (may be eroded in worn shells); metapodial tentacle single or triple; pallial tentacle on left and right — **15**

11. Shell markedly cyrtoconoid, whorls flat-sided, last whorl about 60% or more of height; peristome everted, throat thickened, columella with small tooth opposite umbilical region; labial varix large, white; marine habitats to 35–20‰ — ***Rissostomia membranacea***

    Shell not like this — **12**

12. Last and penultimate whorls with 18–30 thin costae and numerous fine spiral striae; periphery of whorls may be reticulate; protoconch dark purple — ***Rissoa inconspicua***

    Last whorl with 8–25 broad costae, smooth or with microscopic spiral striae in between — **13**

13. Mature shells usually with comma-shaped mark crossing labial varix (juveniles lack mark and varix, periphery of aperture angulate); apex lilac; costae broad-based, with narrow crests, those on last whorl extending to spiral ridge at periphery — ***Rissoa parva***

    Shell never with comma-shaped mark; aperture never angulate — **14**

14. Shell solid; upper whorls of teleoconch flat-sided, spire coeloconoid; costae broad, with broad summit, ten on penultimate whorl; lilac tinge to apex and peristome; south-west shores only — ***Rissoa guerini***

    Shell semi-transparent; whorls tumid, spire slightly crytoconoid; costae do not link with spiral ridge at periphery of last whorl; apex bronze, penultimate and last whorl often with brown transverse bands — ***Rissoa sarsi***

15. Costae strong and flexuous, crossed by narrow, strap-like spiral striae (these may be eroded at the crests of costae); base of last whorl with two strong spiral keels; peristome thick and ridged — ***Manzonia crassa***

    Costae and spiral striae of equal prominence; shell with reticulate pattern — **16**

16. Mature shell with ridges on inside of outer lip; shell rather coarsely reticulate, with strong labial varix; no umbilicus — ***Alvania beani***

    Inside of outer lip not ridged; shell finely reticulate, labial costa slight; with umbilical chink — ***Alvania punctura***

Cingula trifasciata (Montagu) FIG. 12.10

Shell solid, opaque, regularly conical, with six or seven slightly tumid whorls, apex rather blunt. Sutures shallow, slightly channelled; last whorl with spiral striae, often restricted to and more obvious at base. Aperture ear-shaped, with shallow anal sinus where outer lip abuts tangentially to last whorl, peristome somewhat thickened; umbilicus occluded by inner lip, though umbilical groove remains. Commonly yellow-brown, with three spiral brown bands on last whorls, one or two on whorls of spire; var. *rupestris* uniform cream or white; uniform black or dark purple shells occasionally found. Size up to 4 × 2 mm. Body of typical rissoid pattern; mantle edge smooth, with short pallial tentacle on right; metapodial tentacle short, broadly triangular not reaching edge of foot.

Common in silty crevices or under stones from MTL downwards. The var. *rupestris* is confined to crevices and appears to favour exposed shores; it does not appear to mix with banded shells. Distributed from Bay of Biscay to west coast of Norway. On Channel and British North Sea coasts; absent from continental shores of North Sea.

Obtusella intersecta (Wood) FIG. 12.10

Shell small, globose-conical, with blunt tip; spire cyrtoconoid with four or five tumid, semi-transparent whorls. Sculpture of fine spiral striae and prosocline growth lines, visible with × 10 hand lens. Aperture broadly oval with thin peristome; outer lip meets last whorl almost at right angles, with slight anal sinus; inner lip borders distinct umbilical groove and small umbilical chink. Shell uniform buff or pale horn. Size up to 2 × 1.4 mm.

Occurs at LWST and below; locally common amongst algae and on sandy and gravelly bottoms from 10–60 m. Distributed from Iberia to Norway; absent from Baltic and eastern shores of North Sea.

Onoba semicostata (Montagu) FIG. 12.10

Shell solid, opaque, somewhat cylindrical with blunt apex; five or six slightly tumid whorls, with sculpture of spiral striae, and truncated costae confined to apical parts of whorls; costae absent in var. *ecostata*. Aperture small, oval or egg-shaped; peristome thickened, with anal sinus. Inner lip reflected over umbilical groove; umbilicus may be present in young shells. Shell commonly off-white, often with two red-brown bands on last whorl. Periostracal pigmentation may give rise to darker, even black, forms. Size up to 4 × 1.8 mm, most commonly 2 × 1 mm. Body of typical rissoid pattern; mantle edge smooth, with short pallial tentacle on left, and in many animals on right also. Metapodial tentacle narrow, triangular, may extend to edge of foot.

Gregarious, under stones, among corallines and in other silty areas, on rocky shores from near LWM and sublittorally to 100 m; deeper in north of range. Distributed from Mediterranean to Norway; common, especially in summer, on all suitable British shores.

In northern Britain *O. aculeus* (Gould) (Fretter and Graham 1978*b*: 166) may be encountered among weed, not in the silty conditions favoured by *O. semicostata*, and offshore to 200 m. This species differs from *O. semicostata* in its less conspicuous costae (which may be absent) confined to the upper parts of the spire.

Manzonia crassa (Kanmacher, in J. Adams) FIG. 12.10

Shell elongate, conical, white, glossy; with six tumid, semitransparent whorls; apical whorls smooth or finely reticulate, contrasting with strong flexuous costae and spiral striae on teleoconch. Spiral striae accentuated on base of shell, forming two prominent spiral keels; aperture small, oval, with peristome and thickened labial costa (varix); a deep umbilical groove but no umbilicus. Size up to 3.5 × 2 mm. Body of typical rissoid pattern, mantle edge smooth, with short pallial tentacle on right and slightly more prominent one on left. Metapodial tentacle narrow, extends to edge of foot; an additional, smaller projection on each metapodial lobe.

Fairly common among weeds and under stones at LWM; more common sublittorally, to 50 m, on sandy bottoms. Distributed from Mediterranean to Norway; absent from continental coasts of North Sea.

Alvania beani (Thorpe) FIG. 12.10

Shell slightly transparent, markedly cyrtoconoid, with up to seven rather flat-sided whorls; sutures deep. Sculpture of strap-like costae and spiral striae producing conspicuous, rather rectangular reticulations. The segregate species *reticulata* and *calathus* are distinguished by the number of spiral striae on penultimate whorl; six or seven in *reticulata*, four in *calathus*. Aperture pear-shaped, outer lip with well-developed varix crossed by spiral striae, internally thickened and ridged in mature shells; inner lip reflected over umbilical region leaving only slight groove; no umbilicus. Pale orange or cream, often with two darker spiral bands on body whorl. Size up to 3.5 × 2 mm. Body similar to *M. crassa*; metapodial tentacle trilobed.

On weeds and stones in sediment-rich areas; sublittoral to 50 m (deeper in south of range) but occasionally encountered at LWST. Not uncommon. Distributed from Canary Islands and Mediterranean to northern Norway.

Alvania punctura (Montagu) FIG. 12.10

Shell semi-transparent, oval-conical, with six tumid whorls, apex pointed; spire slightly cyrtoconoid. Apical, larval whorls lightly tuberculate, contrasting sharply with decussate pattern of teleoconch; the costate element of decussation lost below periphery of last whorl. Aperture oval or egg-shaped, with peristome and anal sinus; inner lip reflected over umbilical groove leading to narrow, elongate umbilicus; outer lip with slight varix, not thickened or ridged internally. Shell white, cream, or tinged reddish brown, often darkened by conspicuous periostracum; red-brown blotches in three series on last whorl, umbilical region similarly stained. Size up to 3 × 1.8 mm. Body similar to *M. crassa*; metapodial tentacle single.

Locally common among *Laminaria* at LWST and sublittorally on sandy bottoms to 100 m. Distributed from Mediterranean to northern Norway; absent from Baltic and eastern basin of English Channel.

Alvania semistriata (Montagu) FIG. 12.10

Shell solid, oval-conical, with five or six slightly tumid whorls. Sculpture of fine prosocline growth lines, with slight, strap-like spiral striae visible mainly below sutures on spire and toward base of last whorl. Aperture oval, with thickened peristome; outer lip arises tangential to last whorl; inner lip reflected over umbilical groove and basal area of last whorl; umbilicus absent. Shell whitish or buff with two rows of red-brown, comma-shaped marks on last whorl. Size up to 3 × 1.6 mm. Body similar to *M. crassa*, though with a short pallial tentacle on left and a longer one on right; metapodial tentacle trilobed, opercular lobes yellow.

Gregarious. Below stones, in fine weeds and in silted areas of rock pools below MTL; extends sublittorally to 100 m. Distributed from Mediterranean to Norway; on all British coasts, more common in south.

Rissoa parva (da Costa) FIG. 12.11

Shell oval-conical with seven or eight semi-transparent, glossy, slightly tumid whorls. Ornamentation variable: two basic forms may be identified; costate and smooth, the latter often referred to as var. *interrupta*. Uppermost whorls of costate shells rarely ornamented, lower whorls with strong, steep-sided ribs, equal in breadth to intervening spaces, not extending to base of last whorl, terminating at, or slightly below, a peripheral spiral ridge (not conspicuous in all shells). The distribution of costae is somewhat erratic and specimens may show alternation of smooth and costate whorls such that on some costate shells the last whorl may be smooth. Both costate and smooth shells have fine spiral striae on most whorls, most conspicuous between costae of costate shells. Aperture D-shaped, with peristome; often angulate peripherally in young shells and with varix in mature shells. No umbilicus. Colour variable, cream or buff to horn coloured or chocolate brown; apical whorls often tinged violet. On many shells background colour of lower whorls is enhanced by darker spiral streaks and blotches, in most mature specimens a prominent comma-shaped mark present on last whorl, commencing behind and passing through varix. Size up to 5 × 3 mm. Body of typical rissoid pattern; pallial tentacle on right of mantle only; metapodial tentacle well-developed, cylindrical.

Abundant below MTL and sublittorally to 15 m, under stones and on weeds on rocky shores; most commonly associated with corallines and other finely branched algae in which diatoms and detritus collect. The proportion of costate: smooth shells is higher on more sheltered shores, particularly in summer populations. Distributed from Mediterranean to northern Norway; absent from Baltic.

Rissoa inconspicua Alder FIG. 12.11

Shell oval-conical, with five or six semi-transparent, glossy, slightly tumid whorls. Spire profile variable; flat-sided to appreciably cyrtoconoid. Sculpture of fine spiral striae, most prominent at periphery of last whorl; with or without numerous (18–30), fine blunt costae on all but upper three whorls. Interaction of costae and striae produces a fine reticulate pattern, most obvious at periphery of last whorl. A small umbilical groove but no umbilicus. Aperture oval, with peristome and distinct anal sinus. Shell white or horn-coloured, with deep purple apex. Younger whorls often with a series of red-brown streaks below suture; peristome red-brown. Size up to 3 × 2 mm.

On weeds in low shore rock pools, though more common sublittorally on algae and shell gravel to 100 m. Shows limited tolerance of brackish waters. Distributed from the Azores and Mediterranean to Arctic waters; in the Baltic as far as Lubeck Bay.

Rissoa sarsi Lovén FIG. 12.11

Shell semi-transparent, oval-conical, with six tumid whorls; often with globular or inflated body whorl and short spire. Ornamentation of fine sprial striae (not visible without magnification) and irregular growth lines, sometimes with costae on younger whorls. Costae, when present, usually weak, or even absent on youngest parts of last whorl, though a varix may be present. Costae not linked basally to a spiral ridge (cf. *R. parva*); less numerous on penultimate whorl (9–16) than in *R. inconspicua*. Aperture oval, angulate, slightly drawn out adapically; peristome thin, often no more than a glaze where it crosses last whorl. A small umbilical chink. White, cream, or horn-coloured, with apical whorls often darker and sometimes with faint violet tinge (never dark purple/violet as in *R. inconspicua*). Penultimate and last whorls often with red-brown bands (lying between costae in costate shells), which may or may not connect basally to a spiral brown band on last whorl; usually a dark brown streak on columella. Body similar to *R. parva*, though with longer cephalic tentacles and longer, less cylindrical metapodial tentacle.

On rocky shores among weed, especially *Codium* and *Zostera*. Occasionally intertidal but more usually sublittoral to 15 m. Shows some tolerance to brackish waters. Distributed from Mediterranean to northern Norway.

Rissoa guerini Récluz FIG. 12.11

Shell oval-conical, with eight whorls and coeloconoid spire; upper four or five whorls flat-sided, or nearly so, young whorls more tumid. Sculpture of strong costae confined to youngest three or four whorls; number of costae variable on last whorl, costae may fade out toward aperture; ten ribs on each of two preceding whorls. Varix broad, smooth, set a short distance back from outer lip. Fine spiral striae and growth lines give a finely reticulate pattern between costae, absent from upper, non-costate whorls. Aperture oval or D-shaped, with thin-edged peristome and slight umbilical groove. Cream, light horn-coloured or brown with lighter costae. Apical whorls often slate coloured or lilac, may be streaked with reddish brown. Lilac coloration also on peristome and in aperture. Size up to 6 × 3 mm. Body similar to *R. parva*, though pallial tentacle is very short and metapodial tentacle has an expanded base.

Locally common below stones, and on algae and *Zostera* at LWST and below. A southern species distributed along eastern Atlantic coasts between the Canary Islands and south-western coasts of England and Ireland.

Rissoa violacea Desmarest FIG. 12.11

In addition to *R. violacea sensu stricto* this aggregate species includes as segregates *R. lilacina* Récluz, *R. rufilabrum* Alder and *R. porifera* Lovén. All may be distinguished from other *Rissoa* species by the fine reticulation of the shell surface. *R. violacea s.s.* is a Mediterranean species, though the other segregates are represented on British shores.

R. lilacina Récluz: Shell solid, oval-conical, with six or seven slightly tumid whorls, the youngest two or three costate. Ribbing is incomplete on third whorl and apertural view of shell will show costae on last and penultimate whorls only; apart from a wide varix, costae are slight or absent on adapertural part of last whorl, and here a fine reticulate pattern is most prominent. On costate areas of shell reticulation is confined to troughs between costae. On some shells ribbing is slight, requiring careful examination of shell; var. *ecostata* has neither costae nor reticulations. Shell whitish with longitudinal orange-brown stripes which, on costate whorls, lie between costae; apex and peristome with lilac tinge. Size up to 5 × 2.75 mm. Body as in *R. parva*; cephalic tentacles with bright yellow stripe dorsally.

Occasional or locally common on weeds and amongst sandy gravel from LWST down to 50 m. A southern species extending to the western Channel basin, the Welsh coast, western shores of Scotland, and the south and west coasts of Ireland.

R. rufilabrum Alder: Differs only in minor details from *R. lilacina*; costae fewer (11–24 compared to 22–36 in *lilacina*) and, with shell in apertural view, appearing to be confined to last whorl; costae occasionally reduced or absent. Size up to 5 × 2.75 mm.

Habitat as for *R. lilacina*; distributed from Channel Islands to Norway.

R. porifera Lovén: A delicate non-costate version of *R. rufilabrum*; reticulate pattern of depressions prominent, visible on most whorls; varix slight. Horn-coloured, varix white, apex and peristome tinged brown or reddish-lilac. Size up to 5 × 2.75 mm.

Habitat as for *R. lilacina*, from western Ireland to Shetlands, Norway, and in the Kattegat.

Rissostomia membranacea (J. Adams) FIG. 12.11

Shell form is extremely variable in this species and several varieties have been described. Some variability relates in part to environmental conditions, though it is likely that *R. membranacea* is an aggregate species of which some of the major varieties are segregates.

R. membranacea: Shell thin, with eight to nine flat-sided, or very slightly tumid whorls. Last whorl occupies about 60% of total height; apex slightly pointed. Ornament of fine growth lines, microscopic spiral striae, and sinuous costae with shallow profile, generally confined to last whorl (up to ten costae) and extending as far as periphery. Aperture somewhat triangular, with thin, everted peristome and broad, white labial costa; columella often with slight tooth-like prominence opposite umbilical region; inner lip reflected over umbilical groove. Light buff, greenish, or pale horn coloured with longitudinal yellow-brown markings. In costate shells darker markings lie between costae; apical region often tinged with violet. Size up to 9 × 3 mm. Northern Britain, in brackish waters with salinities of 15–20‰.

var. *labiosa* Montagu: More solid with five to seven flattened whorls. Last whorl 66–70% of shell height, with 15–18 costae; tooth on columella pronounced, throat thickened. Southern Britain, in more marine situations (salinity 20–35‰).

var. *octona* Nilsson: Slender shell with long spire of 10–12 tumid whorls; last whorl with ten costae, or none, comprising 40% of shell height. Costae, when present, may be found on all but most apical whorls. Aperture oval without thickening of throat; columella tooth poorly developed. Northern Britain, in brackish waters (salinity 12–16‰).

var. *cornea* (Lovén): Shell small, thin, with five or six slightly tumid whorls; last whorl approx. 60% of shell height, lightly costate. Aperture oval; columella tooth clear; throat thickened. Northern Britain in brackish waters (salinity 6–7‰).

All varieties are associated with *Zostera*, or weeds of similar form, at LWST and below, and are tolerant of a wide range of salinities. Widely distributed from Norway to Canary Islands, extending into the Baltic as far as Rügen Island.

16 BARLEEIDAE

Only one British representative of this family.

Barleeia unifasciata (Montagu) FIG. 12.11

Shell oval-conical, unsculptured, with blunt apex and five slightly tumid whorls. Aperture oval; inner lip reflected over base of last whorl; a small umbilical groove but no umbilicus. Commonly banded brownish-red and white, though may be uniform red, white, or crimson. Size up to 2 × 1.5 mm. Body with prominent bifid snout, and short cephalic tentacles with eyes at base; mantle edge without pallial tentacles. Foot with dark purple opercular lobes; no epipodial tentacles. Operculum dark red, ear-shaped and swollen.

Not uncommon among algae and in rock pools near LWM. Mainly confined to south-west British Isles and Ireland, but recorded from Shetland.

17 CINGULOPSIDAE

Only one British representative of this family.

Cingulopsis fulgida (J. Adams) FIG. 12.11

Shell small, globular, with three or four tumid whorls, last whorl two-thirds of shell height; sutures deep, apex blunt. Aperture large, almost circular, with peristome. Inner lip thickened, reflected over small umbilical groove; many shells with umbilicus. White, cream, or pale horn, with red-brown banding, usually three bands on last whorl, two on penultimate. Young shells mostly uniform dark horn colour. Size up to 1.5 × 0.9 mm. Body with broad, bifid snout; cephalic tentacles short, setose, with eyes at base. Mantle edge without pallial tentacles; no penis in males. Foot with anterior and posterior mucus glands, animals often found suspended from weed or water surface by mucous strings; metapodial tentacle triangular, not reaching edge of foot, operculum yellow, spiral.

In rock pools and on fine weeds, MTL and below; more

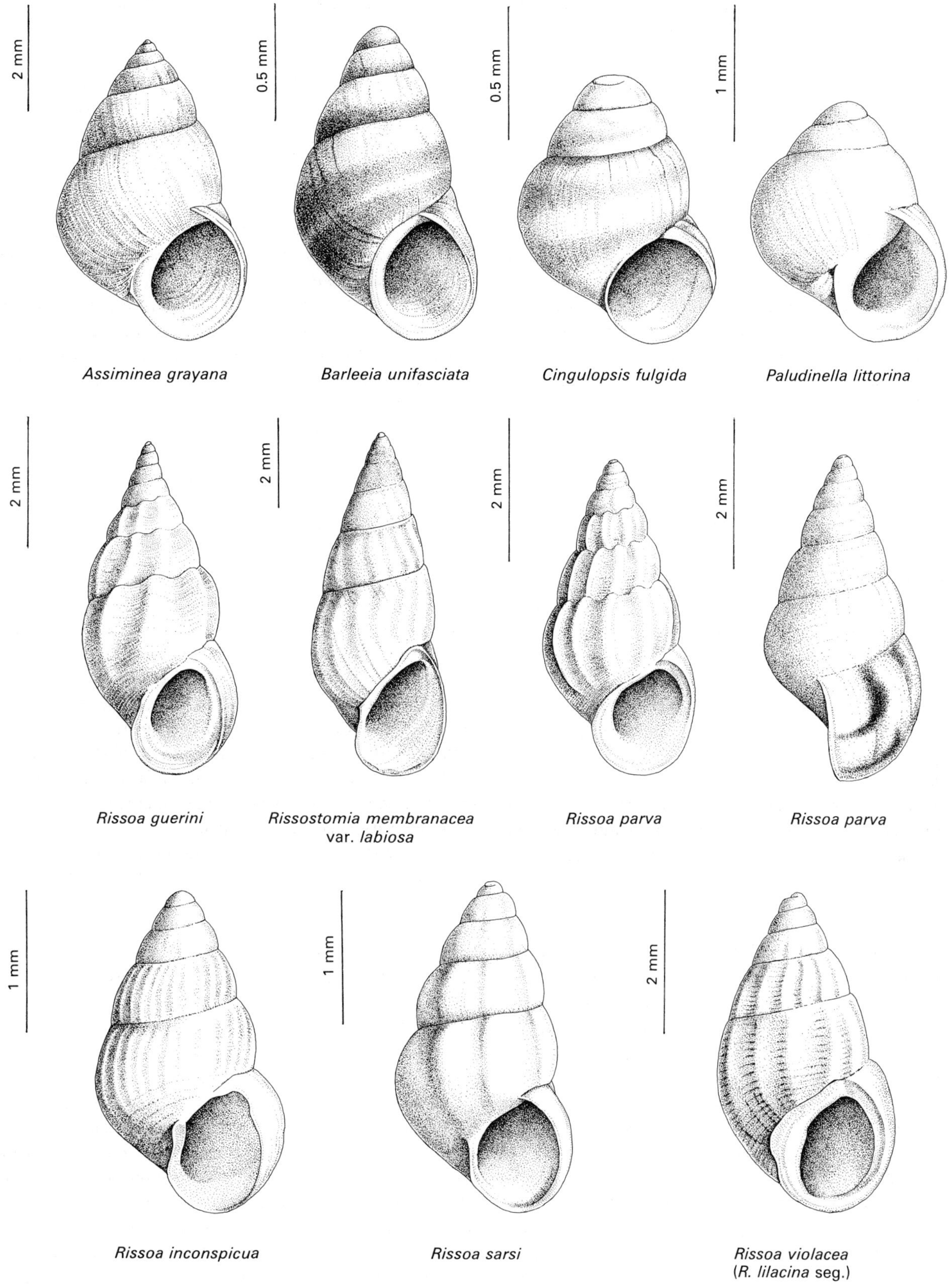
2 mm
0.5 mm
0.5 mm
1 mm
Assiminea grayana
Barleeia unifasciata
Cingulopsis fulgida
Paludinella littorina
2 mm
2 mm
2 mm
2 mm
Rissoa guerini
Rissostomia membranacea var. *labiosa*
Rissoa parva
Rissoa parva
1 mm
1 mm
2 mm
Rissoa inconspicua
Rissoa sarsi
Rissoa violacea (*R. lilacina* seg.)

Fig. 12.11

common in summer. Distributed from west coast of Scotland to Mediterranean. Absent from eastern Channel and North Sea; common on south and west coasts.

18 SKENEOPSIDAE

Only one British representative of this family.

Skeneopsis planorbis (Fabricius) FIG. 12.12

Shell very small (less than 2 mm diameter), semi-transparent, glossy; a helicone of four tumid, rapidly expanding whorls; spire depressed; umbilicus deep, wide. Sculpture of irregular, sinuous growth lines, sometimes with fine spiral lines on upper surface of last whorl. Aperture almost circular, sinuous in profile; with peristome. Red-brown or pale horn coloured, with pale operculum. Size up to 0.75 × 1.5 mm. Snout short and bifid; cephalic tentacles long, cylindrical, bluntly rounded, finely setose at tip. Mantle edges without pallial tentacles. Foot with posterior mucus gland on sole; operculum horny, a concave spiral, with central nucleus.

On fine algae and in coralline pools from MTL to 70 m. Abundant in summer, rare in winter. Often found suspended from weeds or water surface by mucus. Widely distributed, from Azores to Arctic. On all suitable British shores; absent from eastern shores of North Sea and from Baltic.

19 TORNIDAE

Only one British representative of this family.

Tornus subcarinatus (Montagu) FIG. 12.12

Shell small (< 3 mm diameter), nearly circular, with three or four rapidly expanding whorls, spire depressed. Umbilicus deeply excavate so that whorls of spire may be seen from below. Sculpture of keel-like spiral striae: six on last whorl, two on other whorls; also fine spiral lines of tubercles, low flexuous costae and overlying prosocline growth lines. Aperture elongate-oval, in marked prosocline plane, with peristome; small anal spout extends from subsutural spiral keel, the latter marked internally by a groove entering throat. White, possibly with orange tint. Size up to 1.5 × 2.5 mm. Snout and cephalic tentacles long, eyes not discernible; mantle edge with two pallial tentacles on right; gill may protrude from mantle cavity in active animals. Foot with anterio-lateral projections and double edged anterior margin; operculum thin and spiral.

Under boulders in well-oxygenated sand, at LWM and down to 3 m. A southern species extending from Mediterranean to southern shores of British Isles, where it is uncommon.

20 OMALOGYRIDAE

Shell very small (1 mm diameter), a biconcave planospiral of up to four whorls. Aperture circular. Snout broad, bilobed, and ciliate, with eyes at base of each lobe; cephalic tentacles absent. Foot bilobed anteriorly, with median pedal gland opening on sole. No ctenidium.

1. Shell marked only by growth lines and, occasionally, low spiral striae; penultimate whorl abuts aperture slightly below centre **_Omalogyra atomus_**

   Shell with strong transverse costae, often with three spiral ridges; penultimate whorl abuts aperture at centre line **_Ammonicera rota_**

Omalogyra atomus (Philippi) FIG. 12.12

Shell minute, transparent, with two and a half to three rapidly expanding, tumid whorls of almost circular cross-section. Sculpture of irregular growth lines and possibly low spiral ridges. Aperture with peristome; nearly circular in outline, slightly flattened or concave where it abuts penultimate whorl. Periphery of penultimate whorl lies slightly below medial diameter of aperture. Reddish-brown, cream, or yellow. Size up to 0.4 × 1.0 mm. Snout broad, bilobed, ciliate; no cephalic tentacles. Foot bilobed anteriorly; posterior mucus gland on sole. Operculum thin, horny, spiral, with a central nucleus.

On algae, particularly *Enteromorpha* and *Ulva;* common in pools below MTL, often suspended by mucus threads from surface; extends to 20 m. Distributed from Azores to Norway; absent from Baltic and continental shores of North Sea.

Ammonicera rota (Forbes and Hanley) FIG. 12.12

Shell minute, three or four rapidly expanding, tumid whorls of almost circular cross-section. Sculpture of transverse costae but no spiral element; sometimes with three tubercles, above, at, and below periphery which may give a keeled appearance when well-developed. Aperture oval, with peristome; slightly flattened where it abuts penultimate whorl, angulate in keeled shells. Reddish brown or horn coloured, with iridescent throat. Size up to 0.5 × 1.0 mm.

On weeds and in rock pools at LWST; recorded to 25 m on sandy bottoms. Distributed from Mediterranean to Norway; uncommon on British shores, though small size may account for shortage of records. Absent from continental shores of North Sea.

21 RISSOELLIDAE

Shell very small (less than 2 mm), transparent and smooth; short, tumid, and conical, or globular. Umbilicus a shallow chink, or deep and round; operculum with peg-like process articulating with columella. Snout bifid, cephalic tentacles single or bifid. Foot with posterior mucus gland connected by groove to posterior margin of foot. Transparency reveals distinctive dark spots on mantle.

1. Shell short, tumid, conical; aperture < 50% shell height; body whorl ≤ 70% shell height. Tentacles single. Snout deeply bifid, almost tentaculate **_Rissoella diaphana_**

   Shell globular; aperture ≥ 50% shell height; body whorl ≤ 80% shell height **2**

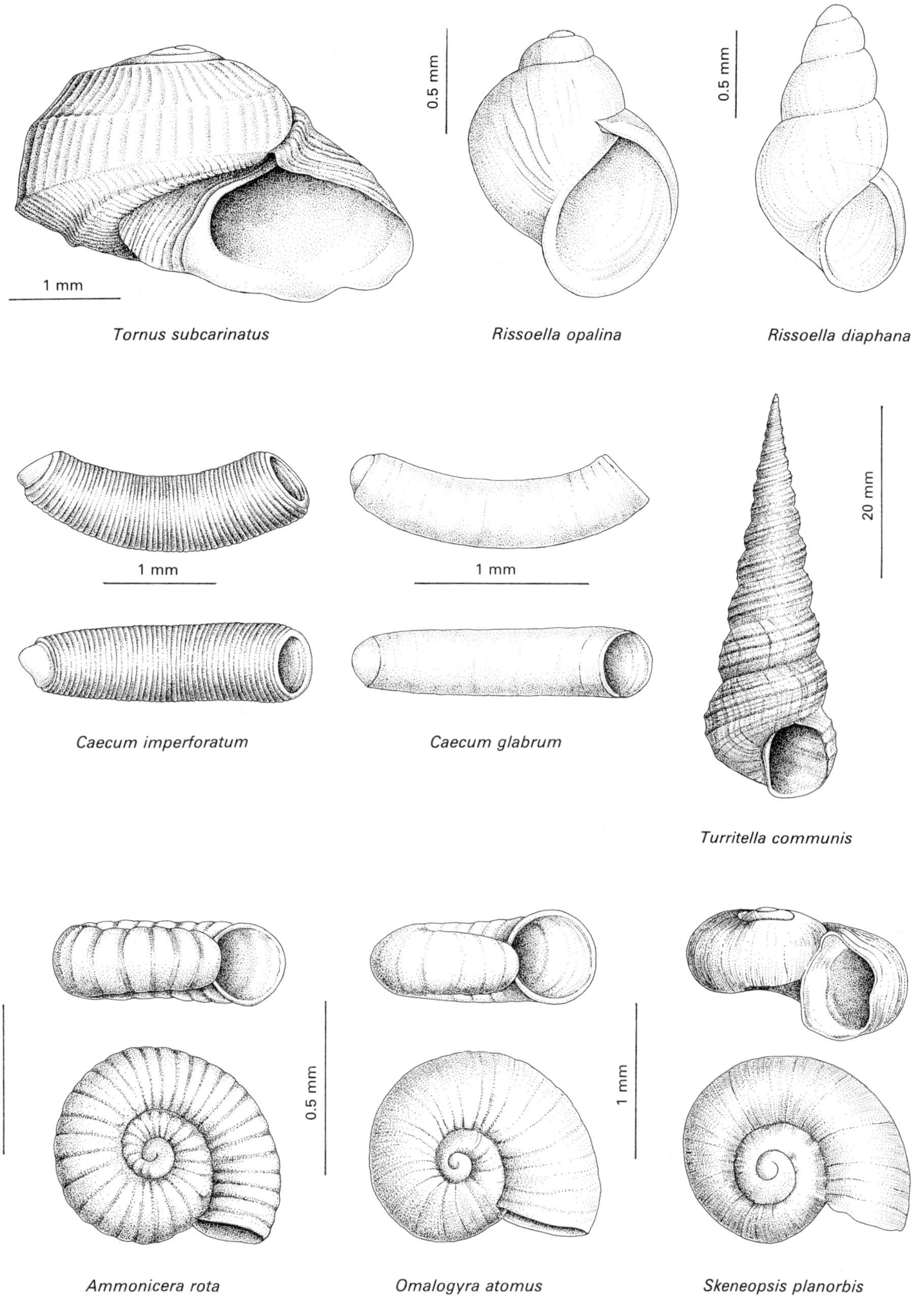
1 mm
Tornus subcarinatus
0.5 mm
Rissoella opalina
0.5 mm
Rissoella diaphana
1 mm
1 mm
Caecum imperforatum
Caecum glabrum
20 mm
Turritella communis
0.5 mm
Ammonicera rota
0.5 mm
Omalogyra atomus
1 mm
Skeneopsis planorbis

Fig. 12.12

2. Shell height greater than breadth; umbilicus a narrow chink; tentacles bifid ***Rissoella opalina***
   Shell height equal to or less than breadth; umbilicus a deep hole; tentacles not bifid ***Rissoella globularis***

Rissoella diaphana (Alder) FIG. 12.12
Shell small, transparent, smooth, fragile; four-and-one-half tumid whorls, slightly stepped at sutures; apex blunt. Last whorl about 70% of shell height. Sculpture of prosocline growth lines. Aperture egg-shaped; inner lip slightly reflected over small umbilical chink. White, viscera visible by transparency in live animals; dark spots on mantle skirt particularly distinctive. Size up to 1.5×0.85 mm. Snout deeply bifid, appearance tentaculate; cephalic tentacles cylindrical, finely setose. Mantle edge without pallial tentacles. Foot with anterior pedal gland opening dorsally, and posterior gland opening ventrally, latter linked by groove to posterior margin of sole. Operculum horny, nucleus close to columellar margin, which bears a peg-like process.

Common in summer among filamentous red algae and in coralline rock-pools between MTL and LWST. Distributed from Mediterranean to Norway; absent from Baltic and most of North Sea.

Rissoella opalina (Jeffreys) FIG. 12.12
Shell small, semi-transparent, smooth; globular, with three or four tumid, rapidly expanding whorls; sutures deep. Aperture almost D-shaped, columella nearly straight, with peristome and small umbilical chink. Pale yellow or horn coloured, with opalescent lustre; three dark spots on mantle show through in live shells. Size up to 2×1.4 mm. Body similar to *R. diaphana*. Snout bifid, slightly lobed; cephalic tentacles bifid, setose.

Locally common in summer on filamentous red algae and in rock-pools near LWST; often with *R. diaphana*. Distributed along western shores of British Isles from Channel Isles to Shetland; absent from North Sea coasts, east Channel, and Baltic.

Rissoella globularis (Jeffreys)
Shell small, transparent, smooth; more globular than *R. opalina*, with three or four tumid, rapidly expanding whorls; sutures deep; breadth equal to or greater than height. Aperture pear-shaped, columella not straight as in *R. opalina*, peristome may project clear of last whorl; inner lip borders deep, round umbilicus. White, nacreous, with dark viscera showing through in live shells; dark spots on mantle not conspicuous. On fine red weed and in coralline rock-pools near LWST; not common. Distributed along Atlantic coasts from France to northern Norway; British records confined to Shetland and north-west Scotland.

22 TURRITELLIDAE
Only one British representative of this family.

Turritella communis Risso FIG. 12.12
Shell tall, sharply pointed, conical, with 16–20 tumid whorls. Sculpture of numerous spiral striae of varying proportions: three (or up to six) usually more prominent than others; sinuous growth lines may interact with spiral striae to give them a beaded appearance. Aperture small, angulate, without peristome; outer lip crenulate. No umbilicus. Buff coloured with lighter striae; may have lilac tinge on base. Size up to 55×16 mm, commonly 30×10 mm. Snout broad, cephalic tentacles short, cylindrical. Mantle edge with numerous pallial tentacles; short, often bifid or trifid on right, longer and more pinnate on left. A deep pallial groove crosses floor of mantle from left, terminating below right cephalic tentacle. (Pallial tentacles and groove associated with filter feeding habit.) Foot small; operculum small, circular, and concave, with numerous marginal bristles.

Locally abundant sublittorally, down to 220 m, on soft bottoms; sometimes at ELWST. Distributed from North Africa and Mediterranean to Lofoten Isles; rare or absent from east Channel and southern North Sea.

23 CAECIDAE
Shell small; young specimens (less than 1 mm) with spiral protoconch from which the body whorl projects as a gently curved tube. Protoconch later breaks off and the truncated body whorl is sealed by a calcareous plug.

---

1. Shell tube annulate, apical plug pointed; operculum flat ***Caecum imperforatum***
   Shell tube smooth, apical plug dome-shaped; operculum convex ***Caecum glabrum***

Caecum imperforatum Kanmacher FIG. 12.12
Adult shell curved, glossy, strongly annulate, apex sealed by eccentrically pointed, calcareous plug. Aperture circular, with thickened rim. Yellow-brown or white. Size up to 4×0.5 mm. Snout flat, deeply bifid; cephalic tentacles long, with setose tips. No pallial tentacles. Operculum thick, flat, circular; polygyrous with central nucleus.

Sublittoral, on sand and sandy-mud, to 250 m. Distributed from Mediterranean to south-west coasts of British Isles. Common, but inconspicuous.

Caecum glabrum (Montagu) FIG. 12.12
Adult shell curved, glossy, without annulations; apex sealed by a dome-shaped calcareous plug. Size up to 2.5×0.5 mm. Body as in *C. imperforatum*. Operculum convex.

Habitat as for *C. imperforatum*. Distributed from Mediterranean to Norway and west coast of Sweden. Common, but inconspicuous.

24 CERITHIIDAE
Only one British representative of this family.

Bittium reticulatum da Costa FIG. 12.13
Shell elongate, conical, up to 16 (usually 10–12) moderately tumid whorls; apex pointed. Sculpture of strong costae and spiral striae, interacting to give oval tubercles: four rows on lower whorls, three on upper whorls; no tubercles on lowest

spiral ridge of each whorl, or on basal ridges of last whorl. One costa, the varix, often more prominent than rest; varix of last whorl close to aperture. Aperture oval, drawn out to a short sinus basally; outer lip crenulate, without anal sinus. Inner lip reflected over columella region and base of last whorl; no umbilicus. Buff to light chestnut-brown; tubercles, varices, and columella lighter. Up to 15 × 6 mm, commonly half this size. Snout long, bifid; cephalic tentacles long, slender with eyes on prominent bulge at base. Mantle edge broadly lobed; foot long, narrow, with doubled-edged anterior margin; large opercular lobe projects posteriorly from below oval operculum.

On soft substrata, associated with *Zostera* and other weeds; from LWST to 250 m. Locally abundant. Widely distributed from Canaries to Lofoten Isles; absent from continental shores of North Sea and east Channel.

B. simplex (Jeffreys), sometimes regarded as a separate species, lacks the varices of *B. reticulatum*, has a thinner shell, and a salmon-pink body (as opposed to pale fawn or brown in *B. reticulatum*). It is perhaps simply a colour variety of *B. reticulatum*.

25 CERITHIOPSIDAE

Shell small, tower-shaped or cyrtoconoid, with many flat-sided or slightly tumid whorls, sculptured by spiral rows of tubercles. Aperture small with short siphonal canal at base; operculum paucispiral. Males lack penis. Only two species likely to be encountered in British waters.

Cerithiopsis tubercularis (Montagu) FIG. 12.13

Shell tall, narrow, conical, up to 14 flat-sided whorls; sutures deep. Sculpture of spiral striae and transverse costae, interacting to give three rows of close set, rounded tubercles (more elongate near aperture); base of last whorl devoid of tubercles but with two spiral ridges. Aperture small, narrow, oval with short siphonal canal; canal wall with /\-shaped notch on abapertural side. Outer lip crenulate, inner lip reflected over columella; no umbilicus. Chestnut-brown, tubercles and columella pale. Size up to 6.5 × 2.25 mm. Head without snout; cephalic tentacles long, slender, setose at tip. Mantle edge simple, drawn out into short siphon on left. Foot with propodial flap behind anterior margin; prominent posterior pedal gland on sole. Operculum thin, spiral, with few turns; opercular lobes large.

On sponges (*Halichondria panicea* and *Hymeniacidon perleve*), particularly where associated with *Lomentaria*, *Corallina*, and other fine red algae; from LWST to 100 m. Most abundant in March and August–September. Distributed from Azores and Mediterranean to Norway; absent from southern North Sea.

Cerithiopsis barleei Jeffreys (Fretter and Graham 1982: 369), is similar to *C. tubercularis* but has a more slender, thinner shell which may be slightly transparent, with sutures more incised. Last whorl is smooth basally. *C. barleei* is pale yellow-brown, and up to 7 × 2 mm. Its habitat is similar to that of *C. tubercularis*; recorded from the sponge *Suberites domunculus*. Not uncommon. More southern in distribution, extending north only as far as south-west coasts of British Isles.

26 TRIPHORIDAE

Shell sinistral, tower-shaped or cyrtoconoid, with many flat-sided whorls, sculptured by spiral rows of tubercles. Aperture squarish, with short siphonal canal; operculum multispiral. Males lack penis.

Triphora adversa (Montagu) FIG. 12.13

Shell sinistral, tower-shaped, with 15 flat-sided whorls; sutures shallow. Sculpture of spiral striae and transverse costae producing spiral rows of tubercles. Last whorl with three rows of tubercles above two or three nodular spiral ridges; tubercles more rectangular as they approach outer lip on abapertural side of last whorl. Penultimate and two or three whorls above with three rows of tubercles, the central row being smaller; upper whorls with two rows. Aperture angular, with small, narrow, almost tubular siphonal canal and slight anal notch. Outer lip crenulate, thin; inner lip reflected over columella; no umbilicus. Chestnut-brown; tubercles, apex, and edge of outer lip often lighter. Size up to 7 × 2 mm. Snout short; cephalic tentacles long, slender with eyes at base. Mantle drawn into siphonal spout on right. Foot with propodial flap above and behind anterior margin; posterior pedal gland opens into a median groove on sole; operculum multispiral.

On, or embedded in, sponges (*Halichondria* and *Hymeniacidon*), under stones, or on weeds from LWST to 100 m. Distributed from Biscay to Norway; absent from southern North Sea and eastern Channel basin.

Triphora pallescens (Jeffreys) (Fig. 12.13) differs from *T. adversa* in that the shell has five rows of tubercles on last whorl; these do not become rectangular as in *T. adversa*; first spiral ridge below tubercles on last whorl bifurcates near outer lip. The body is white in both species but cephalic tentacles are flecked with yellow in *T. pallescens*. Habitat as for *T. adversa*. Distributed from Biscay to western Channel.

Two recently described species, *T. erythrosoma* Bouchet and Guillemot and *T. similior* Bouchet and Guillemot also resemble *T. adversa*. *T. erythrosoma* has four rows of tubercles on the last whorl, red and yellow flecking on head and foot, and yellow tentacles. *T. similior* has black flecks on the head and anterior parts of the foot.

27 APORRHAIDAE

Shell tall, with costae and fine spiral striae. Last whorl in mature shells with prominent spiral rows to tubercles and a palmate extension to outer lip. Last whorl in juveniles with a strong spiral ridge below periphery, and elongate-oval aperture, drawn out basally into a sharp point. Operculum narrow, set across the foot. On sublittoral soft substrata.

1. Uppermost process of palmate extension does not reach shell apex; basal process hooked toward aperture **_Aporrhais pespelecani_**

   Uppermost process of palmate extension projects beyond shell apex; basal process not hooked toward aperture **_Aporrhais serresianus_**

Aporrhais pespelecani (Linnaeus) FIG. 12.13

Shell of mature animals readily distinguished by palmate extension to outer lip; no such extension in shells with less than eight whorls. Spire tall, whorls tumid, with crescentic costae and fine, flat spiral striae; apical whorls finely decussate or with spiral striae only. Last whorl of mature animals with three prominent spiral ridges set with stout tubercles; ridges extend on to palmate outer lip but are not tuberculate at this point. Aperture in mature shells dominated by palmate outer lip; uppermost process fuses with lower part of spire, extending to antepenultimate whorl; basal process curved upward towards aperture (in juveniles this is the only process and resembles a sharply pointed siphonal canal). Cream or sandy coloured, sometimes with purplish stain on back of last whorl; apertural surfaces pearly white. Size up to 42 × 28 mm. Snout long; cephalic tentacles long, cylindrical, with eyes on lateral bulges at base; males with thick, tentaculiform penis behind right tentacle. Foot with double-edged anterior margin; operculum elongate-oval, set across the foot.

Sublittoral; locally common on mud and muddy sand to 180 m. Distributed from Mediterranean to northern Norway and Iceland; not common around Channel Isles and Scilly Isles.

Aporrhais serresianus (Michaud) FIG. 12.13

Shell similar to that of *A. pespelecani*, though generally smaller and more delicate; adult shells distinguished by shape of palmate outer lip, which projects beyond shell apically and does not curve upward basally. White. Size up to 38 × 24 mm.

Sublittoral, on fine mud down to 1000 m. Distributed from Mediterranean to northern Norway and Iceland; in British Isles it occurs off west coast of Ireland, Scotland, Orkney, and Shetland.

28 TRICHOTROPIDAE

Only one common British representative of this family.

Trichotropis borealis Broderip and Sowerby FIG. 12.13

Shell conical, with up to five tumid whorls; covered by a conspicuous periostracum. Sculpture of prominent, chord-like spiral striae of variable dimension, and fine prosocline growth lines. Periostracum on major striae drawn out into stiff bristles. Aperture rather triangular, tapering basally to a V-shaped spout; outer lip thin, crenulate, with periostracal margin; inner lip reflected over columella, forming a narrow umbilical groove. Shell whitish, periostracum light horn coloured. Size up to 11 × 6 mm. Snout short, bifid, with short, dorsally grooved pseudoproboscis; cephalic tentacles long, thin, with eyes on lateral bulges near base. Foot double edged anteriorly and laterally; operculum triangular.

Sublittoral, on stones and shells from hard bottoms; 10–270 m. Locally common. Circumboreal; occurs in northern North Sea, west coast of Scotland, north and west coasts of Ireland.

The trichotropid *Torellia vestita* Jeffreys (Fretter and Graham 1981: 303) has recently been collected from east of the Shetlands. It has a globular, finely reticulate shell with a complete periostracum; periostracal bristles follow the pattern of shell ornamentation. the body has a long, tapering and dorsally grooved pseudoproboscis.

29 CAPULIDAE

Only one British representative of this family.

Capulus ungaricus (Linnaeus) FIG. 12.13

Shell cap-shaped, a rapidly expanding planospiral of up to three whorls; covered by periostracum. Sculpture of numerous spiral striae and fine growth lines; periostracum projects as frilled edge beyond aperture, successive positions of aperture marked by series of similar fringes on last whorl. Aperture large, oval, with finely serrated margin. White, with horn coloured periostracum. Size up to 40 × 25 mm across the aperture and 15 mm high. Snout short, with long, dorsally grooved, non-retractable pseudoproboscis; mantle thick, double-edged. Foot oval, without operculum.

Sublittoral (occasionally at LWST), on stones and shells (especially those of living scallops and *Turritella communis*) down to 850 m. Widely distributed from West Africa to northern Norway; less common in North Sea.

30 CALYPTRAEIDAE

Shell patelliform, with sharp apex, or a depressed hemisphere of variable height; internal partition present. Ciliary feeders, with enlarged mantle cavity and ctenidium; large propodial flap on either side of neck. Foot large, oval, without operculum. Protandrous hermaphrodites.

1. Shell conical, with sharp, central apex; partition tongue-shaped; animals solitary **Calyptraea chinensis**
   Shell cap-shaped, apex subterminal; partition D-shaped; animals usually in chains **Crepidula fornicata**

Calyptraea chinensis (Linnaeus) FIG. 12.14

Shell patelliform, rather depressed, with concave anterior profile; apex with small nipple-shaped beak, often slightly coiled. Sculpture limited to slight, sometimes stepped growth lines, sometimes with small tubercles. Aperture almost circular, with thin, bevelled edge, often reflecting contours of surface to which animal was attached. With a posterior tongue-shaped internal partition. White or yellowish, with pearly white interior. Size up to 15 × 5 mm. Snout short; head elongated behind tentacles, with large propodial folds (cephalic lappets). Mantle edge thickened, with repugnatorial (?) glands; mantle cavity and ctenidium large. Foot large, without operculum.

Sublittoral, on stones and shell gravel associated with soft substrata, to 20 m; under stones at LWST on sheltered shores in south-west England. Distributed from north-west Africa and Mediterranean to west coast of Scotland (Loch Ryan); absent from North Sea and east Channel basin.

Crepidula fornicata (Linnaeus) FIG. 12.14

Shell depressed, elongate cap-shape; aperture applied to substratum, spire set posteriorly and to right; smooth, with irregular

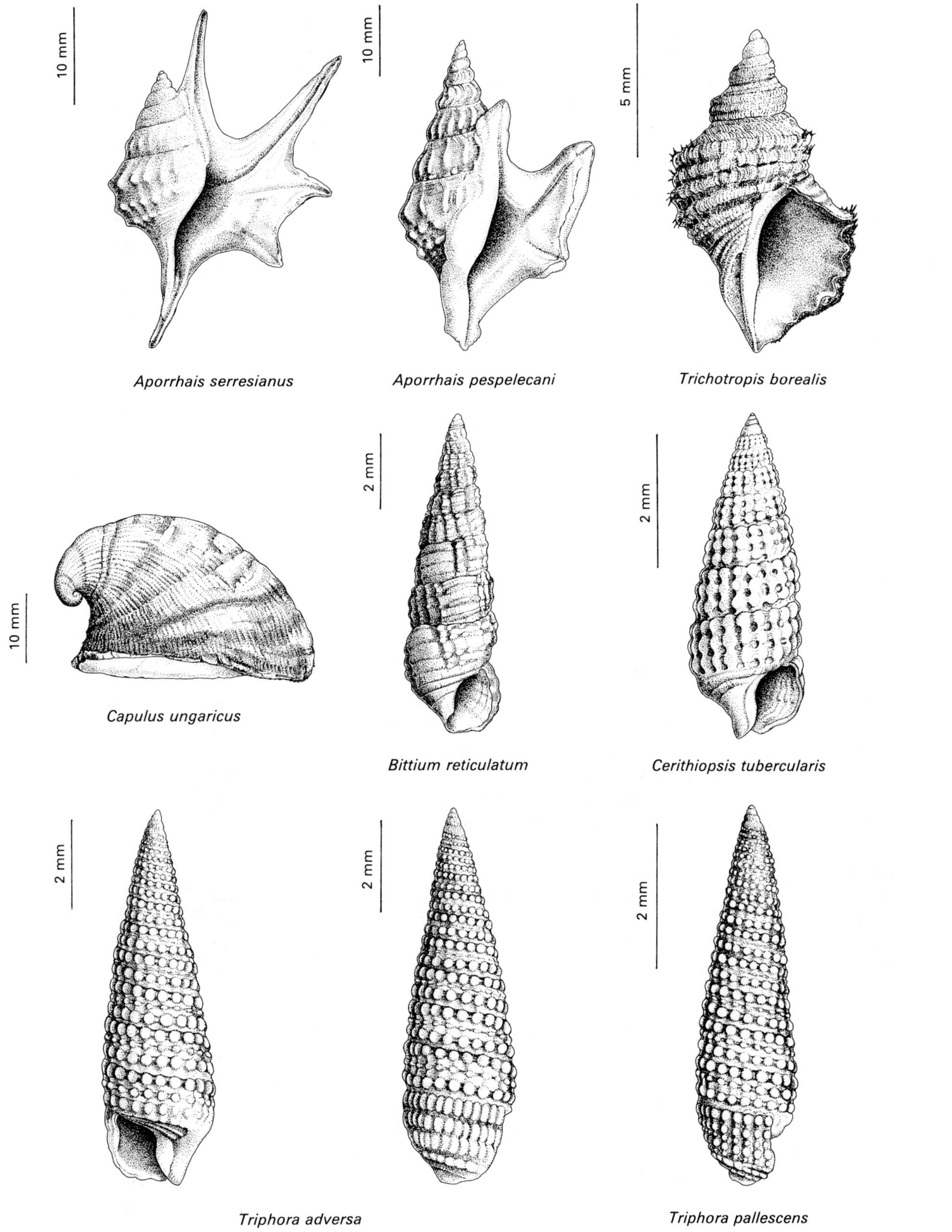
10 mm
Aporrhais serresianus
10 mm
Aporrhais pespelecani
5 mm
Trichotropis borealis
10 mm
Capulus ungaricus
2 mm
Bittium reticulatum
2 mm
Cerithiopsis tubercularis
2 mm
2 mm
Triphora adversa
2 mm
Triphora pallescens

Fig. 12.13

growth lines. Aperture elongate-oval, often slightly concave on spire side; edge thin, moulding to surface to which animal is attached; posteriorly a large shelf-like internal partition. Cream or pinkish, with streaks and blotches of reddish brown; throat chestnut-brown, with white partition. Size up to 25 × 50 mm.

On shells and stones on soft substrata; LWST to 10 m, often cast ashore by storms. The animals form curled stacks of up to 12 individuals; large shells at the bottom, becoming progressively smaller towards the apex of the chain. They are pests of oyster beds. An accidental introduction from America, it is now distributed from Essex along south coast to Bristol Channel, Belfast Lough, and Co. Kerry.

31 LAMELLARIIDAE

Shell wholly or partly covered by papillated mantle lobes; fragile, smooth, with up to three rapidly expanding whorls. No snout; eversible proboscis used to feed on tunicates and coelenterates. Foot with double edged anterior margin; no operculum.

---

1. Shell with thick periostracum; mantle only slightly reflected over shell — **2**

   Shell internal, covered by papillated mantle; resembles dorid sea-slug but with anterior siphonal notch in mantle, smooth tentacles, and no dorsal circlet of gills — **3**

2. Shell with fine, but distinct reticulation; aperture almost circular — ***Velutina velutina***

   Spiral ornament less well-developed; aperture oval; rather rare — ***Velutina plicatilis***

3. Animal up to 20 mm long; dorsal mantle domed, tuberculate, purplish grey with white flecks, occasionally yellowish; shell with high spire — ***Lamellaria perspicua***

   Animal up to 10 mm long; dorsal mantle rather flat, smoother, yellowish with dark flecks; shell with low spire — ***Lamellaria latens***

Velutina velutina (Müller) FIG. 12.14

Shell fragile, semi-transparent, with up to three rapidly expanding, tumid whorls, covered by periostracum; last whorl comprises most of shell; spire low, occasionally in a small depression. Sculpture of fine spiral striae and growth lines, producing reticulate pattern. Aperture almost circular, slightly indented by last whorl; periostracum projects beyond margin of outer lip. Shell white, periostracum dark brown. Size up to 20 × 20 mm, but commonly smaller. Mantle edge smooth, greatly thickened; foot shield-shaped, with double-edged anterior margin; no operculum.

Associated with tunicates, particularly solitary forms such as *Styela*; normally sublittoral, to 1000 m, but may occur at ELWST. Circumboreal, extending down west coasts of British Isles to Mediterranean; deeper in southern parts of range.

Velutina plicatilis (Müller) FIG. 12.14

Shell similar to *V. velutina*; periostracal covering thick, with small bristles, obscuring much of shell contours. Sculpture mainly of irregular growth lines, which may be strong enough to affect periostracum; fine spiral lines may occur. Aperture oval, drawn out basally to right; outer lip thin, flexible, may distort in dried specimens. Periostracum yellowish; shell white or pinkish. Size up to 15 × 12 mm.

Sublittoral, associated with compound ascidians and hydroids on hard substrata; 10–375 m. Not common. Circumboreal, extending into northern North Sea and down west coast of Scotland and Ireland to northern Spain.

Lamellaria perspicua (Linnaeus) FIG. 12.14

Shell permanently covered by mantle; fragile, with two or three rapidly expanding whorls and short spire; last whorl about 95% of shell height. Smooth, apart from growth lines; fine spiral lines on some shells. Aperture rather spoon-shaped, drawn out basally to right; peristome in marked prosocline plane. Colour variable, often grey or grey-lilac flecked with white, but buff or orange forms occur, which when flecked with black may resemble *L. latens*. Shell size up to 9 × 10 mm. Mantle edge thickened, with siphonal groove anteriorly; dorsal surface tuberculate. Foot rather elongate, double-edged anteriorly, without operculum. Body up to 20 × 12 mm.

Fairly common under stones, in pools and on surfaces covered by compound ascidians; LWST down to 1200 m. Distributed from Mediterranean to northern Norway and Iceland; widely distributed around British Isles.

Lamellaria latens (Müller) FIG. 12.14

Shell permanently covered by mantle; fragile, with two rapidly expanding whorls; spire not obvious, profile rather depressed; last whorl equivalent to height of shell. Aperture like that of *L. perspicua* but relatively larger. Shell size up to 5 × 9.5 mm. Body as in *L. perspicua*, mantle smoother and less domed in profile; up to 10 × 6 mm. Tends to be buff with brown or black flecks but this is not a reliable guide to identification.

Habitat and distribution as for *L. perspicua*.

32 ERATOIDAE

Shell convolute or with very short spire; robust, polished, enveloped by mantle lobes in active animals. Aperture long and narrow. No snout; eversible proboscis used to feed on compound ascidians. Foot with posterior pedal gland on sole; no operculum.

---

1. Shell convolute, no spire visible in mature shells; ridged — **2**

   Shell with short spire; smooth or with slight spiral ridges — **3**

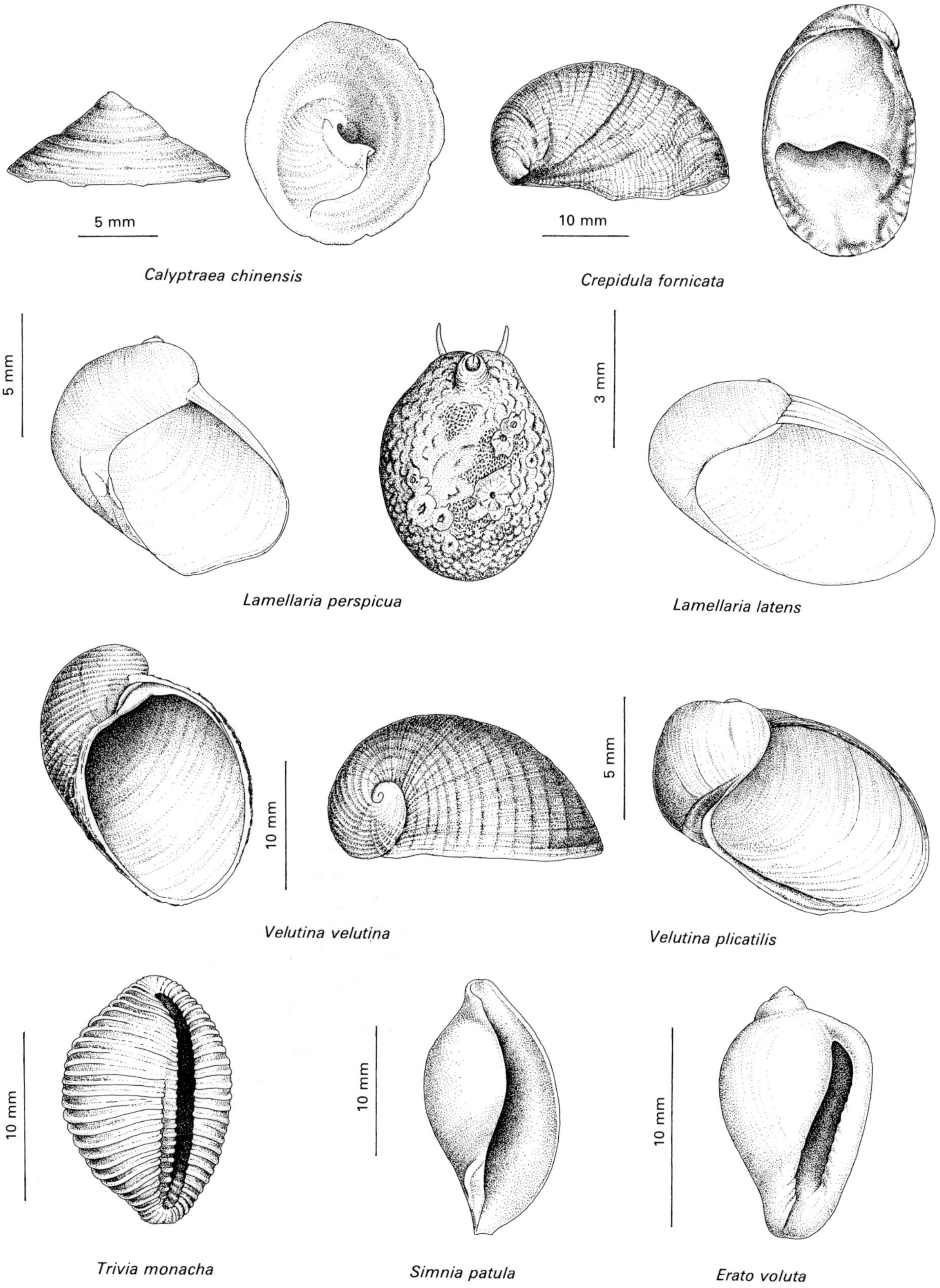
5 mm
Calyptraea chinensis
10 mm
Crepidula fornicata
5 mm
Lamellaria perspicua
3 mm
Lamellaria latens
10 mm
Velutina velutina
5 mm
Velutina plicatilis
10 mm
Trivia monacha
10 mm
Simnia patula
10 mm
Erato voluta

Fig. 12.14

2. Shell with three dark spots on abapertural side; body highly pigmented, back end of foot striped **_Trivia monacha_**

   Shell without spots; body pale, back end of foot yellow **_Trivia arctica_**

3. Shell smooth, thick, narrowing basally to give harp-shaped outline; mantle lobes heavily pigmented **_Erato voluta_**

   Shell slightly ribbed, thin, broad basally, with oblong outline; mantle lobes rather pale **_Trivia juvenile_**

Erato voluta (Montagu) FIG. 12.14

Shell solid, glossy, with three or four slightly tumid whorls; spire short, sutures shallow; last whorl about 80% or more of shell height, harp-shaped in profile. Smooth, with translucent overlay secreted by mantle lobes on most of last whorl and part of spire. Aperture long, narrow, with parallel sides and short siphonal canal; outer lip thick, with internal row of teeth; base of columella grooved. White. Size up to 12 × 7.5 mm. Cephalic tentacles rather short, slender; mantle drawn out into long siphon anteriorly, laterally into large tuberculate lobes which envelop much of shell in active animals. Foot elongate, with double-edged anterior and posterior pedal gland on sole; no operculum.

Sublittoral, associated with ascidians on hard substrata from 20–150 m. Distributed from Mediterranean to Norway; widely distributed around British Isles, more common in southern parts of range.

Trivia monacha (da Costa) FIG. 12.14

Adult shell solid, convolute (spire overgrown by last whorl); egg-shaped, with flattened apertural side. Sculpture of strong, often branching ridges and grooves, those at ends of shell more curved, almost U-shaped. Juveniles (up to 5 mm) with low spire of two or three tumid whorls; smooth at first, developing spiral striae in older shells. Aperture narrow, running whole length of shell; turned to left (i.e. towards swollen parietal region of body whorl) at ends, and ridged on both sides. Juvenile aperture widely open, narrowing adapically, not ridged. Juveniles white, adults pinkish or reddish brown, with three dark spots; apertural side white. Size up to 12 × 8 mm. Cephalic tentacles long, slender; mantle drawn out into long siphon anteriorly, extending laterally over all or most of shell in life. Foot with double-edged anterior margin and posterior pedal gland on sole; females with additional ventral pedal gland. Males with slender, cylindrical penis.

Fairly common on rocky shores, under stones and ledges at LWST; associated with compound ascidians on which it feeds. Distributed from Mediterranean to British Isles, more common in south.

Trivia arctica (Pulteney)

Shell like that of *T. monacha*; lacks pigment spots and usually smaller. Size up to 10 × 8 mm. Body as for *T. monacha* but with flattened, leaf-like penis in males.

Mainly sublittoral, to 100 m in north of range, 1000 m in south; occasionally at LWST, associated with compound ascidians. Distributed from Mediterranean to Norway, more common in north of range.

33 OVULIDAE

Only one British representative of this family.

Simnia patula (Pennant) FIG. 12.14

Shell convolute, elongate-oval, drawn out into apical and basal canals; thin, polished; juvenile shells (< 5 mm long), with short spire. Fine sculpture of low strap-like spiral striae; appears smooth to naked eye. Aperture curved, wider below parietal bulge of last whorl; apical canal short, basal canal wide, bordered on left by ridged columella. White or yellowish. Size up to 22 × 12 mm. Snout short, no proboscis; mantle drawn out anteriorly to a siphon, which projects from basal canal; lateral extensions cover most or all of shell in active animals. Foot large, elongate, with posterior pedal gland on sole and double-edged anterior end; no operculum.

Sublittoral, associated with coelenterates (*Eunicella*, *Alcyonium*, and *Tubularia*) on hard substrata from 15–75 m. Distributed from Spain to west Channel; not common. Occasional records from west Irish and British coasts to Orkney.

34 NATICIDAE

Shell smooth, globular or oval, with low spire and enlarged last whorl; aperture semicircular, umbilicus prominent. Foot enlarged; propodial and opercular regions partially cover shell in creeping animals; operculum large, oligogyrous, horny in British species. Animal ploughs through soft substrata in search of bivalve prey, whose shells it bores. Spawn a characteristic collar or necklace-shaped ribbon, stiffened with sand; commonly known as necklace-shells.

---

1. Outer lip arises tangential to last whorl; sutures slight; umbilicus elongate; last whorl with five spiral rows of brown marks **_Lunatia alderi_**

   Shell not like this **2**

2. Outer lip arises more or less at right angles to last whorl; sutures well-defined; umbilicus rounded; last whorl with subsutural row of brown marks (young shells may have additional, but very faint rows of similar marks) **_Lunatia catena_**

   Shell not like this **3**

3. Outer lip separated from last whorl by deep notch; sutures deep, channelled; columella crossed by deep groove from umbilicus to aperture; shell pale, without pigmented spots **_Lunatia montagui_**

   Outer lip not separated from last whorl by notch; sutures distinct, slightly channelled; groove on columella does not cross inner lip; shell chestnut-brown, tends to retain periostracum **_Lunatia fusca_**

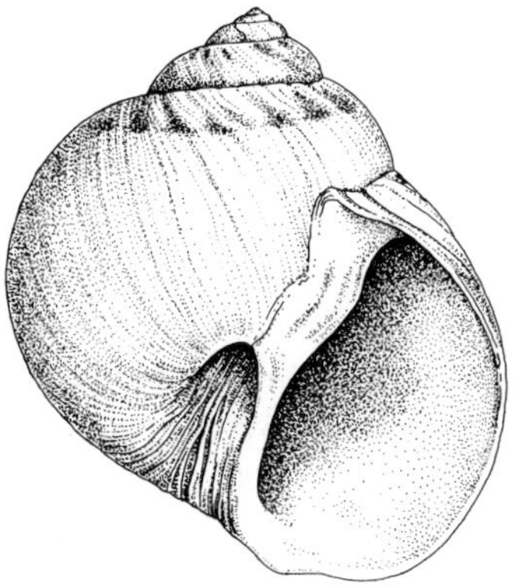

Lunatia catena

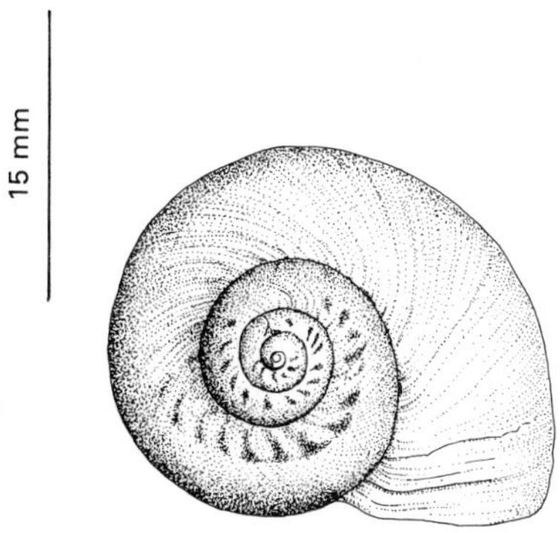

Lunatia catena

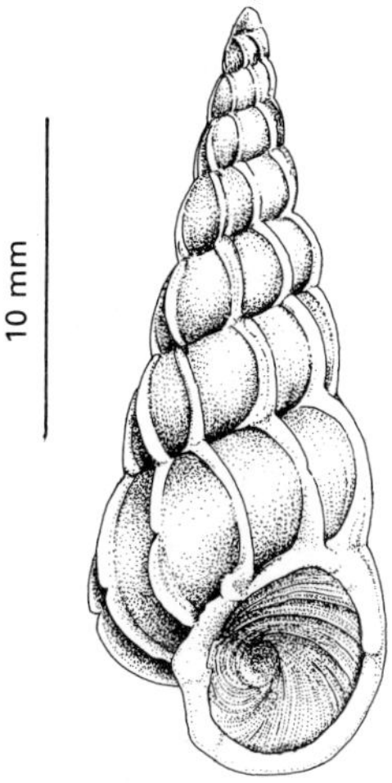

Epitonium clathrus

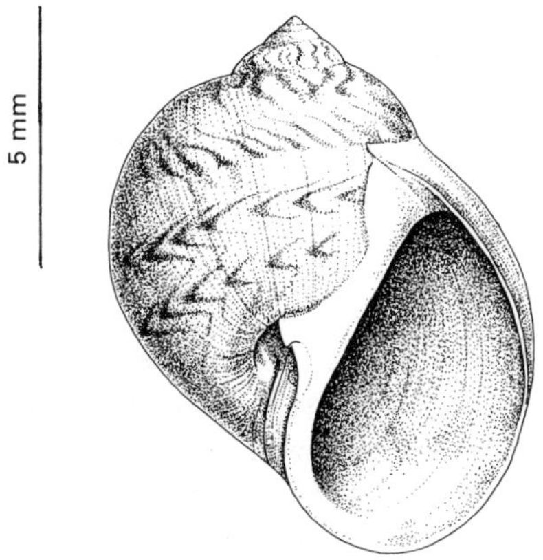

Lunatia alderi

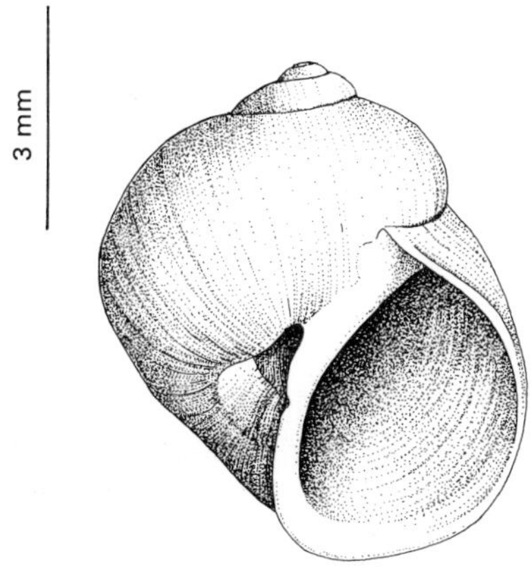

Lunatia montagui

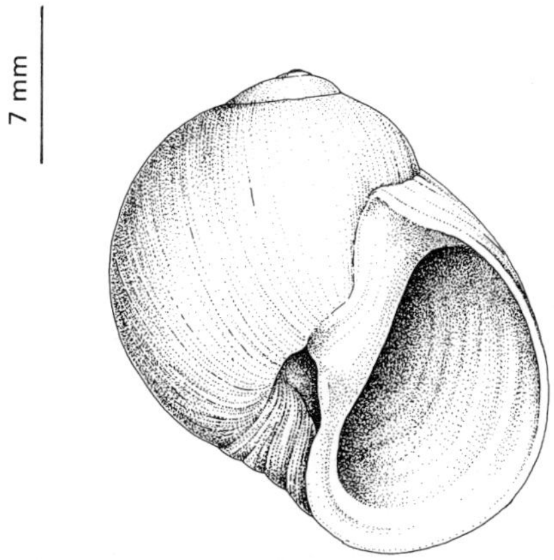

Lunatia fusca

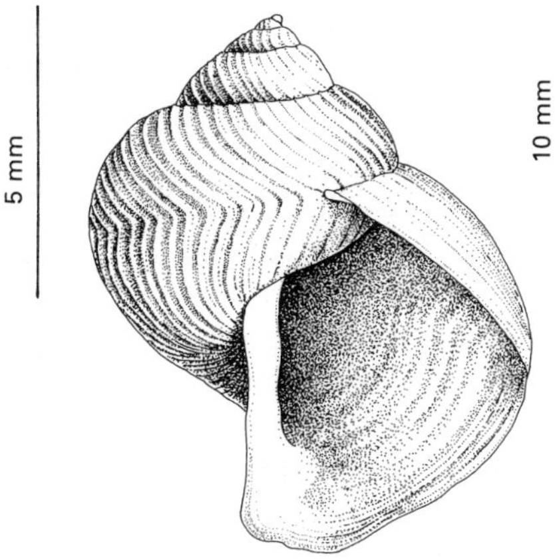

Janthina exigua

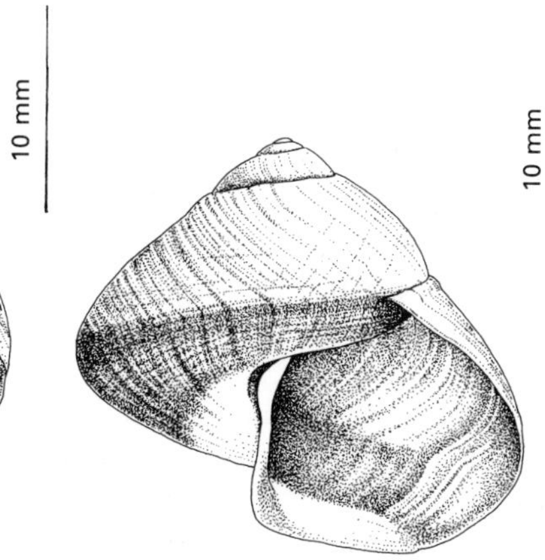

Janthina janthina

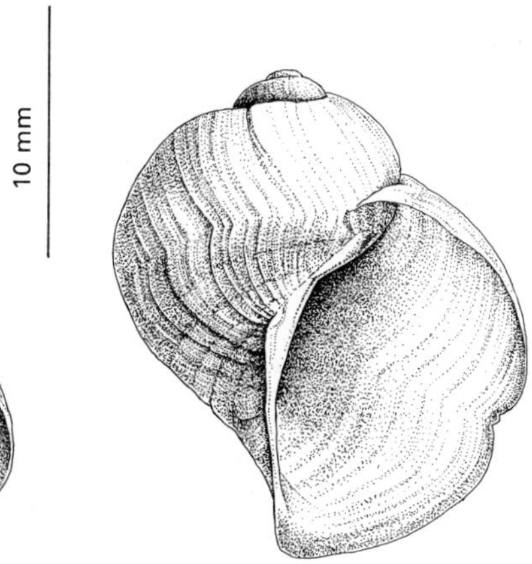

Janthina pallida

Lunatia alderi (Forbes) FIG. 12.15

Shell glossy, globose, with six or seven slightly tumid whorls; last whorl accounts for most of shell height; sutures shallow, spire almost flat-sided. Sculpture of numerous, fine, prosocline growth lines; smooth to naked eye. Periostracum often retained within umbilicus. Aperture large, almost semicircular, with peristome; outer lip arises tangential to last whorl; inner lip partially occludes rather elongate umbilicus, and swells to fill angle between outer lip and last whorl. Buff or light horn-coloured, columella and base pale; five spiral rows of brown marks on body whorl, one row on whorls of spire. Size up to 16 × 14 mm. Snout short, broad, with extensible proboscis; cephalic tentacles long, rather flat. Foot large, with expanded propodial and opercular lobes; in active animals propodium forms a ploughshare, covering anterior of shell, opercular lobes being drawn up around posterior.

On sand and gravel; usually sublittoral from 10 to 50 m (recorded from 2000 m) but occasionally at LWST. Distributed from Mediterranean to Norway; common off all British coasts.

Lunatia catena (da Costa) FIG. 12.15

Shell similar to *L. alderi*. Spire more elevated, with stepped profile and well-defined sutures. Outer lip arises almost at right angles to last whorl; umbilicus deep, round, occlusion by inner lip very slight. Buff, lighter basally and on inner lip, with subsutural row of reddish-brown marks. Younger shells may have additional, but very faint rows of marks on last whorl. Size up to 30 × 30 mm. On sand; LWST to 125 m. Distributed from Mediterranean to Skagerrak; common off all British coasts; absent from coasts of Norway and Sweden.

Lunatia montagui (Forbes) FIG. 12.15

Shell similar to *L. alderi*. Spire rather low, with stepped profile and deeply channelled sutures, producing notch between outer lip and last whorl; inner lip crossed by groove passing from umbilicus to aperture. Light fawn coloured, with darker throat; no spiral rows of brown marks. Size up to 8 × 8.5 mm. Sublittoral, on sandy and muddy bottoms; 15–200 m. A northern species extending down western coasts of British Isles to Mediterranean. More common in north of range.

Lunatia fusca (Blainville) FIG. 12.15

Shell similar to *L. alderi* but with distinct, slightly channelled sutures. Outer lip arises tangential to last whorl, initially slightly concave; umbilicus elongate, partially occluded by inner lip, which is narrowed by groove passing into umbilicus, but not interrupted as in *L. montagui*. Periostracum often retained over large part of shell, which is chestnut-brown. Size up to 25 × 20 mm. Body as in *L. alderi*; anterior margin of propodium rounded.

Sublittoral, on muddy sand to 165 m. Not common. Distributed from Mediterranean to western Channel, west coasts of Ireland and Scotland.

Two other naticids are found rarely: *Lunatia pallida* (Broderip and Sowerby) is a sublittoral, northern species, readily distinguished by its narrow, chink-like umbilicus. *Amauropsis islandica* (Gmelin) is sublittoral, known only from north-east Scotland. Its shell is rather thin and semi-transparent, with periostracum; spire much more pronounced than species of *Lunatia* with stepped profile and channelled sutures; umbilicus absent, or a very small chink

35 EPITONIIDAE

Shell a tall, narrow cone of many tumid whorls; sutures deep or whorls not in contact, connected by thin transverse costae. Aperture small, circular. Animals carnivorous, usually associated with anthozoans; large, eversible proboscis, no snout. Males lack penis.

1. Nine narrow costae on each whorl, aligned with, and fused to, costae of preceding whorl; whorls do not touch, linked only by costae **Epitonium clathrus**

   Costae more numerous, flat or raised; whorls touch, sutures deep **2**

2. Costae flat, deeply undercut on side facing up spire, their adapical ends daubed between costae on whorl above **Epitonium turtonis**

   Costae thin, vertical or curling slightly up spire; with or without triangular spurs at adapical sutures **3**

3. Costae not vertical, crest curling slightly in direction of spire; adapically drawn out into thin, triangular plates **Epitonium trevelyanum**

   Costae vertical, not curling up spire and of constant profile throughout; finely lamellar, numerous (18–22 per whorl); shell not exceeding 12 mm in length **Epitonium clathratulum**

Epitonium clathrus (Linnaeus) FIG. 12.15

Shell a tall glossy cone of up to 15 tumid whorls, circular in section, not meeting at sutures but held together by fusion of thin transverse costae. Crests of costae curling back up spire; aligned to form a series of conspicuous spiral ridges from apex to base. Shell smooth between costae. Aperture almost circular, with rather inflated margin; inner lip reflected over bases of seven costae. Off-white, often with brown spiral bands. Size up to 40 × 12 mm. Cephalic tentacles slender, with eyes at base. Foot with straight, double-edged anterior and median groove on sole; operculum oligogyrous, with depressed nucleus.

Sublittoral, on sandy or muddy substrata to 70 m; may be found at LWST when spawning (long threads of pyramidal capsules). Distributed from Mediterranean to Norway and Kattegat; widely distributed, not uncommon around the British Isles.

E. turtonis (Turton), *E. trevelyanum* (Johnston) and *E. clathratulum* (Kanmacher) are less common that *E. clathrus*. They differ mainly in the characters outlined in the key above and have similar habitat and distribution to *E. clathrus*.

36 JANTHINIDAE

Shell violet, globose, thin and fragile. The animals are pelagic, drifting at the surface of the sea by means of a mucus float secreted by the foot; gregarious, feeding primarily on the pelagic coelenterates *Porpita* and *Velella*.

Janthina janthina Linnaeus FIG. 12.15

Shell thin, fragile, with either tall or depressed globose shape; last whorl with bluntly angulate periphery. Appears smooth, but with numerous regular growth lines (V-shaped on body whorl) and irregular spiral grooves. Aperture angulate at periphery and at base; outer lip with shallow peripheral sinus. Protoconch at an oblique angle to teleoconch. Violet or bluish-purple, darker on base. Size up to 40 × 30 mm. Snout long, with bifid cephalic tentacles half-way along. Foot with broad depression anteriorly from which the float is secreted; no operculum in adults.

Pelagic, with circumequatorial distribution; cast up on western shores of Britain and Ireland after prolonged south-westerly gales.

J. pallida Thompson and *J. exigua* Lamarck (Fig. 12.15) are less common than *J. janthina*. *J. exigua* is readily distinguished by prominent transverse ridges on the shell; it is also darker and smaller (up to 17 × 14 mm) than the other species. The aperture of *J. pallida* is less angulate where it meets the columella; the shell is relatively taller and more globular than *J. janthina*, up to 25 × 23 mm, and the protoconch is not at such an oblique angle to the teleoconch. Both shell and body are pale. All three species have a similar habitat and distribution.

37 ACLIDIDAE

Members of this family are rarely collected in British waters; this may reflect genuine rarity or simply their small size and sublittoral habitat. The shell is less than 6 mm in height, often only 2–3 mm; an elongate, pointed cone of markedly tumid whorls (9–11 in most species), smooth or with peripheral spiral striae. Aperture small, rounded oval; umbilicus a conspicuous chink in most species. Proboscis eversible, no snout; most species without pallial tentacle. Probably carnivorous.

A number of species are described by Fretter and Graham (1982), to which the reader is referred.

38 EULIMIDAE

Shell long, awl-shaped, smooth and highly polished; whorls numerous flat-sided. Animals ectoparasitic on echinoderms, feeding by means of a long, eversible proboscis; foot with large anterior pedal gland. A number of species have been described from Britain, most are uncommon and only three are considered here.

---

1. Shell white **2**

   Shell with two brown spiral bands at periphery of last whorl; never higher than 7 mm ***Eulima bilineata***

2. Long axis of shell straight or nearly so, up to 20 mm high; body white with yellow or orange flecks, and V-shape on snout ***Melanella alba***

   Long axis of shell curved, often markedly so, up to 8 mm high; body white with flecks and streaks of yellow and red ***Vitreolina philippii***

Eulima bilineata Alder FIG. 12.16

Shell a tall narrow cone of up to 13 flat-sided whorls; profile slightly cyrtoconoid. Sutures very slight, paralleled internally and above by false sutures, latter often more conspicuous in live shells. Smooth, with very fine growth lines revealed by reflected light. Aperture narrow, drop-shaped, with outer lip arising tangential to last whorl between two brown spiral bands. Colourless, apart from two brown spiral bands peripherally. Size up to 6 × 2 mm. No snout; long eversible proboscis. Mantle with slight siphonal canal on left and pallial tentacle on right. Foot large, narrow, with double-edged anterior margin into which opens a large pedal gland. Opercular lobes conspicuous, operculum narrow.

On soft substrata from 20–250 m (occasionally at LWST), associated with the ophiuroids *Ophiactis* and *Ophiopholis* but often found free. Distributed from Mediterranean to northern Norway.

Eulima glabra (da Costa) may be encountered off south-west coasts. It is similar to *E. bilineata* but has at least three (up to six) brown bands on last whorl and is up to 10 mm high.

Melanella alba (da Costa) FIG. 12.16

Shell a tall, narrow, highly polished cone; whorls very slightly tumid, profile slightly cyrtoconoid, spire sometimes a little curved. Smooth, with very fine growth lines (seen in reflected light) and varices. Protoconch often lost, leaving 12–14 whorls. Aperture pear-shaped, outer lip shows marked peripheral bulge in side-view; inner lip slightly flared at base. Milky white, body colours showing through shell in life. Size up to 20 × 6 mm. Body similar to *E. bilineata*; lacks pallial tentacle.

On muddy sand and gravel from 16–135 m, possibly associated with holothurians. Distributed from Mediterranean to Norway, more common on western coasts.

Vitreolina philippii (Rayneval and Ponzi) FIG. 12.16

Shell a tall, narrow cone of 10–11 flat-sided whorls; spire bent along its axis; transparent when fresh. Smooth, with occasional very slight growth lines and varices. Aperture narrow, drop-shaped; outer lip with peripheral bulge in side view. Whitish and transparent when fresh, older shells opaque and creamy. Size up to 8 × 2.5 mm.

Common on soft substrata from LWST to 200 m, apparently associated with various echinoderms. Distributed from Mediterranean to Norway; absent from eastern Channel and perhaps also from southern North Sea.

39 STILIFERIDAE

Only one British representative of this family

Pelseneeria stylifera (Turton) FIG. 12.16

Shell globose-conical, surmounted by short pinnacle comprising the three-, or four-whorled protoconch; teleoconch with three rapidly expanding tumid whorls. Appearing smooth but with growth lines and spiral striae, latter more conspicuous towards base. Aperture large, oval, without obvious peristome. Amber when fresh, dead shells white. Size up to 5 × 3 mm. Snout short, broad, with eversible proboscis. Cephalic tentacles broader at base, behind which eye is set. Foot with double-edged anterior margin, large anterior pedal gland and deeply grooved sole; lateral fold extends on to base of shell in active animals.

Ectoparasitic on regular sea urchins; recorded from *Echinus esculentus*, *Psammechinus miliaris*, *Strongylocentrotus droebachiensis*, and *Paracentrotus lividus*. Distributed sporadically from Mediterranean to Norway; not common.

40 MURICIDAE

Shell with costae or spiral striae, or both; where both are present they give rise to conspicuous tubercles. Aperture small, oval, with siphonal canal: often long and in some species forming a tube in mature shells; outer lip thickened, with internal teeth in mature shells. Proboscis eversible; no snout. Foot with accessory boring organ (ABO) anteriorly on sole; operculum horny, with terminal or lateral nucleus. On hard substrata, boring the shell of their prey, often other molluscs, barnacles, and tube worms, by a combined action of radula and ABO. Most lay vase-shaped egg capsules attached to hard surfaces.

1. Shell without spiral striae — ***Boreotrophon truncatus***
   Shell with spiral striae — **2**
2. Shell without transverse costae, often with prominent growth lines; outer lip curves without sharp inflection to tip of short siphonal canal — ***Nucella lapillus***
   Shell with transverse costae; outer lip with marked inflection at base of siphonal canal — **3**
3. Siphonal canal long (about 25% shell height), narrow, almost straight; penultimate whorl with 12 or more ribs — ***Trophonopsis muricatus***
   Siphonal canal long or short, not straight, open or closed in mature shells; penultimate whorl with less than 12 ribs — **4**
4. Spire markedly turreted, ribs with distinct shoulder, which, in mature shells, may be raised adapical to suture; siphonal canal long (about 25% shell height), broad and open in juveniles, closed and somewhat shorter in mature shells; two or three major spiral striae on penultimate whorl, far apart, with minors in between — ***Ocenebra erinacea***
   Shell not like this — **5**
5. Spiral striae numerous (nine to ten on penultimate whorl), approximately equal in width to intervening grooves; siphonal canal open; shell light horn-coloured, body cream; south-east coasts — ***Urosalpinx cinerea***
   Spiral striae less numerous (six to eight on penultimate whorl), broader than intervening grooves; siphonal canal closed in mature shells; shell reddish-buff, body red; south-west coasts — ***Ocinebrina aciculata***

Boreotrophon truncatus (Ström) FIG. 12.16

Shell a tall, sharply-pointed cone of seven tumid whorls, slightly turreted, with deep sutures. Sculpture of sinuous, laminar costae, growth lines and weak microscopic spiral striae; costae variable in number (14–25 on penultimate and last whorls), crests sharp and wave-like, tending to overrun base at leading edge. Aperture oval, with shallow anal sinus and rather short siphonal canal (about 50% of aperture height); distally, the canal bends away from plane of aperture; outer lip thin, smooth internally. Yellowish with pink tinge, throat darker. Size up to 15 × 7.5 mm Distal third of cephalic tentacles narrow, with eyes set on lateral bulges at junction with broader basal portion. Foot with double-edged anterior margin and slight anterio-lateral horns; operculum small, oval.

On stony and gravelly bottoms, from LWST to 200 m. Not uncommon in dredged samples. Circumboreal, extending south into northern North Sea and down western coasts of British Isles to Biscay.

Subfossil specimens of *Boreotrophon clathratus* (Linnaeus) are commonly dredged in British waters; costae are fewer (less than 14 on last whorl), higher and more lamellate than those of *B. truncatus*, with sharply spined shoulders.

Trophonopsis muricatus (Montagu) FIG. 12.16

Shell a tall, narrow cone of seven or eight tumid whorls; sutures deep. Sculpture of costae, spiral striae and fine growth lines; costae numerous (15–20 on penultimate whorl), narrow, interacting with cord-like spiral striae to give a deep reticulation set with raised tubercles; spiral striae numerous (11–15) on last whorl, reducing to five or six on penultimate and two or three on upper whorls of spiral. No spiral striae on sloping subsutural shoulder of each whorl. Protoconch rather inflated, set obliquely on teleoconch. Aperture elongate-oval, with shallow anal sinus and long, straight siphonal canal, about equal to aperture height; outer lip thin, crenulate, and grooved internally. Yellowish, often with spiral brown bands; some shells colourless. Size up to 19 × 8 mm. Body similar to *B. truncatus*; eyes set about half-way along cephalic tentacles.

Sublittoral, on muddy, sandy, and gravelly bottoms from 20–300 m. Not uncommon. Distributed from Mediterranean to south and west coasts of British Isles.

T. barvicencis (Johnston) may occur in dredge samples from stony bottoms off northern Britain. It is distinguished from *T. muricatus* in being markedly turreted, with fewer, rather crinkled, laminar costae (12–14 on penultimate whorl), drawn out into spines at the shoulder.

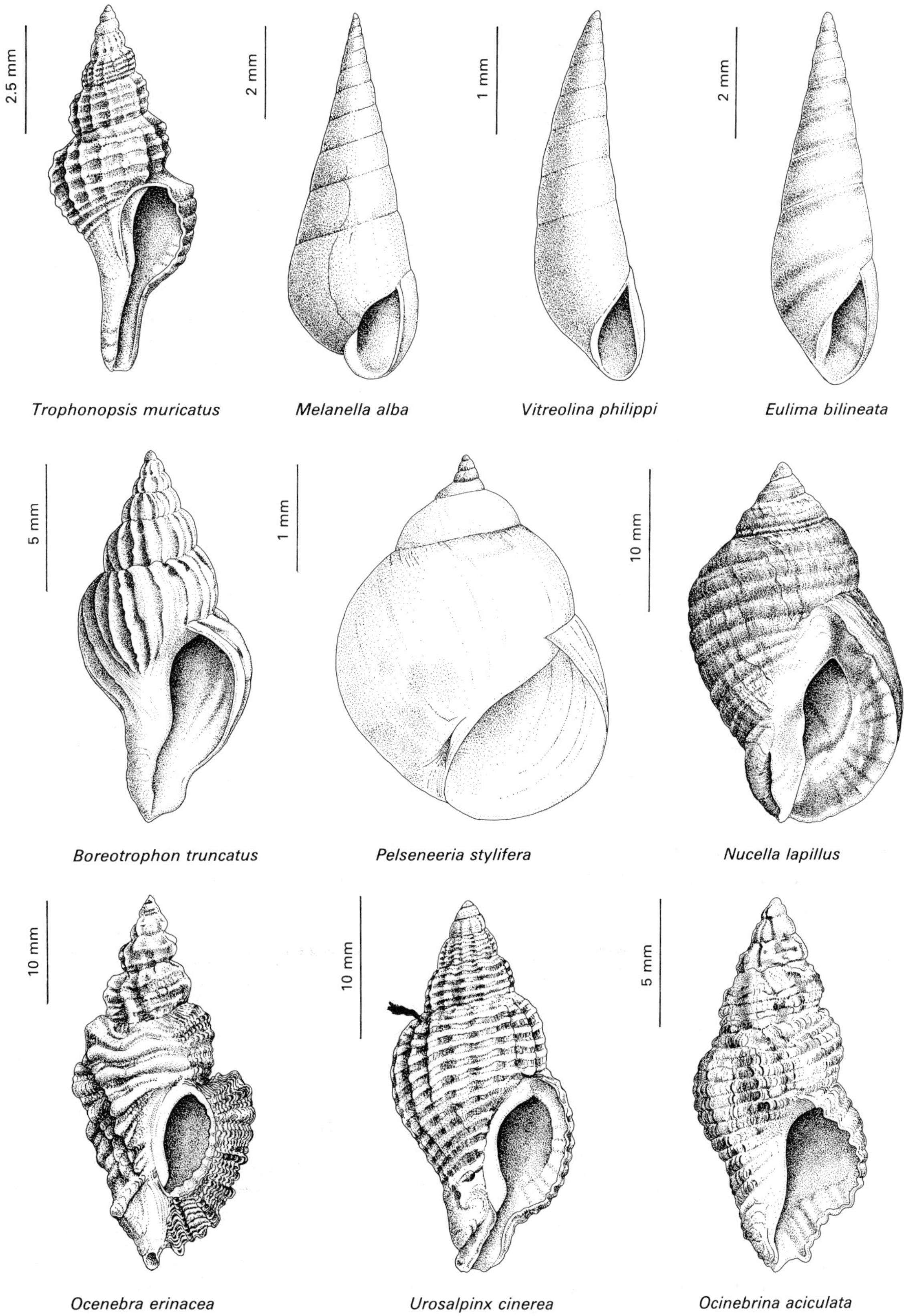
2.5 mm
2 mm
1 mm
2 mm
Trophonopsis muricatus
Melanella alba
Vitreolina philippi
Eulima bilineata
5 mm
1 mm
10 mm
Boreotrophon truncatus
Pelseneeria stylifera
Nucella lapillus
10 mm
10 mm
5 mm
Ocenebra erinacea
Urosalpinx cinerea
Ocinebrina aciculata

Fig. 12.16

Nucella lapillus (Linnaeus) FIG. 12.16

Shell solid, oval-conic, with short, straight-sided or slightly coeloconoid spire; whorls slightly tumid, last whorl large (80% or more of shell height); sutures shallow. Sculpture of broad, low spiral striae and irregular growth lines; last whorl with 11–14 spiral striae, becoming prominent basally as a keel over siphonal canal. Growth lines may produce irregular varices; in juveniles and in animals from sheltered habitats they may be raised to give frilled or imbricate surface to shell (var. *imbricata*). Aperture oval, with short siphonal canal, partially closed by columellar fold; outer lip curves in more or less constant arc from origin to tip of siphonal canal, with little or no inflection at base. In juveniles it is thin and crenulate, later thickening and developing internal teeth. Inner lip reflected over columella and last whorl as a broad glaze. Commonly whitish, with brown or purplish throat but a range of monochrome and striped shell forms also occur. Size up to 42 × 22 mm, occasionally larger in Bristol Channel. Distal two thirds of cephalic tentacles narrow, with eyes at junction with broader basal portion. Foot with double-edged anterior margin, posterior to which is the small pit of the ABO; in females the pit of the ventral pedal gland, used in depositing egg capsules, lies posterior to this; operculum horn-coloured, oval, with basal nucleus.

Abundant on all rocky shores, from MHWNT to MLWST, less common sublittorally to 40 m. Avoids low salinities and excessive weed cover, but tolerant of a wide range of exposure. Widely distributed from Straits of Gibraltar to Arctic.

Urosalpinx cinerea (Say) FIG. 12.16

Shell a tall, sharply pointed cone of seven or eight tumid whorls; sutures deep and sinuous. Sculpture of prominent broad costae (10–12 per whorl), fine growth lines and numerous low, narrow and cord-like spiral striae; costae do not reach base of body whorl; spiral striae do not form prominent keel over siphonal canal. Aperture oval, slightly pointed adapically, extended basally into a short siphonal canal; outer lip thin (thickened internally in mature shells), with sharp inflection at base of siphonal canal; inner lip forming broad glaze over columella and body whorl. Yellowish, off-white, or buff, with darker throat. Size up to 40 × 20 mm, commonly half this. Cephalic tentacles flattened, distal quarter slender, with eyes at junction with broader basal portion. Foot like that of *Nucella*; operculum with lateral nucleus.

On oysters, from lower intertidal to 12 m, hibernating in mud at 7°C or less. An accidental introduction from the eastern U.S.A., presently confined to the Essex and Kent coasts.

Ocenebra erinacea (Linnaeus) FIG. 12.16

Shell a tall, turreted, sharply pointed cone of up to eight tumid whorls; sutures deep and sinuous. Sculpture of strong, buttress costae, cord-like spiral striae and growth lines. Costae with marked shoulder, in mature shells coming to lie adapical to suture; eight or nine costae on penultimate whorl, increasing in number up the spire. Spiral striae of variable size; two or three prominent striae on penultimate whorl, eight or nine on last whorl, with minor striae in between; striae slightly swollen where they cross costae. Growth lines numerous, raised to produce finely laminate surface to shell. Aperture oval, siphonal canal long, equal to aperture height; open and slightly curved in juveniles, closed over in mature shells. Outer lip thin, crenulate in young shells, often with hollow varix behind and internally grooved; thickened internally in mature shells. Cream coloured, often with rusty brown patches. Size up to 50 × 25 mm. Body yellowish-white, with white flecks. Distal third of cephalic tentacles slender, with eyes at junction with broader basal portion. Foot similar to that of *Nucella*.

On rocky shores, in silty crevices and beneath stones from LWST to 150 m; intertidal range increased in summer. Distributed from the Azores and Mediterranean to south and west coasts of British Isles; becoming rare in north.

Ocinebrina aciculata (Linnaeus) FIG. 12.16

Shell a tall, pointed cone of up to seven tumid whorls; sutures sinuous but not deep. Sculpture of broad costae, broad sprial striae and growth lines. Last and penultimate whorls with eight to ten costae, upper whorls of spire with 10–11; costae fade towards base of body whorl. Spiral striae stronger on base of last whorl, forming a spiral keel on siphonal canal; six to eight striae on penultimate whorl, less on higher whorls. Growth lines numerous, raised to produce a finely laminate surface. Aperture oval, with slight anal sinus; siphonal canal short, closed in mature shells. Outer lip thin, crenulate, without varix; thickened and internally fluted in mature shells; shows only slight inflection at base of siphonal canal. Size up to 15 × 10 mm. Body similar to *O. erinacea* but red, flecked with yellow-white spots.

On rocky shores, usually under stones with *Balanus perforatus*; LWST down to 15 m. Distributed from Mediterranean to Scillies and western Channel.

41 BUCCINIDAE

Shell with tall spire, last whorl 60–75% of shell height; adult shell of most species large (commonly 50–100 mm or more); *Chauvetia* 5–6 mm. Sculpture of spiral striae, which may or may not be prominent, with or without costae; when present, costae very prominent, crescentic or orthocline. Aperture commonly large, broadly oval (more elongate in *Colus*); siphonal canal frequently short, never closed. Periostracum on juvenile shells (not *Chauvetia*), often retained, at least in part, on adults.

Proboscis eversible; no snout. Siphon long, commonly projecting well beyond tip of canal in active animals. Foot like that of muricids, but lacks ABO; operculum with terminal or central nucleus. Common on soft substrata.

---

1. Shell with costae and spiral striae **2**

   Shell with spiral striae only, though these may be faint **3**

2. Adult shell large (up to 110 mm), often with periostracum; costae markedly crescentic; spiral striae numerous, of variable size; aperture very broad, without labial varix ***Buccinum undatum***

Adult shell small (up to 5 mm), without periostracum; costae orthocline; four broad, flat, spiral striae on penultimate whorl; aperture narrow, with labial varix ***Chauvetia brunnea***

3. Whorls tumid, with slight subsutural ridge; columella with pronounced spiral ridge (*siphonal fasciole*) extending from base to umbilical region ***Neptunea antiqua***

   Whorls slightly tumid or almost flat-sided, sutures shallow, slightly excavated, but without subsutural ridge; no siphonal fasciole **4**

4. Protoconch with axis of coiling aligned with that of spire, its diameter less than that of next whorl; whorls tumid, spire slightly turreted; periostracum often hispid in unworn shells ***Colus jeffreysianus***

   Protoconch with axis of coiling oblique to that of spire, its diameter less than, equal to, or greater than that of next whorl; whorls tumid or nearly flat-sided **5**

5. Protoconch often bulbous, wider than next whorl; whorls tumid; spiral striae strong (except where eroded), less than 40 on last whorl; siphonal canal rather long, straight, and narrow ***Colus islandicus***

   Protoconch not bulbous, diameter equal to or less than that of next whorl; whorls almost flat-sided; spiral striae less strong, up to 60 on last whorl; siphonal canal rather short, broad, and bent to left ***Colus gracilis***

Colus gracilis (da Costa) FIG. 12.17

Shell tall, fusiform, with eight to ten almost flat-sided whorls, sutures shallow. Sculpture of numerous, narrow, strap-like spiral striae and crescentic growth lines. Striae slightly stronger towards base of shell, about 60 on last whorl, 18 on penultimate. Periostracum thin, often slightly hispid, usually worn away from a triangular area to left of apex of aperture. Protoconch swollen, up to 3 mm diameter, set obliquely on spire. Aperture elongate-oval, pointed apically, with slight anal sinus, drawn out basally into broad, rather short siphonal canal (about 15% of shell height); canal inclined sharply to left. White, with light horn coloured periostracum. Size up to 70 × 28 mm. Cephalic tentacles short, flat, with eyes half-way along. Foot with double-edged anterior margin and slight anterio-lateral horns; females with opening of ventral pedal gland on sole; operculum small, pear-shaped, with terminal nucleus.

On sandy and muddy substrata from 30–80 m; occasionally intertidal in north of range, where it is also more common. Distributed from Norway to Portugal; recorded all round British Isles, rare in southern North Sea and Channel.

Colus islandicus (Mohr) FIG. 12.17

Shell similar to *C. gracilis*, differing only in minor detail. Protoconch bulbous, often of greater diameter than next whorl and set more obliquely on spire; whorls more tumid. Spiral striae often more pronounced, less than 40 on last whorl; triangular area of last whorl lacking periostracum relatively smaller than in *C. gracilis*. Aperture with long narrow siphonal canal; (about 25% of shell height), not markedly inclined to left; outer lip arises at level of spiral stria 10–12 on last whorl. White, with light horn-coloured periostracum. Size up to 150 × 50 mm.

On soft substrata, from 10–3000 m. A northern species extending south to Shetland and the north coast of Scotland; subfossil shells may be encountered in more southern waters.

Colus jeffreysianus (Fischer) FIG. 12.17

Shell similar to *C. gracilis*, differing only in minor detail. Protoconch not swollen, its axis aligned with spire axis; whorls more tumid, spire slightly cyrtoconoid. Spiral striae stronger, less numerous (24–34 on last whorl), often forming fine reticulation with growth lines. Triangular area of last whorl lacking periostracum very much reduced or absent. Periostracum often rather hispid. Aperture like that of *C. gracilis*; outer lip arises at level of spiral stria 8 or 9 on last whorl, siphonal canal relatively longer (about 16–20% of shell height). White, perhaps with bluish or pinkish tinge, periostracum light horn-coloured. Size up to 50 × 25 mm. Body similar to *C. gracilis*; cephalic tentacles longer, less flattened, with eyes nearer tip.

On soft substrata from 30–2000 m. Distributed from Mediterranean to Norway and Iceland; not common around British Isles, though more often encountered off south-west coasts.

Neptunea antiqua (Linnaeus) FIG. 12.17

Shell with tall, coeloconoid spire of seven tumid whorls; protoconch rather bulbous, last whorl about 70–80% shell height; sutures shallow, slightly accentuated by narrow subsutural ridge, more conspicuous on last whorl where it is seen to derive from anal sinus. Periostracum thin, often lost in older shells. Sculpture of numerous, fine spiral striae and flexuous growth lines. A thick spiral keel, the siphonal fasciole, runs from shell base to umbilical region. Aperture broadly oval or egg-shaped, pointed adapically, with small anal sinus; basally a short, broad siphonal canal. Yellowish, sometimes with reddish tinge, throat usually more heavily pigmented. Commonly up to 200 × 50 mm, occasionally up to twice this size. Cephalic tentacles long, distal half slender, eyes at junction with broader basal portion. Siphon does not project far from canal in active animals. Foot large, with double-edged anterior margin and small anterio-lateral horns; females with opening of ventral pedal gland on sole; operculum oval, pointed adapically, with terminal nucleus.

Sublittoral, from 15–1200 m; mainly on soft substrata. Widely distributed from Biscay to Arctic; locally common around much of British Isles, less so in south and off west coast of Ireland.

Buccinum undatum Linnaeus FIG. 12.17

Shell solid, mature individuals especially robust and heavy; periostracum hispid in juveniles, missing from most of shell in adults. Spire of seven or eight tumid whorls, last whorl large (about 70% of shell height); sutures deep. Sculpture of strong crescentic costae, spiral striae and prosocline growth lines. Costae do not extend to base of last whorl; 12–15 on penultimate whorl. Spiral striae numerous, major ridges separated by variable number of minor ridges to form a regular pattern; interaction with growth lines produces fine reticulation. A thick spiral keel, the siphonal fasciole, runs from base of shell to umbilical region. Aperture broadly oval, tapering adapically to a point; outer lip describes a high arc to tip of short, siphonal canal. Inner lip reflected over columella and large area of parietal region of last whorl. Buff, with darker bands above and below periphery; throat and peristome white. Size up to 110 × 68 mm. Cephalic tentacles long, distal portion slender, eyes at junction with broader basal portion. Siphon extends a long way from canal in active animals. Foot large, shield-shaped, with double-edged anterior margin; females with opening of ventral pedal gland on sole; operculum oval, with central nucleus.

On hard and soft substrata, occasionally at LWST but usually sublittoral down to 1200 m. Tolerates brackish water to 15‰. Distributed from Iceland and northern Norway to the Bay of Biscay; common and often abundant around British Isles, except Scilly Isles.

Chauvetia brunnea (Donovan) FIG. 12.17

Shell small, glossy, an elongate cone of five or six tumid whorls; sutures deep. Apex blunt, with swollen protoconch. Sculpture of broad, orthocline costae, strap-like spiral striae and fine orthocline growth lines. Last whorl with nine to eleven costae, the last a varix, not extending to base; penultimate and higher whorls of spire crossed by nine to ten costae. Spiral striae broad, widening slightly over costae, separated by narrow grooves; 10–16 striae on last whorl, those below periphery somewhat narrower, four on whorls of spire. Aperture oval, pointed adapically, with shallow anal sinus and short, wide siphonal canal. Outer lip thin, backed by varix which, in mature shells, is ridged internally. Chestnut-brown, with lighter, horn-coloured grooves between striae. Size up to 5 × 1.5 mm.

Cephalic tentacles long, widely divergent, distal half slender, at junction with broader basal portion. Siphon extends well beyond tip of canal in active animals. Foot narrow, with double-edged, slightly embayed anterior margin and small anterio-lateral horns; operculum oval.

On rocky shores, under stones and in weed at LWST and sublittorally to 60 m. Locally common. Distributed from Mediterranean to shores of south-west England.

42 NASSARIIDAE

Shell oval-conical, whorls rather flat-sided, or tumid, apex sharp. Sculpture of flexuous costae, growth lines, and strap-like spiral striae; often markedly reticulate. With or without distinct periostracum. Aperture with short, deep siphonal canal, never closed; a deep spiral channel runs from tip of canal to umbilical region of columella. Outer lip with labial varix, thickened internally, with row of teeth, in mature shells.

Carrion feeders, with long eversible proboscis and no snout; siphon extends long way from canal and carried aloft in active animals. Foot with well developed anterio-lateral horns and two metapodial tentacles; female with opening of ventral pedal gland on sole; no accessory boring organ (ABO).

1. Adult shell large (up to 30 mm); whorls almost flat-sided; costae and spiral striae of upper whorls of equal height, producing squared, block-like reticulation; usually with complete periostracum **_Hinia reticulata_**

   Adult shell small (less than 15 mm); whorls tumid; costae higher than spiral striae on all whorls, reticulation not block-like or conspicuous; periostracum not obvious **2**

2. Penultimate whorl with eight to ten spiral striae; up to 12 narrow striae on boss below basal spiral channel; dark brown mark on roof of siphonal canal **_Hinia incrassata_**

   Penultimate whorl with four or five spiral striae; five narrow striae on boss below basal spiral channel; no dark brown mark on roof of siphonal canal; not littoral **_Hinia pygmaea_**

Hinia reticulata (Linnaeus) FIG. 12.18

Shell solid, with distinct periostracum; tall, conical, with up to ten almost flat-sided whorls; spire a little cyrtoconoid, slightly stepped at sutures. Sculpture of flexuous costae, strap-like spiral striae and fine growth lines. Costae numerous (20 or more on each whorl), more conspicuous on last whorl and part of penultimate whorl. Spiral striae less numerous (about thirteen on last whorl, five or six on whorls of spire), interacting with costae to produce raised, block-like reticulation, particularly conspicuous on spire. A deep spiral channel runs from base of siphonal canal to umbilical region of columella, below it a boss bearing four to six thin spiral striae. Aperture oval, with slight anal sinus and short, deep, markedly oblique siphonal canal. Outer lip arises tangential to last whorl, curving without inflection to tip of siphonal canal; thickened in mature shells, with internal teeth; often with labial varix. Inner lip spread widely over parietal region of last whorl; toothed or ridged in both columella and parietal regions in mature shells. Cream or buff, often with darker subsutural bands; periostracum imparts a tan colour to live shells, though often discoloured by sulphides in sediments. Size up to 30 × 14 mm. Cephalic tentacles slender, each with short, thickened basal portion bearing eye. Siphon long (about half shell length), held aloft in active animals. Foot long, with double-edged anterior and recurved anterio-lateral horns; posterior tip with slight fork and two metapodial tentacles. Operculum small, oval, with frilled margin.

Common near LWST, in sedimentary areas of rocky shores and sublittorally on soft substrata down to 15 m; often covered in sediment, with only siphon exposed. Distributed from Canaries and Azores to Norway; common on all British coasts.

Hinia incrassata (Ström) FIG. 12.18

Shell solid, tall, conical, with up to eight tumid whorls; periostracum confined to grooves between striae. Sculpture of strong, slightly flexuous costae, strap-like spiral striae, and fine growth lines. Costae more prominent than spiral striae on all

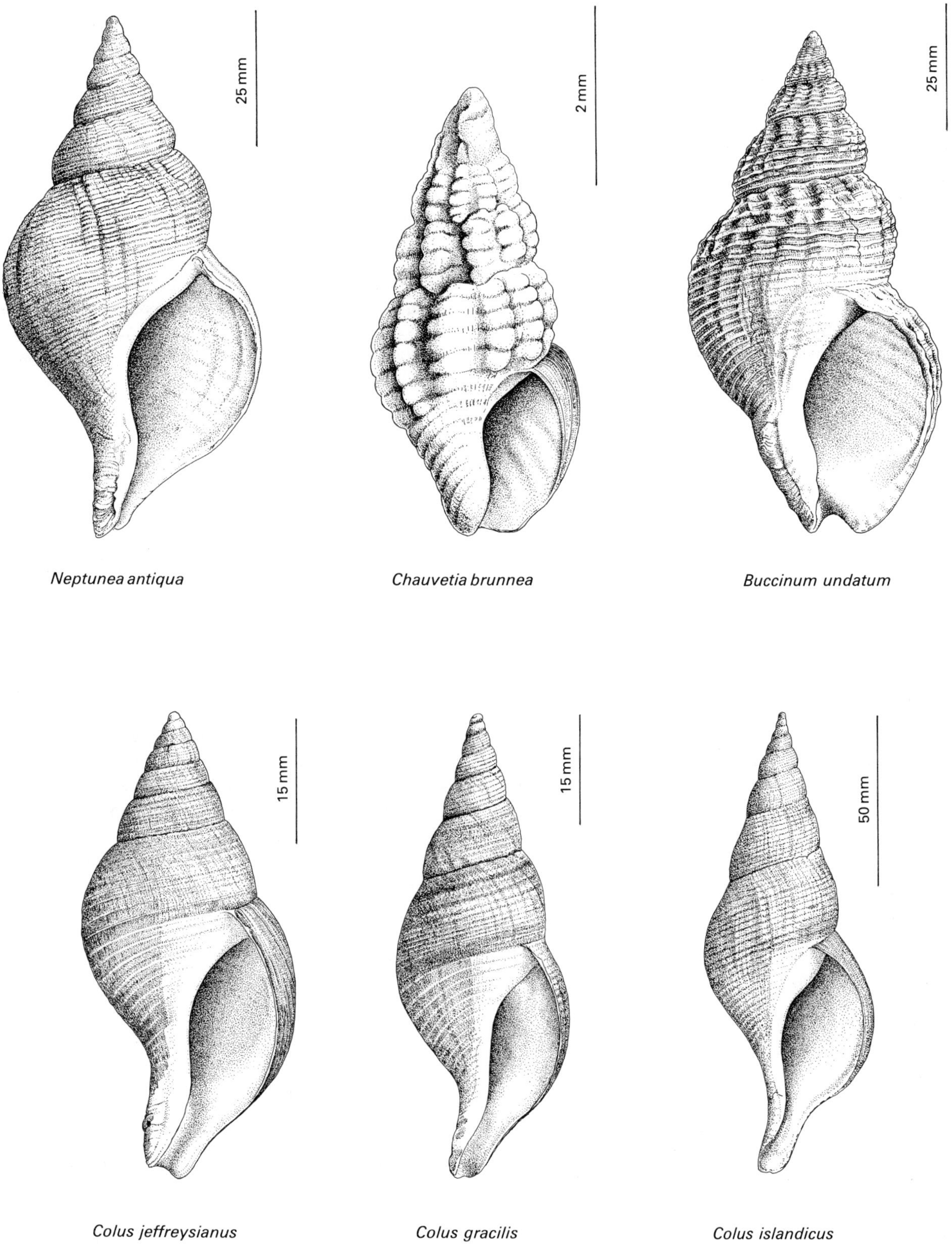
25 mm
2 mm
25 mm
Neptunea antiqua
Chauvetia brunnea
Buccinum undatum
15 mm
15 mm
50 mm
Colus jeffreysianus
Colus gracilis
Colus islandicus

Fig. 12.17

whorls, 12–15 on whorls of spire; last whorl with strong labial varix. Penultimate whorl with eight to ten spiral striae; without marked reticulation. A deep spiral channel runs from base of siphonal canal to umbilical region of columella, below it a boss with up to 12 thin spiral striae. Aperture oval, with slight anal sinus and short, deep, markedly oblique siphonal canal. Outer lip arises at about 45° to last whorl, curving without inflection to tip of siphonal canal; thickened internally in mature shells, with seven teeth below labial varix. Inner lip spread widely over parietal region of last whorl; toothed or ridged in both columella and parietal regions in mature shells. Buff or light tan, often with brown bands below sutures and at base of last whorl; labial varix and columella white, siphonal canal with dark brown mark on roof. Size 12 × 6 mm, occasionally larger.

Common near LWST in silted areas of rocky shores, below stones and in crevices; more abundant in shallow sublittoral, extending to 200 m. Distributed from Mediterranean to northern Norway and Iceland; on all British coasts.

Hinia pygmaea (Lamarck) FIG. 12.18

Shell similar to *H. incrassata*; basal spiral channel shallower, boss below with only five or six spiral striae. Spiral striae less numerous (four or five on whorls of spire), ribs less flexuous, with more pronounced reticulation than in *H. incrassata*; whorls of spire may have varices. Aperture with nine or ten teeth internally on outer lip, inner lip not extending so far over last whorl. Shell colour similar to *H. incrassata*; with brown siphonal canal, but lacking dark mark on roof. Size up to 14 × 8 mm. Body as in *H. incrassata*; siphon and metapodial tentacles longer, anterio-lateral horns on foot more pronounced.

Sublittoral only, on sandy bottoms from 1–1200 m. Not common. Distributed from Mediterranean, north along western coasts of British Isles to Shetland; absent from Irish and North Seas.

43 TURRIDAE

Shell an elongate, often spindle-shaped cone; generally small (up to 15 mm), though some species may reach 25 mm in length. Sculpture variable; strong axial costae and spiral striae, costae and fine spiral element, or spiral striae only. Aperture long and narrow, with siphonal canal and anal sinus. Operculum present or absent. Animal without snout; cephalic tentacles with eyes at varying distance from tip; mantle edge drawn out into a siphon on left.

---

1. Shell with spiral striae only; anal sinus deep **_Teretia teres_**

   Shell with both spiral and costate elements to sculpture **2**

2. Penultimate whorl with 12 or more costae **3**

   Penultimate whorl with less than 12 costae **6**

3. Shell markedly turreted, with broad, sloping subsutural shelf, crossed by costae **_Oenopota turricula_**

   Shell not like this **4**

4. Shell with prominent, broad, minutely granular and concave subsutural band, rarely crossed by costae; costae do not extend beyond periphery of last whorl; spiral striae strap-like; protoconch keeled **_Comarmondia gracilis_**

   Shell with narrow, minutely granular subsutural band; costae extend beyond periphery of last whorl; spiral striae cord-like, nodular, or swollen where they cross costae; protoconch with diamond-shaped reticulation **5**

5. Last whorl with 18–24 narrow, opisthocline costae; spiral striae interact with costae to produce bulbous nodules; outer lip thickened internally, with row of about ten teeth; dark reddish-brown or purplish, with white outer lip **_Raphitoma purpurea_**

   Last whorl with 14–17 orthocline or slightly prosocline costae; spiral striae only slightly inflated on crossing costae; outer lip thickened internally, without teeth; not heavily pigmented, tip of protoconch white **_Raphitoma leufroyi_**

6. Penultimate whorl with 10–11 costae crossed by four to seven narrow, cord-like spiral striae; outer lip thickened internally, with up to nine teeth in mature shells; protoconch with diamond-shaped reticulation; pale, with chestnut-brown spiral striae **_Raphitoma linearis_**

   Shell not like this **7**

7. Costae arise from below narrow, swollen, and often beaded subsutural band (more conspicuous on penultimate and last whorls), much reduced below periphery of last whorl; spiral striae beaded or with scale-like tubercles; outer lip without varix or internal thickening **8**

   Costae arise at sutures and extend to base of last whorl; no subsutural band; spiral striae not beaded, fine and strap-like (may be evident only on upper whorls, not easily seen) or thin, narrow, and numerous (conspicuous or inconspicuous); outer lip with varix, internally thickened **10**

8. Whorls only slightly tumid, profile of shell rather flat-sided; subsutural band prominent, broad, swollen, and beaded; spiral striae numerous, regular (often of two alternating sizes), with beaded tubercles which are not often eroded or scale-like **_Mangelia powisiana_**

   Whorls tumid, profile turreted; subsutural band less well-developed; spiral striae narrow, less regular, tubercles often eroded or scale-like **9**

9. Penultimate whorl with eight to ten costae and five to six major (many more minor) spiral striae; subsutural band narrow with strongly beaded spiral ridge immediately below suture; tubercles on spiral ridges very prominent, scale-like; rarely exceeds 7 × 3 mm **_Mangelia brachystoma_**

   Penultimate whorl with seven to nine ribs and numerous close-set, narrow spiral striae; subsutural band

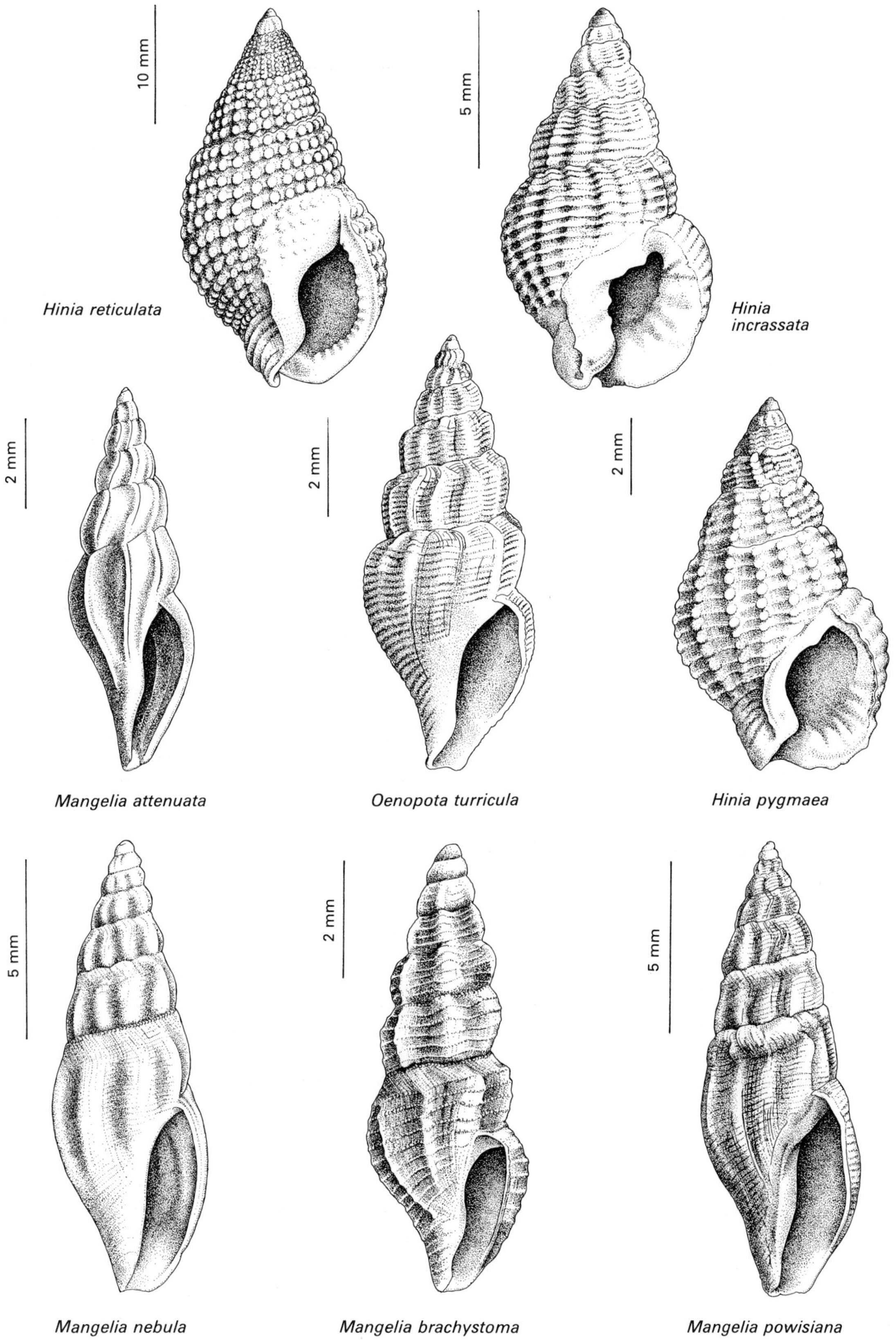
10 mm
5 mm
Hinia reticulata
Hinia incrassata
2 mm
2 mm
2 mm
Mangelia attenuata
Oenopota turricula
Hinia pygmaea
5 mm
2 mm
5 mm
Mangelia nebula
Mangelia brachystoma
Mangelia powisiana

Fig. 12.18

slightly swollen, often beaded but without a prominent, strongly beaded ridge next to suture **_Mangelia nebula_**

10. Shell slender (breadth about 30% of height), with narrow, flexuous costae; spiral striae low, strap-like, and more evident towards apex; anal sinus shallow; siphonal canal rather long **_Mangelia attenuata_**

    Shell not like this **11**

11. Shell markedly turreted; costae narrow, flexuous, angulate in profile; spiral striae numerous, like thin, vertical lamellae, separated by deep grooves of similar width; anal sinus a distinct semicircular notch **_Cytharella smithi_**

    Shell slightly turreted, markedly cyrtoconoid; costae narrow, flexuous, slightly angulate in profile; spiral striae numerous, rather strap-like, not raised into vertical lamellae; anal sinus a distinct semicircular notch **_Cytharella coarctata_**

Oenopota turricula (Montagu) FIG. 12.18

Shell rather elongate, with six or seven turreted whorls; turreting enhanced by sloping, subsutural shelf. Sculpture of costae, spiral striae and growth lines; 12–16 ribs on each whorl, on last whorl fading towards base; costae change direction abruptly on crossing subsutural shelf. Spiral striae numerous, interacting with growth lines to give fine decussation between costae; striae and costae sometimes interact to give bolder decussation especially on upper whorls. Protoconch conspicuous, initially smooth, subsequently developing spiral striae. Aperture elongate, with slight anal sinus at position of subsutural shelf; outer lip thin, inner lip angulate where it extends from base of last whorl to form one wall of siphonal canal. Pale buff or white. Size up to 20 × 9 mm, commonly 12 × 5 mm. Distal one-third of tentacles narrow, eyes at junction with broader basal portion. Siphon extends long way from canal in moving animals; foot with small, oval operculum.

Sublittoral, on sandy bottoms from 20 to 200 m. Circumboreal, more common in northern than in southern waters of British Isles.

Two less common species occur in northern waters. *O. rufa* (Montagu) has a darker shell, with less conspicuous subsutural shelf and a somewhat inflated protoconch. *O. trevelliana* (Turton) has finer and more numerous costae (20–27 on the last whorl).

Mangelia nebula (Montagu) FIG. 12.18

Shell an elongate cone of eight or nine tumid whorls; sutures deep, accentuated by a slightly swollen, often beaded subsutural band. Sculpture of strong, flexuous transverse costae (slightly crescentic on upper whorls), spiral striae and growth lines. Spiral striae numerous with major spirals separating small groups of minors: interaction with growth lines gives them a finely beaded appearance and fine reticulate pattern to shell. Protoconch of three or four whorls, initially smooth, becoming tuberculate before merging with teleoconch. Aperture elongate-oval, with peristome and short, wide siphonal canal. Outer lip markedly prosocline to base of subsutural band where it adopts an opisthocline plane, forming anal sinus. Inner lip thickened, delineated by a shallow groove where it reflects over columella. Light horn coloured or reddish brown. Size up to 14 × 5 mm. Cephalic tentacles with eyes about two-thirds of length from base, tentacle diameter halved distally. Foot shield-shaped, with double-edged anterior; no operculum.

On sand and muddy gravel bottoms from LWST (rarely) to 50 m. Distributed from Mediterranean to Norway; locally common, more frequent in south of British Isles; absent from eastern Channel and southern North Sea.

Mangelia brachystoma (Philippi) FIG. 12.18

Shell an elongate cone of seven or eight tumid whorls; sutures deep, profile rather turreted, with narrow, beaded, swollen subsutural band. Sculpture of strong, flexuous transverse costae, slightly crescentic on upper whorls, spiral striae and growth lines. Spiral striae numerous, major spirals separate small groups of minors; more prominent than in *M. nebula*; interaction with growth lines produces erect, scale-like and conspicuous tubercles. Aperture rather narrower than in *M. nebula*. Pale cream to chestnut-brown, perhaps with lilac tint. Size up to 7 × 3 mm.

Not intertidal; on sand and sandy mud from 4–60 m. Locally common. Distributed from Mediterranean to Norway; off most British coasts; absent from southern North Sea.

Mangelia powisiana (Dautzenberg) FIG. 12.18

Shell an elongate cone of up to ten slightly tumid whorls; sutures deep, bordered below by broad, swollen subsutural band. Spire profile more straight-sided and last whorl curving more gently towards siphonal canal than in *M. nebula*. Sculpture of flexuous transverse ribs, spiral striae and growth lines. Costae much reduced where they cross subsutural band, on upper whorls slightly crescentic. Spiral striae numerous, often alternately broad and narrow; interaction with growth lines gives them a finely beaded appearance and a fine reticulate pattern to shell. Protoconch of two and one-half whorls, initially smooth, becoming tuberculate before merging with teleoconch. Outer lip of aperture less curved than *M. nebula*, opening narrower, almost straight-sided. Pale chestnut-brown, with lighter costae and spiral striae; apex and throat dark chestnut. Size up to 15 × 6 mm, commonly 10 × 4 mm. Eyes closer to tips of cephalic tentacles than in *M. nebula*.

On sand from LWST (rarely) to 100 m. Locally common. Distributed from Biscay to Norway; perhaps confined to west coasts of Britain and Ireland.

Mangelia attenuata (Montagu) FIG. 12.18

Shell a slender, slightly translucent, glossy cone of up to nine tumid whorls (breadth about 30% of height); sutures deep, without subsutural band, profile slightly turreted. Sculpture of narrow, flexuous, buttress-like costae, dipping sharply to sutures; fine, flat spiral ridges and flexuous growth lines visible with hand lens; spiral element often more pronounced towards

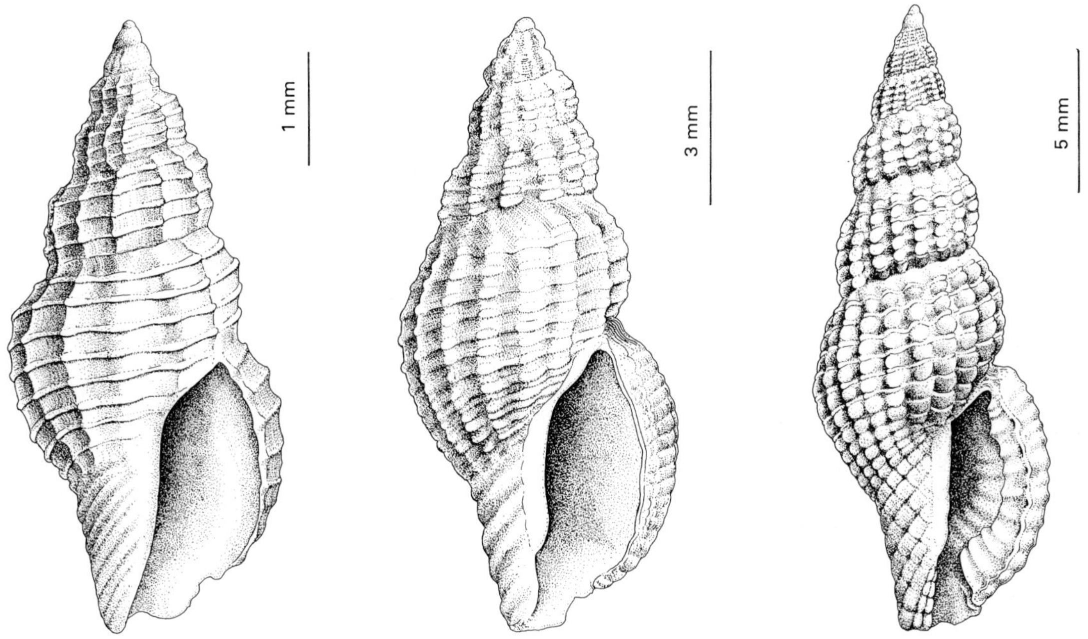

Rhaphitoma linearis *Rhaphitoma leufroyi* *Rhaphitoma purpurea*

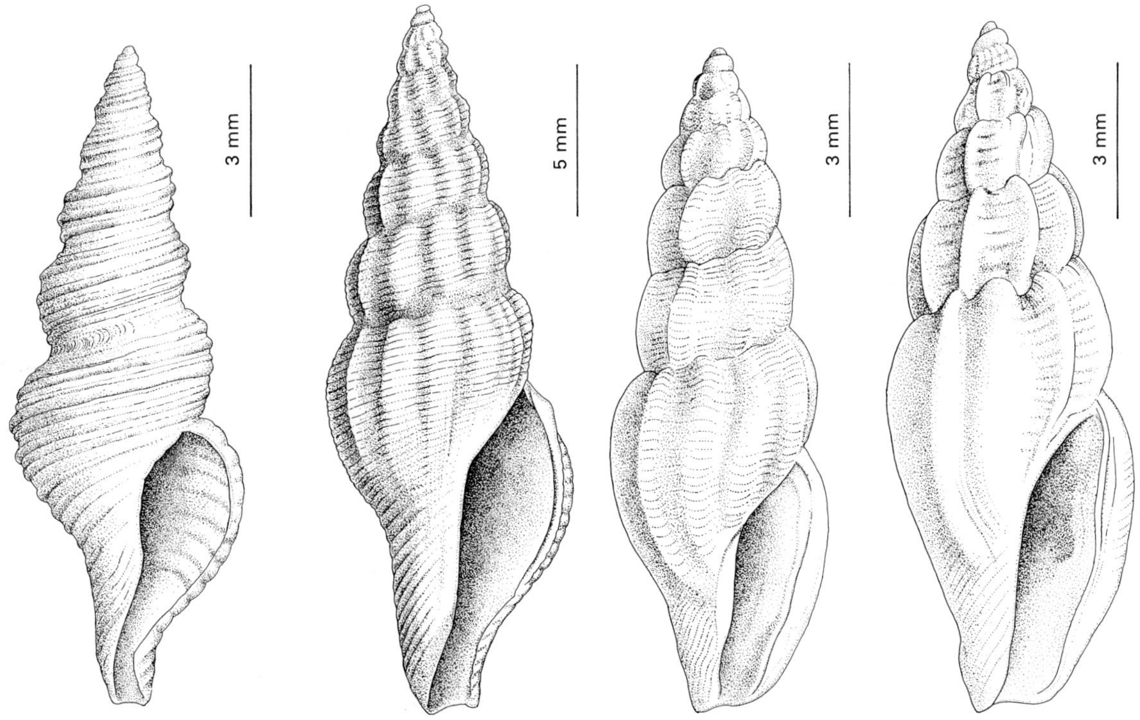

Teretia teres *Comarmondia gracilis* *Cytharella smithi* *Cytharella coarctata*

Fig. 12.19

apex. Protoconch of three and one-half whorls, smooth, becoming reticulate. Aperture long, narrow, almost parallel-sided; anal sinus small, with slightly flared margin; siphonal canal rather long. Outer lip with varix in older shells; inner lip reflected over columellar region, without clear delineation. Light horn-coloured with darker spiral bands; costae milk-white. Size up to 18 × 5.5 mm. Eyes near base of cephalic tentacles.

Sublittoral, on sand or clay bottoms down to 150 m. Distributed from Mediterranean to Northern Norway; sporadic off most British coasts, more common in south-west; absent from southern North Sea.

Cytharella smithi (Forbes) FIG. 12.19

Shell an elongate cone of up to nine tumid whorls; sutures deep, sinuous, without subsutural band; profile slightly cyrtoconoid, turreted. Sculpture of strong, flexuous, buttress-like costae, dipping sharply to sutures, overlain by numerous thin, sharp spiral ridges separated by deep, narrow grooves; no conspicuous interaction between spirals and growth lines. Protoconch of three whorls; smooth, followed by interacting spiral ridges and crescentic costae. Aperture narrow, elongate, almost parallel-sided; outer lip thin, in older shells serrated, with varix; anal sinus semicircular, with everted rim. Cream or light horn coloured with brown spiral band at sutures and variable number of brown spirals on last whorl; costae pale. Size up to 15 × 5 mm, commonly half this size. Eyes nearer tip of cephalic tentacles than on *M. nebula*.

Sublittoral, on soft substrata at 20–50 m. Not common. Distributed from Mediterranean to western coasts of British Isles.

Cytharella coarctata (Forbes) FIG. 12.19

Shell similar to *C. smithi*, or more cyrtoconoid; sometimes more slender, superficially resembling *M. attenuata*; distinguished from latter by deep anal sinus and wider, less elongate siphonal canal. Sculpture of strong, flexuous costae, not dipping sharply to sutures, and numerous fine spiral ridges. Spiral ridges may not be very conspicuous at low magnification; lower, narrower, and more numerous (50 or more on penultimate whorl) than in *C. smithi*, often more pronounced on upper parts of spire. Chestnut-brown, with white base to last whorl or cream with broken spiral bands of chestnut. Size up to 12 × 5 mm, commonly half this size. Eyes half-way up cephalic tentacles.

On sandy bottoms, LWST to 25 m. Distributed from Mediterranean to Norway; may be locally common off western coasts of British Isles.

Comarmondia gracilis (Montagu) FIG. 12.19

Shell an elongate cone of up to eleven (commonly nine or ten) tumid whorls; sutures deep, accentuated by broad, slightly concave subsutural band. Sculpture of broad, opisthocline costae, numerous spiral striae and fine growth lines. Costae do not cross subsutural band nor extend below periphery of last whorl. Spiral striae strap-like, separated by shallow, minutely granular grooves; spiral striae lower and narrower on subsutural band, where granulation is particularly conspicuous. Protoconch of three or four whorls; initially smooth, developing strong peripheral keel, making an abrupt junction with the teleoconch. Aperture elongate-oval, drawn out basally into siphonal canal, which flares slightly and to right at base. Anal sinus a deep, narrow arch. Outer lip thin, crenulate, with a varix set back from margin; internally thickened in older shells, with toothed ridge. Inner lip reflected over columella; in older shells a prominent tooth on base of last whorl at entrance to anal sinus. Light horn coloured, often with darker subsutural band and with paired light and dark band near periphery of last whorl; ribs pale. Size up to 26 × 10 mm. Cephalic tentacles thick, with eyes near tip. Siphon short, not projecting far from canal in active animals. Males with penis behind right tentacle. Foot yellowish with white speckles, with double edged anterior margin and lateral horns; no operculum.

Sublittoral, on muddy or sandy gravel at 7–150 m. Distributed from Mediterranean to Orkneys; on western shores of British Isles, western Channel basin; recorded from Moray Firth but apparently absent from rest of North Sea.

Raphitoma linearis (Montagu) FIG. 12.19

Shell elongate, turreted, slightly cyrtoconoid, with seven to nine tumid whorls; sutures deep, sinuous, with narrow, minutely granular subsutural band. Sculpture of strong costae, spiral striae, and fine growth lines. Costae rather narrow, slightly prosocline (especially on last whorl); about 11–12 on last whorl. Spiral striae narrow, slightly nodular where they cross costae, much reduced on subsutural band; 16–20 on last whorl, four to seven on penultimate. Protoconch of three or four whorls, finely reticulate, with distinctive diamond pattern. Aperture elongate-oval, with short siphonal canal and shallow, semicircular anal sinus. Outer lip thin, crenulate, without a varix though a costa often present at margin; a sharp inflexion at junction with siphonal canal; older shells with a row of nine teeth internally, first and last larger than rest. Cream with lilac tinge to apex or light horn colour; spiral striae often darker horn coloured and visible through outer lip. Size up to 12 × 7 mm. Diameter of cephalic tentacles reduces abruptly about half-way along length; eyes at this point. Siphon on mantle edge extends well beyond siphonal canal in active animals. Foot with double-edged anterior margin and recurved anterio-lateral horns; no operculum.

Occasionally under stones or in pools at LWST; more commonly sublittoral on sand, shell, and gravel bottoms at 10–200 m. Distributed from Mediterranean to northern Norway; occurs off most British coasts but intertidal only on southern shores.

Rhaphitoma leufroyi (Michaud) FIG. 12.19

Shell similar to *R. linearis*. Elongate, markedly cyrtoconoid, profile turreted, with eight or nine tumid whorls; sutures deep, sinuous, with shelf-like, minutely granular subsutural band. Costae and spiral striae on last whorl more numerous (14–17 costae, up to 25 spirals) than in *R. linearis*; spiral striae slightly expanded where they cross costae. Aperture broadly oval with short, wide siphonal canal and shallow anal sinus; no sharp inflexion of outer lip at junction with canal; outer lip thickened internally in older shells but without row of teeth. Light horn coloured, often with chestnut-brown spiral striae; reddish brown patches may be evident, particularly on last whorl; first whorl of protoconch white. Size up to 15 × 6 mm.

Quite common on sandy, shelly, and stony bottoms; LWST to 150 m. Distributed from Mediterranean to northern Norway; widely distributed in British Isles; absent from southern North Sea and east Channel.

Raphitoma purpurea (Montagu) FIG. 12.19

Shell similar to *R. linearis*. Readily distinguished by pattern of shell sculpture; costae opisthocline, narrower, and more numerous (18–19 on penultimate); spiral striae more numerous (>20 on last whorl, >7 on penultimate), interacting with costae to produce prominent bulbous tubercles. Aperture narrower than in *R. linearis*, with 10–11 internal teeth in older shells. Reddish-brown or purplish, often with white outer lip; throat and siphonal canal not white. Size up to 20 × 8 mm. Eyes about two-thirds of way up tentacles; foot white with chalk-white speckles, siphon purplish.

On sand, gravel, and stony bottoms at 10–100 m; at LWST, among weeds (including *Zostera*), under stones, and in rock-crevices on shores in southern British Isles. Not common. Distributed from Mediterranean to northern Norway; occurs in southern North Sea and east Channel.

Teretia teres (Reeve) FIG. 12.19

Shell elongate, slender, with eight or nine tumid whorls; sutures deep, straight, with broad, slightly concave subsutural band marked by low crescentic ridges. Sculpture of strong, cord-like spiral striae and fine growth lines; striae often alternately large and small, absent from subsutural band. Protoconch of up to six whorls, with diamond-shaped reticulation. Aperture oval, drawn out basally into short, open siphonal canal deflected to left. Anal sinus deep, narrow, crescentic terminally, set within subsutural band. Outer lip thin, crenulate; often damaged so that anal sinus appears short. Columella slightly sinuous. White, often with reddish-brown blotches. Size up to 16 × 6 mm, commonly less. Cephalic tentacles long, blunt, with eyes near base. Foot double-edged anteriorly, with anterio-lateral projections; no operculum.

Sublittoral, on sandy bottoms from 30–900 m. Distributed from Mediterranean to Norway; encountered occasionally off most British and Irish coasts; absent from southern North Sea and east Channel.

D. Subclass **Opisthobranchia**

There are more than 3000 species of opisthobranchs around the world, and new ones are still being described. They display a wide range of morphological features, from the shelled bullomorphs and pyramidellids to nudibranchs and sacoglossans, which have lost their shells completely. In the absence of a shell, these two groups, and especially the nudibranchs, have assumed a variety of body forms, colours, and patterns in many different habitats. Evolutionary trends in opisthobranchs have been towards a reduction in the importance of the shell, with the concomitant loss of the operculum, and increasing emphasis on biological and chemical defense. This radiation from early, shelled, burrowing forms probably reflects increasing diversification of feeding, and one sees species, genera, and even families with different, and sometimes very restricted, diets.

The opisthobranchs are divided into nine orders. The burrowing Bullomorpha (Fig. 12.20) usually have an external shell, but this may be internal or absent. An *operculum* is sometimes present. The head forms a flattened *cephalic shield*, sometimes with tentacular processes, and the foot may bear expanded parapodial lobes, or *parapodia*. The Pyramidellidomorpha (Fig. 12.20) are ectoparasites with solid tall-spired shells with the *aperture* in the *body whorl*. The Thecosomata and Gymnosomata are pelagic and are not included here. The Aplysiomorpha (Fig. 12.20), or sea hares, are characterized by enlarged parapodial lobes. The shell is usually internal, or sometimes absent, and is located between the parapodia; the head is elongated, with rolled *oral tentacles* and rolled *rhinophores*. In the Pleurobranchomorpha (Fig. 12.20) the body is flattened, with a *gill* visible on the right side. The shell may be internal, as in British species, external, or absent, and the skin is often able to secrete a strong acid. The head bears an *oral veil* with rolled *oral tentacles* and rolled *rhinophores*. The Acochlidacea are small animals, less than 10 mm long, lacking both shell and operculum; they are found interstitially in sand and gravel, and are not dealt with here. In the Sacoglossa (Fig. 12.20) body form is varied: the shell is external, internal, or absent; they usually possess rolled rhinophores; and the oral tentacles are very small or absent. Finally, the Nudibranchia (Fig. 12.20), lacking a shell, have extremely varied body forms. Flattened oval animals usually have *papillae* or *tubercles* on the *mantle*, which is often referred to as the *dorsum* or *notum*. The *hyponotum* is the underside of the notum, not the foot. Animals may be smooth and limaciform (slug-like), in which case the gills are usually located on the dorsum, often together in a ring or arc around the *anal papilla*. In many nudibranchs finger-like extensions of the body-wall, called *cerata*, are present on the *pallial margin*, the edge between the dorsum and the sides. The cerata contain lobes of the digestive gland and may or may not have *cnidosacs* at their tips. In the Dotoidae the cerata function as gills and some have flap-like basal extensions called *pseudobranchs*. The cerata of the Aeolidiacea do not function as gills, but as defensive organs, the tips bearing cnidosacs. The nudibranch head often bears both oral tentacles and rhinophores; the latter may be smooth, tuberculated, wrinkled, swollen, branched, or lamellated, and are sometimes retractable into simple or elaborate sheaths. The foot is arbitrarily divided into the *metapodium* (posterior) and *propodium* (anterior), the latter often bearing *propodial tentacles*.

All opisthobranchs are hermaphroditic: the reproductive organs of both sexes are present in all adults on the right hand side of the body. Copulation is usually reciprocal, two individuals fertilizing one another. The spawn normally consists of a large number of eggs contained within a gelatinous substance and is laid in ribbons or capsules of many colours, shapes, and sizes. Most species hatch as free-swimming planktonic veligers after an embryonic period that may be as short as five days or as long as fifty. Settlement and metamorphosis of the larvae occur only if specific requirements are met, often the presence of the adult prey, which may be quite limited. For example, *Adalaria proxima* larvae metamorphose only on living colonies of the bryozoan *Electra pilosa*, the preferred adult food (Thompson 1958). Experimental results (Chadwick and Thorpe 1981) indicate that many nudibranchs have specialized feeding requirements and are specific predators of a small number of prey species. This is probably an ecological adaptation to minimize competition for food resources. Body form is related to the type of prey eaten: species that feed on flat encrustations such as ascidians, barnacles, sponges, and some soft corals and bryozoans, have flattened bodies and are generally oval in shape. Nudibranchs feeding on erect prey species (hydroids, sea anemones, and some bryozoans and corals) are more delicate and elongate in shape. The form of the radula is also adapted to the type of prey consumed; prey is cut or torn with the jaws and transferred to the oesophagus by the radula. The structure of the jaws, if present, is a valuable diagnostic feature, as is that of the radula. Most opisthobranchs are carnivorous, although the sacoglossans and aplysiids are herbivorous. The pyramidellids are equipped with a long *proboscis*, armed with a *stylet*, for their ectoparasitic mode of feeding. The hosts are usually tubiculous polychaetes and molluscs, both lamellibranchs and prosobranchs.

The majority of the British Opisthobranchia are epifaunal and found in shallow waters. Some, such as *Acteon tornatilis*, are found on sandy beaches and the majority of the other shelled bullomorphs occur on sandy or muddy bottoms. The larger nudibranchs are easier to find and usually occur low on rocky shores. Small nudibranchs pose greater difficulties because they often form part of the epifaunal community of specialised microhabitats centred on hydroids, sponges, and seaweeds, for example, and are extremely well camouflaged. Sacoglossans and aplysiids are also well camouflaged amongst the algae upon which they feed; together with pleurobranchomorphs, they may be found in clear, shallow rocky habitats, and in rock-pools.

Opisthobranchs should only be preserved after very careful notes are obtained from life on general habitat, species or substrata on which they were found, morphology, sizes, and colours. Accurate drawings are a necessity, and colour slides invaluable. The animals should then be narcotized in 7% aqueous $MgCI_2$ in sea-water; relaxation times varies from minutes to hours depending upon species and size. Specimens are best preserved in 4% sea-water formalin, which retains colour for a longer period than alcohol. However, British species should be easily identifiable with existing literature, and there should be no need to preserve specimens.

REFERENCES

The most recent and complete review of the British Opisthobranchia is contained in Thompson (1976), for the shelled opisthobranchs and sacoglossans, and Thompson and Brown (1984), for the nudibranchs. Alder and Hancock's monograph (1845–1855) is worthy of note for its beautiful colour illustrations. Schmekel and Portmann (1982) include details of habit and ecology for Mediterranean opisthobranchs, many of which also occur in British waters. The parasitic Pyramidellidae are treated by Fretter and Graham (1962) and Graham (1971).

Alder, J. and Hancock, A. (1845–1855). *A Monograph of the British Nudibranchiate Mollusca*. London, Ray Society.

Carefoot, T. H. (1967). Growth and nutrition of *Aplysia punctata* feeding on a variety of marine algae. *Journal of the Marine Biological Association of the U.K.*, **47**, 565–89.

Chadwick, S. R. and Thorpe, J. P. (1981). An investigation of some aspects of bryozoan predation by dorid nudibranchs (Mollusca-Opisthobranchia). In: *Recent and Fossil Bryozoa*, (eds. G. P. Larwood and C. Nielson). Olsen and Olsen, Fredensberg, pp. 51–58.

Eliot, C. N. E. (1910). *A monograph of the British nudibranchiate Mollusca*, Supplementary volume. Ray Society, London.

Fretter V. and Graham, A. (1962). *British Prosobranch Molluscs*. Their Functional Anatomy and Ecology, Vol. 144. Ray Society, London.

Graham, A. (1989). British Prosobranchs. *Synopses of the British Fauna* (new series), no. **2**. 2nd. edition. Leiden: E. J. Brill, for the Linnean Society of London, and the Estuarine and Brackish Water Sciences Association.

Hunnam, P. and Brown, G. (1975). Sublittoral nudibranch Mollusca (Sea Slugs) in Pembrokeshire waters. *Field Studies*, **4** (2), 131–59.

Jeffreys, J. C. (1863–1869). *British Conchology*. Vol. I, (1862); Vol. II, (1863); Vol. III. (1865); Vol. IV, (1867); Vol. V, (1869). van Voorst, London.

Miller, M. C. (1961). Distribution and food of the nudibranchiate Mollusca of the south of the Isle of Man. *Journal of Animal Ecology*, **30**, 95–116

Miller, M. C. (1962). Annual cycles of some Manx nudibranchs, with a discussion of the problem of migration. *Journal of Animal Ecology*, **30**, 545–69

Schmekel, L. and Portmann, A. (1982). *Opisthobranchia des Mitelmeeres: Nudibranchia und Saccoglossa*. Springer-Verlag, Berlin.

Thompson, T. E. (1958). The natural history, embryology, larval biology and post-larval development of *Adalaria proxima* (Alder and Hancock) (Gastropoda: Opisthobranchia). *Philosophical Transactions of the Royal Society*, Series B, **242**, (686), 1–57.

Thompson, T. E. (1976). *Biology of Opisthobranch Molluscs*, Vol. I. Ray Society, London.

Thompson, T. E. (1988). Molluscs: Benthic Opisthobranchs. *Synopses of the British Fauna* (new series), no. 8. 2nd edition. Leiden: E. J. Brill, for the Linnean Society of London, and the Estuarine and Brackish Water Sciences Association.

Thompson, T. E. and Brown, G. H. (1976). *British Opisthobranch Molluscs*. Synopses of the British Fauna (new series), No. 8. Academic Press for the Linnean Society, London.

Thompson, T. E. and Brown, G. H. (1984). *Biology of Opisthobranch Molluscs*, Vol. II. Ray Society, London.

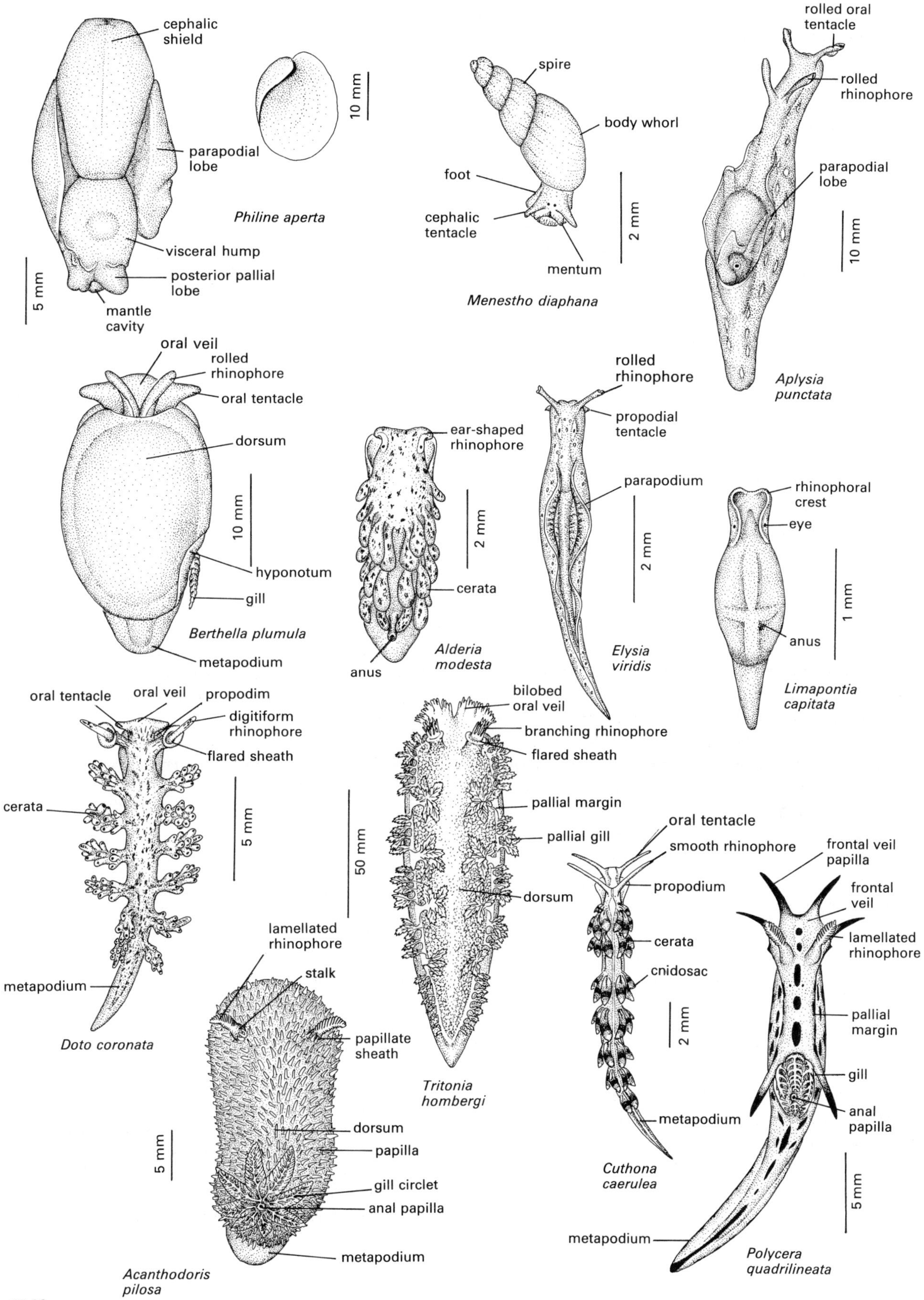
cephalic shield
parapodial lobe
visceral hump
posterior pallial lobe
mantle cavity
5 mm
10 mm
Philine aperta
spire
body whorl
foot
cephalic tentacle
mentum
2 mm
Menestho diaphana
rolled oral tentacle
rolled rhinophore
parapodial lobe
10 mm
Aplysia punctata
oral veil
rolled rhinophore
oral tentacle
dorsum
10 mm
hyponotum
gill
Berthella plumula
metapodium
ear-shaped rhinophore
2 mm
cerata
anus
Alderia modesta
rolled rhinophore
propodial tentacle
parapodium
2 mm
Elysia viridis
rhinophoral crest
eye
1 mm
anus
Limapontia capitata
oral tentacle
oral veil
propodim
digitiform rhinophore
flared sheath
cerata
5 mm
metapodium
Doto coronata
50 mm
bilobed oral veil
branching rhinophore
flared sheath
pallial margin
pallial gill
dorsum
Tritonia hombergi
lamellated rhinophore
stalk
papillate sheath
5 mm
dorsum
papilla
gill circlet
anal papilla
metapodium
Acanthodoris pilosa
oral tentacle
smooth rhinophore
propodium
cerata
cnidosac
2 mm
metapodium
Cuthona caerulea
frontal veil papilla
frontal veil
lamellated rhinophore
pallial margin
gill
anal papilla
5 mm
metapodium
Polycera quadrilineata

Fig. 12.20

KEY TO FAMILIES

1. Shell external — **2**
   Shell internal or absent — **6**

2. Operculum present — **3**
   Operculum absent — **4**

3. Large, up to 25 mm. Spire less than one-sixth shell height. Aperture elongate, obtusely joined to last whorl, length about six-sevenths of last whorl, two-thirds of total shell height — **1. Acteonidae**
   Very small, largest species up to 9 mm, usually 3–4 mm. Spire more than half total shell height. Aperture rounded, acutely joined to last whorl, length about half last whorl, one-third of total shell height (Fig. 12.23) — **7. Pyramidellidae**

4. Cephalic shield with ear-like tentacles (Fig. 12.21) — **3. Retusidae**
   Cephalic shield without conspicuous tentacles — **5**

5. Fragile, bubble-like shell — **8. Akeridae**
   Strong solid shell — **4. Scaphandridae**

6. Shell internal — **7**
   Shell absent — **11**

7. Body consisting of four parts: cephalic shield, visceral hump, paired parapodial lobes (Fig. 12.22) — **5. Philinidae**
   Body not as described — **8**

8. Shell obvious, comprising larger part of animal — **2. Diaphanidae**
   Shell inconspicuous — **9**

9. Animal oval in shape, hard, with an obvious siphon anteriorly. Variously coloured and patterned — **Prosobranchia: Lamellariidae**
   Animal soft, with two sets of anterior tentacles: paired rhinophores and either an oral veil or oral tentacles — **10**

10. 'Sea hares', with enlarged parapodial lobes (Fig. 12.20) — **9. Aplysiidae**
    Slugs with soft, flat bodies, lacking distinct parapodial lobes — **10. Pleurobranchidae**

11. Gills absent. Body tuberculate, tough, black. Anterior tentacles tubular with terminal knobs. South-west rocky shores only — **Pulmonata: Onchidiidae**
    Gills present or absent, body not as described — **12**

12.	Gills situated posteriorly, on right side under mantle (Fig. 12.22)	**6. Runcinidae**
	Gills dorsal, not posterio-lateral, or absent	**13**
13.	Large obvious eyes, usually set in unpigmented region	**14**
	Eyes absent; or if present small, set far back or at posterior base of rhinophores	**16**
14.	Body smooth, elongate, with no projections	**12. Limapontidae**
	Body with lateral parapodial lobes or dorsal cerata	**15**
15.	Conspicuous parapodial lobes (Fig. 12.20)	**11. Elysiidae**
	Cerata arranged in transverse rows	**13. Stiligeridae**
16.	Circlet of gills surrounding anal papilla	**17**
	Gills, when present, not as described	**26**
17.	Gills retractile, completely withdrawing into a deep cavity	**18**
	Gills separately contractile, not completely withdrawn	**22**
18.	Mantle smooth or faintly papillate	**23. Cadlinidae**
	Mantle with conspicuous papillae	**19**
19.	Papillae small and uniform	**20**
	Papillae large, of different sizes	**21**
20.	Mantle velvety; papillae spiculose, with retractile central projection	**27. Kentrodoridae**
	Papillae blunt. Coloured patch present between rhinophores	**24. Rostangidae**
21.	Head with very short oral tentacles. Mantle with blotchy markings of various colours	**25. Archidorididae**
	Head with elongate, finger-like oral tentacles. Mantle with stellate patches	**26. Discodorididae**
22.	Mantle fully developed	**18. Onchidorididae**
	Mantle very reduced, foot obvious in dorsal view	**23**
23.	Body covered with pedunculate, mushroom-like tubercles	**22. Notodorididae**
	Body not as described	**24**
24.	Body smooth, with elongate dorsal processes	**25**
	Body keeled, with reduced mantle skirt	**19. Goniodorididae**
25.	Body with finger-like pallial processes	**19. Goniodorididae pt. 20. Polyceridae**
	Body with single and multiply-forked tubercles	**21. Triophidae**

26.	Pleated sensory organ, *caruncle*, present between rhinophores; fusiform cerata continuous anteriorly	**28. Janolidae**
	No caruncle; cerata, if present anteriorly, not fusiform	**27**
27.	Rhinophores with conspicuous sheaths	**28**
	Rhinophores not as described	**31**
28.	Rhinophores simple, fingerlike, with flared sheaths	**17. Dotoidae**
	Rhinophores divided or lamellated	**29**
29.	Rhinophores branched	**14. Tritoniidae**
	Rhinophores lamellated	**30**
30.	Rhinophore sheaths and dorsal processes arborescent	**16. Dendronotidae**
	Rhinophore sheaths not as described	**15. Lomanotidae**
31.	Rhinophores lamellated, sometimes indistinctly so	**32. Facelinidae**
	Rhinophores wrinkled or swollen	**32**
32.	Rhinophores with up to three bulbous swellings	**33. Favorinidae**
	Rhinophores without swellings	**33**
33.	Propodium extended, forming conspicuous propodial tentacles	**34**
	Propodium not as described	**35**
34.	Body thin, elongated; cerata in clusters	**34. Flabellinidae**
	Body broader; cerata in close packed rows	**29. Aeolidiidae**
35.	Oral tentacles absent	**30. Embletoniidae**
	Oral tentacles present	**36**
36.	Cerata swollen, flattened or tuberculate	**31. Eubranchidae pt.**
	Cerata fusiform, clavate or digitiform	**31. Eubranchidae pt.** **35. Tergipedidae**

1 ACTEONIDAE

External shell stout, relatively solid, up to eight whorls; able to accommodate animal when retracted. Operculum present. Cephalic shield with two pairs of lobes. One British species.

Acteon tornatilis (Linnaeus) FIG. 12.21

Shell solid, glossy, ridged transversely, with longitudinal lines sometimes visible; opaque, varying shades of pink with white bands, each with a narrow darker pink edge on either side; two white bands on body whorl, one on each of other whorls. Up to 25 mm long; aperture two-thirds of shell height, a peg on inner columella. Operculum triangular, transparent, amber colouring and fine lines diffusing from edge. Body creamy white, large and fleshy, completely retractable into shell. Foot and head divided anteriorly; head with large lobes, foot with small, blunt propodial tentacles.

Acteon tornatilis burrows in clean, rather fine but not silty, sand, below MTL and in the shallow sublittoral. It has been observed to feed on the polychaetes *Lanice conchilega* and *Owenia fusiformis*. It is most easily captured on a rising tide, when the animals burrow along the surface leaving a characteristic trail in the sand.

2 DIAPHANIDAE

Shell external or internal, colourless, fragile, several whorls. No operculum. Cephalic shield and propodium with short tentacular processes.

Colpodaspis pusilla M. Sars FIG. 12.21

Shell internal but obvious; large, bubble-like, completely covered by mantle, up to 3.2 mm long. Body white with opaque white speckling extending beyond mantle edge and shell at both ends. The posterior lobe is an extension of the mantle, not the foot: the foot is completely hidden, Y-shaped, with the propodium bilobed. Total length up to 5 mm.

C. pusilla can be found offshore among hydroids and erect bryozoans. It is widely distributed off the west coasts of northern and southern England and Ireland.

3 RETUSIDAE

External shell rather fragile, able to accommodate whole animal. Operculum absent. Cephalic shield with rounded or pointed, posterio-lateral, tentacular processes.

1. Shell up to 10 mm long, aperture shorter than last whorl. Cephalic tentacles rounded ***Retusa obtusa***

   Shell up to 5 mm long, aperture larger than last whorl. Cephalic tentacles pointed **2**

2. Shell cylindrical, may have a protruding spire. Tapering, pointed cephalic tentacles ***Retusa truncatula***

   Shell elliptical, aperture much longer than last whorl, spire not protruding. Swollen, pointed cephalic tentacles ***Reluse umbilicata***

Retusa obtusa (Montagu) FIG. 12.21

Shell fragile, translucent white, glossy, and sculptured; variable in shape, usually elongated, not rounded like *Acteon*. Aperture usually smaller than last whorl; spire may or may not be present. Up to 10 mm. Body whitish, extending beyond shell only anteriorly. Cephalic shield square, slightly indented in mid-line anteriorly; posteriorly, two rounded tentacles point backwards and outwards. Foot very small, squat, shield-like shape in ventral view, approximately same size as head, indented anteriorly. Total length up to 15 mm.

R. obtusa is a burrower in mud or muddy sand, living below the surface in the first few centimetres, and feeds on *Hydrobia ulvae* (Pennant). It can be found by sieving the mud; in finer substrata, the flat-topped variety *pertenuis* only reaches 3 mm in length. Generally distributed all around the British Isles on sheltered muddy shore, locally abundant in favourable localities. Occurs from lower shore to 300 m.

Retusa umbilicata (Montagu) FIG. 12.21

Shell up to 4 mm, translucent whitish, glossy, fragile; elliptical, with spiral striations, aperture usually larger than last whorl and spire height combined. Spire flattened or indented. Body white. Cephalic shield indented medially, cephalic tentacles flattened, stout, and somewhat pointed. Total length up to 6 mm.

Found in fine muddy sand; sublittoral, to 2000 m. Predominantly northern distribution.

Retusa truncatula (Bruguière) FIG. 12.21

Shell up to 5 mm, translucent white or yellowish, glossy; cylindrical, with longitudinal striations. Body white. Cephalic shield indented anteriorly at mid-line, cephalic tentacles pointed. Foot elongate, with rounded corners, indented anteriorly. Total length up to 7 mm.

Burrows in sand, muddy sand, and mud; lower intertidal to about 50 m. Feeds on *Hydrobia ulvae* and perhaps foraminiferans and small lamellibranchs. All British coasts, except extreme south-east.

4 SCAPHANDRIDAE

External shell stout and rounded, usually unable to accommodate entire animal. Operculum absent. Cephalic shield with or without tentacular processes; parapodial lobes present.

1. Shell large, 45 mm. Inflated, pyriform; aperture three to four times wider at base than apex ***Scaphander lignarius***

   Small shell, less than 15 mm, aperture of almost constant width **2**

2. Shell elongate, cylindrical; aperture as long as last whorl ***Cylichna cylindracea***

   Shell bubble-like; aperture longer than last whorl ***Roxania utriculus***

Cylichna cylindracea (Pennant) FIG. 12.21

Shell cylindrical, solid, glossy, opaque, whitish with areas of brown; up to 13 × 4.5 mm, about three times longer than broad. Body white to pale buff. The only visible portions are the cephalic shield, indented anteriorly, and smooth, short, rounded tentacles posteriorly. Foot a little larger than head, with very small parapodial lobes between head and shell.

C. cylindracea is a burrower in fine sand, found in the top 30 mm. Rarely on the lower shore; sublittoral, to 1500 m. All British coasts.

Roxania utriculus (Brochi) FIG. 12.21

Shell solid, glossy, translucent, white to pale yellow, usually with brownish markings; sculpture of transverse lines of dots and dashes, crossed by finer longitudinal lines. Up to 15 mm, length about 1.5 times width; too small for animal to withdraw

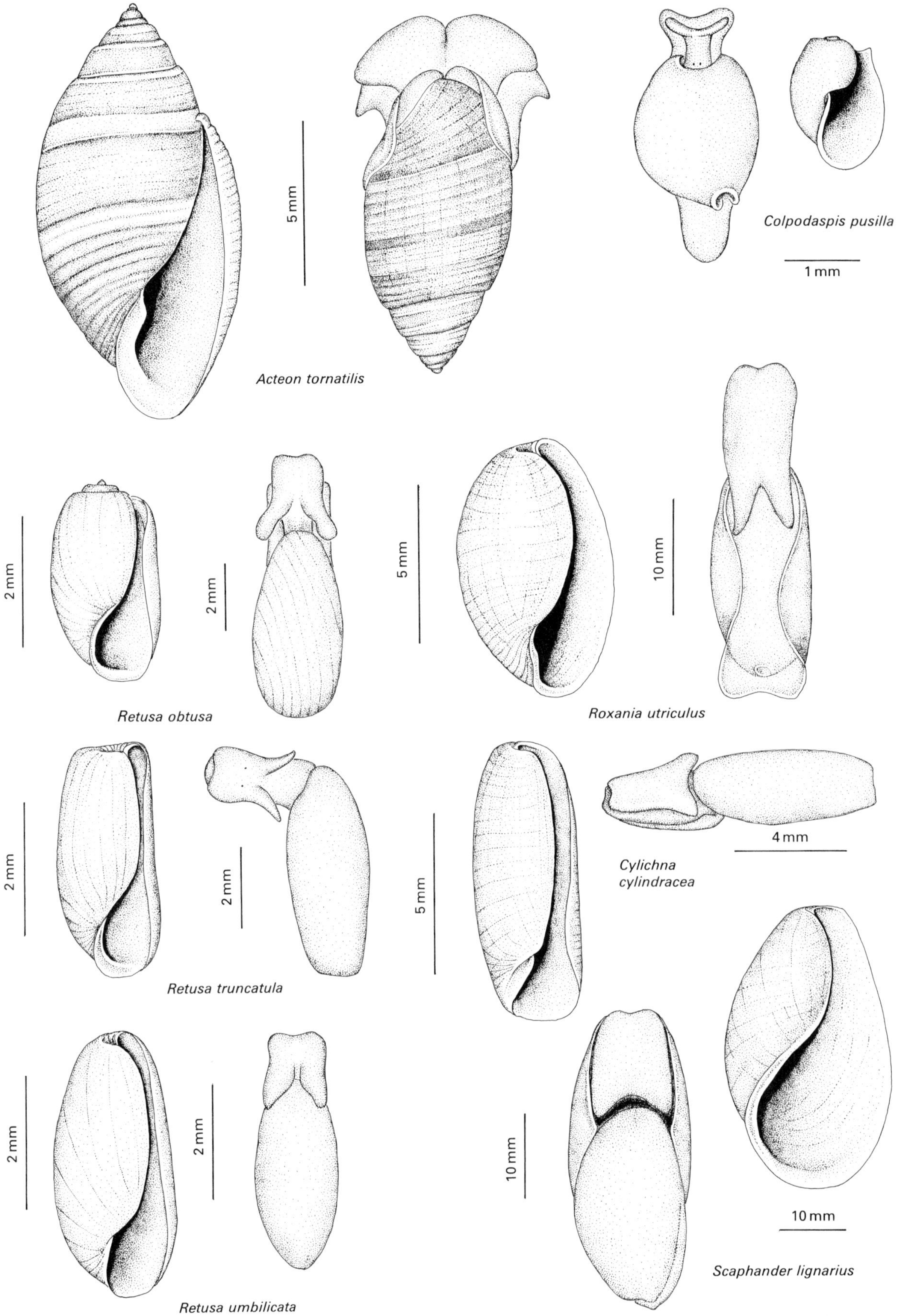
5 mm
Colpodaspis pusilla
1 mm
Acteon tornatilis
2 mm
2 mm
5 mm
10 mm
Retusa obtusa
Roxania utriculus
2 mm
2 mm
5 mm
4 mm
Cylichna cylindracea
Retusa truncatula
2 mm
2 mm
10 mm
10 mm
Scaphander lignarius
Retusa umbilicata

Fig. 12.21

into completely. Body grey-white with yellowish-brown tinge; only cephalic shield, enlarged parapodial lobes, and posterior mantle lobe visible. Head indented anteriorly; produced posteriorly into smoothly rounded tentacular processes which point straight back, overlapping shell, not flaring outwards.

In muddy sand; sublittoral, to 1500 m. Northern Britain and west coast of Ireland.

Scaphander lignarius (Linnaeus) FIG. 12.21

Shell solid, glossy, opaque, green, yellow, or brown; sculpture of spiral lines, crossed by longitudinal lines, with pigment distributed along some of the lines. Up to 45 mm. Body white or orange. Head very short, squat, indented anteriorly; tentacular processes smooth, short, conical. Parapodial lobes extending from anterior end of head shield to about middle of shell; pallial lobe also visible. Total length up to 60 mm.

In muddy sand; sublittoral, to 700 m. *S. lignarius* is a burrower, preying on infaunal worms and lamellibranchs. Common, but absent from east coast of Britain.

5 PHILINIDAE

Shell internal, with large aperture. Parapodia large and fleshy but not meeting mid-dorsally. Posterior pallial lobe conspicuous, extending beyond the metapodium.

1.	Cephalic shield elongate, cylindrical	**2**
	Cephalic shield not as described	**3**
2.	Body up to 20 mm. Shell cylindrical	***Philine scabra***
	Body up to 7 mm. Shell rounded	***Philine catena***
3	Body translucent white, or cream with white spots, up to 70 mm.	***Philine aperta***
	Body light yellow with red-brown specks, up to 5 mm	***Philine punctata***

Philine aperta (Linnaeus) FIG. 12.22

Body up to 70 mm, about twice as long as wide, translucent, white to pale yellow, with white dots, quadripartite. Cephalic shield longer than wide, indented posteriorly, may or may not be indented anteriorly. Parapodial lobes fleshy, rounded on outer sides, moulding to form of head and visceral hump. Shell large, up to half body length, open, inflated, covering visceral hump, enclosed by folds of mantle. Mantle cavity widely open posteriorly, indented dorso-medially.

A burrower in sand and muddy sand, feeding on *Pectinaria*, *Echinocyamus*, foraminiferans, and small infaunal lamellibranchs and gastropods. Lower shore and shallow sublittoral, common. All British coasts.

Philine catena (Montagu) FIG. 12.22

Body cream to pale brown or buff, with specks of dark brown, orange, and white; dark brown present only on dorsal surfaces. Usually less than 7 mm long. Cephalic disc not indented anteriorly; mantle cavity not gaping posteriorly, extending only a short distance beyond the metapodium.

Intertidal and sublittoral, on rocky shores; diet unknown. Reported from the Irish Sea and Bristol Channel, and recently off Aberdeen.

Philine punctata (Adams) FIG. 12.22

Body very small, up to 5 mm, pale yellow with red-brown specks. Cephalic shield rounded posteriorly, slightly angular anteriorly. Posterior quarter very rounded, with gaping mantle cavity. Parapodial lobes less fleshy than in *P. aperta*, and body parts more clearly demarcated. Distinguished from *P. aperta* mainly by its much smaller size.

Lower shore, in pools and on silty sand; offshore to 240 m. Not uncommon; found all around the British Isles but predominantly off north east and north coasts, also western Ireland.

Philine scabra (Müller) FIG. 12.22

Body white, sometimes with dark specks, up to 20 mm. Cephalic shield very long and narrow, indented anteriorly. Posterior part of body rendered elliptical by shell. Shell cylindrical, bottom edge serrated.

In sand and muddy sand, offshore. Not widely reported; recent records suggest a northern distribution but there are early records (prior to 1951) for the English Channel and the Irish Sea.

6 RUNCINIDAE

Shell reduced or absent. Operculum absent. Cephalic shield united with mantle, no tentacular processes. No parapodial lobes. Body smooth and elongated. One British species.

Runcina coronata (Quatrefages) FIG. 12.22

Body smooth, elongate, up to 6 mm. Colour pattern produced by variations in pigment concentration: body brown with white speckles; head separated by a light collar lacking brown pigment but with increased white, a similar oval patch at posterior end of mantle. Brown colour not extending to mantle edge, which is translucent with orange tinge. Cephalic shield fused with mantle, indented anteriorly, no tentacles. Foot translucent cream, visible on each side of head and neck; eyes barely visible on edge of brown pigment. Metapodium extending beyond mantle by about one-quarter to one-third body length, with central band bordered on each side, just beyond mantle, by patch of white spots; border translucent, with very pale orange tinge. Three gills protrude beneath posterior edge of mantle on right side of body.

Intertidal and shallow sublittoral, in clear coralline rock-pools; may be abundant in favourable conditions. *Runcina coronata* has been observed feeding on *Codium*. English Channel, west coasts, northwards to Orkney.

7 PYRAMIDELLIDAE

Small animals, with spiral conical shell, into which soft parts may be completely withdrawn, closed by an operculum. Protoconch sinistral, rest of shell dextrally coiled. A portion of the foot, the *mentum*, projects anteriorly between head and propodium, slightly indented in mid-line. Tentacles are not of typical nudibranch form (oral or propodial): they have a concave surface which may be moved to face a variety of directions. Eyes located between tentacles: a diagnostic difference with prosobranchs (on stalks lateral to tentacles). These opisthobranchs have no radula, but a probosics armed with a stylet, specialized for a parasitic mode of life. They are most often found in association with molluscs and tubicolous polychaetes; recent studies have shown that though they are not host-specific, they may exhibit some preferences. Perhaps as many as 40 species of pyramidellid occur around the British Isles but most are rare, or only infrequently recorded.

1. Shell a slender spire of ten or more whorls **2**
   Shell with less than ten whorls, rather short and fat **3**

2. Shell with strong, straight costae, with concentric grooves between ***Turbonilla crenata***
   Shell with thick, sinuous oblique costae, no concentric sculpture ***Turbonilla elegantissima***

3. Shell with well-marked costae; spiral ridges on abapical portion of body whorl ***Chrysallida spiralis***
   Shell smooth, or with some fine spiral striae or grooves, but without costae **4**

4. Each whorl with well-marked spiral grooves on abapical portion ***Menestho divisa***
   Whorls smooth, or with fine striae only **5**

5. Shell small, comprising just three whorls and the protoconch ***Menestho diaphana***
   Shell of more than three whorls **6**

6. Last whorl smoothly rounded, without a keel **7**
   Last whorl with a faint keel, appearing angular in apertural view **9**

7. Shell slender, up to six whorls, with last whorl comprising less than half total shell height. Columellar tooth well-marked, extending as ridge into aperture ***Odostomia plicata***
   Shell rather fat, of five whorls or less. Last whorl comprising about half total shell height **8**

8. Columellar tooth weakly developed; umbilicus distinct ***Odostomia scalaris***
   Columellar tooth strongly developed, extending as ridge into aperture; umbilicus indistinct ***Odostomia eulimoides***

9. Shell small, four whorls only. Last whorl keeled at broadest point. Aperture oval ***Odostomia turrita***
   Shell with up to five whorls. Last whorl angular at broadest point but not distinctly keeled. Aperture quadrate ***Odostomia unidentata***

Chrysallida spiralis (Montagu) FIG. 12.23

Shell usually long and slender, sometimes rather short and fat, comprising four whorls and the protoconch, size up to 2.8 mm; white, solid, and deeply ridged. Abapical part of last whorl with spiral grooves between flat ridges. Aperture thickened, rather square, with slightly flared lip; tooth on columella variably developed, most prominent on shells with thickened apertural rim.

Found all around Great Britain except south-east corner.

Menestho diaphana (Jeffreys) FIG. 12.23

Shell fusiform, slender, up to 3.4 mm; comprising three whorls and the protoconch; smooth, white and rather shiny, with some very fine striations on abapical part of last whorl. Aperture elongate, thickened, columellar tooth variably developed, umbilicus large.

M. diaphana is an obligate parasite of the sipunculid *Phascolion strombi* and has been reported from scattered localities around the British Isles.

Menestho divisa (Adams) FIG. 12.23

Shell tall with elongate spire, maximum size 5.0 mm; with six whorls and a protoconch; creamy yellow to white, glossy. All whorls with incised spiral lines abapically and faint growth lines; top of whorls flat at sutures. Aperture elongate-oval, columellar tooth small, blunt, variably developed; umbilicus well-marked, deep.

Found chiefly on northern coasts.

Odostomia eulimoides Hanley FIG. 12.23

Shell with long conical spire, maximum size 5.2 mm, with five whorls and protoconch; body whorl comprising approximately half total shell height. Shell white to yellowish-white, no sculpture; weak growth lines visible on last whorl; whorls flat-topped at sutures. Aperture elongate-oval, with thickened outer lip; columellar tooth well-developed, extending as ridge into aperture; umbilicus shallow, obscured by columella but visible on broken shells.

Parasitic on a large number of lamellibranchs, and also the prosobranch *Turritella communis*. Scattered localities around the British Isles.

Odostomia plicata (Montagu) FIG. 12.23

Shell with tall slender spire, maximum size 3.8 mm, with six whorls and the protoconch; white, smooth, and glossy, with faint growth lines. Sutures well-marked, whorls flat-topped. Aperture broadened forming a drop shape, with gently curved, thickened outer lip. Columellar tooth well-marked, extending as ridge into aperture; umbilicus small, sometimes obscured by columellar lip.

All around British Isles except south-eastern coasts.

Odostomia scalaris MacGillivray FIG. 12.23

Shell with short fat spire, maximum size 4.4 mm, with five whorls and the protoconch; white, shiny, and smooth. Whorls

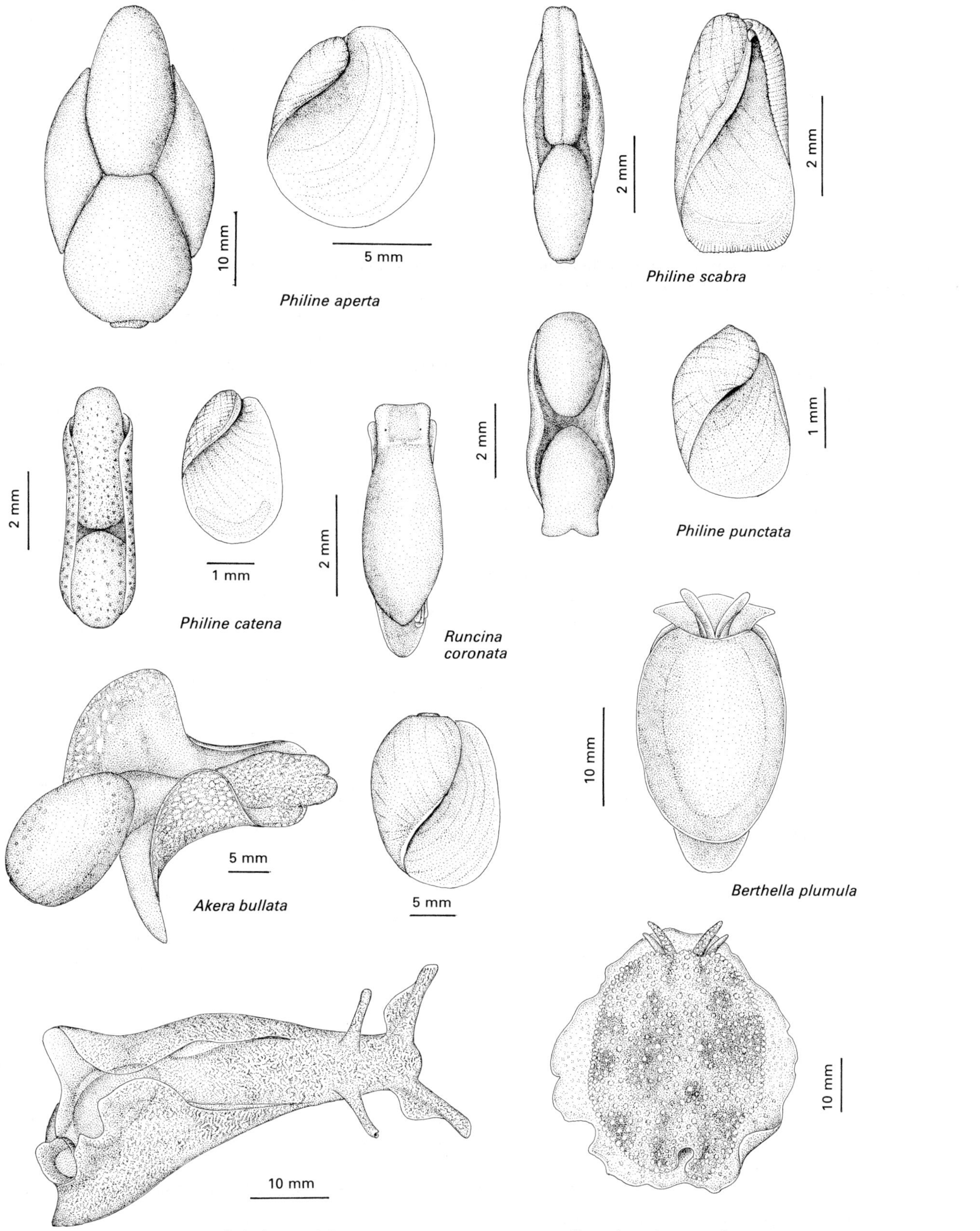
10 mm
5 mm
Philine aperta
2 mm
2 mm
Philine scabra
2 mm
1 mm
Philine catena
2 mm
1 mm
Runcina coronata
2 mm
Philine punctata
5 mm
Akera bullata
5 mm
10 mm
Berthella plumula
10 mm
Aplysia punctata
10 mm
Pleurobranchus membranaceus

Fig. 12.22

rounded, curving into sutures basally, with a few faint growth lines. Aperture oval, almost as long as last whorl, with flattened outer lip. Columellar tooth weakly developed; umbilicus small but distinct.

All British coasts. Hosts are many prosobranchs and several lamellibranchs, especially *Mytilus edulis*.

Odostomia turrita Hanley FIG. 12.23

Shell with tall, blunt spire, maximum size 2.8 mm, with four whorls and the protoconch; shiny, opaque white to yellow, with greyish tinge towards spire. Faint growth lines present; whorls flat-sided, last whorl keeled at broadest point, appearing angular. Aperture oval; columellar tooth and umbilicus distinct.

Reported from scattered localities along the Channel, and on northern coasts; host unknown.

Odostomia unidentata (Montagu) FIG. 12.23

Shell with tall spire, maximum size 3.8 mm, with five whorls and the protoconch; white to off-white or greyish, smooth except for few growth lines on last whorl, no keel but slight angle to last whorl. Whorls dipping very slightly into sutures. Aperture elongate, rather angular; columellar tooth well-developed and conspicuous.

All British coasts.

Turbonilla crenata (Brown) FIG. 12.23

Shell with elongate, slender spire; maximum size 8 mm, with 11 whorls and protoconch; creamy white to yellow, banded with yellowish brown. Whorls marked with strong longitudinal, closely spaced costae, and concentric grooves betweeen costae; fairly straight-sided, sutures deep and angular; a series of concentric grooves on abapical part of last whorl. Protoconch obvious, smooth, slightly bulbous. Aperture rather squared, lip thickened; in old, thickened, individuals lip may be out-turned. Umbilicus small, usually obscured by columellar lip.

Recorded from western and northern coasts.

Turbonilla elegantissima (Montagu) FIG. 12.23

Shell with slender, elongate spire, maximum size 9 mm with ten whorls and protoconch. Whorls almost straight-sided with deep, flat sutures; marked with thick, flat, slightly sinuous, oblique costae, indistinct on first whorl, prominent thereafter; abapical portion of body whorl smooth with few very faint growth lines; no concentric grooves between costae. Protoconch smooth and rather bulbous. Aperture rather square, thickened in older shells; no umbilicus. Deep orange operculum.

Lower shore and subtidally, feeding on sedentary worms, such as *Audouinia* and *Amphitrite*. Chiefly west coast of Britain.

8 AKERIDAE

With oval external shell of several whorls, too small to accommodate whole animal. Operculum absent. Large parapodia, used for swimming, which meet mid-dorsally. One British species.

Akera bullata (Müller) FIG. 12.22

Shell large, bubble-like, fragile, up to 40 mm; glossy, white to pale brown, up to six whorls forming a rather flat spire. Body up to 60 mm, pale grey to orange, often streaked anteriorly with blotchy lines of purplish-brown, covered with white and dark spots. Head indented anteriorly, expanded laterally to form flattened tentacular lobes. Parapodial lobes enlarged, used for evasive swimming action; covering lateral and dorsal areas of shell when at rest.

In and on soft fine mud, in sheltered bays formerly favoured by *Zostera*. Herbivorous. Lower shore to 370 m, all British coasts.

9 APLYSIIDAE

Shell internal or absent. Head elongated, with rolled oral tentacles and rolled rhinophores. Parapodial lobes present.

Aplysia punctata (Cuvier) FIG. 12.22

Shell fragile, transparent, pale amber; situated betweeen parapodia and visible through mantle foramen. Body up to 300 mm, olive-green, brown, or purplish-black in adults, pinkish-red in juveniles, with blotches or veins of darker pigment, interspersed with white or grey; colour very variable. Body elongate; head with paired oral tentacles, and paired, rolled rhinophores; eyes just anterior to rhinophores. Parapodia characteristic: widely spaced anteriorly, joined quite high up posteriorly; margins dark brown or blackish in adults, with white line on very edge; in juveniles violet with white edge. Anterior and lateral edges of propodium, edges of oral tentacles and tips of rhinophores marked with dark pigment. Metapodium smooth, quite high but without a keel, evenly pointed.

Aplysia punctata is a herbivore and its growth rate is directly related to the algal species eaten (Carefoot 1967). *Phocamium*, *Enteromorpha*, and *Ulva* are the three most beneficial, and *Delesseria* and *Laminaria* among the least. However, preferred food is not necessarily that most readily available. Colour may be related to diet. Common off all British coasts; can be found in shallow water and juveniles, in particular, may occur in weedy rock pools.

10 PLEUROBRANCHIDAE

Internal shell. Oral tentacles smooth; oral veil obvious. Rhinophores rolled. Gill present as rachis with double row of tubercles, situated on right side between foot and mantle.

1. Mantle brown, with patches of darker brown, papillate **Pleurobranchus membranaceus**

   Mantle pale yellow to orange, smooth **Berthella plumula**

Pleurobranchus membranaceus (Cuvier) FIG. 12.22

Shell thin, fragile, transparent, flattened and widely open, colourless to pale amber, up to 50 mm. Body up to 120 mm, pale brown with patches of dark brown between soft, conical, retractile mantle papillae. Mantle much smaller than foot, may lift and curl forming an exhalant siphon. Foot smooth, indented at head region. Head situated between foot and mantle, flattened,

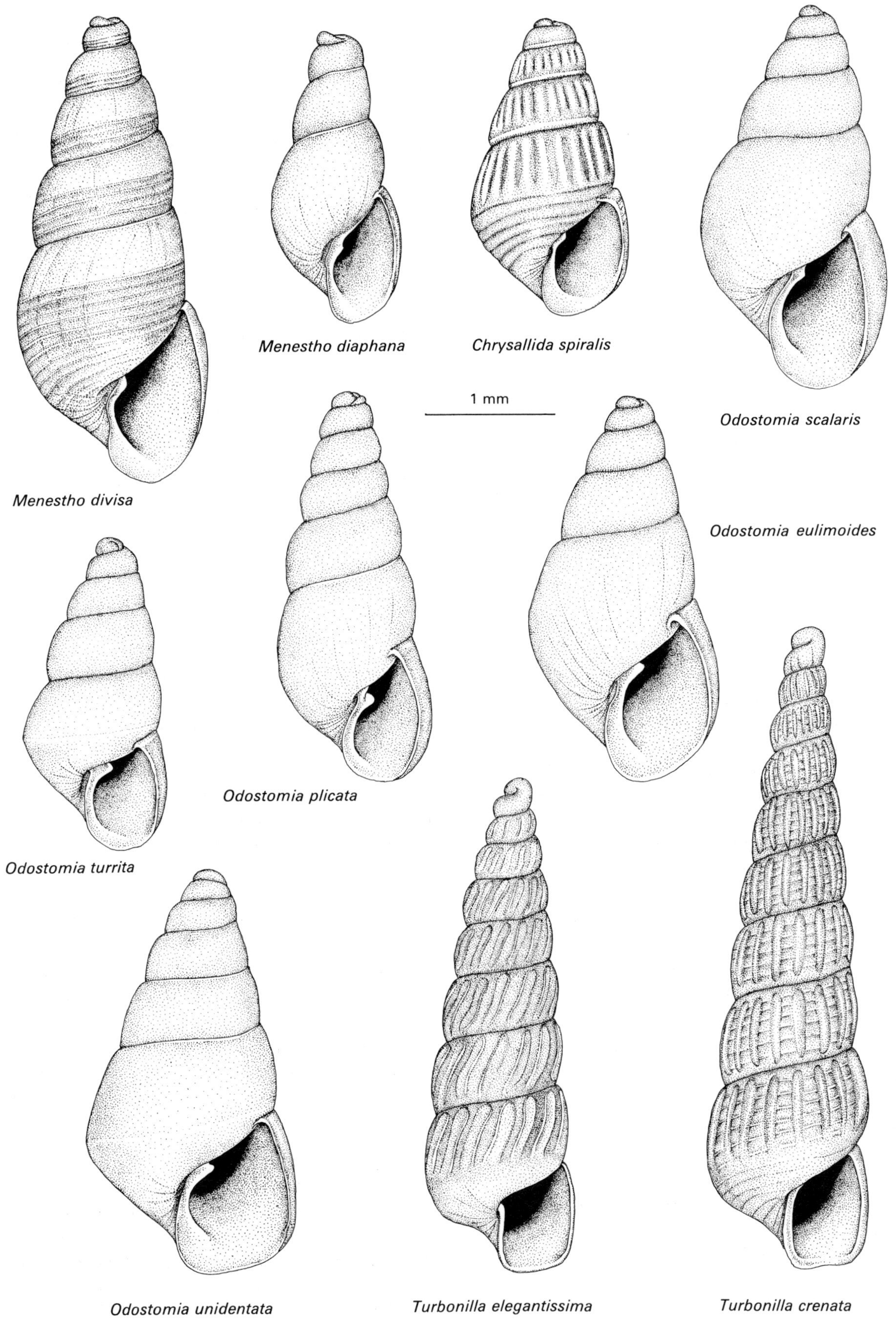
Menestho divisa
Menestho diaphana
Chrysallida spiralis
Odostomia scalaris
1 mm
Odostomia eulimoides
Odostomia plicata
Odostomia turrita
Odostomia unidentata
Turbonilla elegantissima
Turbonilla crenata

Fig. 12.23

with smooth, elongate oral tentacles; rhinophores smooth, rolled, protruding from beneath anterior end of mantle. A posterior metapodial gland, of unknown function, develops on the dorsal side of the animal at a pre-sexual stage, when body length is approximately 40 mm. The epidermis of the mantle contains acid-gland cells which secrete sulphuric acid on disturbance.

Lower shore and shallow sublittoral, to 70 m, in clean rocky areas where it preys on ascidians. May be common at certain, unpredictable, times of the year. Western coasts, from the Channel to Orkney.

Berthella plumula (Montagu) FIG. 12.22

Shell large, up to 30 mm long, completely covered by mantle, which extends just a little beyond it, all around. Body up to 60 mm, smooth, cream to orange, often with reticulate markings in middle of mantle. Head flat, large oral veil lying between propodium and mantle; oral tentacles slightly rolled, rhinophores enrolled. Mantle edge may be raised on right side, above gill, to facilitate respiration. A metapodial gland present dorsally, appearing when body is about 30 mm long.

A specialized predator of compound ascidians, commonly observed on *Botryllus*; under stones and in rock-pools, low shore and sublittorally to 10 m. Common off all British coasts.

11 ELYSIIDAE

Shell absent. Body smooth, elongate, with parapodial lobes. Oral tentacles absent. Rolled rhinophores. One British species.

Elysia viridis (Montagu) FIG. 12.24

Body elongate, tapering, up to 45 mm; colour varying according to diet, from vivid green to brown or bright red, with tiny brilliant spots of blue, red, and green. Parapodia extending along almost whole body length, often with patches of white, and with green chloroplasts visible. Propodial tentacles conspicuous; rhinophores rolled, opening ventrally, distal ends usually white. Black markings sometimes present, white patches may surround eyes. Anus dorso-lateral, below right rhinophore.

A distinctive sacoglossan, *Elysia viridis* is commonly found on algae, especially *Codium fragile*, in shallow water, and rock-pools. The chloroplasts derived from *Codium* are retained undamaged by *Elysia* and continue to photosynthesize, to the benefit of the slug. All British coasts.

12 LIMAPONTIIDAE

Shell absent. Small, less than 10 mm in length, no dorsal papillae or parapodia. Smooth, simple body form. Cephalic tentacles present in varying forms or absent.

1. Head without ridges or tentacles. In salt-marshes, on *Vaucheria* — ***Limapontia depressa***
   Head with crests or tentacles. In coralline rock-pools — **2**
2. Head with rhinophoral crests, bearing eyes on their sides — ***Limapontia capitata***
   Head with smooth, tapering rhinophores; eyes in pale areas behind rhinophores — ***Limapontia senestra***

Limapontia capitata (Müller) FIG. 12.24

Length commonly about 4 mm, rarely up to 8 mm; dark brown or black, velvety. Rhinophoral crests forming dorso-lateral flaps, pale externally and around the eyes. Pale areas also on head, metapodium, and sometimes middle of dorsum. Anal and renal openings situated to right of mid-line, close together.

Common, found in clean coralline rock-pools from MTL to the lower shore, feeding on species of *Bryopsis*, *Cladophora*, *Enteromorpha*, and *Chaetomorpha*, algae whose frond diameter approximates to the width of the foot of the slug. Species of algae with much larger or smaller diameters cannot be eaten. All British coasts.

Limapontia depressa (Alder and Hancock) FIG. 12.24

Body smooth, up to 6 mm; dark, often with tiny pale spots on dorsum and white patches around eyes. The var. *pellucida* Kevan is bright yellow with pale areas on sides of head, occasional bluish-white and black specks on dorsum, with greenish lobes of digestive gland visible through skin. No tentacles; body shape more flattened than in other *Limapontia* species; anal and renal openings widely spaced, dorsal, close to mid-line. Metapodium short.

On salt marshes, feeding on *Vaucheria*. Common on all British coasts.

Limapontia senestra (Quatrefages) FIG. 12.24

Body smooth, up to 6 mm; marbled olive-brown to black, with paler metapodium and rhinophores, and pale patches around eyes. Digitiform tentacles present, lacking in juveniles shorter than 2 mm. Anal and renal openings located near mid-line, close together. Ventrally, propodium indented on mid-line and head bilobed. Distinctive bursal swelling on right side. The var. *corrugata* Alder and Hancock has a wrinkled dorsum.

Intertidal, in coralline rock pools, feeding on *Cladophora rupestris* and *Enteromorpha*. All British coasts.

13 STILIGERIDAE

Shell absent. Body elongated, with cerata (lacking cnidosacs). Oral tentacles lacking. Rhinophores present as flattened, ear-like (*Alderia*), rolled (*Hermaea*), or digitiform (*Stiliger*) structures.

1. Rhinophores absent, or forming short ear-shaped projections. Ratio of foot-width to body-length 1:3. Short swollen cerata; anus located posteriorly — ***Alderia modesta***
   Not as described — **2**
2. Rhinophores rolled, with open slit on dorsal side. Ratio of foot-width to body-length 1:20. Body white, tinted rose-pink anteriorly. Cerata lanceolate and faintly tuberculate — ***Hermaea bifida***
   Rhinophores rolled, with open slit on ventral side. Ratio of foot-width to body-length 1:8. Body greenish white. Cerata cylindrical, smooth, rounded at tip — ***Hermaea dendritica***

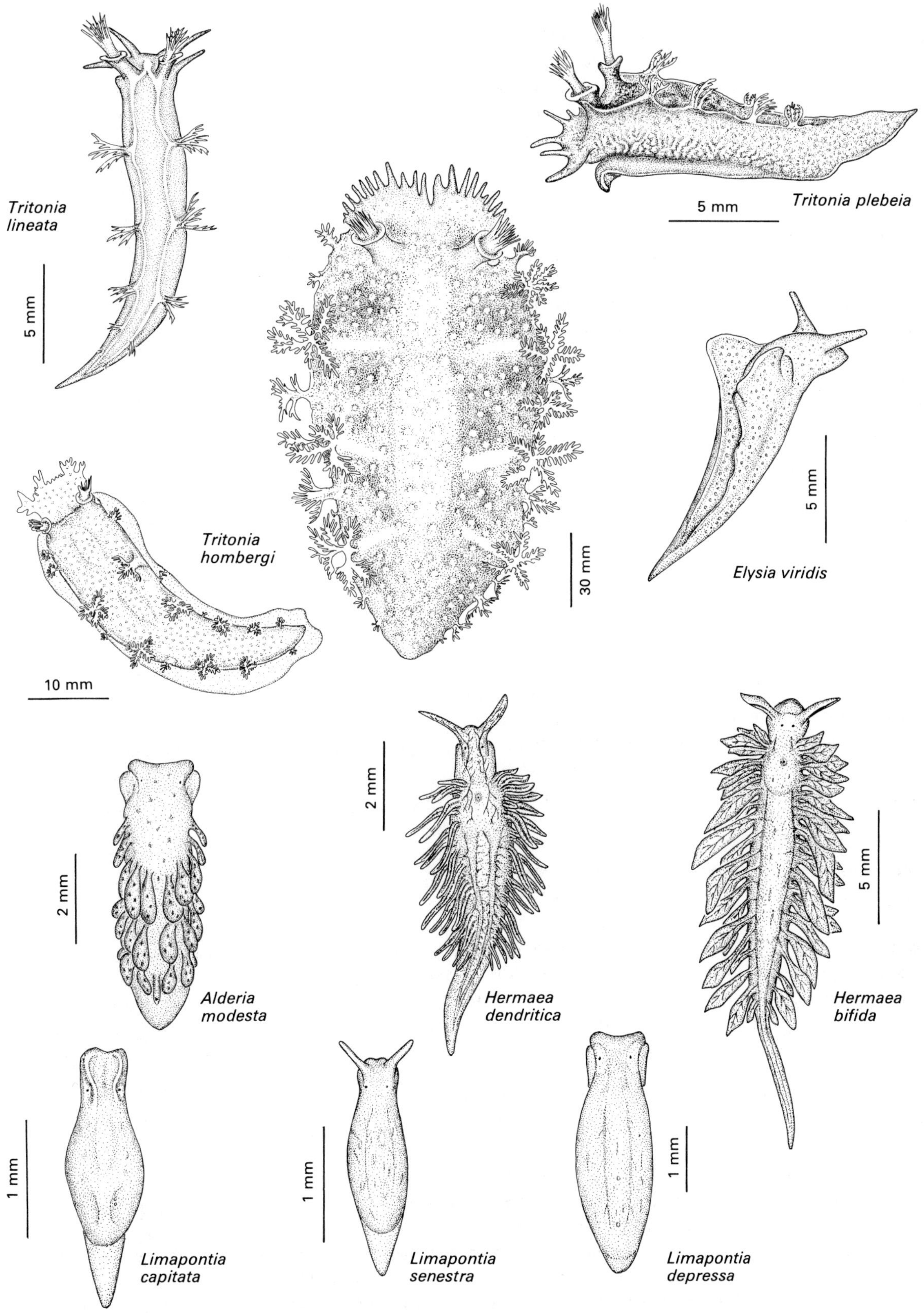
Tritonia lineata
5 mm
Tritonia plebeia
5 mm
Tritonia hombergi
10 mm
30 mm
Elysia viridis
5 mm
Alderia modesta
2 mm
Hermaea dendritica
2 mm
Hermaea bifida
5 mm
Limapontia capitata
1 mm
Limapontia senestra
1 mm
Limapontia depressa
1 mm

Fig. 12.24

Alderia modesta (Lovén) FIG. 12.24

Body flattened, up to 10 mm, pale fawn with patches of white, green, and brown. Up to seven transverse paired rows of broad, round-topped cerata, pulsating regularly to circulate blood; rows very difficult to count. Anal papilla posterio-dorsal, on mid-line.

In salt marshes, on damp mud clear of brackish water, associated with species of *Vaucheria*. All British coasts.

Hermaea bifida (Montagu) FIG. 12.24

Body very long and thin, up to 20 mm, ratio of foot-width to body-length 1:20; white with rose blush on rhinophores and head. Cerata dorso-lateral in 14 indistinct rows, up to three per row; lanceolate, faintly tuberculate, pointed distally. A conspicuous pair of brilliant red-brown digestive gland channels visible through body wall, extensions running into and branching within each ceras. Propodium slightly bilobed; head large, flattened, ventrally divided into five parts. Rhinophores enrolled, with dorsal opening. Anal papilla located on mid-line, anterior to pericardium.

H. bifida is found on red algae such as *Delessaria* and *Griffithsia*, rarely on the shore, more commonly in shallow water offshore. Reported from west coasts of Britain and Ireland, and the east coast of Scotland.

Hermaea dendritica (Alder and Hancock) FIG. 12.24

Body up to 11 mm, ratio of foot-width to body-length 1:8; colour variable, usually greenish white, with green or brown digestive glands branching into cerata, across head and into rhinophores. Digestive gland consists of two parts, anterior and posterior, pericardium devoid of tributaries posteriorly. Cerata smooth, elongate, smoothly pointed, with white tips; eight rows on each side, up to four per row. Rhinophores enrolled, with ventral opening. Propodium simple, lobed; mouth region small in comparison with *H. bifida*. Metapodium flattened and smoothly pointed.

Intertidal and shallow sublittoral, on green algae, particularly *Codium tomentosum* and *Bryopsis plumosa*. All British coasts, but more frequent on west.

14 TRITONIIDAE

Body elongated, quadrangular in cross section. Pallial gills present. Oral veil often bilobed and produced into digitiform processes. Rhinophores issuing from dilated sheaths and with branched subterminal processes.

---

1. Oral veil bilobed, with numerous digitiform processes. Very large, up to 200 mm **Tritonia hombergi**

   Oral veil not obviously bilobed; with fewer, longer processes. Smaller, up to 34 mm **2**

2. Body translucent white with opaque white dorsal markings **Tritonia lineata**

   Body not as described **3**

3. Body pale, white to brown, with green and brown mottling. Oral veil rounded, with six processes, more or less equal length **Tritonia plebeia**

   Body rose to salmon-pink. Oral veil flattened, with six processes of unequal length, outer two being longest **Tritonia nilsodhneri**

Tritonia hombergi (Cuvier) FIG. 12.24

Juveniles translucent white with opaque white pigment on dorsum, darkening with age to dark purplish-brown mottling; with lighter ventral surface. Oral veil, in both juvenile and adult, large, bilobed, with many digitiform processes; translucent white, or peach with denser white or coloured spots. Rhinophores with conspicuous trumpet-shaped sheaths, thick, branching into feathery processes, tips tinted greenish or yellowish-brown, much darker than rest of coloration. Mantle with soft tubercles and translucent white or peachy feathery gills; larger primary gills paired, with up to six branches, flexed towards mid-line, secondary gills with fewer branches, projecting laterally. Gill number increases with age. Foot translucent, with white spots.

This is the largest British nudibranch and can attain a length of 200 mm. It is always found in association with the soft coral *Alcyonium digitatum*. Shallow sublittoral, and offshore to 80 m; *T. hombergi* occurs off all British coasts but is more common in the north.

Tritonia lineata Alder and Hancock FIG. 12.24

Body up to 34 mm; white, tinted rose from internal organs. A brilliant line of opaque white on each side of dorsum, linking rhinophores and gills, meeting in a long point posteriorly. Oral veil produced into four long tapering processes, distal halves of which are white. Rhinophores as for *T. hombergi*, tips sometimes tinged brown. Gills paired, four to six on each side, consisting of two main branching stems; increasing in size anteriorly, white. Foot translucent, lobed anteriorly, pointed posteriorly.

Diet is uncertain, but *T. lineata* has been found on *Kirchenpaueria pinnata*, gorgonians, and small octocorals such as *Sarcodictyon catena*. Shallow sublittoral, to 40 m; less common than *T. hombergi*, localized, but occurring off all British coasts.

Tritonia nilsodhneri (Marcus)

Body slender, up to 34 mm, rose to salmon-pink. White areas around gills and rhinophores; gill bases and rhinophoral sheaths pale pink to white. Rhinophores branched, rather than feathery, darker pink. Oral veil produced into six short digitiform tentacles of unequal length, the outer one on each side being the longest. Gills paired, up to eight on each side. Foot irregularly broader than mantle, lobed anteriorly, pointed posteriorly.

T. nilsodhneri feeds on gorgonians, especially *Eunicella verrucosa*, from which it derives its rose-pink coloration. Rare, occurring only in south-west Britain and off western Ireland, down to 40 m.

Tritonia plebeia Johnston FIG. 12.24

Intermediate in form between *T. hombergi* and *T. lineata*. British specimens up to 30 mm; body white to pale yellow or brownish, with green and brown mottling. Oral veil translucent white with brown speckling; with six short processes of approximately equal length, one or two often broken off. Mantle flatter than in other *Tritonia* species, pallial margin conspicuous, usually paler, with darker brown mottling on either side. Gills, up to ten pairs, translucent whitish or yellowish.

Feeds on the soft corals *Alcyonium digitatum* and *Paramuricacea placonis*, as well as the gorgonian *Eunicella verrucosa*. Found in clear water, shallow sublittoral to 129 m, all around the British Isles, though there are no records for the south-east corner.

15 LOMANOTIDAE

Elongated; conspicuous pallial rim produced laterally, with wavy margin. Cerata present. Bilobed oral veil with two pairs of digitiform tentacles. Rhinophores swollen and lamellated, issuing from dilated sheaths.

1. Body up to 34 mm; drab, with epidermal pigmentation of brown and white spots and speckles. Cerata never curved or wrinkled **_Lomanotus marmoratus_**

   Body up to 94 mm; colourless, or with uniform coloration, not speckled, tips of cerata may be yellow. Longest cerata often with curved tips, and gill-like wrinkles in the curves **_Lomanotus genei_**

Lomanotus genei Verany FIG. 12.25

Body up to 94 mm; colourless, or brightly coloured, pale orange-red, deep crimson, or purple; unicolorous, without spots, cerata may be tipped with yellow or orange. Cerata comparatively short and flat (cf. *L. marmoratus*), often curved at the ends and bearing a series of longitudinal wrinkles on the under surface.

Shallow sublittoral, feeding on *Nemertesia ramosa*. All British coasts.

Lomanotus marmoratus (Alder and Hancock) FIG. 12.25

Body up to 34 mm; colour very variable, dark grey to white, yellow, or crimson, spotted with white, brown, and gold. Dorso-pallial rim expanded and divided on each side, forming a series of fleshy, fan-shaped gills; unevenly produced edge extends anteriorly around front of animal, above which is the oral veil, with two pairs of elongated processes. Rhinophores swollen, lamellated; sheath rims with finger-like processes. Posterior to last pair of gills, dorsopallial rim expanded into an undulating membrane; tail lashed from side to side during swimming escape reaction.

Feeds on calyptoblastic hydroids such as *Nemertesia antennina*. Offshore, to 40 m; reported from localities off all British coasts.

16 DENDRONOTIDAE

Body limaciform with four to eight pairs of pallial cerata. Oral veil with two to five pairs of digitiform tentacles, usually subdivided. Long, swollen, lamellated rhinophores, sheaths with branched rims.

One British species.

Dendronotus frondosus (Alder and Hancock) FIG. 12.25

Body up to 100 mm. Colour varying with age; juveniles to 4 mm pale; from 4–30 mm, uniform or mottled white to orange or reddish-brown, often in contrasting bands and stripes; pattern fading in specimens larger than 30 mm, becoming rather drab. Dorso-pallial rim not as conspicuous as in tritonids but produced into richly arborescent gills: up to nine pairs of large gills, with smaller ones in between. Rhinophore sheaths and anterior margin of head similarly arborescent.

A large nudibranch, aptly named. Juveniles feed on calyptoblastic hydroids, such as *Hydrallmania*, *Obelia*, and *Sertularia*, adults on larger gymnoblastic species, especially *Tubularia indivisa* and *T. larynx*. Found in shallow inshore waters off all British coasts.

17 DOTOIDAE

Body with tuberculate, swollen pallial cerata; usually a single row down each side; *pseudobranchs* (small gill-like excrescences) on the mesial face of the largest cerata in some species. Rhinophores long and tapering. Sheaths tall and flared.

1. Tubercles on cerata with coloured distal spots **2**

   Not as described **6**

2. Body very densely pigmented with blackish-red spots of varying shapes. Rhinophore sheaths trumpet-shaped, speckled inside. Pseudobranchs present as well as crests on head **_Doto dunnei_**

   Body pale with streaks or mottling **3**

3. Distal tubercular spots red, crimson, maroon, or purple **4**

   Distal tubercular spots brown or black **5**

4. Large dark red to purple red spots; body with red streaks or blotches. Five to eight pairs of cerata **_Doto coronata_**

   Small bright red spots; body with evenly distributed red spots missing in a line between the cerata. Four to five pairs of cerata. Squared oral veil **_Doto eireana_**

5. Black or brown spots. Lateral papillae present between cerata, each tipped with a black spot. Prominent crests running anteriorly from rhinophores **_Doto pinnatifida_**

   Black spots, except on terminal tubercle. Pseudobranchs present. Transverse rows of black-tipped tubercles on dorsum between pairs of cerata; lateral longitudinal row of tubercles outside cerata **_Doto tuberculata_**

6. Head with prominent crests running from rhinophores forward **7**
   Head not as described **8**
7. Small head. Large pseudobranchs present on the cerata; scattering of brown or black spots on cerata. Rhinophore sheath rims white ***Doto millbayana***
   Pale fawn to orange body; rhinophores ochre. Form, colour, and size of cerata variable ***Doto fragilis*** complex
8. Large obvious oral veil. No spots on tubercles or cerata. Trumpet-shaped rhinophore sheaths ***Doto lemchei***
   Fawn with dense speckling of brown or black. Crenulated rhinophore sheath rims with longer anterior sides ***Doto cuspidata***

Doto coronata (Gmelin) FIG. 12.20

Body up to 12 mm; translucent white or pale yellow, with crimson, red, maroon, or purple pigmented spots coalescing to form streaks or blotches covering much of dorsum, sometimes concentrated at bases of cerata and often parts of their sides as well. Some specimens lack pigment around and between cerata, often with a dark maroon streak in the centre of each stalk. Rhinophores translucent white with spots of opaque white. A confusingly variable species; general characteristics are: relatively large head with, at most, only slight crests; long, smooth, digitiform rhinophores, sheaths either trumpet-shaped with smooth rims slanting upwards from median side, or with a tongue-like protrusion of anterior margin; five to eight pairs of cerata, each with three to four rings of tubercles, a round, dark red or maroon spot on each tubercle. Oral veil dilated into distinct lateral lobes, with smooth contours. Propodium lobed, metapodium elongate, mantle fusing into it.

Intertidal and sublittoral, to 180 m; common in shallow waters off all British coasts. *D. coronata* feeds on a wide range of gymnoblastic and calyptoblastic hydroids, including *Dynamena pumila, Clava multicornis, Obelia* spp., *Hydrallmania falcata, Sertularia* spp., *Tubularia larynx*.

Doto cuspidata Alder and Hancock FIG. 12.25

Body up to 14 mm, fawn and crimson, with dense speckling of dark brown or black. Up to ten pairs of cerata, each with four to six concentric rows of triangular, unpigmented, white-tipped tubercles. Rhinophore sheaths crenulated, with longer processes on anterior sides. Oral veil dilated, with sharply projecting lateral lobes. Small, white, epidermal tubercles on dorsum.

Feeds on *Nemertesia ramosa*; sublittoral, to 148 m. Not common; reported from south Cornwall, Bristol Channel, and Irish Sea.

Doto dunnei Lemche FIG. 12.25

Body up to 25 mm; fawn-yellow to brownish, with very dense pigmentation of irregular maroon to black spots. Seven or eight pairs of contrasting light brown cerata, each with seven rings of five to eight tubercles, pyriform, decreasing in size basally. Tubercles lighter coloured than cerata, with distinct spot at tip. Wing-like pseudobranchs on each ceras, with three to five leaves on each side. Rhinophores very long, rounded distally; sheaths wide and trumpet-shaped, speckled within, more so on outer anterior side. Oral veil with distinct rounded lateral flaps notched basally; head with crests. Propodium simply rounded, not lobed, notched on mid-line.

In shallow water, on *Kirchenpaueria pinnata*. Western shores of Britain and Ireland.

Doto eireana Lemche FIG. 12.25

Body up to 7 mm; whitish with evenly distributed red spots on dorsum and upper sides, a distinct spot-free line between cerata on each side. Four or five pairs of cerata, the fifth pair, when present, small and closely placed to fourth; three or four rings of tubercles on each ceras, four to six tubercles per ring, each with a red spot. Oral veil squared, with squared lobes. Rhinophores twice length of sheaths, with scattered white spots, particularly concentrated distally; sheaths not flared, longer anteriorly, with few spots.

A small species of the '*coronata* group', found subtidally on *Sertularia operculata*. Recorded from south Cornwall, Bristol Channel, south-west Scotland, and western Ireland.

Doto fragilis (Forbes) FIG. 12.25

Three forms are recognized, perhaps representing three separate species. Typical form up to 26 mm, brownish-yellow body, sometimes with orange tinges. Up to nine pairs of fusiform cerata, unspotted, stalk darker than body, with up to seven concentric rows of pale tubercles. Rhinophores elongate, digitiform, three times length of sheaths, ochre with cream to yellow tips; sheaths flared, with white, slightly wavy rims. Oral veil with paired lateral lobes. Found on species of *Nemertesia*.

Giant form long and thin, up to 34 mm, body orange-brown with distinctive translucent white foot and metapodium. Cerata rounded and compressed; tubercles more numerous, tightly packed, cream-coloured; additional clumps of tubercles scattered between cerata and towards mid-line. Rhinophores digitiform, only twice length of sheaths; flared, crenulate rhinophores sheaths, orange-brown with paler rims. On *Halecium muricatum*.

Smallest form pale creamy orange, with overall rather uniform coloration. Cerata less rounded than in giant form, but less elongate than typical form, slightly darker than body; tubercles in about five concentric rows, each with pigmented end. Rhinophores ochre with white tips, sheaths half length of rhinophores. Found on *Halecium halecinum*.

All three forms occur commonly around the whole British coast, from the lower shore to about 200 m.

Doto lemchei Ortea and Urgorri FIG. 12.25

Body up to 10 mm, translucent pale fawn with darker brown pigment mottling anterior to rhinophores, between rows of cerata and sparsely on sides. Five, sometimes six, pairs of cerata,

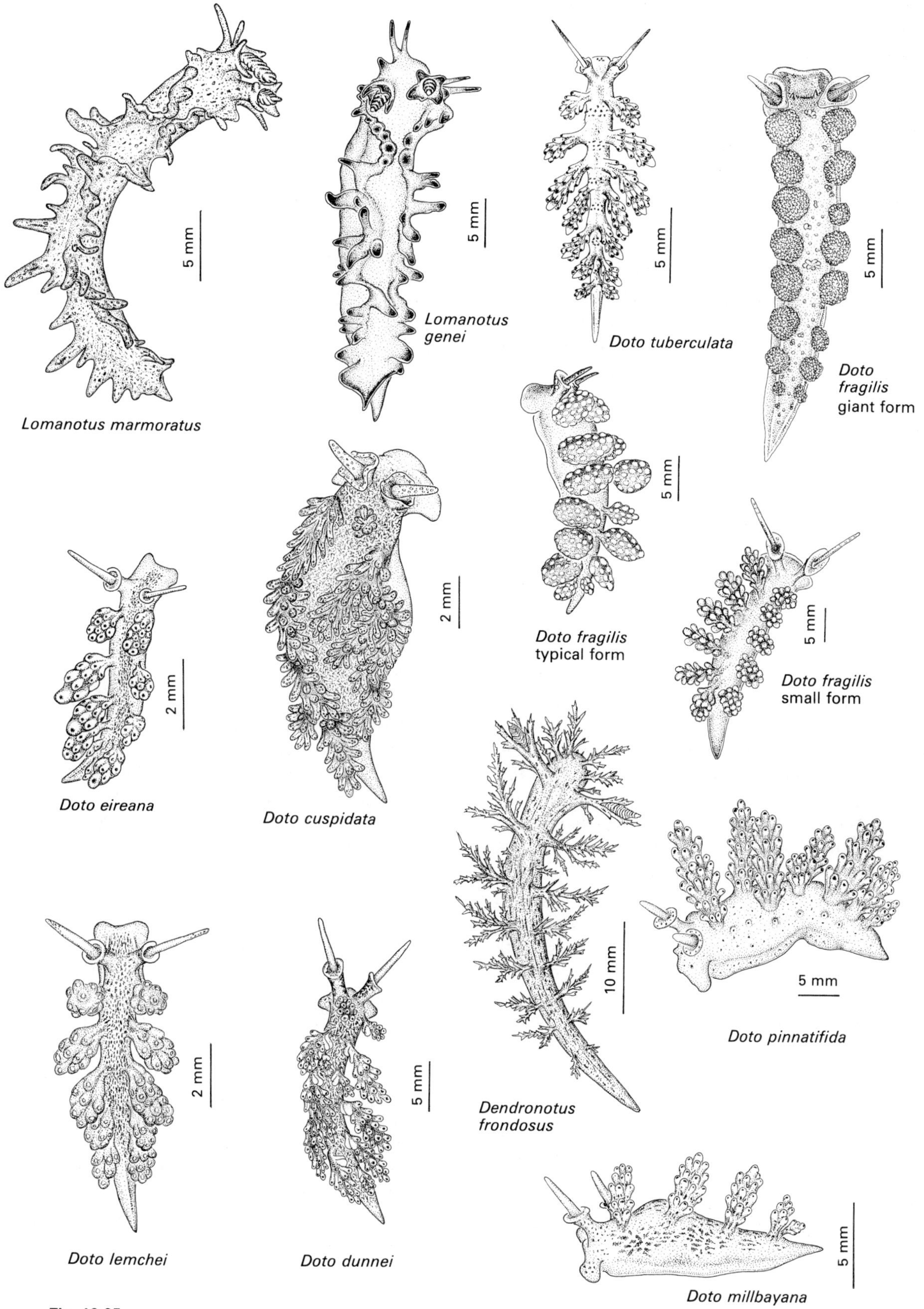
5 mm
Lomanotus marmoratus
5 mm
Lomanotus genei
5 mm
Doto tuberculata
5 mm
Doto fragilis giant form
5 mm
Doto fragilis typical form
2 mm
Doto eireana
2 mm
Doto cuspidata
5 mm
Doto fragilis small form
10 mm
Dendronotus frondosus
5 mm
Doto pinnatifida
2 mm
Doto lemchei
5 mm
Doto dunnei
5 mm
Doto millbayana

Fig. 12.25

with three to five rings, each of seven tubercles, per ceras; tubercles short and blunt, no pigment spots, but with characteristic clusters of white glands. No tubercles on body. Rhinophores about twice length of sheaths, pale, with scattered white spots distally, concentrated at rounded tips; sheaths trumpet-shaped. Head rather elongated, with two faint crests and two large, oval, anterio-lateral flaps.

Seems to feed exclusively on *Aglaophenia tubulifera*, on which it also lays its spawn. Recorded to date from south Cornwall, Scilly Isles, Lundy, west Cork, and the east coast of Northern Ireland.

Doto millbayana Lemche FIG. 12.25

Body up to 14 mm, whitish, digestive gland showing yellowish through body wall, brown in the cerata; scattered blackish-brown spots on bases of rhinophore sheaths and on cerata, particularly abundant on lateral sides of cerata. Distal ends of rhinophores, and rims of sheaths, white. Six pairs of cerata each with four or five rings of five to eight low, rounded tubercles; well developed pseudobranchs, each with five processes in feather-like arrangement, spotted at tips. No papillae on body. Oral veil with very small, almost semicircular flaps, directed anterio-laterally. Prominent head crests. Foot with a slight transverse furrow anteriorly, but no distinct corners or propodial tentacles. Propodium broader than metapodium, which tapers sharply to acute end.

Feeds on *Plumularia setacea*, a small hydroid sometimes associated with *Nemertesia ramosa*, which is also occasionally eaten. Deposits flat, broad, winding spawn ribbon. Not common. Recorded around the Cornish peninsula, and from west coast of Ireland, and Orkney. However, it will probably be found to occur wherever its prey occurs, which should be searched carefully.

Doto pinnatifida (Montagu) FIG. 12.25

Body up to 29 mm; cream or pale fawn to pale orange, with brown to black mottling down centre. Up to ten pairs of cerata, unicolorous with body; tubercles arranged in five or six concentric rows, long, conical, spread apart, nearly all tipped with a dark brown or black spot. Lateral papillae between larger cerata, also tipped black. Rhinophores translucent, with white flecks along most of length; sheaths flared, rims with black spots, sometimes undulate. Oral veil smoothly rounded. Head region appearing concave due to white speckled lateral ridges extending from rhinophore bases.

Shallow sublittoral, feeding on *Nemertesia antennina*. On western coasts of Britain and Ireland, may be locally common.

Doto tuberculata Lemche FIG. 12.25

Body up to 19 mm; pale fawn, orange or greenish-yellow, with finely dispersed darker brownish speckling on dorsum. Foot, ceratal tubercles, and rhinophores without speckling. Usually eight pairs of cerata, slightly compressed, swollen in middle, each with three or four rings of simple or complex, short, conical tubercles, all, except terminal one, tipped with a black dot. Pseudobranch, in larger specimens, forming a ridge with five small lobes, the largest lobes each with a distal spot. Two to four black-tipped, raised tubercles in rows across body between each two pairs of cerata. Pallial border just visible as a yellowish crest running longitudinally between cerata, fading towards last pairs. Foot rounded anteriorly, slightly notched medially, expanding and then narrowing to merge with lateral margins, thereafter narrower than body. Oral veil with a small rounded lobe on each side. Rhinophores pigmented white distally, about three times length of sheaths, which are flared, with undulating edge.

Shallow sublittoral, on *Sertularella gayi* and *Abietinaria abietina*. South-west England and north-west Ireland.

18 ONCHIDORIDIDAE

Phanerobranchiate dorids (gills separately contractile), flattened and oval with mantle skirt usually larger than foot. Rhinophores lamellated, no sheaths. Feeding on encrusting bryozoans.

1.	Brightly coloured dorsum: white with central red region and yellow submarginal band	***Onchidoris luteocincta***
	Body pigment not as described	**2**
2.	Body uniformly pigmented	**3**
	Body with patches or spots of colour	**5**
3.	Body usually >20 mm, varying in colour, with long, soft, pointed papillae; rhinophore sheaths papillate	***Acanthodoris pilosa***
	Body <17 mm, yellow or white	**4**
4.	Dorsum usually white, with stout flattened tubercles; rhinophores and gills colourless; gills in horsehoe around anal papilla	***Onchidoris muricata***
	Body usually yellow, with fusiform tubercles which elongate at periphery; gills and lamellate portion of rhinophores usually darker; gills form circlet around anal papilla	***Adalaria proxima***
5.	Body pale brown with scattered orange spots	***Onchidoris depressa***
	Dorsum pale with coloured blotches	**6**
6.	Body large, up to 40 mm, with brown blotches; up to 29 gills	***Onchidoris bilamellata***
	Body small, <8 mm, with variously coloured dorsal patches; peculiarly shaped patch of different colour around rhinophore; up to 12 gills	***Onchidoris sparsa***

Acanthodoris pilosa (Abildgaard in Müller) FIG. 12.26

Body commonly about 30 mm, but up to 70 mm; white, pale grey, or brown, to dark purplish-black or charcoal grey. Dorsum covered with long, soft, pointed papillae, of uniform length. Rhinophores with long conical stalks, the distal, lamellate ends posteriorly directed; rhinophore sheaths low, papillate. Up to

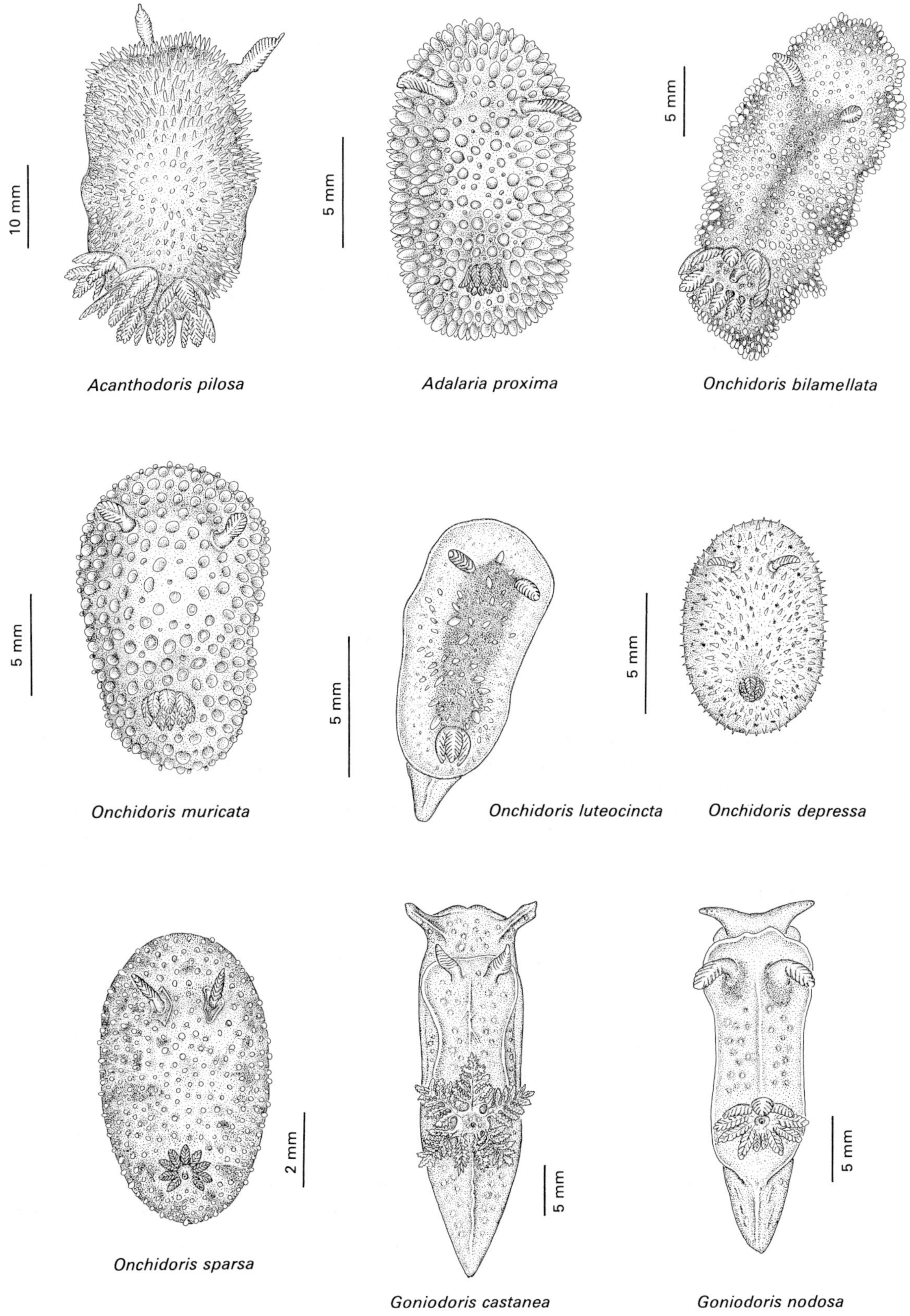
10 mm
Acanthodoris pilosa
5 mm
Adalaria proxima
5 mm
Onchidoris bilamellata
5 mm
Onchidoris muricata
5 mm
Onchidoris luteocincta
5 mm
Onchidoris depressa
2 mm
Onchidoris sparsa
5 mm
Goniodoris castanea
5 mm
Goniodoris nodosa

Fig. 12.26

nine large gills encircling anal papilla. Metapodium visible posterior to mantle.

Intertidal and shallow sublittoral, usually found on its bryozoan prey: *Alcyonidium* spp., *Flustrellidra hispida, Callopora* spp. Common on all British coasts.

Adalaria proxima (Alder and Hancock) FIG. 12.26

Body up to 17 mm, white or yellow. Dorsum covered with club-like papillae, becoming elongate towards periphery. Rhinophores digitiform, diverging and directed posteriorly; stalk very translucent, lamellae often darker yellow than dorsum. Up to 12 gills encircling anal papilla, often darker than dorsum; frequently with tubercles within the gill circlet. The gills are rather small, especially in comparison with *Acanthodoris pilosa*. Often confused with *Onchidoris muricata*, which has rather uniform, mushroom-shaped, mantle tubercles, which do not change form towards the periphery. Rhinophores are blunter and the digestive gland extends further to the anterior in *A. proxima*. *O. muricata* is usually white, though pale yellow specimens occur, specially in northern localities.

Feeds on a variety of encrusting bryozoans, especially *Electra pilosa*; intertidal and shallow sublittoral, to 60 m. South Cornwall, Bristol Channel, Irish Sea, western Ireland, and north coasts of Britain.

Onchidoris bilamellata (Linnaeus) FIG. 12.26

Body up to 40 mm, dull white with brown blotches of varying intensity, decreasing towards periphery; with hand lens pigment is seen to consist of wavy lines around and between tubercles. Dorsum with rough, club-like tubercles, becoming more elongate towards margins. Rhinophores more densely pigmented basally than distally; lamellae rough, uneven, with irregular ends. Gill circlet relatively large; gills speckled brown around edges; tubercles inside gill circlet.

Feeds on barnacles; intertidal and shallow sublittoral, to 20 m. All British coasts.

Onchidoris depressa (Alder and Hancock) FIG. 12.26

Body small, flattened, elliptical, up to 9 mm; pale brown with scattered orange spots, those close to mid-line with central black point. Mantle spiculose, with soft, evenly spaced, long pointed tubercles. Oral veil without tentacles. Rhinophores tending to curve posteriorly.

Intertidal and shallow sublittoral, to 15 m; said to feed exclusively on the bryozoan *Schizomavella linearis*. Reported from scattered localities around the whole British Isles.

Onchidoris luteocincta (M. Sars) FIG. 12.26

Body up to 11 mm, dorsum white, with red central region and yellow submarginal band. Tubercles sharply conical. Metapodium extending beyond dorsum, with obvious keel and crenulated edge marked in opaque white; tip often yellow.

One of the most colourful of the British dorids and rather conspicuous; intertidal and shallow sublittoral, feeds on the bryozoans *Smittoidea reticulata, Cellepora pumicosa*, and *Crisia*.

Onchidoris muricata (Müller) FIG. 12.26

Body up to 14 mm, white, or more rarely pale yellow. Mantle noticeably spinulose, with abundant, stout, flattened, spiculose tubercles. Rhinophores colourless. Up to 11 colourless gills in a horseshoe around anal papilla. This species is frequently confused with *Adalaria proxima* but is generally paler in colour, with more pointed rhinophores; the tubercles of *O. muricata* are more flattened than in *A. proxima*, and the digestive gland, in ventral view, is situated farther to the posterior. The radula of *O. muricata* has only two lateral plates, while that of *A. proxima* has ten.

Lower shore and shallow sublittoral, to 15 m, feeding on encrusting bryozoans. All British coasts, but more common in north.

Onchidoris sparsa (Alder and Hancock)

Body up to 8 mm, oval; pale brown with pink, red, orange, and/or brown patches. Mantle with widely spaced, translucent, mushroom-shaped tubercles, simetimes with basal black rings and terminal black spots. Rhinophores straight, pointed, broadly lamellate, with brown spots; sheaths arising from area of dorsum conspicuously different in colour from rest and bearing three or four large tubercles. Nine or ten small, simply pinnate gills, usually white, encircling the anal papilla.

Sublittoral; feeds on the bryozoans *Cellepora pumicosa* and *Porella concinna*. Predominantly northern distribution.

19 GONIODORIDIDAE

Phanerobranchiate, soft and limaciform (may be broad – *Okenia*) with reduced mantle which often has long papillae. Rhinophores unsheathed, lamellate, sometimes with long papillae. Feeding on encrusting bryozoans and ascidians.

---

1. Dorsum keeled, without conspicuous projections **2**

   Dorsum not as described **3**

2. Fawn to black ground colour, with creamy white raised spots scattered over the whole animal ***Goniodoris castanea***

   Translucent milky white ground colour; dorsum with luminescent white spots, yellow tinges around margins and on rhinophores and oral tentacles ***Goniodoris nodosa***

   White, with crimson dorsal blotches and yellow submarginal band ***Onchidoris luteocincta*** (p. 000)

3. Broad, elongate mantle projections issuing from a raised rim around rhinophores and gills, longest anteriorly. Gills feathery, numerous. Colourful: white, red, and yellow ***Okenia elegans***

   Elongate, slender and smooth dorsal projections: posteriorly, a ring or a pair beside three tripinnate gills; anterior projections issuing from rhinophore bases **4**

4. Ring of pallial processes white with brilliant orange or yellow tips. Paired, long processes protruding from rhinophore bases **_Ancula gibbosa_**

   Creamy white, with a pair of flaps on each side of gills and a digitiform process issuing from each rhinophore base **_Trapania pallida_**

Goniodoris castanea Alder and Hancock FIG. 12.26

Body up to 38 mm, fawn to red-brown, or even black. Mantle very reduced, smaller on all sides than foot; anteriorly extending just beyond rhinophores, posteriorly disappearing on each side of gills; edge curled upwards, especially laterally. Mantle and dorsal surface of foot with keel, usually white. Head with angular oral tentacles, each with a keel or ridge extending posteriorly over head and disappearing beneath mantle. Surface of mantle, hyponotum, foot and head with scattered, creamy white, raised spots. Head brownish, distal ends of tentacles white. Rhinophores irregularly lamellated, darkening distally but with white tips. Up to nine tripinnate gills around prominent anal papilla; colour variable, generally brownish with white edges and some white speckling on rachides.

A predator of the compound ascidians *Botryllus schlosseri* and *Botrylloides leachi* on which it is well camouflaged. Sometimes on large solitary species such as *Ascidia mentula*. Offshore, to 90 m; south-east and western coasts, not common.

Goniodoris nodosa (Montagu) FIG. 12.26

Body up to 27 mm, translucent white with almost luminescent opaque white spots. Mantle reduced, showing foot but less so than in *G. castanea*; edges curled, translucent, but often tinged yellow at margin. Areas around rhinophores, gill circlet, anterior foot, most of oral tentacles and proximal region of gills without spots. Notum with median keel. Metapodium shortened, yellow distally, with median keel. Rhinophores evenly lamellated, with faint yellow tinge and white spots. Oral tentacles smooth, with increased density of white pigment at tips but otherwise yellowish. Up to 13 gills, fat and fleshy.

All British coasts, shallow inshore waters to 120 m, common. Juveniles can be found on encrusting bryozoans such as *Alcyonidium gelatinosum* and *Callopora dumerilii*, adults on compound ascidians such as *Botryllus schlosseri, Dendrodoa grossularia*, and *Diplosoma listerianum*.

Okenia elegans (Leuckart) FIG. 12.27

Body up to 80 mm. Mantle much reduced, but produced into tentacles around whole of margin; typically dark red, marginal processes red or orange with yellow or white tips. A mid-dorsal row of processes posterior to rhinophores, terminating with a pair anterior to gills. Two long, slender tentacles anteriorly, size and colour of which vary. Foot broad, translucent white or pink, with yellow edge. Rhinophores same colour as mantle at base, paler just proximal to lamellae, becoming dark red on distal third, with yellow terminal knob. Up to 21 gills, white basally, grading from red to yellow distally.

One of the more conspicuous British species, usually found in association with its prey, solitary ascidians such as *Ciona, Molgula* spp., and *Polycarpa* spp.

Ancula gibbosa (Risso) FIG. 12.27

Body up to 33 mm long, translucent cream or white. Rhinophores and gills with coloured tips; in most specimens tips of pallial processes and rhinophoral processes are bright yellow or yellow-orange, rarely white; in some specimens just rhinophores and pallial processes are tipped with creamy yellow. Rhinophores with crinkly stalks, cylindrical in region of lamellae, swollen medially, each with ten lamellae, flattening distally, ending in a long blob; two elongate processes issuing from rhinophore stalk, not from head. Three pinnate gills, encircled by 14 pallial processes. Oral tentacles elongate, finger-like, directed anterio-laterally. Metapodium sometimes with a yellow tip, sometimes with a median row of yellow pigment. Foot margins may be coloured yellow.

This species is known to eat the compound ascidians *Diplosoma listerianum, Botryllus schlosseri*, and *Botrylloides leachi*. It has been found among erect bryozoans and may feed on these as well. Under rocks in shallow water, offshore to 110 m; all British coasts.

Trapania pallida Kress FIG. 12.27

Body up to 20 mm; uniformly cream or white, with iridescent white patches distally on oral tentacles, rhinophores, rhinophoral processes, gills, pallial processes, and posteriorly on the mantle. Morphologically similar to *Ancula gibbosa* but with a thick digitiform process on each rhinophore, long oral tentacles, obvious propodial tentacles, and a wing-like flap on each side of the gill region.

T. pallida is thought to feed on entoprocts, such as *Loxocalyx*, and is found on the sponges and bryozoans to which its prey is attached. Sublittoral; reported from north-west Ireland and south-west England.

20 POLYCERIDAE

Soft, limaciform phanerobranchiates. Pallial margin often with elongated papillae laterally and anteriorly. Rhinophores lamellated, sometimes sheathed. Feeding on encrusting and erect bryozoans.

---

1. Frontal veil with digitiform or tubercular papillae **2**

   Frontal veil smooth, dorsum with black and yellow spots **_Thecacera pennigera_**

2. Frontal veil papillae long and digitiform **3**

   Frontal veil papillae short **5**

3. Frontal veil papillae smooth, tapering. With paired, smooth, digitiform processes flanking gills **4**

   Frontal veil papillae feathery, mantle with paired digitiform processes **_Limacia clavigera_**

4. Usually four, sometimes six, frontal veil papillae. Dorsal oval blotches of orange on slightly raised mounds **_Polycera quadrilineata_**

   Minimum of eight frontal veil papillae. Yellow-orange blocthes, if present, not on mounds **_Polycera faroensis_**

5. Body bright orange with blue spots **_Greilada elegans_**

   Body brownish green, with round, yellow tubercles on dorsum **_Palio dubia_**

Polycera faeroensis Lemche FIG. 12.27

Body up to 45 mm; translucent white, with bright yellow pigment covering most of frontal veil papillae and distal ends of pallial processes. Pallial margin between head and gills usually yellow edged; metapodium with yellow crest, merging at tip with yellow foot margins. Distal portion of rhinophores and tips of gills and propodial tentacles also yellow; dorsum sometimes with yellow patches. Usually eight frontal veil papillae, sometimes with smaller, additional papillae. One pallial process on each side of gills, usually simple, rarely compound with two or three points. Gills simple, five large, two small; posterior, surrounding anal papilla. Rhinophores with large stalks and bulbous lamellated region, with a relatively small, yellow, terminal knob.

Despite variation in coloration, *P. faeroensis* is immediately distinguished from *P. quadrilineata* by its larger number of frontal veil papillae. Feeds on encrusting and erect bryozoans, from shallow sublittoral to 35 m. South and west coasts, most common in extreme south-west.

Polycera quadrilineata (Müller) FIG. 12.27

Body up to 39 mm; translucent white with patches of yellow and orange, of varying shade, typically with a row of oval blotches along the mid-line, merging posteriorly with a line along the mid-line of the distal metapodium; occasionally with black streaks, blotches or fine spots on the dorsum. Pallial margin edged with yellow-orange, usually with spots on flank. Edge of foot also usually dotted with colour. Four frontal veil papillae, rarely more, coloured yellow. Pallial process on each side of gills elongate, pointed, longer and thinner than in *P. faeroensis*; up to 11 pinnate gills. Distal tips of pallial processes, and gill rachides, yellow. Rhinophores with thick stalks, scarcely narrower than lamellate portions, which are held angled to stalks, terminated with cylindrical knob; distal portion of rhinophore yellow.

Occurs intertidally, and sublittorally to about 60 m, on encrusting Bryozoa. All British coasts. Some specimens may have the body almost entirely black, or grey with black speckles.

Palio dubia (M. Sars) FIG. 12.27

Body up to 30 mm, but commonly less than 20 mm, inconspicuous; yellow to dark greenish-brown, with numerous pale yellow or fawn tubercles. Foot light-coloured, produced anteriorly into digitiform propodial tentacles. Rhinophores fawn, equally divided into stalk and conical lamellate region, latter with darker spots of pigment at base. Five white, pinnate gills; pallial margin on each side of gills produced into a series of backwardly projecting lobes. Pallial margin and lobes pale coloured. Body with high profile, almost square-sectioned in anterior view; pallial margin extending around base of rhinophores on each side, and anteriorly, almost vertically not quite meeting medially at the head.

Sublittoral, to 100 m, but sometimes found in rock pools. Feeds on encrusting Bryozoa. Locally distributed, all British coasts.

Palio nothus (Johnston) has been reported from shore and shallow sublittoral habitats all around the British Isles. It is distinguished from *P. dubia* by its more oval head, and by its distinctly club-like dorsal processes. Thompson and Brown (1984) and Thompson (1988) discuss the morphological differences between the two species.

Greilada elegans Bergh FIG. 12.27

Body up to 43 mm; spectacularly coloured, bright orange with iridescent blue spots (rarely with white spots). Frontal veil produced anteriorly into small digitiform papillae. Pallial margin tuberculate, of contrasting colour, most usually white, fusing posterior to gills to form a keel; metapodium also keeled, with just a few blue tubercles, blue mounds on flanks also continuing onto metapodium. Rhinophores large, lamellate portion angled to stalk, same colour as body. Up to seven large, feathery gills, anterior to anal papilla, each white edged, with blue dots on inner surfaces. No elongated pallial processes adjacent to gills.

Greilada elegans feeds on the bryozoan *Bugula*, especially *B. turbinata*. It has been reported only from south-west England and the west coast of Ireland, from shallow sublittoral habitats.

Thecacera pennigera (Montagu) FIG. 12.27

Body up to 30 mm, white with dense, small black and yellow spots, and larger, irregular orange blotches. Three to five pinnate gills, posterior to which are two long, white-tipped digitiform processes. Rhinophores lamellate, with distinctive smooth-rimmed sheaths produced posteriorly, and with lateral, median openings. Head rounded, lacking oral tentacles but with small, paired propodial tentacles.

An inconspicuous species known to feed on the bryozoan *Bugula plumosa*, on which it also lays its spawn. Shallow sublittoral. rare, though locally abundant; recent records are from Dover, south Cornwall and the Bristol Channel.

Limacia clavigera (Müller) FIG. 12.27

Body up to 18 mm, white with yellow or orange dorsal papillae; rhinophores, gills, processes and cerata also tipped with yellow or orange. Pallial margin produced to form a paired series of projecting, pallial cerata; head fringed with similar, smaller processes with feathery ends. Rhinophores long, lamellated, with short stalks issuing from low sheaths with thickened rims; oral tentacles long, conspicuous, with dorsal grooves. Three to five pinnate gills situated anterior to anal papilla. Foot white except at tip where marginal and median pigment bands merge.

Shallow sublittoral, to 80 m, on encrusting bryozoans, including *Membranipora, Electra, Cryptosula, Callopora*, and *Schizoporella*. All British coasts, except for the extreme south-east.

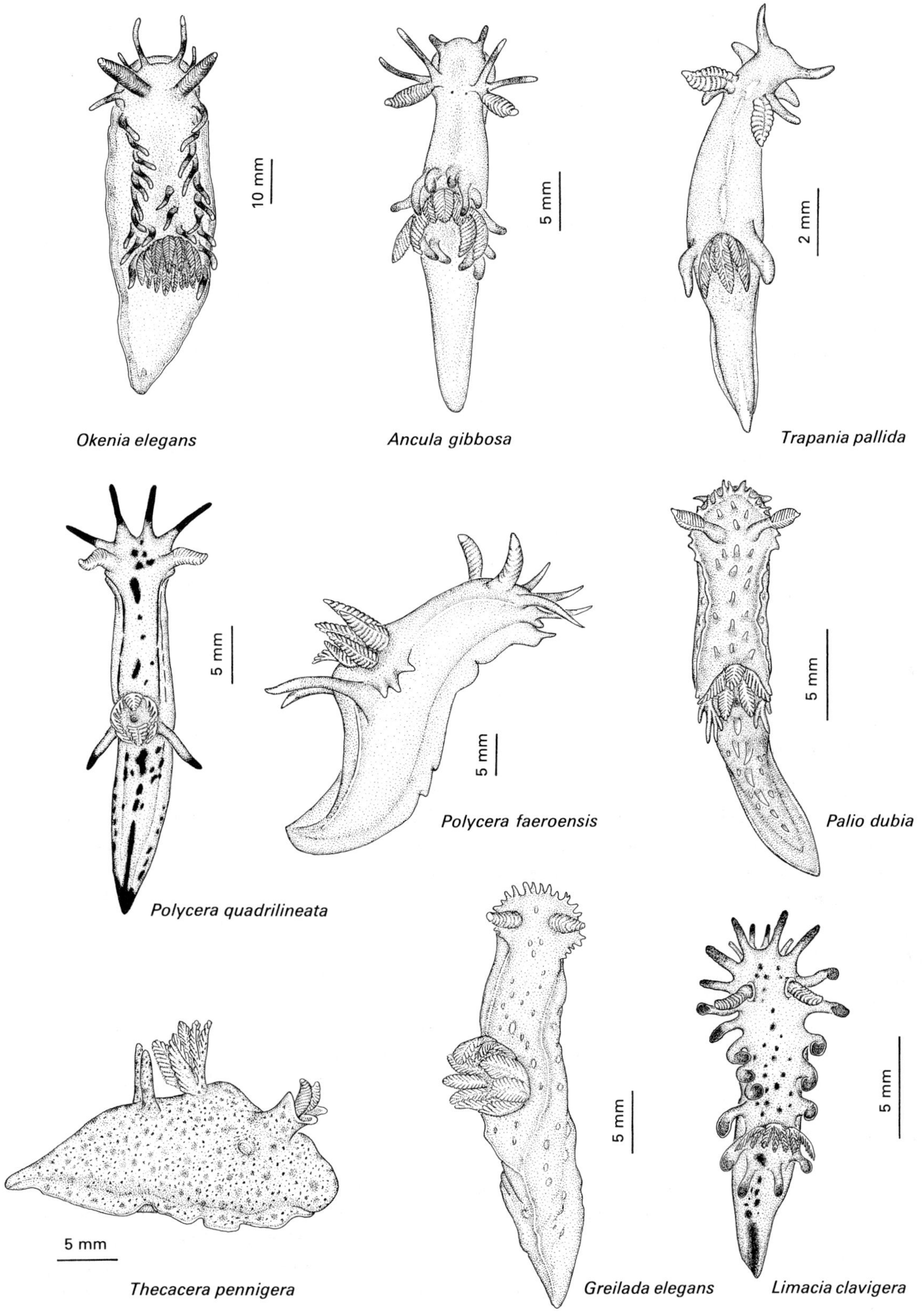
10 mm
Okenia elegans
5 mm
Ancula gibbosa
2 mm
Trapania pallida
5 mm
Polycera quadrilineata
5 mm
Polycera faeroensis
5 mm
Palio dubia
5 mm
Thecacera pennigera
5 mm
Greilada elegans
5 mm
Limacia clavigera

Fig. 12.27

21 TRIOPHIDAE

Phanerobranchiate, limaciform, soft; mantle reduced to form tuberculated ridge, continuous anteriorly; tubercles often arborescent. Similar processes sometimes occur on flanks and dorsum. Rhinophores lamellated; low simple sheaths. Bipinnate or tripinnate gills forming arc around anal papilla. Feeding on encrusting bryozoans and sponges. One British species.

Crimora papillata Alder and Hancock FIG. 12.28

Body up to 35 mm; translucent white, with numerous pale yellow to bright orange, forked papillae, distributed around frontal veil and pallial margin, and on the dorsum, flanks and metapodium. Rhinophores tapering to smooth points, translucent white basally with orange lamellae; sheaths low, distinct, with thickened rims. Three to five pinnate gills; white with orange tips, to almost completely orange.

Sublittoral, to 30 m, on the foliaceous bryozoans *Flustra foliacea* and *Chartella papyracea*. South-western England and western Ireland, locally abundant.

22 NOTODORIDIDAE

Phanerobranchiate, limaciform, firm; mantle reduced and indistinct. Dorsal surface tuberculate. Rhinophores smooth; sheaths raised (*Aegires*) or expanded (*Notodoris*). Simple, bi- or tripinnate gills form an arc around the anal papilla. Feeding upon encrusting bryozoans and sponges. One British species.

Aegires punctilucens (D'Orbigny) FIG. 12.28

Body up to 20 mm; fawn with white flecking and brown spots, with numerous pedunculate tubercles, each light brown at tip with dark brown or black central spot. Ocelli distributed between tubercles: oval, light brown discs with borders of dark brown spots and iridescent turquoise or blue centre. Rhinophores smooth, blunt, light brown with white flecks, tips white with dark brown transverse band; sheaths with tuberculate rims. Three pinnate gills anterior to anal papilla, white with brown markings.

Sublittoral, to 100 m, among clumps of the calcareous sponge *Leucosolenia*. All British coasts except south-west.

23 CADLINIDAE

Cryptobranchiate (gills retractile simultaneously into common branchial pit), broad with ample mantle skirt. Rhinophores lamellate without sheaths. Propodium bilaminate. Feeding upon encrusting sponges. One British species.

Cadlina laevis (Linnaeus) FIG. 12.28

Body up to 32 mm; translucent white, with opaque white (rarely yellow) edge to mantle. Dorsum with opaque, conical tubercles of varying size; mantle skirt ample, with distinctive, glistening, white or yellow subepidermal glands. Rhinophores short, tapered, white, often with yellow tips; sheaths low with crenulate rims. Five, rarely six or seven, tripinnate gills.

Intertidal and shallow sublittoral, feeding on sponges (including *Hemimycale columella*). Predominantly northern but recorded from all British coasts.

24 ROSTANGIDAE

Cryptobranchiate; broad, ample mantle skirt. Dorsum finely tuberculate. Rhinophores lamellate. Propodium bilaminate, anterior one notched medially. Feeding upon encrusting sponges. One British species.

Rostanga rubra (Risso) FIG. 12.28

Body up to 15 mm (in Mediterranean specimens larger), but commonly less; bright red, with scattered black spots and a rough, yellowish patch between rhinophores. Mantle coarsely tuberculate. Head with short, lateral oral tentacles; rhinophores short and swollen, cream coloured, often with red spots on rachides. Up to ten short, pinnate gills.

Intertidal and shallow sublittoral, feeding on red sponges such as *Microciona atrasanguinea* and *Ophlitaspongia seriata*. West coasts only.

25 ARCHIDORIDIDAE

Cryptobranchiate; broad, ample mantle. Dorsum tuberculate, some tubercles coalescing. Rhinophores lamellated. Sheaths sometimes present. Propodium bilaminate. Feeding upon encrusting sponges.

Archidoris pseudoargus (Rapp) FIG. 12.28

Body up to 120 mm long, oval, firm; variously mottled and blotched with yellow, green, brown, and red. Mantle skirt ample, hiding foot; coarsely tuberculate, tubercles densely spaced, rounded and of varying sizes. Head with vestigial oral tentacles; rhinophores short, conical, pale yellow. Up to nine tripinnate gills, translucent, with brown pinnules

Juveniles tend to be nondescript creamy yellow, but always with purple-brown patch of pigment between the tubercles, anterior to the gills. Dark specimens may be confused with *Discodoris planata* (below), which is distinguished, however, by white, stellate dorsal markings, and small but distinct oral tentacles.

The most common, and perhaps familiar, of the British nudibranchs, popularly termed the Sea Lemon. Common on lower shore, and offshore to 300 m; feeds mainly on the sponge *Halichondria panicea* or, at greater depths, on other encrusting sponges. All British coasts.

26 DISCODORIDIDAE

Cryptobranchiate; broad, ample mantle. Dorsum tough, tubercles spiculose. Rhinophores lamellated. Head with digitiform oral tentacles, bilaminate, anterior lamina usually notched. Feeding upon encrusting sponges.

Discodoris planata (Alder and Hancock) FIG. 12.28

Body up to 65 mm, oval; predominantly brown, reddish- or purplish-brown. Dorsum covered with small spiculose tubercles and up to 12 white or pale, stellate patches of acid-secreting glands, in a roughly paired series. Head with small, but well-developed oral tentacles. Rhinophores short, pointed, mottled light and dark brown, often with opaque white or golden tips; sheaths tuberculate, with pale rims.

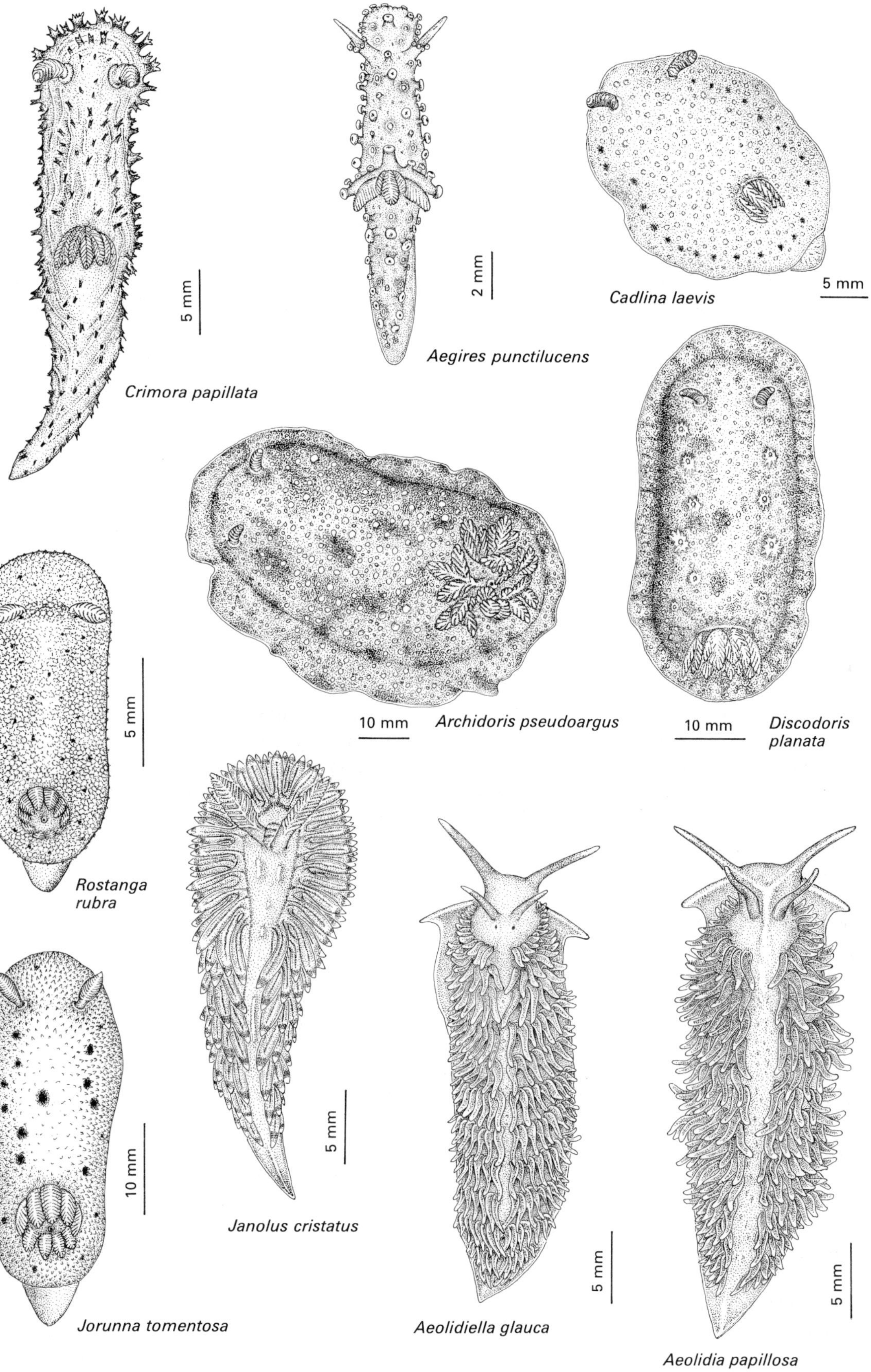

5 mm
Crimora papillata
2 mm
Aegires punctilucens
Cadlina laevis
5 mm
10 mm
Archidoris pseudoargus
10 mm
Discodoris planata
5 mm
Rostanga rubra
10 mm
Jorunna tomentosa
5 mm
Janolus cristatus
5 mm
Aeolidiella glauca
5 mm
Aeolidia papillosa

Fig. 12.28

Shallow sublittoral, feeding on sponges, including *Hemimycale columella*. Western coasts only, uncommon.

27 KENTRODORIDIDAE

Cryptobranchiate; broad, ample mantle skirt. Dorsum finely tuberculate. Rhinophores lamellated. Head with short oral tentacles; propodium bilaminate, anterior one notched. Feeding upon encrusting sponges. One British species.

Jorunna tomentosa (Cuvier) FIG. 12.28

Body up to 55 mm, elongate oval, posterior portion of foot visible in dorsal view; colour variable, cream, fawn, or sandy brown, almost always with a loosely paired series of dark brown blotches, as well as speckles of light brown. Mantle covered with small, uniform, spiculose tubercles, with a soft, velvety texture; rhinophore and gill pockets slightly raised. Head with slender oral tentacles; rhinophores darker or lighter than mantle, with brown flecks, but always white tipped. Up to 17 tripinnate gills, pale coloured.

Lower shore and shallow sublittoral, feeding on a number of sponges including *Halichondria panicea* and species of *Haliclona*. All British coasts, widespread and common.

28 JANOLIDAE

Elongated, with tapering tail. Cerata present; digitiform, continuous around anterior. Rhinophores lamellate (*Janolus*) or wrinkled and papillate (*Proctonotus*).

Janolus cristatus (Delle Chiaje) FIG. 12.28

Body up to 80 mm, tapered posteriorly, with thickly distributed lateral cerata, continuous around the whole of the head; transparent, with white patches on dorsum; cerata with brown central streak of digestive gland, tipped with bluish-white, often with brilliant metallic or iridescent sheen. Rhinophores large, light brown or orange, with white tips; linked basally by a wrinkled, light brown or orange, sensory caruncle.

Offshore, to 40 m; feeding on the bryozoans *Bugula* (including *Bugula turbinata*), *Callopora*, *Cellaria*. A beautiful and conspicuous species; locally common on all British coasts, except the southeast.

29 AEOLIDIIDAE

Nudibranchs with cerata in rows; cerata contain hepatic lobes of the digestive system and cnidosacs at the tips, where they store unfired nematocyts obtained from their prey. Propodial and long, agile, oral tentacles present. Rhinophores smooth (British species). Feeding on actinians (British species).

1. Body large (up to 120 mm); colour variable, with a white band spreading over head to oral tentacles; rhinophore tips pale. A median dorsal zone lacking cerata **_Aeolidia papillosa_**

   Body up to 40 mm; no white pattern on head; rhinophore tips pigmented white. More substantially covered with cerata **_Aeolidiella glauca_**

Aeolidia papillosa (Linnaeus) FIG. 12.28

Body up to 120 mm, though commonly less; wide, flattened, tapered posteriorly, with prominent oral tentacles and rhinophores, and abundant, closely-set cerata; mid-dorsal region lacking cerata. Colour very variable: cream, fawn, brown, orange pink, or dark purple-brown; usually a distinctive, pale V-shaped patch on head, between rhinophores and oral tentacles. Propodial tentacles elongate, prominent. Rhinophores and oral tentacles speckled with brown, tipped with white; rhinophores usually darkest towards base. Cerata elongate, straight or with recurved ends, always tipped with white.

The largest North Atlantic aeolid, *A. papillosa* is widespread and common on all British coasts, on the shore and in the shallow sublittoral. It feeds on sea anemones, including species of *Actinia*, *Anemonia*, *Metridium*, *Sagartia*, and *Tealia*.

Aeolidiella glauca (Alder and Hancock) FIG. 12.28

Body up to 40 mm, broad, with prominent oral tentacles, and numerous cerata arranged in distinct rows; translucent white, pale yellow, grey-brown, or orange-pink, depending on diet. Dorsum, cerata, and tentacles covered with white spots or blotches. Cerata with creamy white tips, a pinkish subterminal tinge, and pale brown to greenish-brown digestive gland.

On the lower shore, and in the shallow sublittoral, feeding on various species of actinians including *Sagartia elegans*. All British coasts.

Very similar to *Aeolidiella glauca* are *A. sanguinea* (Norman) and *A. alderi* (Cocks). The first has so far only been recorded from the west coast of Ireland, and is characteristically orange or red. *A. alderi* is characterized by a white ruff of cerata just posterior to the rhinophores; in these the digestive gland extends only a little way into the cavity of the cerata, and the white cnidosacs are much larger.

30 EMBLETONIIDAE

Flattened aeolidaceans with swollen cerata in single series down each side of body. Ceratal tips rounded (British species); cnidosacs absent and replaced by terminal pads of nematocysts. Smooth rhinophores; dilated bilobed oral veil; propodial tentacles absent. Diet uncertain. One British species.

Embletonia pulchra (Alder and Hancock) FIG. 12.29

Body up to 7 mm, an atypical aeolid with few cerata, a bilobed oral veil, no oral tentacles; translucent white or pink with white epidermal speckling or blotches. Cerata in a single row of up to seven on each side. Digestive gland varying from pale yellow to orange, red and brown. Rhinophores smoothly tapering, widely spaced.

Shallow sublittoral; inconspicuous and frequently overlooked. *E. pulchra* is thought to feed on common shallow water hydroids, such as *Hydrallmania*, *Nemertesia*, *Dynamena*, and *Tubularia*. All British coasts.

31 EUBRANCHIDAE

Aeolidaceans with swollen cerata of varying shapes. Arranged in simple or branched rows. Smooth rhinophores and oral tentacles; propodial tentacles absent. Feeding on hydroids.

1. Cerata inflated **2**
   Cerata smooth and digitiform, or knobbly, up to ten rows, olive green or brown, sometimes banded ***Eubranchus cingulatus***
2. Cerata flattened, not in rows; each with white tip and yellow subterminal ring ***Eubranchus tricolor***
   Cerata not as described **3**
3. Cerata not obviously in rows, white with orange or yellow subterminal ring; rhinophores and oral tentacles orange tipped ***Eubranchus farrani***
   Not as described **4**
4. Maximum of ten cerata per side, sparse, urn-shaped, with brown pigment usually forming interrupted bands ***Eubranchus exiguus***
   Many cerata, in closely packed rows; white, mottled with brown and orange, with yellow subterminal ring ***Eubranchus pallidus***

Eubranchus cingulatus (Alder and Hancock) FIG. 12.29

Body up to 29 mm, though commonly much smaller; greyish brown with white spots, and variably developed spots, blotches or large patches of olive-green or brown, particularly noticeable between bases of cerata. Dorsum with up to ten rows of long, digitiform cerata; each row with up to 13 cerata. Slightly inflated, and appearing knobbly when contracted, but not distinctly tuberculate (cf. *E. doriae*). Tips of cerata transparent, with white subterminal band, followed by two or three incomplete bands of olive-green or brown; white spots usually present, digestive gland extensions white, buff, pale yellow, or pale brown. Rhinophores about twice length of oral tentacles; both tipped with white, with subterminal band of olive-green or brown, followed by sparse or dense speckles of white and olive or brown.

Eubranchus cingulatus feeds on the hydroid *Kirchenpaueria pinnata*. Records are at present unreliable through confusion with *E. doriae* (Trinchese) (Thompson and Brown 1976; Schmekel and Portmann 1982: both as *E. cingulatus*). This species is smaller than *E. cingulatus* (up to 8 mm long), similar in coloration, but with inflated, tuberculate cerata, in up to seven rows, each with at most eight cerata.

Eubranchus exiguus (Alder and Hancock) FIG. 12.29

Body up to 10 mm, translucent greyish- or yellowish-white with speckles or patches of olive-brown on oral tentacles, head, rhinophores, dorsum, and cerata. Opaque white dots scattered all over body. Metapodium long and flattened with definite edge between foot and mantle; translucent white with olive-brown marginal pigmentation. Rhinophores almost smooth, white-tipped, with olive-brown pigment, which may form a distinct band. Oral tentacles smooth, about half length of rhinophores, with similar pigmentation. Cerata few, usually ten or less, not in rows; variable shape, usually urn-like, but never knobbly; each with clear tip above olive-brown bands, tending to be reddish, or more brown, than other pigmented areas. Up to three colour bands present on each ceras. Hepatic lobes pale, often colourless.

This small, delicate species occurs on *Obelia* spp, *Laomedea flexuosa*, and other small, shallow water calyptoblastic hydroids, often together with *Tergipes tergipes*. All British coasts.

Eubranchus farrani (Alder and Hancock) FIG. 12.29

Body up to 20 mm, slender, with a maximum of five cerata, inflated and not flattened. This is a variable species, often confused with *E. tricolor*, and occurs in five colour forms; intermediate forms perhaps occur and juveniles may be completely colourless except for the hepatic contents:

1. White or greyish-white, with ends of cerata, rhinophores and oral tentacles suffused with orange or orange-yellow; tips clear, with white glands below. Scattered patches of orange and orange-yellow on dorsum and head, and sometimes on cerata.
2. Chocolate-brown, with or without orange bands on cerata, oral tentacles and rhinophores. Tips of cerata translucent, usually patches of orange on dorsum, white glands not usually visible.
3. Golden, yellow, or orange; cerata with translucent tips and white glandular rings. Body usually pale; cerata rather slender.
4. Translucent white; cerata with white cnidosacs and tips; hepatic lobes pale brown, buff or orange, clearly visible. With sparse or dense white speckling; rhinophores and oral tentacles often with orange pigment, cerata and body may have a few orange patches.
5. Translucent, with no coloured pigmentation; rhinophores and oral tentacles translucent, cerata with white tips. Brownish hepatic lobes visible in cerata.

E. farrani feeds on calyptoblastic hydroids, such as *Obelia*, attached to kelps; occurs in shallow, sublittoral water off all British coasts.

Eubranchus pallidus (Alder and Hancock) FIG. 12.29

Body up to 23 mm; translucent greyish-white, with markings varying from yellow-orange to orange, crimson, orange-brown, and dark brown; small white spots always present, sometimes forming larger patches. Cerata inflated, arranged in up to ten rows of seven; with pale or translucent tips, and subterminal white bands, often interspersed with, or partially obscured by, a yellow or gold ring; covered with superficial white specks and orange, red or brown patches or spots. Propodium rather wider than in other *Eubranchus* species. Oral tentacles translucent with white or clear tips, speckled with white; rhinophores about twice as long as oral tentacles, band of orange to red present subterminally.

Pale specimens occur and should not be confused with paler varieties of *E. farrani*. Unlike the latter, *E. pallidus* occurs offshore, on shell gravel and stones, feeding on various calyptoblastic and gymnoblastic hydroids; adults are found most often on *Tubularia*, juveniles on smaller hydroids attached to *Tubularia*. All British coasts.

Eubranchus tricolor Forbes FIG. 12.29

Body up to 45 mm, pale yellow or greyish-white. Cerata flattened, inflated, not arranged in rows; extensions of digestive gland so numerous (up to 150) that the cerata are flattened, with no pattern; unicolorous with body, with subterminal ring of yellow or orange-yellow, white cnidosac extending from tip to just beyond yellow ring. Digestive gland extensions smooth, brown or red-brown, usually darker brown or violet at tip, just below cnidosac. Rhinophores smooth, tapering, suffused with pale yellowish-brown, except at tips.

Offshore, to about 80 m, on various hydroids, especially *Obelia geniculata*. All British coasts.

32 FACELINIDAE

Nudibranchs with regular rows of tapering cerata. Long lamellated rhinophores, lamellae as rings or meeting obliquely. Propodial tentacles and long mobile oral tentacles both present. Propodium bilaminate, anterior lamina notched. Varied diet.

1. Typically with blue iridescence over head and cerata; subterminal white on cerata extending basally on anterior side. Digestive lobules darker, or more reddish, at distal end **_Facelina coronata_**

   Translucent white body, with rose tinge. Cerata long, overlapping, not in clusters. White patch between rhinophores **_Facelina bostoniensis_**

   White speckling on dorsum and cerata; the latter tipped with white. **_Facelina annulicornis_**

Facelina bostoniensis (Couthouy) FIG. 12.29

Body up to 55 mm, translucent white, tinged with rose-pink; a white patch between rhinophores, others may be present on head and pericardium; oesophagus usually clearly visible, red. Oral tentacles up to half body length, distal third sometimes white; rhinophores lamellated, tips with white pigment. Cerata arranged in loosely defined clusters, anterior ones overlapping the posterior; translucent, with distinct subterminal white areas, hepatic lobe light to dark brown, or greenish-brown, clearly visible. Foot wider than body.

Usually sublittoral, on *Tubularia*; intertidal on *Clava multicornis*, but will feed on a variety of hydroids. All British coasts except extreme south-east.

Facelina coronata (Forbes and Goodsir) FIG. 12.29

Body up to 38 mm; translucent white with pink tint, usually with a beautiful blue iridescence on head and cerata. Rhinophores half length of oral tentacles, light to dark brown, with ring-like lamellae; anterior surfaces, and distal parts of oral tentacles, with superficial white pigmentation. Red oesophagous always visible posterior to rhinophores. Cerata in distinct rows, with gap between rows one and two, iridescent blue with white subterminal bands, extending basally on anterior side. Digestive lobules clearly visible, thin or thick, light brown to dark brownish-red, darkening or reddening distally. Foot little wider than body.

Intertidal and shallow sublittoral, on various hydroids, especially *Tubularia indivisa*. All British coasts, except south-east.

Facelina annulicornis (Chamisso and Eysenhardt)

Up to 43 mm in length. Rhinophores obliquely lamellated. Similar to other *Facelina* species but distinguished by white spots generally distributed on dorsum and cerata, which are tipped with white.

33 FAVORINIDAE

Similar to Facelinidae but with cerata in arcs or horseshoe shapes, not rows. Rhinophores smooth or with one to three bulbous swellings. Recurved propodial tentacles; large flexible oral tentacles present. Diet varied.

1. Rhinophores with three bulbous swellings near base; white pigment on oral tentacles extending on to head and dorsum as a streaky line **_Favorinus blianus_**

   Rhinophores with subterminal swelling and brown pigment below swelling; white pigment on propodial tentacles and head, forming a rhombus shape posterior to rhinophores **_Favorinus branchialis_**

Favorinus blianus Lemche and Thompson FIG. 12.30

Body up to 30 mm, pale straw-yellow. Propodial tentacles long, tapered, curved posteriorly, with superficial white pigment; oral tentacles long, tapering, with white pigment continuous on to head, and extending posteriorly as a streaky line on the dorsum, and on to the metapodium. Rhinophores yellowish, with three bulbous swellings near base, a dark brown stripe on posterior side broadens over each swelling. Cerata digitiform, curved towards centre of body, arranged in horseshoe pattern, the first placed very close to rhinophores; white tipped, with white streaks distally on anterior side, hepatic lobes pale yellow brown.

On lower shore, and subtidally to about 35 m; it is known to feed on *Tubularia larynx* and the eggs of other nudibranchs. Western coasts of Britain and Ireland.

Favorinus branchialis (Rathke) FIG. 12.30

Body up to 15 mm, delicate, slender; white, digestive gland varying in colour according to prey. Cerata long and tapering, white tipped with subterminal band, often olive-green or brown, covered with white spots; up to six clusters of six or seven cerata, clusters V-shaped anteriorly and approximately linear posteriorly. Propodial tentacles digitiform, curving posteriorly; oral tentacles long and graceful, subterminal half white. Rhinophores slender, smooth, with translucent white, elongate tip, a cream or pale yellow subterminal swelling (rarely two), and

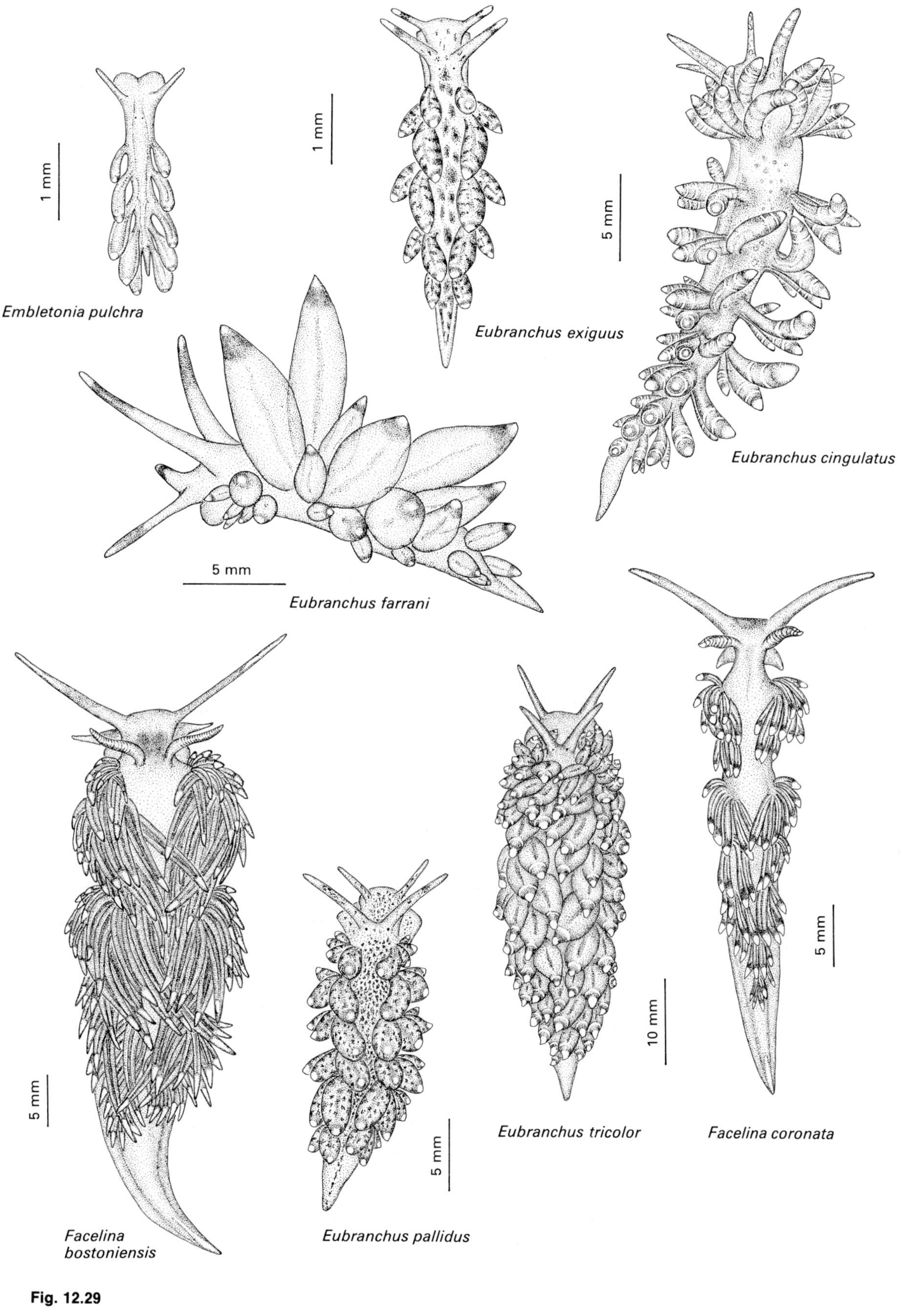
1 mm
Embletonia pulchra
1 mm
Eubranchus exiguus
5 mm
Eubranchus cingulatus
5 mm
Eubranchus farrani
5 mm
Facelina bostoniensis
5 mm
Eubranchus pallidus
10 mm
Eubranchus tricolor
5 mm
Facelina coronata

Fig. 12.29

brown basal pigmentation. White pigment between oral tentacles, and dorsal part of propodial tentacles, extends posteriorly between rhinophores, widens to rhombus shape posterior to rhinophores, and narrows again continuing as a line down the dorsum to the metapodium. The distribution of superficial brown and white pigment varies; juveniles may lack both pigment and rhinophoral swellings.

Lower shore and shallow subtidal, to about 20 m; usually feeding on eggs of other opisthobranchs, but also on calyptoblastic hydroids such as *Obelia geniculata* and *Sertularia argentea*. On western coasts, extending farther to north and south than *F. blianus*.

34 FLABELLINIDAE

Aeolidaceans with cerata in rows, either on a notal ridge or on slight expansions of the pallial ridge (British species), which separates the sides from the dorsum. Rhinophores wrinkled or smooth (British species), distinct propodial tentacles; long, mobile oral tentacles. Feeding on hydroids.

1.	Body (excluding cerata) purple	***Coryphella pedata***
	Body (excluding cerata) translucent white	**2**
2.	Body with opaque white lines	**3**
	Body not as described	**4**
3.	Opaque lines on dorsum, sides, oral tentacles, rhinophores, usually on cerata	***Coryphella lineata***
	Opaque line down dorsum	***Coryphella verrucosa***
4.	Cerata long, tapering, with white tips; rhinophores smooth	***Coryphella pellucida***
	Not as described	**5**
5.	Body length up to 15 mm, narrow subterminal band on cerata; well-developed pallial ridge	***Coryphella gracilis***
	Body length up to 50 mm, broad subterminal band on cerata; pallial ridge not obvious	***Coryphella browni***

Coryphella browni Picton FIG. 12.30

Body up to 50 mm, pellucid white, with an interrupted streak of white pigment on dorsum from posterior cerata to tip of metapodium. Oral tentacles long, rhinophores wrinkled, both streaked with white on distal third; propodial tentacles pointed and grooved. Up to eight clusters of cerata in transverse rows: seven precardial and 15 postcardial, each row with up to five cerata. Only the first two groups of cerata are clearly separated by a pallial ridge, which disappears between the other clusters. A broad white subterminal band encircles the cnidosac on each ceras (cf. *C. gracilis*); digestive gland varies from bright red to chocolate brown.

Frequently in large numbers on the hydroid *Tubularia indivisa*, often with *C. lineata*. All British coasts.

Coryphella gracilis (Alder and Hancock) FIG. 12.30

A small *Coryphella*, maximum body length 15 mm; translucent white, more opaque when developing ova are present. Up to six clusters of cerata, all originating from a distinct pallial ridge; each ceras with a thin band of white pigment subterminally (cf. *C. browni*), digestive gland varies from bright red to bright green. Propodial tentacles rather thick; oral tentacles smooth, rhinophores slightly wrinkled, both streaked distally with white. Metapodium also with white streak.

C. gracilis is known to feed on *Eudendrium ramosum*, and possibly feeds also on *Clytia johnstoni* and *Halecium* spp. In shallow waters, all British coasts.

Coryphella lineata (Lovén) FIG. 12.30

Body up to 50 mm, translucent white. Oral tentacles long, smooth, rhinophores slightly longer, wrinkled; both colourless except for a thin white stripe running from white tips. Stripes from oral tentacles meet on the head and continue posteriorly along the dorsum to the tip of the colourless metapodium. Lateral white stripes originate anterior to rhinophores and unite with median stripe on metapodium. Propodium notched, transversely grooved; propodial tentacles moderately long. Cerata elongated, arranged in five to eight clusters on a distinct pallial ridge; anterior clusters with two parallel rows of cerata, posterior comprising single rows, or even single cerata. Each ceras with a subterminal ring of white extending basally as a longitudinal stripe, the anterior one bearing the thickest; white pigment sometimes poorly developed. Digestive gland lobes reddish-orange or reddish-brown, more rarely dark brown, crimson, or green.

Shallow sublittoral, between 20–40 m, preferred diet is *Tubularia indivisa*. Common on all British coasts.

Coryphella pellucida (Alder and Hancock) FIG. 12.30

Body up to 40 mm, slender, translucent white. Distinguished principally by its large number of cerata, up to seven per row. Cerata long, tapering, with crimson digestive gland lobes and white tips, no subterminal band. Oral tentacles and rhinophores long, smooth, tapering; white tipped, often with streaks of white extending basally. Foot as narrow as upper body, with recurved propodial tentacles; propodium grooved transversely. Anterior lamina notched medially.

Offshore, to 120 m, feeding preferentially on *Tubularia indivisa*. Northern coasts only.

Coryphella pedata (Montagu) FIG. 12.30

Body up to 48 mm; distinguished by its violet/magenta coloration. Cerata arranged in a maximum of seven clusters: first group of parallel rows, middle clusters in horseshoe shapes, and the posterior comprising single rows. All clusters arising from expansions of the pallial ridge, typically fewer cerata than in other British coryphellids. Each ceras with white subterminal band, digestive gland lobe pinkish-orange to red. Oral tentacles and rugose rhinophores darker violet than body, with white tips. Propodial tentacles rather long, tapering, recurved; paler than oral tentacles and rhinophores, sometimes tipped with white.

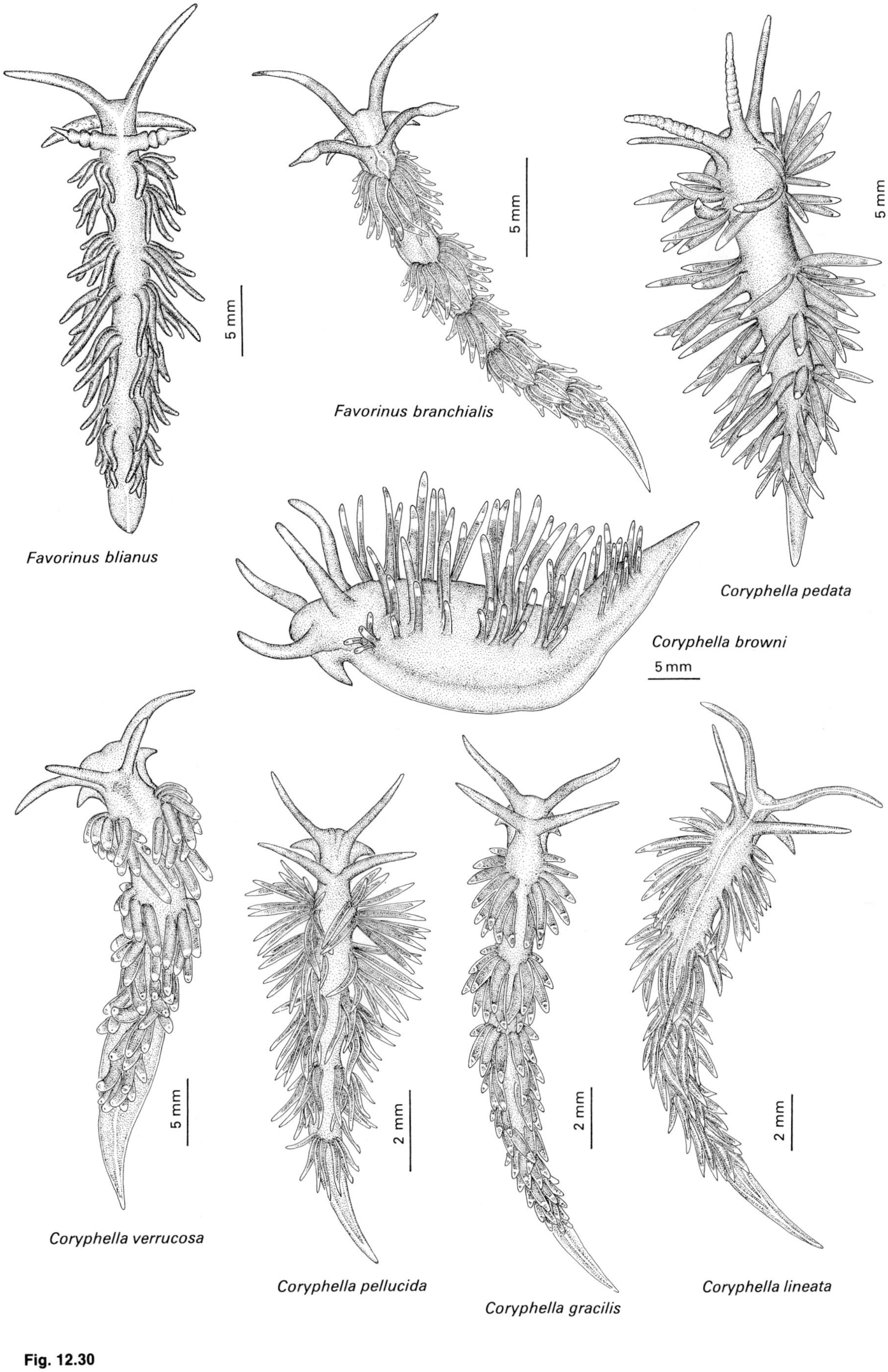
Favorinus blianus
5 mm
Favorinus branchialis
5 mm
Coryphella pedata
5 mm
Coryphella browni
5 mm
Coryphella verrucosa
5 mm
Coryphella pellucida
2 mm
Coryphella gracilis
2 mm
Coryphella lineata
2 mm

Fig. 12.30

Shallow sublittoral, to 40 m, usually on the hydroid *Eudendrium ramosum*. Common on all British coasts.

Coryphella verrucosa (M. Sars) FIG. 12.30

Body up to 35 mm, translucent white. Cerata stout, in five to seven clusters on an expanded pallial ridge, which is absent between clusters. Ceratal clusters more or less evenly spaced (cf. other British coryphellids) and quite tightly packed; each ceras with white subterminal band, digestive gland lobes light brown to crimson or maroon. A white dorsal stripe extends from pericardium to tip of metapodium. Rhinophores rugose, shorter than oral tentacles; both with white distal streaks. Propodium notched medially and transversely grooved, the groove extending onto the short, stout propodial tentacles.

Sublittoral, feeding on *Tubularia indivisa* and other hydroids, and on the scyphistomae of *Aurelia aurita*. A boreo-arctic species, occurring in the North Sea and on the coasts of Scotland.

35 Tergipedidae

Small aeolidaceans with cerata in even rows (reduced to single row in *Tergipes*); cerata clavate, digitiform, or fusiform. Propodial tentacles reduced or absent; oral tentacles smooth (reduced in *Tenellia*). Feeding on hydroids, *Cuthona nana* especially on *Hydractinia echinata*.

---

1. Head rounded, with short oral tentacles **Tenellia adspersa**
   Head with long oral tentacles **2**

2. Cerata sparse: arising singly down each side **Tergipes tergipes**
   Cerata in obliquely transverse rows **3**

3. Rhinophores and oral tentacles suffused with lemon yellow; cerata with pale subterminal band, second band transparent, green, turquoise, or blue, and third band yellow, orange, or red **Cuthona caerulea**
   Not as described **4**

4. Rhinophores banded with orange, crimson, olive, or brown **5**
   Rhinophores otherwise pigmented **7**

5. Rhinophores with crimson band and white streak; oral tentacles with solid white streak **Cuthona rubescens**
   Rhinophores and oral tentacles banded with orange, red, olive, or brown **6**

6. Tentacles with olive or brown bands and white tips; no pigment on body other than white speckling **Cuthona amoena**
   Tentacles with orange or red bands; crescent-shaped patches of orange, red, or brown between rhinophores to first cerata (sometimes one between oral tentacles and rhinophores) **Cuthona foliata**

7. Rhinophores with white pigmented ends **8**
   Rhinophores orange or brown, darker than body, darkening distally **Cuthona gymnota**

8. Digestive gland, within cerata, dark green with black specks **Cuthona viridis**
   Digestive gland pink to brown **9**

9. Cerata numerous, extending in front of rhinophores; club-shaped with white tips; body suffused with pink posteriorly **Cuthona nana**
   Cerata fewer, clavate, colourless tip with white subterminal ring; digestive gland brown, possibly with a violet tinge **Cuthona concinna**

Cuthona amoena (Alder and Hancock) FIG. 12.31

Body up to 10 mm; with scattered, small, creamy white spots, often on raised tubercles, especially on head, between rhinophores, and on prominent pericardial swelling; with faint brown or olive-brown mottling. Rhinophores smooth, with opaque white tips and an olive or brown band (rarely two) subterminally, its width equal to its distance from tip; oral tentacles short, pigmentation similar to rhinophores. Three rows of cerata anterior to pericardial swelling, five rows posteriorly; each ceras with scattered reddish brown and white spots, darkening basally; small cnidosac visible at tip, usually a subterminal ring of red-brown spots. Digestive gland lobes sandy brown to olive green, often darkest basally.

Sublittoral, to 40 m, on *Halecium halecium* and *H. beani*. All British coasts.

Cuthona caerulea (Montagu) FIG. 12.31

Body up to 26 mm, translucent white to pale greenish, often with opaque white pigment in a streak, running dorsally and dividing around pericardial swelling. Smooth oral tentacles and longer rhinophores suffused with yellow, concentrated towards tips. Cerata bulbous or cylindrical, but always tapering to a point; up to 12 rows, with six per row; terminal cnidosac opaque white; often with an uneven white subterminal ring, always followed by a red, orange, or yellow band, and often then a blue, turquoise, or green band; a third coloured band matches the colour of the first but fades towards base of ceras. Digestive gland lobe brilliant green, shades of blue, brown, or almost black.

Sublittoral, to 270 m, on various hydroids, especially *Sertularella polyzonias* and *S. gayi*. On western coasts of Great Britain and Ireland.

Cuthona concinna (Alder and Hancock) FIG. 12.31

Body up to 12 mm, rarely larger; translucent white or cream, often with yellowish tinge, generally drab and inconspicuous. Rhinophores and oral tentacles translucent or colourless, with white tips. Cerata clavate, up to ten rows with maximum of five per row. Tip containing cnidosac colourless, white subterminal ring may be present, sometimes forming indistinct streak

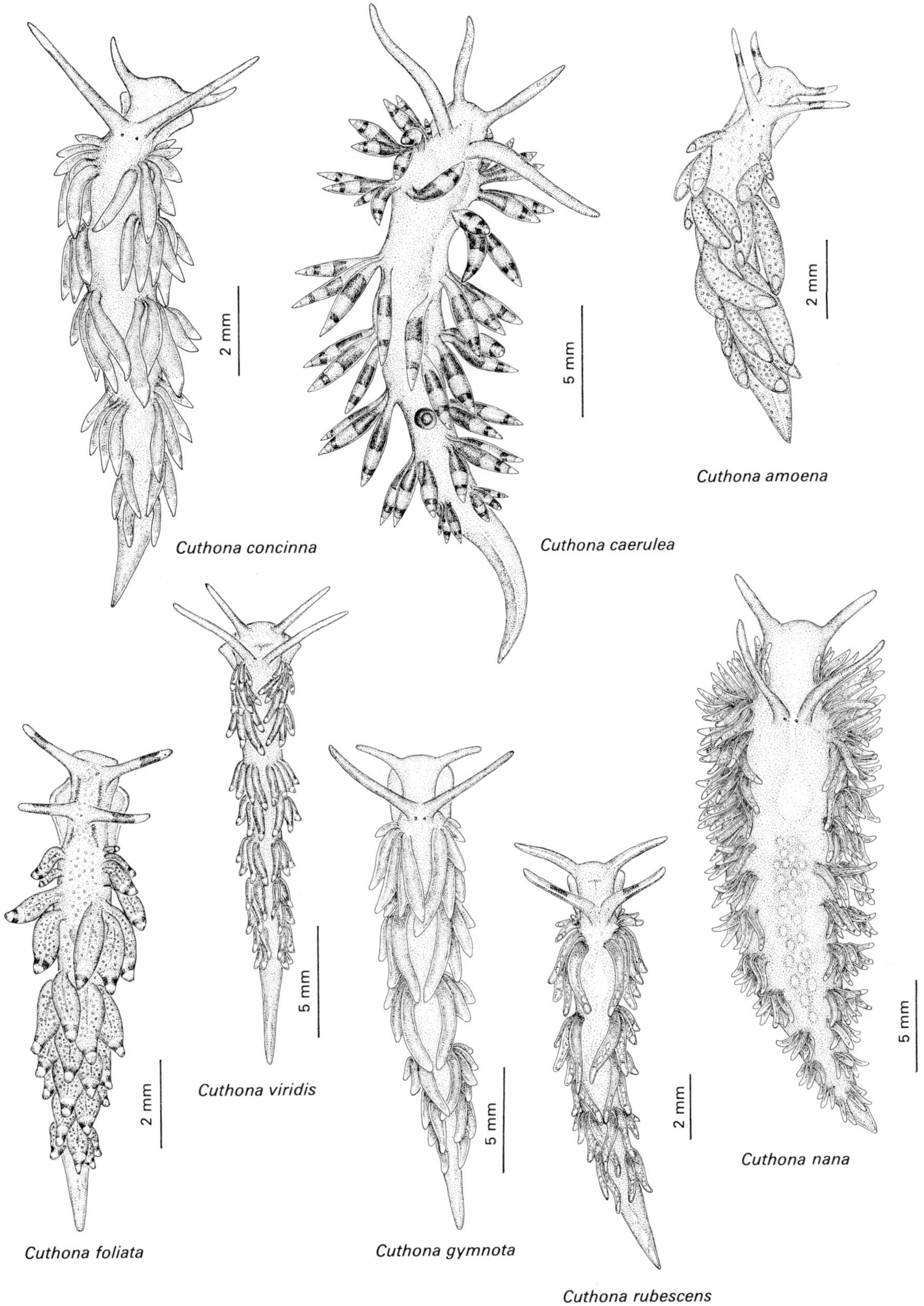
2 mm
Cuthona concinna
5 mm
Cuthona caerulea
2 mm
Cuthona amoena
2 mm
Cuthona foliata
5 mm
Cuthona viridis
5 mm
Cuthona gymnota
2 mm
Cuthona rubescens
5 mm
Cuthona nana

Fig. 12.31

towards base of ceras; digestive gland lobe dull brown, often with violet tinge.

Shallow sublittoral, to 25 m; diet includes *Laomedea gelatinosa, Obelia longissima, Sertularia cupressina*, and *S. argentea*. Predominantly northern distribution.

Cuthona foliata (Forbes and Goodsir) FIG. 12.31

Body up to 11 mm; yellowish, with distinctive crescentic patches of orange, red or brown posterior to rhinophores, often with a red or orange patch on each side of head between oral tentacles and rhinophores, and on dorsum posterior to prominent pericardial swelling. Rhinophores short and stubby, banded with orange or red; oral tentacles with less distinct banding, speckled with white dots, extending on to head. Cerata fusiform, in seven to eight rows, each of up to five cerata; subterminal yellow speckling, fading towards base, digestive gland lobe green, brown, or yellow-brown.

Intertidal and shallow subtidal, feeding on calyptoblastic hydroids, especially *Abietinaria abietina, Sertularella polyzonias*, and *S. gayi*. All British coasts.

Cuthona gymnota (Couthouy) FIG. 12.31

Body up to 22 mm, translucent white or colourless, or tinted orange or rose. Rhinophores deeper orange or brown, fading basally; oral tentacles may be pale or dark. Cerata digitiform or fusiform, rather stout; up to 12 rows, with up to seven cerata in the largest rows close to the pericardium. Each ceras with large white cap, and subterminal band of dull white, yellow, or orange; digestive gland lobe pinkish-orange to red-brown. Metapodium short and rounded.

Lower shore and shallow subtidal, to 30 m; preferred prey is *Tubularia indivisa* but on more exposed shores feeds on *T. larynx*. All British coasts, common.

Cuthona nana (Alder and Hancock) FIG. 12.31

Body up to 28 mm, translucent white sometimes suffused with pink posterior to pericardial swelling. Head broad; oral tentacles short, digitiform, often white distally; rhinophores smooth, bluntly pointed, also tending to whiten distally. Cerata abundant, more than 20 rows arranged in up to 16 clusters; club-shaped, white tipped, digestive gland lobes brown in starved specimens, otherwise tending to pink.

Intertidal and sublittoral, to 30 m, in sheltered waters; feeds predominantly on *Hydractinia echinata*. Recorded from scattered localities on all British coasts.

Cuthona rubescens Picton and Brown FIG. 12.31

Body up to 12 mm; translucent white with white specks, distinguished from *C. amoena* by white streaks on upper surfaces of oral tentacles, and by smooth, crimson-banded rhinophores, also streaked with white. Cerata cylindrical with narrow bases and rounded tips; up to 11 rows with maximum of six in longest rows; crimson basally, with white spots, often forming a ring at the level of the cnidosac.

Sublittoral, 10–30 m, probably feeds on *Halecium halecinum*. Uncommon; recorded from coasts of Devon, Pembrokeshire, Donegal, Orkneys, and Mull.

Cuthona viridis (Forbes) FIG. 12.31

Body up to 30 mm, semi-translucent white or pale yellow. Oral tentacles and rhinophores white distally. Cerata elongated, cylindrical, in up to nine rows of seven; cnidosacs white or pale yellow, almost filling tip of each ceras, a subterminal ring of white or pale yellow, streaking anteriorly in larger specimens. Diagnostic feature is the dark green digestive gland lobes, interspersed with black specks.

Intertidal, and offshore to 100 m. Feeds on a number of hydroids, including *Abietinaria abietina, Sertularella* spp., *Nemertesia* spp. and, intertidally, *Clytia johnstoni*, perhaps with a preference for *Sertularella rugosa*. All British coasts, except extreme south-east.

Tenellia adspersa (Nordmann) FIG. 12.32

Body up to 9 mm, pale yellow to dull brown with black speckling of varying density. Head domed and spatulate, oral tentacles, if present, very short; long oral tentacles typical of other British tergipedids lacking. Rhinophores smooth, tapered to blunt points, white distally. Cerata fusiform, in up to six rows, each of up to three cerata; pale yellow to orange; cnidosacs distinct close to tips; sometimes with subterminal band of white dots.

Intertidal and shallow sublittoral; euryhaline, often in harbours, estuaries and canals. Feeds on a wide range of hydroids, including *Laomedea longissima, L. loveni, Cordylophora lacustris, Protohydra leuckarti*. Rare, the few British records are from the Bristol Channel and from Fleet and Portishead, on the south coast, but it is a small and generally inconspicuous animal which may prove to have a wider distribution.

Tergipes tergipes (Forskål) FIG. 12.32

Body up to 8 mm; translucent white, with distinct greenish digestive gland; streaks of red brown originating on sides of head, at bases of oral tentacles, run posteriorly on each side; similarly coloured band extends from base of rhinophores on each side to meet lateral line. Up to eight, rarely ten, fusiform cerata in two single, longitudinal series; anterior two paired, successive ones alternating; each with relatively large white tip, reddish-brown subterminal ring overlying cnidosac.

Shallow sublittoral, to 20 m, on small calyptoblastic hydroids growing on kelps or moored structures. All British coasts, common.

E. Subclass **Pulmonata**

The pulmonate gastropods are primarily adapted to terrestrial and freshwater habitats, perhaps having evolved from a number of estuarine or supralittoral stocks. Pulmonates exhibit detorsion, lack ctenidia, and display a major modification of the mantle cavity which functions as a *pulmonary sac* or *lung*. The opening of the mantle cavity is reduced as a contractile *pneumostome*. A shell may be present, spirally coiled and large enough to accommodate the whole of the animal, or small and cap-like,

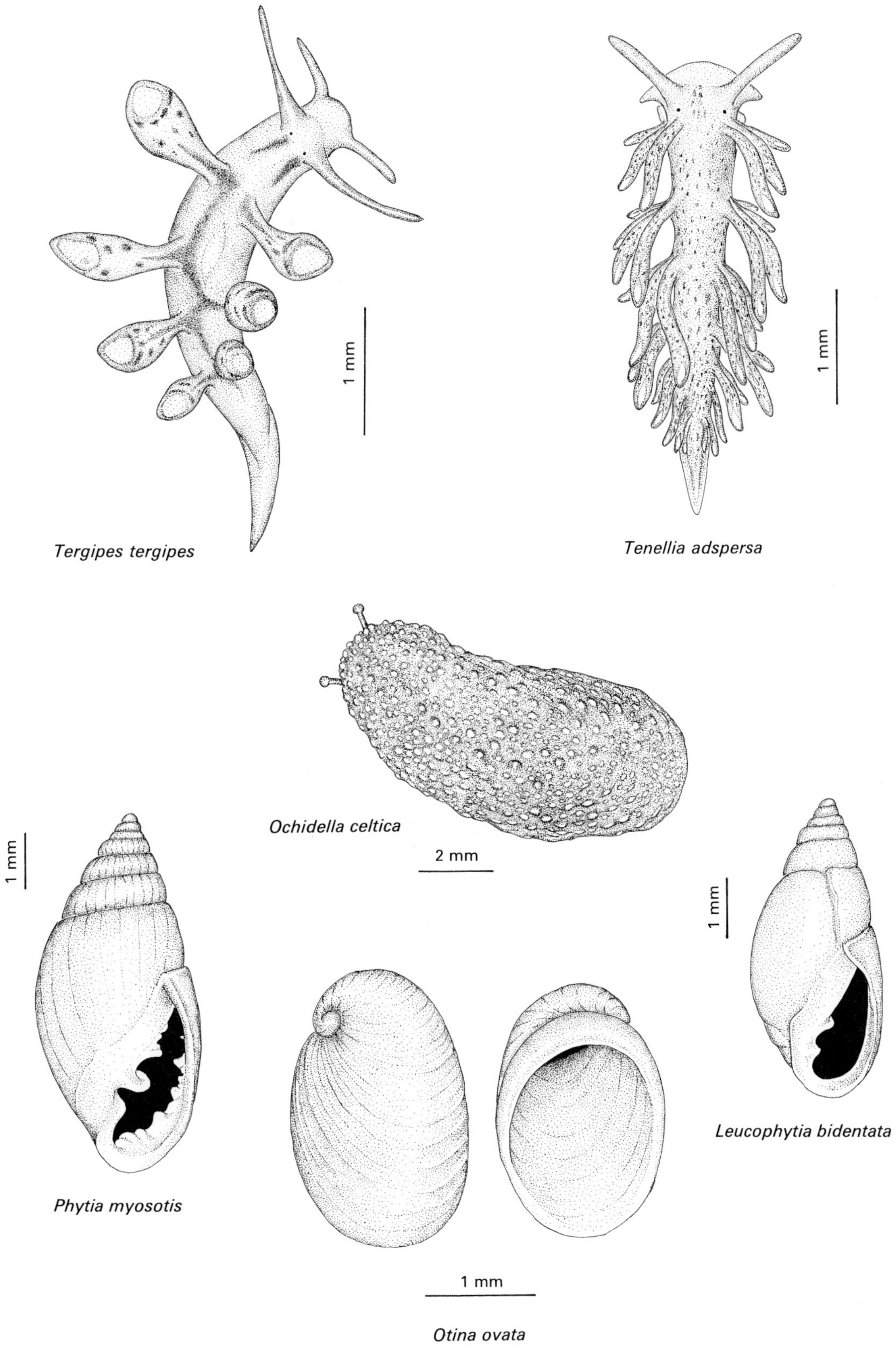
1 mm
Tergipes tergipes
1 mm
Tenellia adspersa
Ochidella celtica
2 mm
1 mm
Phytia myosotis
1 mm
Leucophytia bidentata
1 mm
Otina ovata

Fig. 12.32

and covering only the posterior portion of the body. Certain genera possess a large limpet-like external shell, while others have a small internal shell. Shelled pulmonates do not have an operculum. In a majority of land slugs a shell is lacking altogether.

Some groups of pulmonates have returned to marine habitats, and a few species may be found on rocky coastlines in all parts of the world. Snail-like, limpet-like, and slug-like genera occur, which may be confused with similar prosobranch or opisthobranch genera. A few representatives of marine pulmonate snails and slugs occur commonly on British shores.

1 ELLOBIIDAE

Shell fusiform, spirally coiled, with large last whorl and thickened, toothed aperture.

1. Shell slender; aperture with two teeth on columella, outer lip untoothed **_Leucophytia bidentata_**

   Shell broadly oval; aperture with three or more teeth on columella, outer lip with three or more teeth **_Phytia myosotis_**

Leucophytia bidentata (Montagu) FIG. 12.32

Shell rather solid, glossy white, slender fusiform; up to 6 mm, with six or seven whorls, the last whorl comprising about two-thirds of total shell height. Smooth, with no visible sculpture. Aperture elongate oval; outer lip thin, slightly flared abapically; inner lip thickened, with two strong folds on columella, projecting into aperture as well marked teeth.

Upper shore, in crevices and among seaweeds, particularly on sheltered estuarine shores. Reported from scattered localities around southern half of Britain, and on all Irish coasts. May be locally common.

Phytia myosotis (Draparnaud) FIG. 12.32

Shell thin, rather short and broad, greyish-brown; up to 8 mm, with seven whorls, the last whorl comprising about three-quarters of total shell height, smooth. Aperture oval, with distinctly thickened inner and outer lips, each bearing at least three distinct teeth.

Upper shore, particularly in estuaries, among seaweed, saltmarsh plants and rotting debris. Southern and western coasts only, locally common.

2 OTINIDAE

Shell cap-shaped, with minute spire of two whorls and a broad, shallow last whorl.

Otina ovata (Brown) FIG. 12.32

Shell minute, oval, up to 2 mm long, consisting largely of a shallow last whorl, and completely covering the body of the animal; reddish-brown to purple, smooth, with faint concentric growth lines.

Upper shore, in barnacle tests and among short, dense weeds such as *Lichina pygmaea*. Predominantly southern in distribution; perhaps locally common, but inconspicuous.

3 ONCHIDIIDAE

Pulmonate sea-slugs, lacking a shell, the head and foot hidden by a tuberculate mantle. Head with a pair of retractile tentacles bearing eyes at tips. Pneumostome posteriorly situated, close to anus.

Onchidella celtica (Forbes and Hanley) FIG. 12.32

Body solid and fleshy, oval, domed, up to 12 mm long, deep olive to black. Mantle covered with large, coarse, well-spaced tubercles, obscuring head and foot completely when the animal is motionless. Two short, thick tentacles, bearing eyes at tips, visible anteriorly when the animal is crawling.

Between tidemarks on exposed rocky shores, active during periods of total emersion, crawling on small algae. Retreats to crevices, upshore, as tide rises. Coasts of Cornwall and south Devon, locally abundant.

13 MOLLUSCA II
BIVALVIA AND CEPHALOPODA

PHYLUM MOLLUSCA

F. Class **Bivalvia**

THE bivalve molluscs are largely sedentary animals. They may be firmly attached to fixed substrata, or hidden in cracks and crevices, and some are found as borers in a variety of rocks, wood, or organic carbonates. Many simply burrow into fine or coarse bottom deposits, some lying unattached in the very top layers, others inhabiting more permanent burrows. With a few exceptions, such as the scallops and the file shells, mobility is limited to movement within the burrow or the immediate environment. Most bivalves are microphagous feeders although two small groups, the Protobranchia and the Septibranchia, are specialized, respectively, as particulate deposit feeders and carnivores. Adaptive radiation of this homogenous group is thus expressed largely in shell form.

Externally the bivalve body appears very simple compared to that of other molluscs (Fig. 13.1). There is no definite head; the gut and visceral mass are enclosed laterally by two flaps of body wall, the *mantle*; the space between the flaps is referred to as the *mantle cavity* and its most conspicuous contents are a pair of extensive, and often elaborately folded, *ctenidia* or gills. The mantle secretes the two lateral valves of the shell; its role is two-fold, producing both the calcium carbonate component and the organic matrix, *conchiolin*, on which the former is deposited. Dorsally, an uncalcified portion of conchiolin forms an elastic *ligament* which may be external or partly or wholly internal with respect to the valves. The elasticity of the ligament tends to force the valves apart and closure is achieved by antagonistic musculature, usually arranged as paired, internal *adductor muscles*. The outer surface of the shell is covered by a horny *periostracum*; this may be thin and smooth, or coarse, and often produced as periostracal spines. The muscular *foot* of bivalves is typically wedge-shaped. In active burrowers it plays an important role in probing and digging into the substratum, and in the highly specialized, fast burrowing Solenidae it forms a complex and functionally diverse organ. The two edges of the mantle flaps may be fused to a greater or lesser extent, and posteriorly they are often produced into tubular *siphons*. The presence, type, and degree of development of the siphons reflect the mode of life of the animal. They are short in shallow burrowing cockles, longer in the Veneracea; the dorsal, *exhalant*, and ventral, *inhalant*, tubes may be separate, elongate, and highly flexible in deposit feeding species, or partly or wholly fused, and in static deep burrowers may be relatively rigid.

Morphological features of the shell are widely used in classification and are the most useful features for the purposes of identification (Fig. 13.1). The bivalve shell is typically bilaterally symmetrical (*equivalve*), although in many families one valve is frequently smaller than the other (*inequivalve*). Growth proceeds by the deposition of calcite along the free edges of each valve and the oldest part of the valve is dorsal in position, where it forms a convex, and often prominent, *umbo*. The umbones of each valve may be spaced apart, with a dorsal *cardinal area* between; if close together, the cardinal area may be divided into an anterior *lunule* and a posterior *escutcheon*. However, many bivalves lack defined cardinal areas altogether. Each valve may be symmetrical about a mid-line drawn dorso-ventrally through the umbo (*equilateral*), more frequently the umbo is situated in front of or behind the mid-line (*inequilateral*). The two valves articulate along the dorsal *hinge line*, which frequently bears a number of interlocking teeth. These may be situated immediately below the umbo (*cardinal teeth*) or to either side of it (*lateral teeth*), and both types may be present in the same species. The *ligament* is always posterior to the umbones; it may be external to the hinge line or partly or wholly internal. The internal element is usually accommodated in a pit, or spatulate depression (*chondrophore*). The external surface of the valve may bear concentric rings or ridges marking successive growth increments; these should not be confused with concentric sculpture, of ridges or grooves, which is also present in many species. Radiating striations or *ribs*, often produced into sharp spines, may also occur. The internal surface of the valve displays some useful taxonomic features. Most prominent are the *adductor muscle scars*, one anterior and one posterior, the shape and relative proportions of these are often important. The attachment of the mantle to the inner surface of the valve is marked by another scar, the *pallial line*, extending between the two adductor scars. Posteriorly, this may show an indentation (*pallial sinus*) indicating the position of the retracted siphons in the closed shell.

In using the following keys allowance must be made for the fact that the more delicate morphological features of the shell may be worn or broken. It is also necessary to orientate the specimen correctly so that left and right valves may be distinguished; thus, the *pallial sinus* (when present) and the *ligament* are always *posterior* in position, *umbones* tend to be inclined *anteriorly* and, when *muscle scars* are unequal, the *anterior* is the smaller. The most highly modified shells will still be difficult to orientate and notes on these are introduced where appropriate in the text. Similarly, additional morphological features are displayed in Fig. 13.1, or introduced and defined in the keys and family descriptions. The nomenclature employed here, for the most part, follows that of Howson (1987).

REFERENCES

Fischer-Piette, E. (1975). Révision des Venerinae. *Mémoires du Muséum National d'Histoire Naturelle, Paris (Zoologie), Serie A*, **93**, 1–64.

Fischer-Piette, E. (1977). Révision des Cardiidae. *Mémoires du Muséum National d'Histoire Naturelle, Paris (Zoologie), Serie A*, **101**, 1–212.

Fischer-Piette, E. and Delmas, D. (1967). Révision des mollusques lamellibranches du genre *Dosinia* Scopoli. *Mémoires du Muséum National d'Histoire Naturelle, Paris (Zoologie), Serie A*, **47**, 1–91.

Fischer-Piette, E. and Metevier, B. (1971). Révision des Tapetinae (Mollusques, Bivalves). *Mémoires du Muséum National d'Histoire Naturelle, Paris (Zoologie), Serie A*, **71**, 1–106.

Howson, C. M. (1987). *Directory of the British Marine Fauna and Flora.* Marine Conservation Society, Ross-on-Wye.

Madsen, F. J. (1949). Marine Bivalvia. *Zoology of Iceland*, **4** (63): 1–116.

Seed, R. (1974). Morphological variations in *Mytilus* from the Irish coast in relation to the occurrence and distribution of *M. galloprovincialis Lmk. Cahiers de Biologie marine*, **15** (1), 1–25.

Tebble, N. (1976). *British bivalve seashells*, 2nd edn. Her Majesty's Stationery Office, Edinburgh.

Turner, R. D. (1966). *A survey and illustrated catalogue of the Teredinidae.* Museum of Comparative Zoology, Harvard, Cambridge, Mass.

KEY TO FAMILIES

1. Hinge line of shell with numerous small, identical teeth, alternating with sockets (Taxodont: Figs. 13.2, 13.3) **2**
   Hinge line with or without cardinal and/or lateral teeth, but not taxodont **6**

2. Shell equilateral, almost circular in outline. Light or dark reddish-brown, typically with zigzag patterns **4. Glycymeridae**
   Shell inequilateral, triangular, ovate, or irregular in outline **3**

3. Shell drawn out into a short, angular spout posteriorly; without a gape **2 Nuculanidae (*Nuculana minuta*)**
   Shell without a spout; if with angular ridges developing from the umbones, then with a ventral gape **4**

4. Ligament external, extending across a broad, grooved cardinal area between widely spaced umbones. Hinge teeth in a continuous series **3. Arcidae**
   Ligament internal, in a small indentation on the hinge line, separating the teeth in each valve into a posterior and an anterior series **5**

5. Shell minute (<5 mm), inequilateral, the posterior end slightly upcurved. Teeth in approximately equal numbers in the anterior and posterior series. Small pallial sinus present **2. Nuculanidae (*Portlandia philippiana*)**
   Shell often exceeding 10 mm, roughly triangular. Teeth in markedly unequal series, up to twice as many anteriorly as posteriorly. No pallial sinus **1. Nuculidae**

6. Hinge line of shell drawn out into projecting ears on one or both sides of the umbones. Typically fan-shaped, with radiating ridges (Fig. 13.6) **7**
   Hinge line not forming paired ears **9**

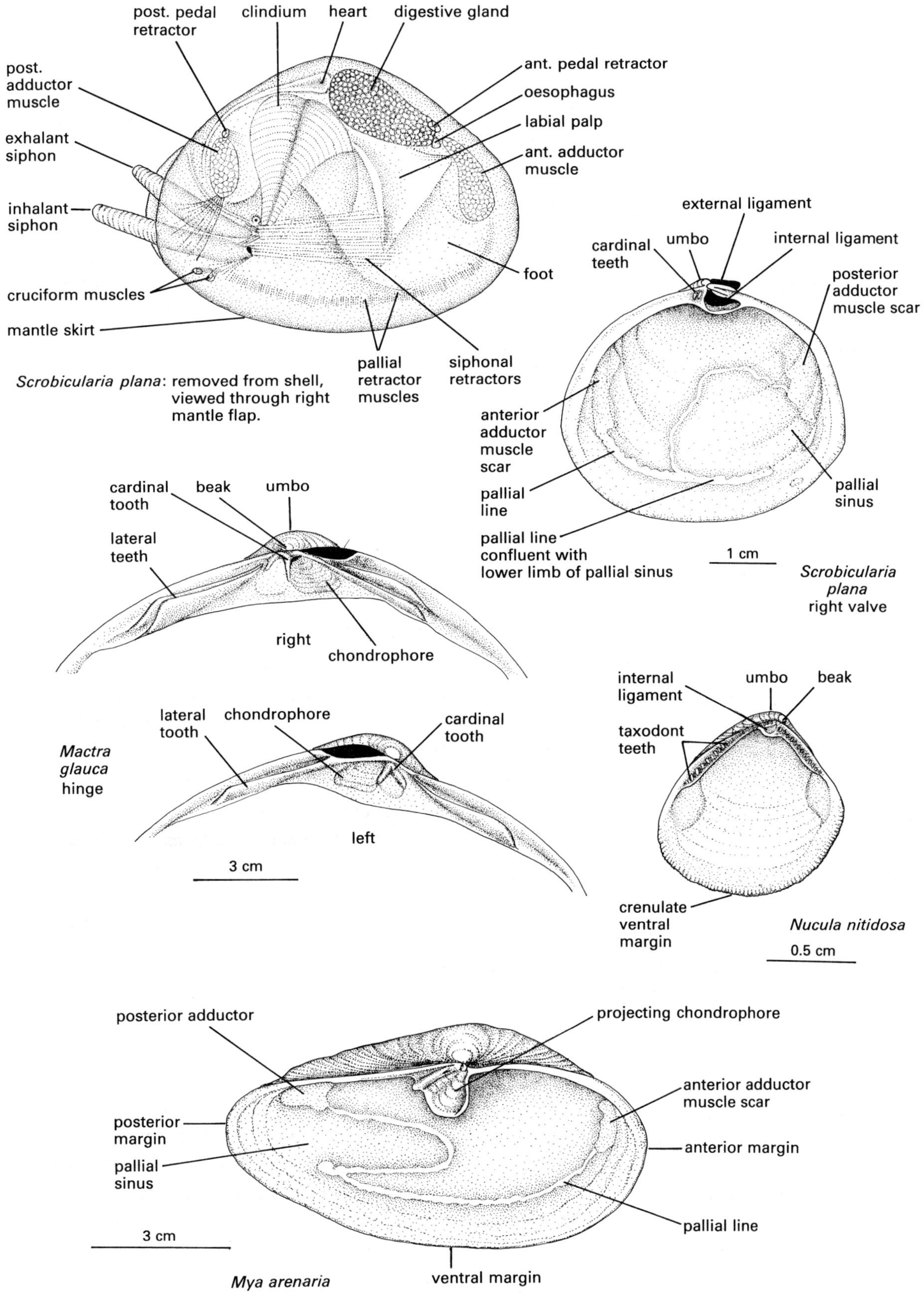
post. pedal retractor
clindium
heart
digestive gland
post. adductor muscle
exhalant siphon
inhalant siphon
cruciform muscles
mantle skirt
ant. pedal retractor
oesophagus
labial palp
ant. adductor muscle
foot
pallial retractor muscles
siphonal retractors
Scrobicularia plana: removed from shell, viewed through right mantle flap.
external ligament
cardinal teeth
umbo
internal ligament
posterior adductor muscle scar
anterior adductor muscle scar
pallial line
pallial sinus
pallial line confluent with lower limb of pallial sinus
1 cm
Scrobicularia plana right valve
cardinal tooth
beak
umbo
lateral teeth
right
chondrophore
lateral tooth
chondrophore
cardinal tooth
Mactra glauca hinge
left
3 cm
internal ligament
umbo
beak
taxodont teeth
crenulate ventral margin
Nucula nitidosa
0.5 cm
posterior adductor
projecting chondrophore
anterior adductor muscle scar
posterior margin
anterior margin
pallial sinus
pallial line
3 cm
ventral margin
Mya arenaria

Fig. 13.1

7. Shell markedly inequilateral; hinge line as long as the greatest length of the shell, largely comprising the posterior ear. Fragile and brittle, without marked radiations **7. Pteriidae**

   Shell equilateral, or only slightly inequilateral, although ears may be of unequal size; with radiating sculpture, often very strong **8**

8. Shell length half, or less, of depth; outline elongate-oval, ears small **11. Limidae**

   Shell length about equal to depth; curved edge of margin forming at least a semicircle, ears prominent **10. Pectinidae**

9. Each valve with one adductor muscle scar only, left valve with or without associated byssal scars. Shell inequivalve, often irregular in outline with a corrugated surface **10**

   Each valve with at least two adductor scars **11**

10. Shell thin and brittle, rarely exceeding 60 mm. Right valve flat, with a large hole close to the umbo. Left valve convex, the single round adductor scar adjacent to, or partly fused with, one or two small byssal scars **5. Anomiidae**

    Shell solid, often very thick and rugose, frequently exceeding 60 mm. Right valve flat, without a hole, left valve convex; each with a single, elliptical adductor scar, often deeply recessed **9. Ostreidae**

11. Hinge teeth absent. Ligament reduced, internal. Shell equivalve, inequilateral, and irregular in outline, frequently gaping. Anterior dorsal margin reflected in front of the umbones (Fig. 13.16); accessory plates present dorsally or at the tips of the siphons. Borers in wood, rock, peat, etc. **12**

    Not with the above association of characters **14**

12. Valves elongate, oval, or quadrate, gaping to a varying extent along the anterior, ventral, or posterior margins. Each valve with a slender, fragile process (the *apophysis*) curving out from beneath the umbo (look for stump if broken) **38. Pholadidae**

    Valves highly modified; short and globular, each formed from several distinct lobes **13**

13. Dorsal accessory plate present, apophyses absent. Shell gaping anteriorly, closed posteriorly **39. Teredinidae (*Xylophaga dorsalis*)**

    Long slender apophyses present, dorsal plates absent. Shell gaping anteriorly and posteriorly **39. Teredinidae**

14. Hinge line without teeth or chondrophore **15**

    Hinge line with either or both teeth and chondrophore **21**

15. Shell markedly inequilateral; umbones anterior, terminal or subterminal **16**

    Shell only slightly inequilateral, or almost equilateral; umbones close to mid-line, rarely subterminal above a large, oval ventral gape **17**

16. Shell broadly triangular, large, up to 300 mm long, with conspicuous anterior ventral and posterior gapes. A series of septa close to the umbo, on the inner surface of the shell **8. Pinnidae (*Atrina fragilis*)**
Shell subtriangular, oval, or bean-shaped. No septa; without anterior or posterior gapes **6. Mytilidae**

---

17. Shell inequilateral, umbones subterminal, a broad oval gape extending the whole length of the shell. Boring in limestone, sandstone, or organic concretions **37. Gastrochaenidae**
Shell not markedly inequilateral. If ventral gape present, then valves almost equilateral **18**

---

18. Shell with distinct radiating ribs **19**
Shell with concentric sculpture only **20**

---

19. Shell elongate-oval, hinge line straight; anterior and posterior margins rounded, with distinct ventral gape **20. Galeommatidae (*Galeomma turtoni*)**
Shell oval or rhomboidal, hinge line curved, without a gape **6. Mytilidae (*Crenella* and *Rhomboidella*)**

---

20. Shell orbicular, with prominent umbones. A conspicuous umbonal ridge extending to ventral margin **15. Thyasiridae (*Thyasira flexuosa*)**
Shell elongate-oval to quadrate, umbones less prominent. No umbonal ridge **20. Galeommatidae (*Devonia perrieri*)**

---

21. Posterior of shell drawn out as a blunt projecting spout (Fig. 13.17) **43. Cuspidariidae (*Cuspidaria*)**
Shell regular or irregular in outline but without a posterior spout **22**

---

22. Right valve flat, left valve convex. Pallial line consisting of a series of small, separate muscle scars. Ligament internal **42. Pandoridae**
Not with the above association of characters **23**

---

23. Shell very elongate, straight or gently curved, with dorsal and ventral margins almost or quite parallel. Umbones and ligament anterior, subterminal in position **33. Solenidae**
Not with the above association of characters **24**

---

24. Pallial line with a posterior sinus; rarely, this is small and difficult to distinguish **25**
Pallial line without a sinus **38**

---

25. Shell subtriangular with truncate posterior margin. Right valve more convex than left, enclosing it, leaving a broad border of right valve covered only by a fringe of left periostracum. Left valve with a deep pit adjacent to the chondrophore, right valve with a corresponding tooth. Pallial sinus small **35. Corbulidae**
Not with the above association of characters **26**

---

26. Ligament external only, no chondrophore **27**
Ligament internal and external, chondrophore present **33**

27. Right valve with one cardinal tooth, left valve with two, or none. All teeth liable to be worn. Shell fragile, often irregular in outline **36. Hiatellidae**
Not with the above association of characters **28**

28. Shell elongate, cylindrical, with anterior and posterior gape. Cardinal teeth peg-like: two right and one left, or two left and one right **32. Solecurtidae**
Not with the above association of characters **29**

29. Two cardinal teeth in each valve. Shell often thin and fragile. Cruciform muscle scars present; pallial line often confluent with lower border of pallial sinus **30**
Three cardinal teeth in each valve. Shell typically thick and solid, subtriangular. No cruciform muscle scars; pallial line usually distinct from border of pallial sinus **24. Veneridae**
Three cardinal teeth in left valve, two in right. Shell thin, circular or oval. No cruciform muscle scars; pallial line distinct from border of pallial sinus **25. Petricolidae**

30. Hinge line with cardinal and lateral teeth, the latter conspicuous in the right valve, but in the left appearing simply as slender ridges close to the margin **31**
Hinge line with cardinal teeth only **32**

31. Inner margin of shell crenulate or finely serrate **28. Donacidae**
Inner margin of shell smooth **29. Tellinidae**

32. Shell almost circular, colour variable but often in concentric bands. Intertidal only, often abundant in soft estuarine deposits **29. Tellinidae (*Macoma balthica*)**
Shell elongate-oval, anterior end not regularly semicircular, often keeled or ridged posteriorly. Sublittoral, occasionally on lower shore. Not estuarine **31. Psammobiidae**

33. One or more cardinal teeth present **34**
No cardinal teeth **36**

34. Left valve with two cardinal teeth joined to form a single $\wedge$-shaped structure. Two anterior and two posterior lateral teeth may be present in right valve. No cruciform muscle scar **35**
Left valve with one or two cardinal teeth, but not joined. Right valve without lateral teeth, or with single, small anterior and posterior teeth. Cruciform muscle scars present **30. Scrobiculariidae**

35. Shell subtriangular or elliptical, without a gape. Umbones on or close to mid-line **26. Mactridae**
Shell elongate, gaping at both ends. Umbones in anterior half **27. Lutrariidae**

36. Internal ligament supported by a projecting, spatulate chondrophore in the left valve and a smaller, concealed chondrophore in the right valve. Shell with a posterior gape **34. Myacidae**

Internal ligament supported by a concave chondrophore inset into the hinge line of each valve. Shell without a posterior gape **37**

---

37. Each valve with a short crack in the calcification extending posteriorly from the umbo **40. Periplomatidae**

Valves without a crack **41. Thraciidae**

---

38. Shell globular, equivalve. Umbones prominent, enrolled, curving away from the hinge line, Cardinal and lateral teeth parallel to hinge line **22. Glossidae**

Not with the above association of characters **39**

---

39. Shell equivalve; valves strongly convex, with radiating ridges often well-developed. Ligament external **23. Cardiidae**

Radiating ribs rarely developed, sculpture of concentric lines or ridges. Ligament external or internal **40**

---

40. Shell oval or circular, often large; heavy and solid, with prominent umbones. Periostracum thick, brown or black. Ligament external. Three cardinal teeth in each valve **21. Arcticidae**

Ligament external or internal. Three cardinal teeth in one valve only, or both valves with less than three **41**

---

41. Anterior adductor muscle scar larger than posterior. Ligament external **42**

Both muscle scars of similar size. Ligament external or internal **43**

---

42. Anterior adductor scar comma-shaped, completely fused with the pallial line along its anterior edge. Shell smooth, often silky, sculptured with very fine concentric lines between distinct growth stages **14. Diplodontidae**

Anterior adductor scar elongate-oval, separated from the pallial line along a varying proportion of its anterior edge. Shell matt, with concentric lines or ridges **13. Lucinidae**

---

43. Ligament external. Lateral teeth absent. At least two cardinal teeth in each valve, three may be present in left valve. Shell typically subtriangular, solid **12. Astartidae**

Ligament internal. Lateral teeth may be present. Cardinal teeth variable, never more than two in each valve. Shell typically fragile and small (<12 mm) **16. Kelliidae**

17. Lasaeidae

18. Leptonidae

19. Montacutidae

---

1 NUCULIDAE

Shell subtriangular or oval in outline, smooth, tumid; equivalve, inequilateral, with umbones well to posterior of mid-line. Hinge line with taxodont teeth in two distinct series, anterior more numerous than posterior. Ligament internal, in an elliptical chondrophore, below and just in front of beaks. Adductor muscle scars about equal; pallial line faint, without a sinus.

1. Inner margin of shell smooth — ***Nuculoma tenuis***
   Inner margin of shell with small, distinct crenulations — **2**
2. Shell decussate: distinct at the hinge line and forming coarse, transverse corrugations anteriorly across the lunule — ***Nucula sulcata***
   Shell not, or only faintly decussate; lunule smooth, without transverse ridges — **3**
3. Radiating bands of colour (red-brown to grey) running from umbones to margin. With posterior hinge line vertical, anterior hinge line curving upwards — ***Nucula hanleyi***
   Radiating striae present, but colour bands, when present, concentric — **4**
4. Periostracum very glossy, often with bold concentric bands of grey or greenish-yellow. With posterior hinge line vertical, anterior hinge line curved about a horizontal line — ***Nucula nitidosa***
   Periostracum dull or matt. With posterior hinge line vertical, anterior hinge line curved above the horizontal — ***Nucula nucleus***

Nucula nitidosa Winckworth — FIG. 13.2

Shell white or grey with bluish growth lines; periostracum very glossy, olive or yellow-olive, often with concentric bands of light yellow. Fine radiating striations and fainter concentric lines visible with hand lens. Commonly 10 mm long, but up to 13 mm. Anterior hinge line curving about the horizontal when the posterior is vertical; 20–30 hinge teeth anteriorly, 10–14 posteriorly; lunule and escutcheon poorly defined. Margin of shell finely crenulate.

On silt and fine sand. Common, and often abundant, off all British coasts, down to 100 m. Distributed from Norway south to the Mediterranean and west Africa.

Nucula nucleus (Linnaeus) — FIG. 13.2

Shell white or grey; periostracum matt, light yellow, light brown, or greenish yellow, with darker or lighter concentric bands. Fine radiating striations visible with hand lens, with coarser concentric lines. Lightly decussate posteriorly. Commonly less than 10 mm long, but may reach 12 mm. Anterior hinge line curved above the horizontal when the posterior is vertical; 16–25 hinge teeth anteriorly, 10–14 posteriorly; lunule poorly developed, escutcheon distinct with strongly pouting hinge line. Margin of shell crenulate.

On coarse sand and fine gravel, often common. Off all British coasts, to about 150 m. Distributed from Norway south to the Mediterranean, West Africa, and the coasts of South Africa.

Nucula sulcata Bronn — FIG. 13.2

Shell white or grey; periostracum matt, light yellow to brown, often with darker patches, and brown or black incrustations about hinge line. Fine radiating striations and coarser concentric lines give a decussate effect, well-marked at hinge line. Commonly about 15 mm long, but often exceeding 20 mm. Anterior hinge line curving about the horizontal when the posterior is vertical; 20–30 hinge teeth anteriorly, 10–14 posteriorly; lunule crossed by strong transverse corrugations, escutcheon distinct, with depressed border and pouting hinge line. Margin of shell crenulate.

On mud and sand, from 10–200 m. Mostly off west coasts. Distributed from northern Norway to the Mediterranean and West Africa.

Nucula hanleyi Winckworth — FIG. 13.2

Shell white or grey; periostracum smooth and shiny but without a high gloss, yellow-brown or grey-brown, with darker or lighter bands of colour radiating from umbones. Faint radiating striations and some concentric lines, occasionally lightly decussate posteriorly. Commonly about 10 mm long, but reaching 16 mm. Anterior hinge line curved gently above the horizontal when the posterior is vertical; 15–25 hinge teeth anteriorly, 10–12 posteriorly; lunule poorly defined, escutcheon distinct, with hinge line slightly pouting. Margin of shell crenulate.

On fine shell-gravels, offshore to 40 m. A southern species, occurring in the English Channel and off western coasts as far north as the Isle of Man. Distribution extends south to the Mediterranean.

Nuculoma tenuis (Montagu) — FIG. 13.2

Shell white or grey, thin and fragile; periostracum glossy, greenish-yellow to brown. Faint radiating striations visible with hand lens, and slightly bolder concentric lines. Commonly about 10 mm long, often less. Anterior hinge line curving well above the horizontal when the posterior is vertical; 14–18 hinge teeth anteriorly, 6–10 posteriorly; lunule poorly defined, escutcheon strongly pouting. Margin of shell smooth.

On muddy deposits, offshore, perhaps to the edge of the Continental Shelf. Reported mostly off northern coasts of Britain. Distribution circumboreal; present in the Mediterranean.

2. NUCULANIDAE

Shell subtriangular or irregularly oval in outline, often slender and rostrate posteriorly; equivalve, inequilateral, umbones anterior to mid-line. Hinge line with taxodont teeth in two distinct series, separated by a small chondrophore supporting internal ligament. Adductor muscle scars about equal; pallial line faint, with a small sinus.

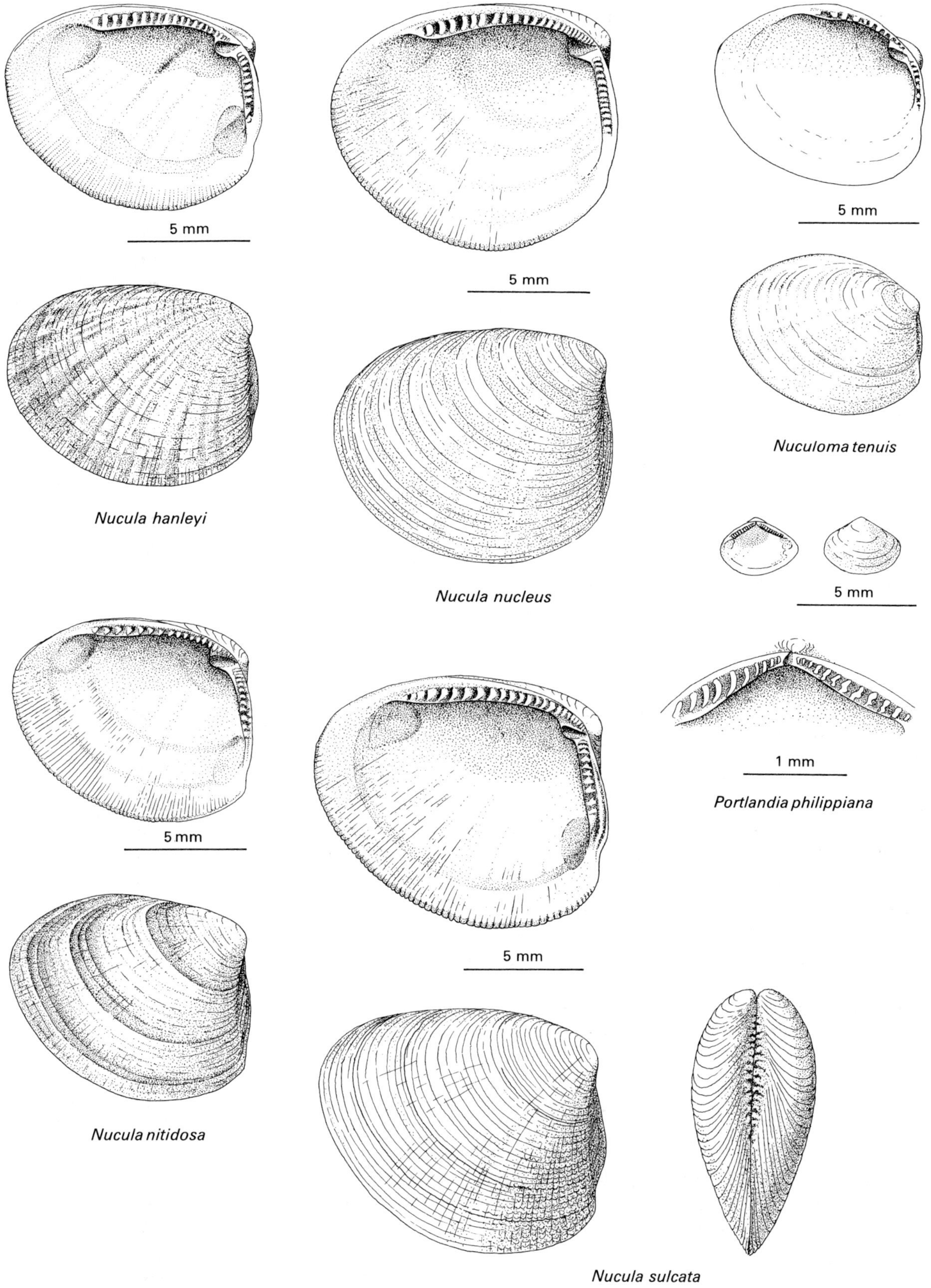
5 mm
5 mm
5 mm
Nuculoma tenuis
Nucula hanleyi
5 mm
Nucula nucleus
1 mm
Portlandia philippiana
5 mm
5 mm
Nucula nitidosa
Nucula sulcata

Fig. 13.2

Nuculana minuta (Müller) FIG. 13.3

Shell slender, elongate, rounded anteriorly, posteriorly drawn out as a short, truncate spout; off-white or greyish, periostracum dull, light brown to olive-brown, with darker concentric bands. Sculpture of bold, concentric ridges; on each valve, a raised ridge extends from umbone to posterior border, delimiting lower edge of truncate rostrum. Up to 20 mm long, but usually about 15 mm. Hinge with 15–18 teeth anteriorly, 18–22 posteriorly. Lunule indistinct, escutcheon prominent, extending from umbones to tip of rostrum, with fine longitudinal striations.

On muddy sand and gravel, from 10 m to 200 m. Reported from scattered localities off all British coasts, local but sometimes very common. Distribution circumboreal; in the north-east Atlantic extending south as far as the English Channel.

Portlandia philippiana (Nyst) FIG. 13.2

Shell rounded anteriorly, slightly produced posteriorly, tumid, umbones prominent and rounded; off-white, periostracum glossy, pale, yellowish brown or yellowish green. Surface smooth, except for a few faint concentric lines, growth stages clear. Very small, rarely exceeding 5 mm in length. Lunule and escutcheon indistinct, grading smoothly into shell margin. Hinge with 10–13 teeth anteriorly and 12–16 posteriorly.

Recorded off north and west coasts only, in deep water. Distribution circumboreal; present in Mediterranean.

3 ARCIDAE

Shell thick and strong, tumid, oval, quadrate or irregular in outline; equivalve, inequilateral, umbones anterior to mid-line, widely separated. Ligament external, extending between umbones across a broad, grooved, diamond-shaped cardinal area. Hinge line straight, with a continuous series of identical teeth, alternating with sockets. Adductor muscle scars about equal. No pallial sinus.

1. Shell elongate, angular. Hinge line straight, its junction with posterior margin marked by a sharp step. Pronounced ventral gape present. ***Arca tetragona***

   Shell roughly oval. Hinge line straight, grading gently into posterior margin. No byssal gape ***Striarca lactea***

Arca tetragona Poli FIG. 13.3

Shell elongate, very irregular; anterior end deeper than posterior, umbones prominent and widely spaced, each with a sharp, raised ridge running from the beak to posterior margin. Up to 50 mm long. Cardinal area broadly triangular, with distinct chevron ridges, partly obscured in older specimens. Colour off-white or yellow, periostracum brown. Sculpture of fine radiating ridges; older shells worn to a rough, chalky white surface, coarse and rugose between umbones. Hinge line with 40–50 teeth, intersecting the rounded anterior margin at right angles, forming a sharp step with concave posterior margin. A small ridge extends from posterior end of hinge line to umbones. Ventral margin irregular and variable, usually convex, occasionally concave, finely crenulate, with a distinct byssal gape ventral to umbones. Inner surfaces dull brown in smaller shells, off-white in older shells, with bluish, pale brown, or green blotches. Adductor scars prominent, often in relief.

Sessile, attached by a thick green byssus; in crevices, among rocks and shell debris, often encrusted with other organisms, or abraded by the substratum. From the lower shore to about 100 m, off all British coasts. Distributed from Norway south to the Mediterranean and west Africa.

Striarca lactea (Linnaeus) FIG. 13.3

Shell very tumid, up to 20 mm long but often smaller; umbones just in front of mid-line, prominent, blunt, and enrolled. A thick ridge extending from beaks to anterior and posterior margin on each valve. Cardinal area a slender diamond shape with fine chevron striations, ligament deep brown. Shell off-white or yellowish; periostracum coarse, brown, forming a distinct fringe ventrally. Sculpture of fine radiating and concentric ridges, giving a decussate effect on the central areas of each valve, but with only radiating ridges prominent at anterior and posterior margins. Growth stages visible as raised ridges. Hinge line with 40–50 teeth. Anterior and posterior margins rounded. Inner surface white, adductor scars well-marked.

From the lower shore to below 100 m, on shell-gravels and coarse bottoms. Off southern and western coasts from the English Channel to the Clyde Sea area. Distributed south to the Mediterranean, West Africa, and the coasts of South Africa.

A third, rarer, species, *Bathyarca petunculoides* (Scacchi) occurs off the northern and western shores of Scotland. Most similar to *S. lactea,* it is distinguished by its small size (up to 5 mm) and by the presence of three or four larger, accessory teeth at each end of the hinge line.

4 GLYCYMERIDAE

Shell equivalve, equilateral, umbones situated on mid-line. Ligament external, in a narrow, elongate cardinal area. Hinge teeth in a continuous series in young shells, separated by a smooth area below umbones in older individuals. Adductor muscle scars about equal, no pallial sinus.

Glycymeris glycymeris (Linnaeus) FIG. 13.3

Shell thick, robust; almost circular in outline, anterior hinge line curving more abruptly downwards than posterior. Up to 65 mm long. Umbones central, scarcely inclined anteriorly; cardinal area very elongate. Surface dull; brown, yellow, or light purplish-red, either uniformly coloured or in irregular concentric zigzags on a cream background; periostracum dark brown, coarse, velvety to touch, usually remaining as a broad band around margin. Sculpture of fine concentric lines and radiating striae, growth stages clear. Hinge line with 6–12 teeth in each of two series, on either side of a smooth area below umbo; in young shells these are continuous. Inside of shell glossy, white, brown, or reddish, often with irregular patterns; margin crenulate. Adductor scars and pallial line well-marked.

A shallow burrower in fine shell-gravels, or sandy/muddy gravels offshore to about 100 m, common off all British coasts.

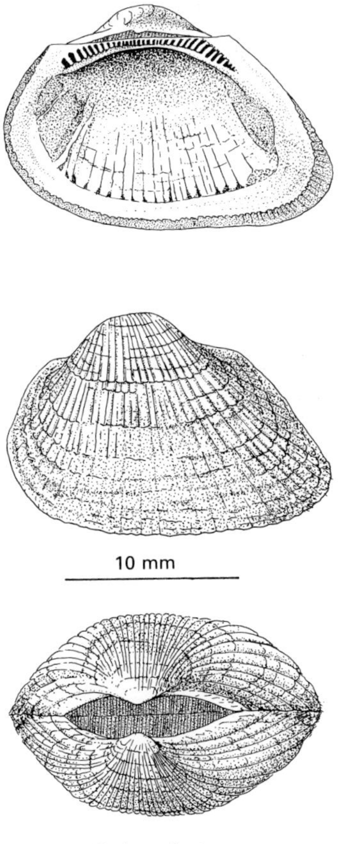

Striarca lactea

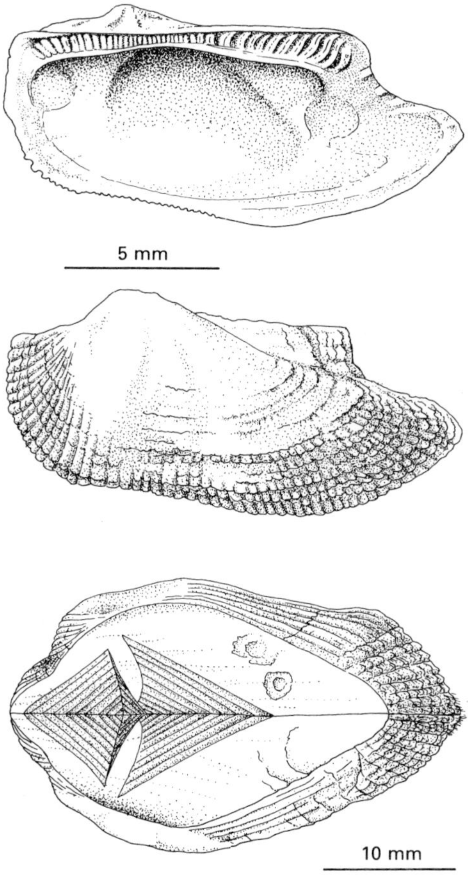

Arca tetragona

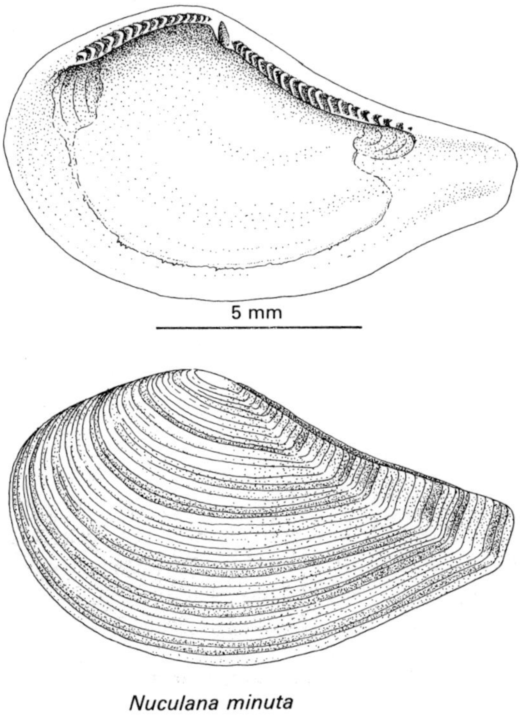

Nuculana minuta

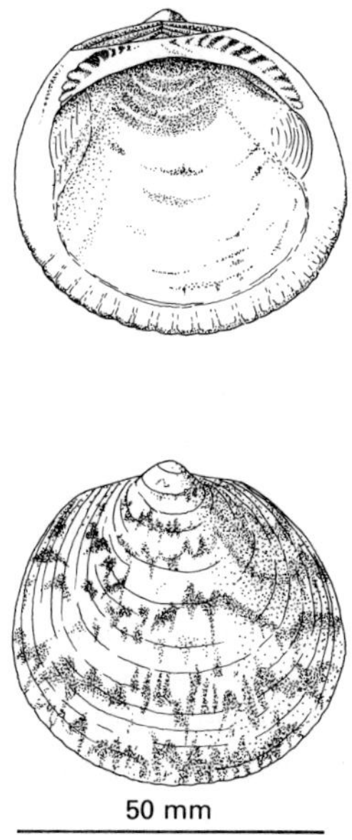

Glycymeris glycymeris

Fig. 13.3

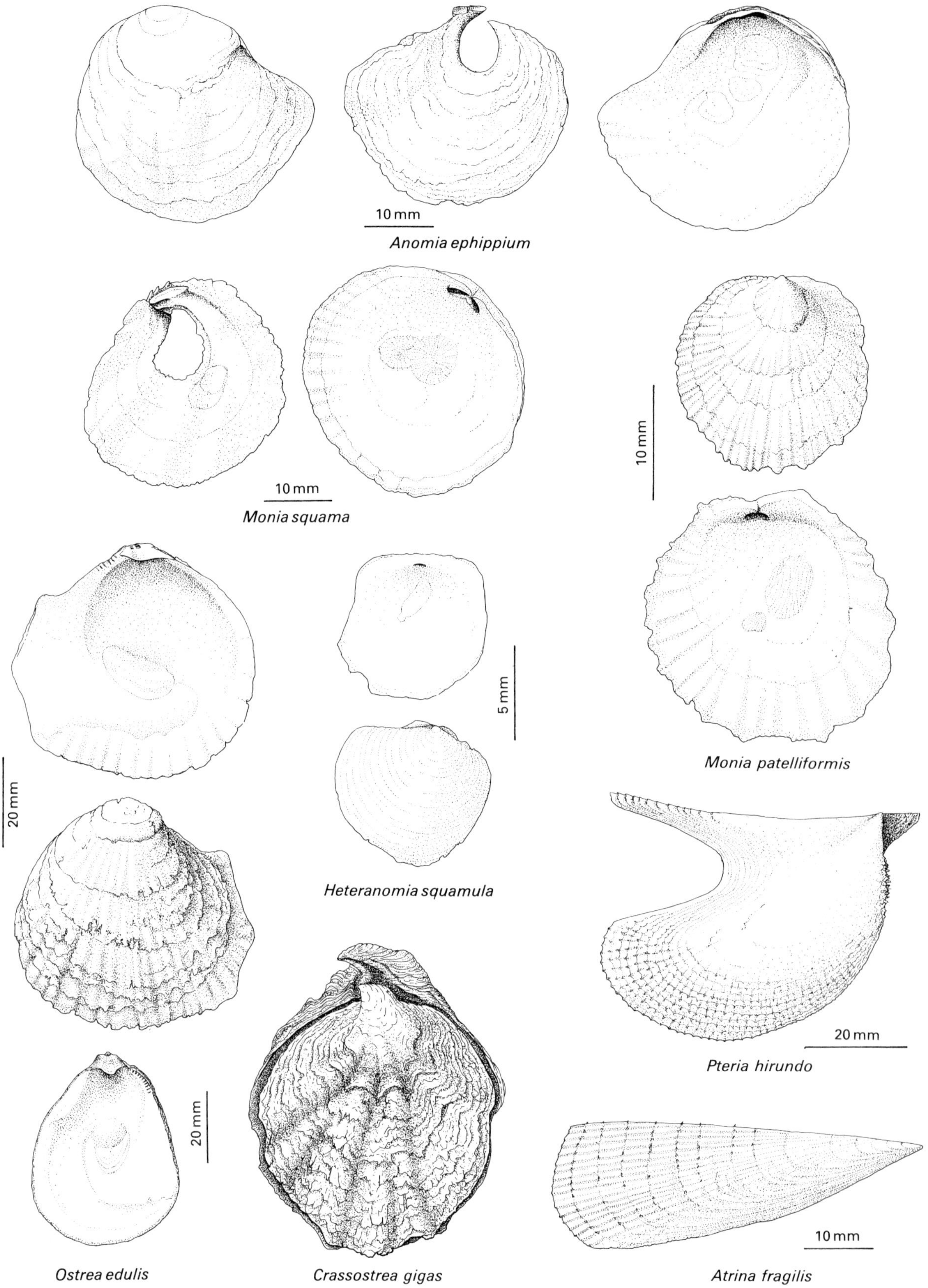
10 mm
Anomia ephippium
10 mm
Monia squama
10 mm
20 mm
5 mm
Monia patelliformis
Heteranomia squamula
20 mm
Pteria hirundo
20 mm
10 mm
Ostrea edulis
Crassostrea gigas
Atrina fragilis

Fig. 13.4

Dead shells often accumulate in vast deposits in some areas, such as the western end of the English Channel. Distributed from Norway south to the Mediterranean and west Africa.

5. ANOMIIDAE

Shell thin and brittle, approximately circular in outline, often irregular; attached by a byssus passing through a hole in right valve, with left valve uppermost. Inequilateral, umbones rounded, indistinct. Inequivalve; left valve convex, right valve flat or concave. Right valve with a single adductor muscle scar; left valve with a single adductor scar, adjacent to, or fused with, one or two byssal attachment scars. No hinge teeth. Ligament internal, in a deep pit below umbone of right valve. Inner surface of shell nacreous, glossy, no pallial line.

---

1. Left valve with three closely spaced scars in the concavity of the shell **_Anomia ephippium_**

   Left valve with two distinct scars, or two partly fused **2**

2. Shell very small, rarely exceeding 10 mm. Scars in left valve smooth-surfaced, closely juxtaposed, forming an obliquely elongate shape **_Heteranomia squamula_**

   Shell larger, up to 40 mm long. Scars in left valve with striated surface, fused or discrete **3**

3. Left valve with two discrete scars **_Monia patelliformis_**

   Left valve with two fused scars, forming a dumb-bell shape **_Monia squama_**

Anomia ephippium Linnaeus FIG. 13.4

Shell brittle, fragile, roughly circular, the posterior margin frequently bluntly produced; often rather irregular when attached to an uneven substratum. Rarely larger than 60 mm. Left valve convex, often deeply cupped and overlapping right valve; with three distinct scars on inner surface, in an oblique line, in concavity of shell. Right valve thin, flat, or irregularly concave, with elliptical byssal aperture dorsally; single adductor scar postero-ventral to byssal aperture. Colour white or grey, occasionally with a pink or bluish tinge. Sculpture of wavy concentric lines, often raised as irregular scales. Growth stages usually clear, except in older shells when they may be obscured by incrustations or, on the left valve, worn smooth.

Distributed from the lower shore to at least 150 m, and perhaps to the edge of the Continental Shelf. Attaches to various hard substrata, including other molluscs, and to algal holdfasts. Widespread and common off all British coasts. Occurs in Mediterranean; distributed from Iceland to West Africa, and to the South Atlantic Islands.

Monia patelliformis (Linnaeus) FIG. 13.4

Shell thin, circular or oval, sometimes irregular; white or grey, occasionally with patches or streaks of pale brown. Usually less than 40 mm long. Umbones marginally anterior to mid-line, set back from hinge. Left valve shallowly convex, inner surface white, or pale brown with a bluish tinge; two distinct scars, a larger, oval byssal scar immediately below umbo, and a circular, postero-ventral muscle scar. Both scars with striated surface, visible with hand lens. Right valve flat or slightly concave, byssal aperture large, oval or drop-shaped. Sculpture of left valve consisting of coarse concentric ridges and numerous radiating ribs, margin crenulated; growth stages usually clear and often forming prominent ridges. Right valve with fine concentric lines, often forming coarse scales close to margin, but without ribs. Surface of upper (left) valve often obscured by encrusting organisms.

Attached to hard substrata, offshore to at least the edge of the Continental Shelf. Widespread off all British coasts; distributed from Norway to the Mediterranean.

Monia squama (Gmelin) FIG. 13.4

Shell thin, fragile, almost circular; off-white, usually darkened or discoloured by incrustations. Up to 40 mm long. Umbones low and indistinct. Left valve shallowly convex, often almost flat; inner surface glossy, white or greenish, often with bluish tints or darker patches of green. The larger, dorsal, byssal scar fused with a smaller, ventral adductor scar, forming a single obliquely orientated, dumb-bell shaped scar, with a wrinkled or striated surface. Sculpture of fine, closely spaced concentric lines, often raised into small scales, and numerous delicate, radiating ridges, margin finely crenulate. Growth stages unclear. Right valve thin and translucent, closely adpressed to substratum, byssal aperture drop-shaped, with fine concentric lines but no ribs.

Distributed from the lower shore to about 100 m, attached to hard substrata, including other molluscs. Reported from west Scotland, Isle of Man, and south-west British Isles. Ranges from Norway south to the English Channel.

Heteranomia squamula (Linnaeus) FIG. 13.4

Shell very thin and delicate, approximately circular, almost flat; colour off-white. Rarely attaining 15 mm long, usually much less. Left valve only slightly convex, umbo distinct, forming a short cone; inner surface smooth, white; oval, dorsal byssal scar and ventral adductor scar closely adpressed, slender, obliquely orientated, with smooth surfaces. Right valve very thin, translucent. Both valves with numerous, fine, concentric lines, visible with hand lens.

On hard substrata and algae; occurs off all British coasts, ranges from Arctic waters south to the Bay of Biscay.

6. MYTILIDAE

Shell elongate, equivalve; inequilateral, beaks at anterior end, terminal or subterminal; rarely almost equilateral. Periostracum usually conspicuous, typically darker than shell, with or without spines. Hinge line without teeth, rarely with small crenulations, continuous with shell margin. Ligament internal, or external in a narrow groove and inconspicuous. Adductor muscle scars unequal, anterior scar reduced. No pallial sinus. Attached by a diffuse, fibrous byssus.

1. Shell smooth, with or without concentric or radiating lines, but without raised ribs **2**

   Shell with two series of radiating ribs, separated by a smooth area extending from the umbones to the ventral margin **7**

2. Umbones terminal, at the extreme anterior end of shell **3**

   Umbones subterminal, with anterior margin extending beyond the beaks **4**

3. Umbones rounded, not downturned. Ventral margin convex or pouting. Mantle edge typically light yellow-brown ***Mytilus edulis***

   Umbones beaked or pointed, downturned. Ventral margin slightly concave posterior to umbones, the two shell edges forming a narrow, flattened area. Mantle edge typically purple or violet ***Mytilus galloprovincialis***

4. Hinge line of shell with an oval, ridged area close to umbones (visible with hand lens). Periostracum hairy, covered with smooth spines. Usually less than 20 mm long ***Modiolula phaseolina***

   Hinge line of shell smooth. Periostracum with or without spines. Frequently exceeding 20 mm **5**

5. Periostracum with fringes of broad serrated spines, particularly well-developed posteriorly. Shell light yellowish-red, with darker yellow-brown periostracum ***Modiolus barbatus***

   Periostracum without spines, or with smooth spines in small individuals. Shell white, yellow with darker red rays, or blue/purple **6**

6. Shell white, bluish, or purple. Periostracum light yellow-brown in juveniles, deepening to rich mahogany with paler umbones in older specimens; with a patch of rough scales posteriorly, bearing a fringe of smooth spines in small individuals ***Modiolus modiolus***

   Shell off-white to light yellow, with distinct variously developed rays or zigzags of light red. Periostracum light yellow-brown, smooth, without spines ***Modiolus adriaticus***

7. Periostracum light brown or olive in smaller specimens, deep mahogany or almost black in larger shells. Posterior series of 40–60 fine ribs. Maximum length about 50 mm. Offshore ***Musculus niger***

   Periostracum horn-coloured or greenish, with darker patches of green, brown, or purplish-red. Usually fewer than 40 ribs posteriorly. Maximum length about 20 mm. Frequent on lower shore **8**

8. Umbones very prominent, shell appearing quite tumid. Dorsal margin downcurved towards posterior. Up to 18 ribs anteriorly, 20–35 posteriorly ***Modiolarca tumida***

   Umbones distinct, but not prominently so, shell more slender. Dorsal margin upcurved, posterior end as deep as mid-line **9**

9. Shell elongate-oval. Periostracum light or dark greenish brown, with streaks and blotches of green and reddish-brown, often with red-brown chevrons along hinge line. 20–30 ribs posteriorly ***Musculus costulatus***

   Shell oval to rhomboidal. Periostracum pale green or brown, often unicolorous, or with a few darker patches. 30–40 ribs posteriorly ***Musculus discors***

Mytilus edulis Linnaeus FIG. 13.5

Shell brittle, oval, pyriform, or subtriangular; terminal umbones situated at apex of anterior end. Length variable: commonly 50–100 mm, although many populations rarely reach 30 mm, and largest shells may attain 150–200 mm. Colour light horn to blue or purple, dark purplish blue when older; small specimens often marked with pale blue and brown rays on a translucent ground; periostracum thick, deep blue to black, occasionally dirty brown, glossy where new, often abraded over much of the anterior end. Umbones prominent, with raised umbonal ridges curving posteriorly and ventrally. Sculpture of fine concentric lines, growth stages clear. Inner surface of shell white below ligament, progressively blue and dark, towards a deep purplish black posterior margin. Mantle edge of live specimens typically yellow-brown.

Attached by a byssus issuing ventrally, close to umbones, *M. edulis* occurs in dense beds from upper shore and into the shallow sublittoral. Both size and shape are greatly influenced by environmental factors.

Widespread and common on all British coasts, ranges from Arctic waters south to the Mediterranean.

Mytilus galloprovincialis (Lamarck) FIG. 13.5

Shell brittle, oval, subtriangular, or pear-shaped, umbones prominent, pointed, and slightly curved ventrally. Posterior to umbones ventral margin is slightly concave and both shell edges form a flattened area. Colour blue to deep purplish black, periostracum light brown to blue-black; without the rays of colour typical of *M. edulis*. Sculpture of fine concentric lines, growth stages clear. Inner surfaces white beneath umbones, becoming bluer and darker posteriorly. Mantle edge of live specimens typically purple.

This species is readily confused with *M. edulis* and many authorities have considered that only one species is present in British waters. However, Seed (1974) has carried out an extensive survey of mussel populations around the British Isles and has shown that *M. galloprovincialis* is present intertidally around south-west England, South Wales, and southern and western Ireland. From Cherbourg it extends westward along the coasts of France and south to the Mediterranean. The most useful distinguishing features of *M. galloprovincialis* are the beaked, downturned umbones, the degree of flattening of the ventral

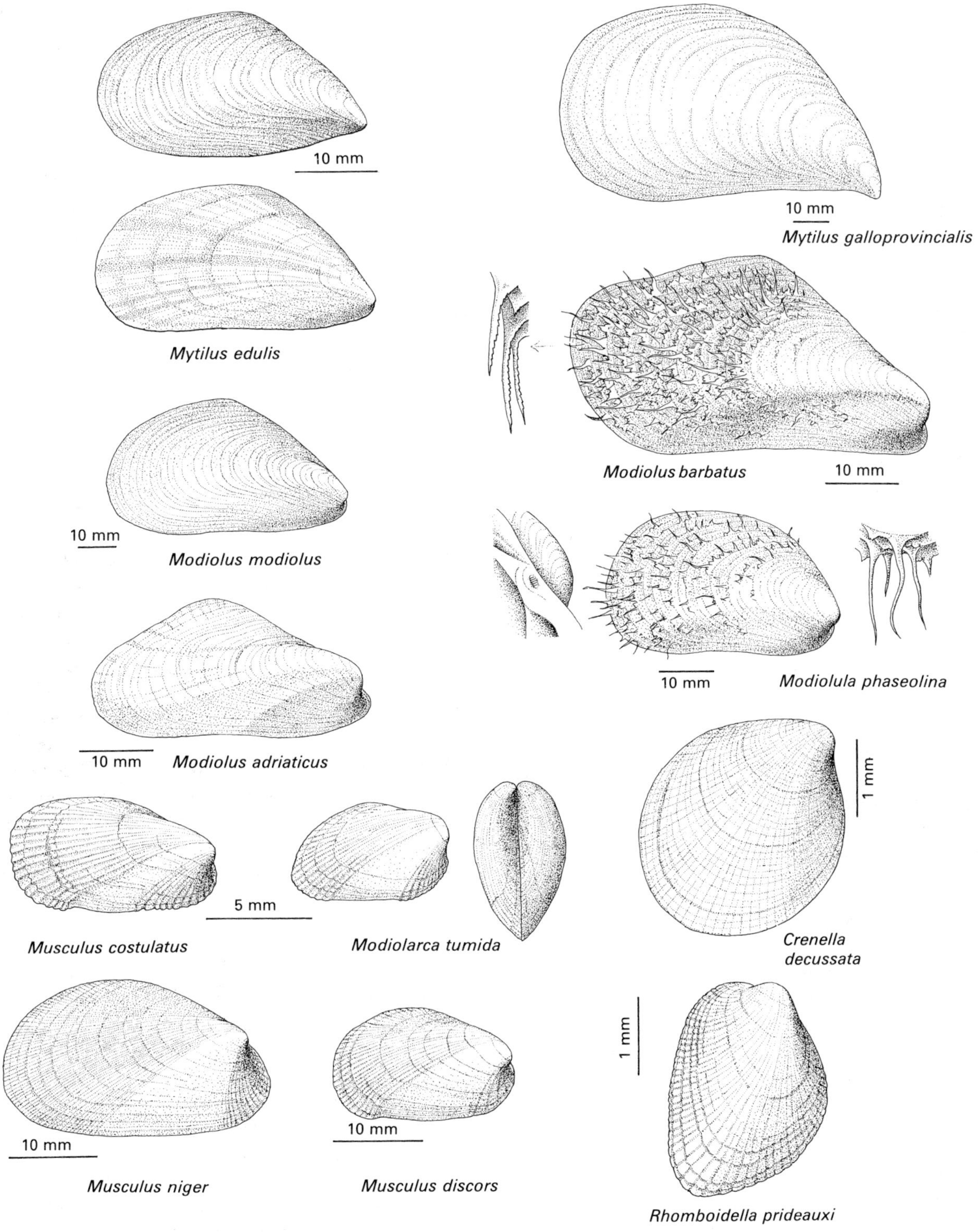

Fig. 13.5

margin and the colour of the mantle edge. However, both species of *Mytilus* display great variation in shell morphology through the interaction of ontogenetic and environmental factors, and old specimens may be almost impossible to identify with certainty. As mixed populations of both species are found over much of the northern range of *M. galloprovincialis*, the possibility of hybridization may add a further complication.

Modiolus modiolus (Linnaeus) FIG. 13.5

Shell thick, tumid, irregularly oval or rhomboidal in outline;

umbones anterior but subterminal, with a projecting, rounded, anterior margin; dorsal margin markedly convex, ventral margin slightly concave. Umbonal ridges well-developed. Size variable, commonly 50–100 mm, frequently more or less. Shell bluish white to slate-blue, darkening in larger specimens; periostracum very glossy, light horn to deeper mahogany or dark brown in old shells, usually with a lighter yellow-brown strip along the umbonal ridges. Young specimens with numerous long, smooth, slender periostracal spines, frequently with attached detritus, lost as the shell grows, represented only by a roughened area at posterior end. Sculpture of fine concentric lines and ridges, growth stages often clear. Inner surface of shell white, sometimes with a faint bluish tinge towards posterior margin.

Distributed from lower shore to about 150 m, on coarse grounds; offshore it may form immense aggregations. Off all British shores, south to the Bay of Biscay.

Modiolus barbatus (Linnaeus) FIG. 13.5

Shell thin and brittle, elongate-oval; umbones subterminal, above a curved anterior margin. Hinge line at a steep angle to umbones, ventral margin slightly concave; umbonal ridges very prominent, rounded, almost parallel with ventral margin. Rarely larger than 60 mm long. Shell yellowish white to light yellow or reddish brown; periostracum glossy, yellow, reddish brown, or light mahogany, lighter along umbonal ridges, produced into dense fringes of flattened, serrated spines. These are present in even the largest specimens, though often abraded at anterior end, and may trap quantities of silt and detritus. Sculpture of fine concentric lines; growth stages clear where not obscured by periostracum. Inner surface of shell pale bluish white, with tinges of red or light purple.

Distributed from lower shore to about 100 m, on coarse grounds. Southern and western, north to Yorkshire on the east coast and the Clyde in the west, ranging south to the Mediterranean and north-west Africa.

Modiolus adriaticus (Lamarck) FIG. 13.5

Shell thin and fragile, elongate-oval, appearing irregularly triangular when viewed from inner surface. Ventral margin very slightly concave, sometimes convex; umbones subterminal, anterior margin protruding well beyond beaks. Up to 40 mm long. Shell light yellow, often with a reddish tinge, with radiating streaks of light red, sometimes faint and occasionally absent in smallest specimens. Periostracum smooth and very glossy, light yellow or yellow-brown, darkest on each side of umbonal ridges. Sculpture of fine concentric lines, growth stages distinct. Inner surface white or very pale pink, with the external pale red rays visible, sometimes with streaks or patches of pale purple.

Offshore, to about 100 m, uncommon; predominantly southern in distribution. Distributed south to the Mediterranean and the Canary Isles.

Var. *ovalis*, reported from a few localities in the south-west, from Weymouth to the Helford River, is distinguished by a darker periostracum with green, brown, or purple rays. It grows up to 50 mm long.

Modiolula phaseolina (Philippi) FIG. 13.5

Shell thin, fragile, tumid; broadly oval, with a concave ventral margin, appearing bean-shaped, up to 20 mm long. Umbones subterminal. Shell light yellow to light purple; periostracum horn, light-brown, or yellow-brown, paler along umbonal ridges, raised into broad fringes of smooth, slender spines, often absent about umbones and anterio-ventral margin. Sculpture of fine concentric lines. Inside of shell bluish white to light purple; small oval protuberance on hinge line, close to umbones, with series of transverse ridges—visible with hand lens in the largest specimens.

Distributed from lower shore downwards; reported from most British coasts. Distributed from northern Norway to the Mediterranean and north-west Africa.

Musculus costulatus (Risso) FIG. 13.5

Shell thin, elongate oval, dorsal margin curving gently above umbones, posterior end appearing almost quadrate. Umbones subterminal, distinct. Up to 13 mm long. Shell off-white, periostracum variably coloured: light or dark greenish brown, with patches of deep brown, reddish brown, or purple, umbones frequently darker; often with broad streaks or zigzags, particularly anteriorly, and typically with reddish-brown chevrons along hinge line. Ventral and posterior margins often marked with a broad band of deep green. About 10 broad, radiating ribs anterior to umbones, 20–30 posteriorly—thin dorsally, but broad and flat close to posterior margin. Umbonal ridges without ribs. Fine concentric lines also present, not markedly decussate where they intersect ribs; growth stages clear. Margin strongly crenulate in ribbed areas. Inner surface of shell pearly, sometimes tinted brown, external sculpture visible.

On rocky shores, lower shore and shallow sublittoral. Mostly south and west in distribution; ranging south to the Mediterranean and north-west Africa.

Musculus discors (Linnaeus) FIG. 13.5

Shell thin, fragile, oval to rhomboidal; dorsal margin curving well above umbones, posterior broadly rounded, not tapering. Umbones subterminal, distinct. Rarely exceeding 13 mm long. Shell light yellow-brown, periostracum light brown, with a pale green tinge, often darker along ventral margin. Sculpture of 8–12 broad ribs anteriorly, 30–50 more slender ribs posteriorly, and a few fine concentric lines; not decussate, margin of ribbed areas crenulate. Inner surface of shell pearly, sometimes with a bluish tinge, external sculpture visible.

On rocky shores from mid-tidal zone into shallow sublittoral. Widespread and common off all British coasts. Distribution circumboreal, in the north-east Atlantic ranging south to the Mediterranean and Madeira.

Musculus niger (Gray) FIG. 13.5

Shell thin, brittle, broadly oval in outline, ventral margin more or less straight. Umbones subterminal, anterior margin extending well beyond beaks; umbonal ridges not pronounced, grading gently into rest of shell. Up to 50 mm long. Shell light tan to deeper brown, with a paler greenish area ventrally between the two ribbed areas; periostracum light brown or olive to deep brown, or almost black in large specimens. About 12–15 delicate radiating ribs anteriorly, 40–60 posteriorly, fine concentric lines also present; ribbed areas with finely crenulate margins, appearing decussate when viewed with a hand lens, growth stages clear. Inside of shell pale bluish white, with some darker tinges; external sculpture visible.

Occurs offshore, on sandy mud and fine gravel. Northern, extending south to the Yorkshire coast.

Modiolarca tumida (Hanley) FIG. 13.5

Shell thin, fragile, oval, dorsal margin downcurved, tapering posteriorly. Umbones and umbonal ridges very prominent, giving the two valves a very tumid appearance. Anterior margin almost straight. Rarely reaching 20 mm long. Shell off-white to light yellow-brown, periostracum horn-coloured, or light green, occasionally with darker brown or red-brown markings. With 15–18 broad ribs anteriorly, 20–35 posteriorly; fine concentric lines present, most conspicuous on prodissoconch. Inner surface of shell white, external sculpture visible.

Occurs on the lower shore and in the shallow sublittoral, among algal holdfasts, attached to undersides of shells and stones, or frequently embedded in the tests of large tunicates, particularly *Ascidia mentula* (Müller) and *Ascidiella aspersa* (Müller). Widespread and common, on rocky shores, off all British coasts. Present in Mediterranean.

Crenella decussata (Montagu) FIG. 13.5

Shell oval, tumid; almost equilateral, umbones more or less on mid-line with beaks curving anteriorly. Small, rarely exceeding 3 mm in length. Shell white, periostracum light yellow or brown. Sculpture of numerous radiating ribs, extending over whole shell surface, intersecting with equally numerous concentric lines, giving a finely decussate appearance; prodissoconch apparently smooth, with exceptionally fine sculpture. Growth stages clear, but not prominent. Inner surface white, external sculpture visible, adductor scars indistinct. Margin crenulate, crenulations extending along hinge line and resembling small teeth, developed below umbo as small, grooved callus.

Offshore, on sand and fine gravel. Reported from the Bristol Channel northwards, absent from the south coasts of the British Isles.

Rhomboidella prideauxi (Leach) FIG. 13.5

Shell rhomboidal, tumid; inequilateral, umbones anterior to mid-line with beaks directed anteriorly, posterior margin sloping rapidly away from umbones, becoming convex ventrally, anterior margin strongly convex. Shell white, periostracum light yellow. Sculpture of numerous radiating ribs and fewer concentric lines, not noticeably decussate; growth stages distinct, ridged, forming raised scales where they intersect with ribs. Prodissoconch with fine concentric lines but no ribs. Inner surface white, external sculpture visible, adductor scars indistinct. Margin crenulate, continuous with a small ridged callus below umbo.

From lower shore downwards; south coast and Channel Isles only. Distributed south to the Mediterranean and the Canary Isles.

7 PTERIIDAE

Shell thin, fragile, equivalve; inequilateral, obliquely oval, umbones anterior, hinge line straight, grossly enlarged, as long as the greatest length of the shell, forming a short anterior and a long posterior wing. Ligament external, in a slender cardinal area. A single tooth in the left valve. A single adductor muscle scar.

Pteria hirundo (Linnaeus) FIG. 13.4

Shell thin and very brittle, valves almost flat; pale brown, white, or translucent. Obliquely oval along an anterior-dorsal/posterior-ventral axis, with umbones well to anterior. Umbones distinct, separated by slender cardinal area. Hinge line straight, produced into prominent ears on each side of umbones, posterior being about five times as long as anterior. Margin deeply concave below posterior ear, convex ventrally, with narrow anterior ventral gape. Periostracum light brown, with narrow, discontinuous streaks of light reddish-brown radiating from umbones. Sculpture of irregular concentric corrugations, forming rough scales close to margin, particularly prominent anteriorly, and closely spaced near to ears.

Lives offshore, attached by a byssus issuing from the anterior gape. On mud, sand, and fine gravel. Reported from all British coasts but predominanty southern and western. Distributed south to the Mediterranean.

8 PINNIDAE

Shell equivalve, inequilateral, triangular; gaping posteriorly and along antero-ventral margin. Numerous close-packed ridges, *septa*, beneath umbones at anterior end of shell, but no hinge teeth. Anterior adductor scar small, close to umbones; posterior larger, close to middle of shell.

Atrina fragilis (Pennant) FIG. 13.4

Shell fragile, thin. Equivalve; inequilateral, roughly triangular. Up to 300 mm long. Dorsal margin straight, ligament thin, inconspicuous, lying along anterior half of margin; ventral margin straight anteriorly, becoming concave or irregular posteriorly, grading into rounded posterior margin. Valves gaping posteriorly and along anterio-ventral margin. Colour light yellow-horn to darker brown, occasionally with few darker patches. Periostracum thin. Sculpture of fine concentric lines and variable number (<20) of smooth, wavy ridges radiating from umbones, raised dorsally and posteriorly into delicate, flat spines. Interior of shell glossy, anterior adductor scar small, hidden beneath umbone, posterior large and distinct, in middle of the valve. A number of vertical ridges, or septa, closely packed, may be seen beneath umbones at anterior end of shell.

Offshore, in mud, sand, or fine gravel; lives with the pointed anterior end deeply buried, and attached by a byssus emerging from the ventral gape. Predominantly southern and western.

9 OSTREIDAE

Shell thick, rugose; inequivalve, left (lower) valve convex, often forming a deep bowl and frequently overlapping the right valve, usually cemented to substratum when small, right valve flat or slightly concave. Inequilateral, umbones anterior to mid-line. Juveniles with small taxodont teeth on each side of umbones, absent in adults. Ligament internal, thick; inner surface nacreous, with single adductor scar, elliptical, distinct, and often recessed.

---

1. Left valve sculptured with coarse concentric grooves and ridges, often raised to form scales, and ten or more fine ribs, worn and indistinct on older parts of shell; right with concentric sculpture but no ribs. Adductor scar white, off-white, or cream **_Ostrea edulis_**

   Both valves with coarse concentric grooves and few (six or seven) very prominent ribs. Adductor scar deep purple or brownish **_Crassostrea gigas_**

Ostrea edulis Linnaeus — FIG. 13.4

Shell thick and solid, but often brittle and flaking easily when dead. Shape variable: typically oval or pear-shaped, or circular when young, becoming irregular, freely influenced by substratum and often distorted. Colour off-white, yellowish, or cream, with light brown or bluish concentric bands on right valve. Up to 100 mm long, rarely larger. Left valve often deeply cupped, with coarse concentric sculpture, frequently developed as flat scales, and crenulated bands marking growth stages; numerous irregular, radiating ribs give a crenulated effect to margin; surface often abraded, or encrusted. Right valve usually more brittle than left, with well-marked concentric sculpture but lacking ribs. Inner surface of both valves pearly, white or bluish grey, often with darker blue areas marking inclosures of detritus; adductor scar elliptical or comma-shaped, smooth-surfaced, white or yellowish white.

Widely distributed around the British Isles, from the lower shore to about 80 m, as indigenous populations or introduced stocks. Occurs naturally from Norway south to the Mediterranean.

Crassostrea gigas (Thunberg) — FIG. 13.4

Shell thick and very rugose; shape variable, usually elongate oval, its greatest dimension being dorsoventral, up to 180 mm, but often less. Umbones prominent and often enrolled. Left valve deeply cupped, with coarse concentric sculpture, growth stages prominent, raised and frilled, forming flat scales; few, usually about six, very bold raised ribs, imparting grossly crenulate, saw-toothed, appearance to margin. Right valve similarly sculptured, ribs corresponding to channels in opposite valve. Colour off-white, yellowish, or bluish grey, often with darker purplish patches. Inner surfaces of both valves pearly, white or bluish white, external sculpture extending about one-third of its length into the shell; adductor scar distinct, deep purplish-blue or brown.

Introduced to south-east and south-west coasts, breeding only sporadically. Dead shells may occur elsewhere. The American oyster, *C. virginica* (Gmelin) is deep purple or reddish brown in colour but unlike *C. gigas*, the shell lacks bold ribs and a crenulate margin.

10 PECTINIDAE

Shell thick and strong, tyuically oval or circular, occasionally irregular and distorted, usually with bold, radiating ribs. Equivalve or inequivalve, one valve often flat or less concave than the other. Hinge line straight, produced as short, projecting ears on each side of umbones. Equilateral, or inequilateral with anterior ears longer than posterior. Hinge line without teeth, although occasionally infolded along its edge to form articulating surfaces. Ligament internal, in a well-marked chondrophore. One adductor scar only.

---

1. Both shell valves irregular or distorted; ribs often worn, or obscured by concretions. Right valve strongly convex, usually cemented to the substratum in adults **_Chlamys distorta_**

   Both valves regular, without distortion in shape or outline **2**

2. Right valve strongly convex, left valve flat; each with up to 17 broad, rounded ribs. Large, up to 150 mm long **_Pecten maximus_**

   Both valves shallowly convex **3**

3. Anterior ear of each valve much longer than posterior **4**

   Anterior ear of right valve with a byssal notch; otherwise ears of both valves essentially similar **6**

4. Shell sculpture of very fine radiating lines, occasionally prominent at edges of shell as ridges: 30–40 in number, up to 10 of which may be particularly broad and prominent, variably developed, but not regular over whole surface. Margin of shell semicircular. Often with coloured bands, rays, and zigzags **_Palliolum tigerinum_**

   Sculpture of raised ribs extending from umbones to margin, always prominent. Margin of shell forming a semi-ellipse **5**

5. Up to 35 radiating ribs, of regular width, each with numerous spines, particularly erect near margin, where they may overlap edge. Byssal notch in right anterior ear deep and conspicuous, with distinct teeth on the lower side **_Chlamys varia_**

   Each valve with 35–50 radiating ribs, of irregular width; spines less frequent, often small. Byssal notch in right anterior ear shallow, not particularly conspicuous **_Chlamys distorta_** juv.

6. Each valve with about 20 radiating ribs, without spines and smooth on the crest, but separated by transverse corrugations **_Aequipecten opercularis_**

   Sculpture of fine radiating lines and concentric striations, with 3–10 (typically seven) smooth, flattened ribs **_Pseudamussium septemradiatum_**

Pecten maximus (Linnaeus) FIG. 13.6

Shell thick and solid; inequivalve, left valve flat, right valve strongly convex, slightly overlapping the left along its margin. Equilateral, with a small byssal notch in the right anterior ear. Ears equal. Right (convex) valve off-white, yellowish, or light brown, often with bands or spots of darker pigment, left valve light pink to reddish brown; periostracum dirty brown. Up to 150 mm long, slightly longer than deep, both valves with bold, radiating ribs, about 16 in number, distinctly quadrate in section on left valve, but more rounded on right. Finer radiating ridges on both ribs and intervening grooves, concentric sculpture of corrugated lines, growth stages clear. Ears with numerous fine ridges. Inner surfaces of shell nacreous, often with patches or streaks of dark or light chestnut, external sculpture discernible over most of the shell.

Attached by a byssus when small, later free and capable of swimming. Lives in depression in sand or fine gravel with left valve uppermost, offshore to about 100 m. Widespread off all British coasts; distributed from Norway to the Atlantic coast of Spain.

Aequipecten opercularis (Linnaeus) FIG. 13.6

Shell thin, brittle. Inequivalve, left valve more convex than right. Equilateral, with ears more or less equal; anterior ear of right valve slighty larger than posterior, overlapping that of left valve and with distinct fold along hinge line, below it a small, rounded byssal notch, finely dentate along lower edge. Up to 90 mm long, frequently less, as long as deep, ventral margin constituting more than a semicircle. Colour very variable: light pink to brown, orange, or yellow, usually marked with bands, zigzags, rays, and spots of darker or lighter shades, right valve often paler than left, frequently with coloured markings on a creamy ground. Sculpture of about 20 bold, radiating ribs, finer ridges, and concentric corrugated grooves, ears similarly sculptured; growth stages unclear; margin very crenulate.

Attached by byssus when young, later unattached, lives with the left valve uppermost and swims freely and swiftly. Occurs in dense aggregations offshore, occasionally between tidemarks, to 100 m, on sand and fine gravel. Widespread and common off all British coasts. Distributed from Norway south to the Mediterranean and Canary Isles.

Chlamys distorta (da Costa) FIG. 13.6

Shell thick, irregularly oval. Inequivalve: right valve with small byssal notch in anterior ear, cemented to the substratum in later life, becoming irregular, and more deeply convex than the left, upper valve. Up to 50 mm long. Colour variable, from dull white or grey, to yellow, red or brown, sometimes in irregular patterns. Small individuals, attached only by the byssus, with 35–50 regular, radiating ribs on each valve; larger, cemented shells with up to 70 coarse, spiny ribs on each valve, usually worn or partly obscured by encrusting organisms; margin crenulate.

C. distorta lives attached to hard substrata, from low on the shore, offshore to at least 100 m. Widely distributed, and common on all British coasts. Distributed from northern Norway and Iceland, south to West Africa, and in the Mediterranean.

Chlamys varia (Linnaeus) FIG. 13.6

Shell thin, more or less oval. Inequivalve: right valve with a distinct, rounded byssal notch in anterior ear, with fine teeth on lower border. Inequilateral: both valves with anterior ear more pronounced, twice as long as posterior. Up to 60 mm long. Colour very variable, from off-white, through yellow and orange to brick-red, purple, or brown, often with bands or patches of darker or lighter colour. Very young stages usually unicolorous. Sculpture of 25–35 bold ribs, and a few corrugated concentric lines, raised into prominent spatulate spines along ribs, most pronounced close to margin. Ears similarly ribbed. Growth stages usually clear. Inner surface of shell glossy, often coloured same as exterior, sculpture visible, margin crenulate.

Attached by byssus when young, but may become free later; often grows enclosed by algal holdfasts. Widespread and common off all British coasts, from lower shore to about 100 m. Present in the Mediterranean, distributed south to west Africa.

Palliolum tigerinum (Müller) FIG. 13.6

Shell thin, broadly oval, the curved ventral margin constituting almost a semicircle. Almost equivalve, right valve with narrow byssal notch in anterior ear, few poorly developed teeth on ventral border. Inequilateral: both valves with prominent anterior ear and negligible posterior ear. Rarely exceeding 25 mm in length. Colour very variable, from white or cream through light reddish-brown to purple, in radiating or concentric bands or zigzags, or in irregular blotches, or both combined. A common form has a broad median band extending from umbone to margin. Left valve frequently more highly coloured and patterned than right, which is often unicolorous. Sculpture of very fine concentric and radiating lines, visible only with hand lens, shell otherwise appearing smooth; numerous fine ribs may develop in later growth stages, up to 50 in number, often of unequal width, margin finely crenulate as a result. Growth stages clear but not prominent. Anterior ears of both valves with about six bold ridges.

Distributed from lower shore to about 100 m, most frequent on coarser ground. Widespread and common off all British coasts. Distributed from Norway to the Atlantic coasts of Spain, Portugal, and Morocco.

With the development of ribs *P. tigerinum* may appear similar to *Pseudamussium septemradiatum* (below), but is easily distinguished from the latter by the ears. Two other, rarer, species of *Palliolum* may also be confused with *P. tigerinum*. *P. striatum* (Müller) is a small species, rarely larger than 20 mm, occurring offshore on western coasts. Like some specimens of *P. tigerinum* it may have numerous fine ribs, but is distinguished by the presence of small spines on the ribs and by the byssal notch, which is more prominent than in *P. tigerinum*. *P. furtivum* (Lovén), occurring rarely off southern and western coasts, is recognized by its complete absence of ribs and by the byssal notch, which has numerous fine teeth extending from the lower border of the notch along the suture between ear and shell, almost to the umbone. *Similipecten similis* (Laskey), another rare species, also lacks ribs but has concentric sculpture continuous between ears and valve. It has no teeth on the byssal notch.

Pseudamussium septemradiatum (Müller) FIG. 13.6

Shell thin and fairly brittle, almost circular. Equilateral. Almost equivalve, right anterior ear with very small byssal notch. Up to 50 mm long. Left valve dull brick-red, brown, or purple, with white or cream spots and blotches; right valve pale brick-red with spots in small specimens (< 15 mm), becoming paler as animal grows. In large specimens right valve is largely creamy white with reddish umbones, colour extending, and fading, down the ribs. Sculpture of fine radiating striations and coarser concentric lines (occasionally spinous); each valve with three to ten, most frequently five to seven, evenly spaced, broad rounded ribs, increasing rapidly in width from umbone to margin. The right valve, in addition, may have a group of four or five slender ribs close to anterior margin. Ears with about six ribs each, and concentric corrugations. Growth stages clear.

Lives with the left valve uppermost; occurs offshore to about 200 m, reported from the west coast of Scotland, and the coasts of Northumberland and Durham. Distributed from Norway to the Mediterranean and north-west Africa.

11. LIMIDAE

Shell thin, delicate, typically elongate-oval along a dorsoventral axis. Equivalve. Equilateral; or inequilateral with anterior marginally larger than posterior and curvature of anterior margin more pronounced. Valves often gaping. Umbones spaced apart, with distinct cardinal area between. Sculpture of fine radiating ribs. Usually dull coloured. Ligament internal, in well-marked chondrophore. No hinge teeth. One indistinct adductor muscle scar only.

1. Shell perfectly equilateral, tumid, umbones prominent above short, triangular ears. Sculpture of delicate radiating ridges — ***Limatula sulcata***
   Shell inequilateral, anterior margin more prominent than posterior — **2**
2. Outline of shell bean-shaped. With delicate radiating ribs, without marked gape — ***Limaria loscombi***
   Shell irregularly oval, with large anterior and posterior gape, exposing fringe of brilliant red or orange pallial tentacles — ***Limaria hians***

Limaria hians (Gmelin) FIG. 13.6

Shell thin, fragile, obliquely oval along an anterior dorsal-posterior ventral axis. Inequilateral, anterior ears distinctly more prominent than posterior, sharply pointed; elliptical, dorsal gape on anterior margin and narrower gape along whole of posterior margin. Up to 25 mm long. Colour off-white, often stained dirty brown by surroundings. Sculpture of numerous radiating ribs (approx. 50) and coarse concentric lines; ears, and parts of valves close to them, with concentric lines only. Margin serrated, growth stages clear. Hinge line straight, cardinal area broadly lozenge-shaped.

Limaria hians lives on coarse bottoms, from lower shore to about 100 m. Thick fringe of orange or red pallial tentacles protrudes all around margin of shell. The animal is able to swim; when settled it uses its byssus threads to construct a nest of gravel and shell debris. Reported off southern and western coasts only. Extends south to the Mediterranean and the Canary Isles.

Limaria loscombi (Sowerby) FIG. 13.6

Shell thin, fragile, tumid; approximately bean-shaped in outline, anterior dorsal margin markedly more convex than posterior. Inequilateral; ears small, triangular, anterior slightly more prominent than posterior, with a very small byssal notch ventrally. Valves not gaping. Rarely reaching 20 mm in length, generally less. Colour white, translucent. Sculpture of numerous (> 50) fine radiating ribs, of varying width, and finer concentric lines. Ears without ribs. Growth stages clear. Hinge line straight, with a slender, fusiform cardinal area.

Occurs offshore, from 10–200 m, on fine muddy gravel or sand. Constructs a nest with its byssal threads. Reported from southern and western coasts only; distributed south to the Mediterranean and north-west Africa.

Limatula sulcata (Leach, in Brown) FIG. 13.6

Shell thin, fragile, regularly oval about a dorsoventral axis; both valves strongly convex. Equilateral; ears short and triangular. Rarely larger than 13 mm. Colour off-white to cream. Sculpture of about 30 fine, radiating ribs, and concentric lines, giving a finely decussate effect on the most convex areas of the shell. Ribs absent from ears and anterior and posterior edges of shell. Ventral margin finely crenulate. Umbones prominent, rounded, with slender cardinal area between. Growth stages clear.

Occurs on fine bottoms, mud and sand, from 20 m and deeper. Reported from west Scotland and the Isle of Man. Distributed from Norway south to the Mediterranean and west Africa.

Two other equilateral species of Limidae are found in British waters: *Limatula subauriculata* (Montagu) is similar to *L. sulcata* but is smaller (< 10 mm) and more slender, with fewer ribs. It has been reported offshore, on western coasts only. *Limea sarsi* (Lovén) is more broadly oval, with a coarse sculpture of bold ribs and concentric lines; it rarely exceeds 3 mm in length and is known only from the northern North Sea.

12 ASTARTIDAE

Shell thick, strong, oval or subtriangular; equivalve, inequilateral, umbones just anterior to mid-line. Sculpture usually prominent, typically concentric lines or ridges; periostracum thick. Two cardinal teeth in right valve, two or three in left valve. True lateral teeth absent. Ligament external. Adductor muscle scars about equal; pallial line distinct, without sinus.

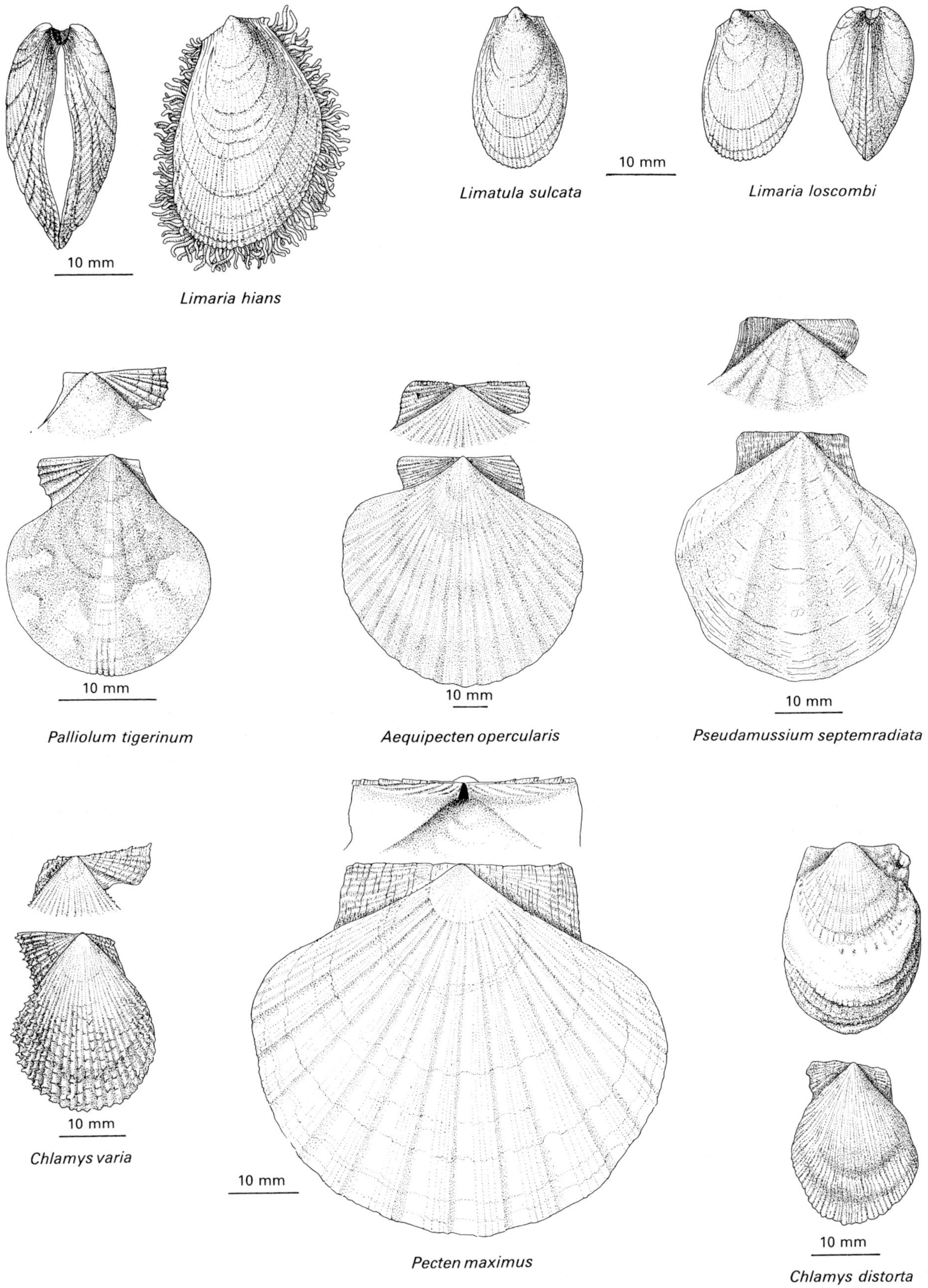
10 mm
Limaria hians
Limatula sulcata
10 mm
Limaria loscombi
10 mm
Palliolum tigerinum
10 mm
Aequipecten opercularis
10 mm
Pseudamussium septemradiata
10 mm
Chlamys varia
10 mm
Pecten maximus
10 mm
Chlamys distorta

Fig. 13.6

1. Shell dorso-ventrally elongate, triangular, rarely exceeding 3 mm in length ***Goodallia triangularis***
   Shell more broadly oval, up to 30 mm long **2**

2. Inner margin of shell crenulate; sculpture of about 20 bold concentric ridges. Up to 30 mm long ***Astarte sulcata***
   Inner margin of shell smooth; sculpture of 30–40 finer concentric ridges. Up to 15 mm long ***Tridonta montagui***

Astarte sulcata (da Costa) FIG. 13.7

Shell thick and robust, broadly oval, prominent umbones imparting subtriangular outline. Usually white (rarely pink) with thick, mid- to dark-brown periostracum, often encrusted and stained black about umbones. Up to 30 mm long. Lunule and escutcheon distinct, deep; ligament visible in anterior half of escutcheon. Sculpture of bold concentric ridges, growth stages indiscernible. Left valve with two prominent cardinal teeth below umbones and a third, slender, indistinct and often abraded, extending below ligament. Right valve with very thick anterior cardinal tooth and slender one posteriorly; additionally, slender ridge below escutcheon may be taken for posterior lateral tooth. Inner surface of shell dull white, often with pink or brown tinge; adductor scars and pallial line very clear, inner shell margin distinctly crenulate (rarely, smooth).

In mud, or fine gravels with sand, shell, or mud, offshore. Present off all British coasts; distributed from southern Arctic waters to the Mediterranean and north-west Africa.

Tridonta montagui (Dillwyn) FIG. 13.7

Shell broadly oval, umbones centred on mid-line, prominent, giving subtriangular outline. Up to 15 mm long. Colour white, with a thick, light or dark brown periostracum. Lunule and escutcheon distinct, elongate, ligament occupying half length of latter. Sculpture of 30–40 raised concentric ridges, growth stages unclear. Right valve with two cardinal teeth: anterior broad, ridged or bifid, posterior indistinct. Left valve with two equal-sized teeth anteriorly; posteriorly, the third is poorly developed and may appear to be absent. Inside of shell smooth, white; margin smooth, adductor scars and pallial line very indistinct.

Found offshore to about 100 m, in sand and fine gravel. A circumboreal species, most frequent off north and west coasts, extending no farther south than the Bay of Biscay.

The northern species *T. elliptica* (Brown) resembles *Astarte sulcata*, but is distinguished by its smooth margin. It is larger than *T. montagui*, with more prominent umbones and always fewer than 30 concentric ridges. *T. borealis* (Schumacher), reported from the northern North Sea, is broadly oval in shape with concentric lines rather than raised ridges.

Goodallia triangularis (Montagu) FIG. 13.7

Shell triangular, its greatest dimension being along its dorso-ventral axis; umbones prominent, situated on mid-line. About 3 mm long. Colour white, overlain by light yellowish to brown periostracum. Lunule and escutcheon indistinct, shallow. Sculpture of numerous fine concentric lines, closely spaced. Growth stages distinct. Right valve with two cardinal teeth, anterior broad and triangular, posterior slender. Left valve with two approximately equal-sized cardinals, posteriorly a third one appears as a ridged projection, like a lateral tooth. Inner surface white, margin with distinct beadlike crenulations.

Occurs offshore, to at least 100 m, in sand and fine gravel. Widespread off all British coasts, distributed south to the Mediterranean and north-west Africa.

13 LUCINIDAE

Shell equivalve; inequilateral, umbones just anterior to mid-line. Hinge line with cardinal and lateral teeth; ligament external, often deeply inset, lunule and/or escutcheon typically well-developed. Pallial line without sinus. Adductor scars unequal: anterior larger, often partly or largely separated from pallial line. Shell usually white and dull.

1. Anterior adductor scar very elongate and narrow, touching the pallial line only at its anterior end. Shell with numerous, fine, closely spaced, concentric ridges ***Lucinoma borealis***
   Anterior adductor scar fused with the pallial line along about half of its anterior edge. Shell with concentric lines or ridges **2**

2. Shell margin strongly convex anterior to umbones. Escutcheon absent. Sculpture of fine concentric lines ***Loripes lucinalis***
   Shell margin shallowly concave anterior to umbones. Escutcheon distinct, bordered by raised, crenellated margins ***Myrtea spinifera***

Lucinoma borealis (Linnaeus) FIG. 13.7

Shell thick, strong, tumid, roughly circular; colour dull white, with a thin light-brown periostracum. Up to 40 mm long. Umbones small but distinct, just in front of mid-line; lunule pyriform, escutcheon fusiform, deep. Ligament clearly visible. Left valve with a bifid anterior and a single posterior cardinal tooth, right valve with single anterior and bifid posterior. Each valve with single anterior and posterior lateral teeth, weakly developed, often worn and indistinct. Sculpture of numerous fine concentric ridges, growth stages clear; surface often abraded in older shells. Inner surface white, pallial line and adductor muscle scars distinct; anterior scar up to five times as long as broad, only its narrow anterior end attached to pallial line.

In sand and gravel from the lower shore to about 100 m. Ranges from Norway to the Mediterranean, present on most British coasts.

Loripes lucinalis (Lamarck) FIG. 13.7

Shell approximately circular, almost equivalve; white or cream, periostracum with very pale beige tints. Up to 20 mm long. Umbones distinct, medially situated but curving anteriorly beyond mid-line. Lunule short and deep, no escutcheon; liga-

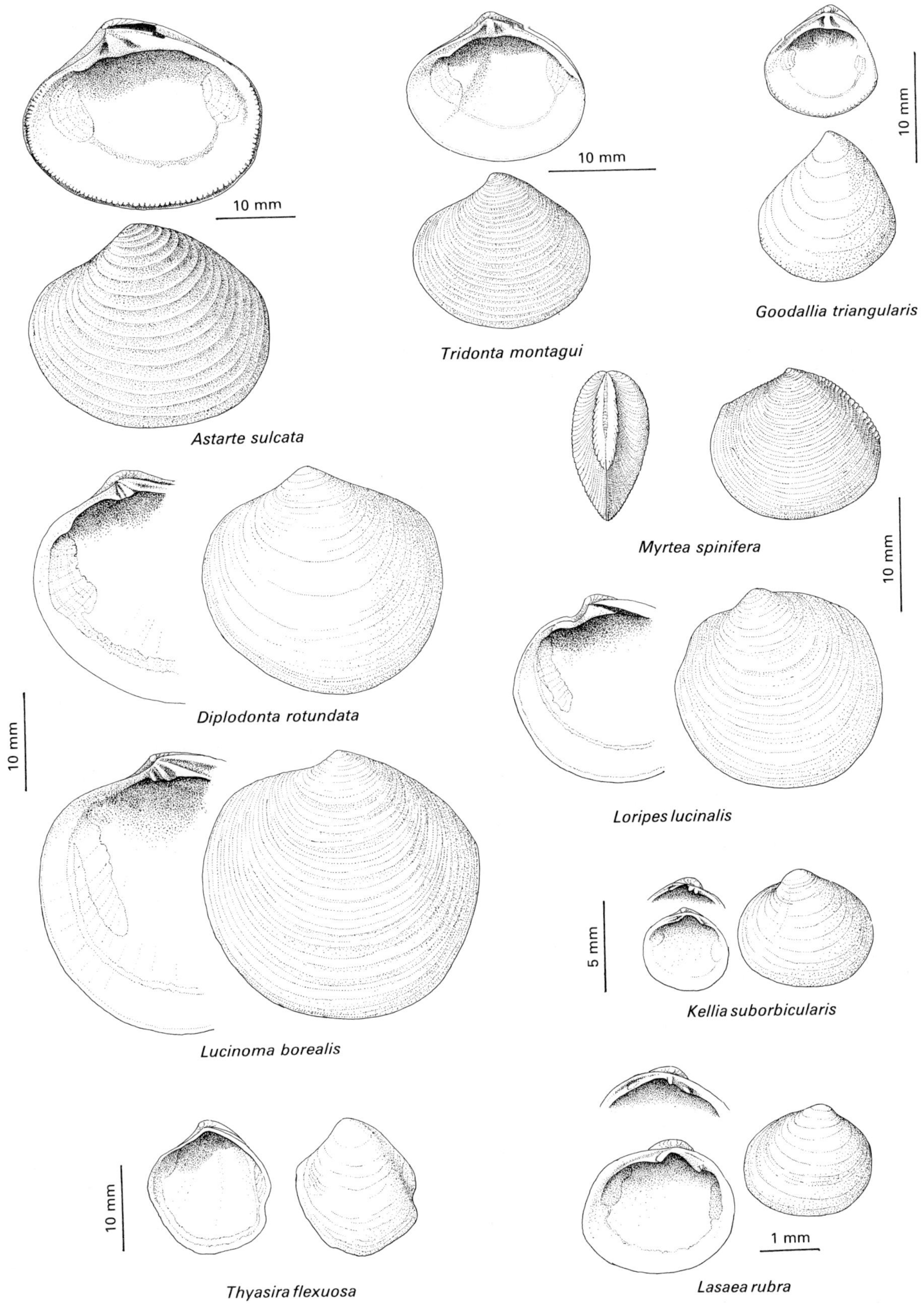
10 mm
Astarte sulcata
10 mm
Tridonta montagui
10 mm
Goodallia triangularis
Myrtea spinifera
10 mm
Diplodonta rotundata
10 mm
Loripes lucinalis
Lucinoma borealis
5 mm
Kellia suborbicularis
10 mm
Thyasira flexuosa
1 mm
Lasaea rubra

Fig. 13.7

ment deeply inset and hidden. Left valve with bifid anterior and single posterior cardinal teeth, right valve with single anterior and bifid posterior. Each valve with single anterior and posterior lateral teeth, often indistinct. Sculpture of numerous fine, concentric lines, growth stages clear. Inner surface white; pallial line distinct, anterior adductor scar elongate-oval, fused with the pallial line along the dorsal one-third of its anterior border.

In sand and fine gravel, from the lower shore to at least 150 m. Reported from the Channel, off the Northumberland coast, and from the Isle of Man; distributed south to the Mediterranean and west Africa.

Myrtea spinifera (Montagu) FIG. 13.7

Shell thick, oval; beaks just anterior to mid-line, pointed and extending anteriorly. White or cream with an indistinct, pale beige periostracum. Up to 25 mm long. Dorsal margin anterior to umbones concave, with a distinct oval lunule; ligament visible in deep fusiform escutcheon bordered laterally by erect, flared, and crenellated rims. Sculpture of numerous fine, closely spaced, concentric ridges; growth stages unclear. Right valve with single cardinal tooth, and single anterior and posterior lateral teeth. Left valve with two cardinal and single anterior and posterior lateral teeth, but in addition ridges may be developed along hinge line and appear as extra laterals. Inner surface of shell white, pallial line distinct; anterior adductor scar elongate-oval, roughly dumb-bell shaped, touching the pallial line along its anterior ventral border.

In mud and muddy sands, offshore to about 100 m. Distributed on western coasts from south Devon to the Shetland Isles, rarely elsewhere. Ranges from Norway to the Mediterranean and the Azores.

14 DIPLODONTIDAE

Shell equivalve; inequilateral, umbones anterior to mid-line. Hinge line with cardinal and lateral teeth, ligament external. Pallial line without sinus. Adductor muscle scars subequal, anterior being the larger.

Diplodonta rotundata (Montagu) FIG. 13.7

Shell thin, brittle, tumid; irregularly oval, posterior half deeper than anterior. Colour white, periostracum light yellowish brown; surface smooth or silky in unworn specimens. Up to 30 mm long. Lunule indistinct; escutcheon well-marked, deep, ligament distinct. Left valve with bifid anterior cardinal tooth and a single posterior cardinal tooth; right valve with single anterior and bifid posterior. Both valves with single anterior and posterior lateral teeth. Sculpture of very fine concentric lines; growth stages clear, often coarse in larger specimens. Inner surface of shell white, lustrous; anterior adductor scar dorso-ventrally elongate, comma-shaped, closely applied to the pallial line.

In sand and gravel, offshore to about 70 m. Reported from the English Channel and the south and west Irish coasts, south to the Mediterranean and west Africa.

15 THYASIRIDAE

Shell equivalve, inequilateral. Ligament external. Hinge line without teeth. Typically with fine concentric sculpture, growth stages indistinct. Adductor scars about equal. Pallial line unclear, without sinus.

Thyasira flexuosa (Montagu) FIG. 13.7

Shell thin and fragile, broadly oval, tending to be irregular. Hinge line concave anteriorly, convex posteriorly; umbones on mid-line, beaks clearly directed anteriorly. Lunule heart-shaped, shallow; escutcheon elongate, well-marked, hinge line pouting. Dull white, with pale yellowish brown periostracum; up to 10 mm long. Sculpture of fine concentric lines, growth stages visible but not conspicuous. Smooth ridge extends postero-ventrally from umbones to margin of each valve, marked on either side by distinct groove; lower groove more conspicuous and causes deep indentation where it reaches posterior margin. Hinge teeth absent, although hinge line of right valve may be thickened below umbo, forming short callus, or peg. Inner surface white; pallial line broad, ill-defined, adductor scars indistinct.

Offshore, to the edge of the Continental Shelf; in muddy sand. Present off all British coasts and often common; distributed from Norway to the Azores, extends into the Mediterranean.

Two other species of *Thyasira* occur off the north and west coasts of Scotland, *T. ferruginea* (Winckworth) and *T. croulinensis* (Jeffreys). Although similar in outline to *T. flexuosa*, they differ principally in lacking the umbonal ridge and the posterior indentation in the margin.

KELLIIDAE, LASAEIDAE, LEPTONIDAE, MONTACUTIDAE

1. Shell thin and flat, fragile, approximately square in outline, outer surface densely punctate **_Lepton squamosum_**

   Shell often thin and fragile, but distinctly convex and broadly oval in outline. Without punctulations **2**

2. Right valve with one cardinal tooth, left valve with two; one posterior lateral tooth in each valve, but no anterior lateral **_Kellia suborbicularis_**

Cardinal teeth present or absent; anterior lateral teeth present in one or both valves, but no posterior laterals **3**

Cardinal teeth present or absent; one anterior and one posterior lateral present in each valve **4**

---

3. Shell with fine radiating ridges, most distinct about the umbones. White. Single anterior lateral tooth in each valve, other teeth absent ***Montacuta substriata***

Shell with concentric lines only. White, usually stained with red deposit. Left valve with one anterior lateral tooth, right valve with one cardinal tooth ***Tellimya ferruginosa***

---

4. Shell with fine radiating striations. Each valve with single cardinal tooth, and single anterior and posterior laterals ***Epilepton clarkiae***

Shell with fine concentric lines only, or with few striations irregularly developed about the umbones **5**

---

5. Shell elongate-oval, with low umbones, cardinal teeth absent from both valves. Distributed from the lower shore to the outer continental shelf ***Mysella bidentata***

Shell plump, broadly oval, with prominent umbones. Left valve with single cardinal tooth. Between tidemarks only, attached by byssus ***Lasaea rubra***

---

16 KELLIIDAE

Shell equivalve, inequilateral, small. Ligament internal and external; hinge with cardinal teeth, typically small and indistinct, and poorly developed lateral teeth. Pallial line broad, ill-defined, without sinus.

Kellia suborbicularis (Montagu) FIG. 13.7

Shell tumid, umbones just in front of mid-line. Broadly oval, almost circular; white, with pale horn or yellowish periostracum. Up to 10 mm long. Sculpture of fine concentric lines, growth stages distinct. Lunule broad and shallow, escutcheon scarcely discernible. Right valve with a single cardinal tooth below beak of umbo, and single posterior lateral; left valve with two cardinal teeth below beak, and single posterior lateral. Small external ligament visible posterior to beaks; in each valve position of internal ligament is maked by embayment in hinge line posterior to cardinal tooth/teeth. Inside of shell white or translucent; pallial line broad, ill-defined, adductor scars about equal, not distinct.

Distributed from the lower shore to at least 100 m; occurs in holes and crevices, within old shells, barnacle tests, etc. Most frequent on southern and western shores. Distributed from Norway south to the Mediterranean and West Africa.

17 LASAEIDAE

Shell equivalve, inequilateral, small. Ligament internal; hinge line with cardinal and lateral teeth, few and small. Adductor scars about equal, pallial line without sinus.

Lasaea rubra (Montagu) FIG. 13.7

Shell thin and fragile, obliquely oval, tumid, both valves strongly convex; inequilateral, umbones posterior to mid-line. Rarely exceeding 3 mm in length. Colour white, with concentric bands of pale pink or red, darkest towards margins; periostracum light yellow-brown. Sculpture of numerous, closely spaced, fine, concentric lines, a few faint radiating striae may be present; growth stages clear, raised. Right valve with single anterior and posterior lateral teeth, left valve with one small cardinal tooth and single anterior and posterior laterals; in both valves position of the internal ligament is marked by deep embayment below hinge line and broad plate below posterior lateral tooth. Inner surface white, adductor scars relatively large, pallial line broad and ill-defined.

An intertidal species, common on rocky shores all around the British Isles. Attached by byssus, it is found in crevices, among *Fucus* clumps or in *Lichina* tufts; often present in very large numbers. Distributed from Norway to the Mediterranean and the Canary Isles.

The venus shell *Turtonia minuta* is the same size as *Lasaea rubra*, occupies similar habitats and superficially resembles it. However, *Turtonia* may be distinguished by the group of three cardinal teeth present in each valve.

18 LEPTONIDAE

Small round shells, thin and generally rather fragile. Equivalve; equilateral or just inequilateral. Ligament internal. Hinge with one cardinal tooth centrally, and with anterior and posterior lateral teeth. Adductor scars about equal. Pallial line indistinct, without sinus.

Lepton squamosum (Montagu) FIG. 13.8

Shell very thin and fragile, roughly square in outline, umbones small, just anterior to mid-line. Up to 12 mm long. White, translucent, periostracum thin and inconspicuous. Sculpture of fine concentric lines, most cospicuous close to distinct growth stages; surface of both valves marked with dense punctulations, reminiscent of a brachiopod. Hinge plate distinct; right valve with one minute cardinal tooth and single anterior and posterior lateral teeth, left valve with small cardinal tooth and single anterior and posterior lateral teeth. Inside of shell translucent, with external punctulation showing through. Adductor scars about equal, pallial line diffuse.

In silty sand and fine gravel from ELWS into shallow sublittoral; this species lives in or close to burrows of callianassid shrimps (*Upogebia* spp.). Distributed from Norway to the Mediterranean; in British waters, most frequent off the southern and western coasts. *Lepton nitidum* (Turton) is very similar to *L. squamosum* but is almost circular in shape; it occurs offshore to perhaps the edge of the Continental Shelf, but is not commensal.

Epilepton clarkiae (Clark) FIG. 13.8

Shell thin and fragile, inequilateral, umbones posterior to mid-line; obliquely oval, rarely reaching 2 mm in length. Colour yellowish white, periostracum indistinct, visible as darker bands about the growth stages. Sculpture of very fine concentric lines and irregular, radiating striations; growth stages clear. Each valve with one cardinal tooth and single anterior and posterior lateral teeth; distinct in right valve, less so in left. Adductor scars and pallial line indistinct.

Distributed from the lower shore into the sublittoral, in association with various species of sipunculid. Reported from south and west coasts of the British Isles; present in Mediterranean.

The coin shells *Neolepton sulcatulum* (Jeffreys) and *N. sykesi* (Chaster) have similar dentition to *E. clarkiae*. Both are almost circular in outline, with bold concentric lines, and in British waters are known only from the Channel Isles.

19 MONTACUTIDAE

Small, fragile shells. Equivalve; inequilateral. Hinge line with thin anterior and/or posterior lateral teeth, cardinal teeth often absent. Ligament internal. Adductor scars about equal. Pallial line diffuse, without sinus. Many commensal species.

Tellimya ferruginosa (Montagu) FIG. 13.8

Shell thin and fragile, elongate-oval, umbones posterior to mid-line; up to 8 mm long. White; periostracum thin, light yellowish, usually covered with thick, granular, rusty red deposit. Sculpture, when visible, of fine concentric lines and few radiating striae. Right valve with single cardinal tooth, left valve with one anterior lateral. Inside of shell dull white, sometimes tinted purple. Adductors scars clear, anterior slightly larger, pallial line wide.

This bivalve lives commensally in the burrows of the sea urchin *Echinocardium cordatum*. Widespread and common off all British coasts; distributed from Norway to the Mediterranean and north-west Africa.

Montacuta substriata (Montagu) FIG. 13.8

Shell thin and fragile, obliquely oval, umbones posterior to mid-line; rarely exceeding 3 mm long. White, with inconspicuous pale brown periostracum. Ventral margin often sightly concave medially. Sculpture of fine concentric lines and few low, radiating ribs. Growth stages clear. Each valve with single anterior lateral tooth only. Inner surface of shell white, adductor scars clear, anterior slightly larger, pallial line wide and diffuse.

M. substriata lives in association with sea urchins, usually attached to the anal spines. Recorded from various localities, associated with *Spatangus purpureus* and *Echinocardium flavescens*. Distributed from ELWS, offshore to about 100 m. Ranges from Norway to the Mediterranean.

Mysella bidentata (Montagu) FIG. 13.8

Shell thin and fragile, oval, umbones posterior to mid-line; up to 3 mm long. White or translucent, periostracum light brown or olive. Sculpture of fine, closely spaced, concentric lines; growth stages clear. Each valve with single anterior and single posterior lateral tooth. Inside of shell white, adductor scars about equal, distinct; pallial line broad.

From ELWS to about 100 m, in muddy sand or fine gravel, in crevices of dead oyster valves, in the burrows of the sipunculid *Golfingia*, or associated with the ophiuroid *Acrocnida brachiata*. Widespread around Britain, it ranges from Norway to West Africa and the Mediterranean.

20 GALEOMMATIDAE

Small shells. Equivalve, inequilateral. Ligament internal. Hinge line without teeth. Adductor scars about equal. Pallial line without sinus. Many commensal species, attaching by byssus, or by modified foot.

Galeomma turtoni Sowerby FIG. 13.8

Shell thin and brittle, umbones anterior to mid-line. Oblong-oval, slightly deeper posteriorly than anteriorly; up to 15 mm long. Colour white or pale brown. Hinge line straight, without teeth, umbones low, prodissoconch smooth and prominent. Anterior and posterior margins downcurved, ventral margin shallowly convex, with a wide ventral gape. Valves regularly convex or with a deep dorsoventral channel along the mid-line. Sculpture of numerous fine radiating ribs and concentric ridges, intersecting to give regularly decussate effect. Growth stages indistinct. Inner surfaces dull white, adductor scars and pallial line indistinct.

Distributed from the lower shore into the shallow sublittoral, attached by a byssus or freeliving. Occurs on south-western shores and around the Channel Isles; ranges south to the Mediterranean and West Africa.

Devonia perrieri (Malard) FIG. 13.8

Shell thin and fragile, umbones situated close to posterior end. Elongate, almost rectangular in outline, up to 5 mm long. White, translucent, with pale brown tints, periostracum thin, light brown. Posterior margin with narrow gape. No hinge teeth. Sculpture of fine concentric lines, growth stages clear.

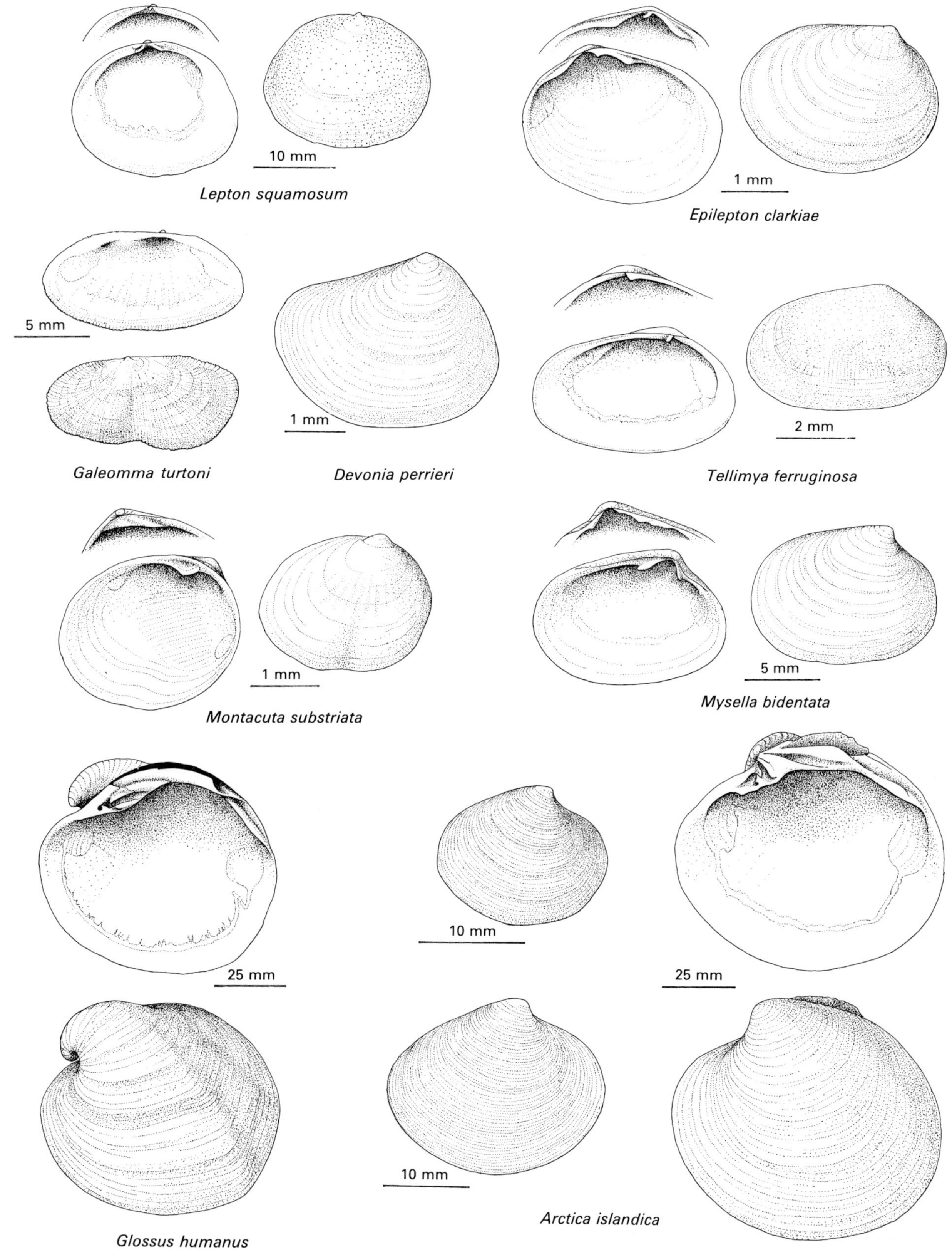
10 mm
Lepton squamosum
1 mm
Epilepton clarkiae
5 mm
1 mm
2 mm
Galeomma turtoni
Devonia perrieri
Tellimya ferruginosa
1 mm
Montacuta substriata
5 mm
Mysella bidentata
25 mm
10 mm
25 mm
10 mm
Glossus humanus
Arctica islandica

Fig. 13.8

A little-known species, *D. perrieri* lives attached to the holothurian *Leptosynapta*. Recorded from Salcombe, Falmouth, and Shetland, on *L. inhaerens*, and from Cullercoats on *L. bergensis*.

21 ARCTICIDAE

Shell equivalve; inequilateral, umbones anterior to mid-line. Periostracum thick, conspicuous; ligament external. Cardinal and lateral teeth present. Pallial line without sinus.

Arctica islandica (Linnaeus) FIG. 13.8

Shell thick and strong, broadly oval, with prominent umbones, up to 120 mm long; dull white with thick periostracum: glossy chestnut-brown in smaller individuals, deep greenish brown to black in large specimens, typically brittle and peeling readily in largest shells. Anterior hinge line strongly curved, lunule ill-defined; escutcheon shallow, occupied by thick, arched, brown or black ligament. Sculpture of concentric lines and few irregular grooves, growth stages distinct. Right valve with three prominent cardinal teeth and single posterior lateral tooth; triangular pit in front of anterior cardinal, surrounded by small knobs and ridges. Left valve with three cardinals and one posterior lateral, anterior cardinal continuous with series of small ridges and denticulations. Inner surfaces smooth, glossy; adductor scars distinct, posterior slightly larger than anterior. Ventral margin crenulate.

Young specimens appear rather different from older ones, the periostracum tending to be smoother and lighter in colour; umbones are relatively less prominent and the shells appear to have a more oval and less tumid shape.

In sand and muddy sand; offshore, perhaps to the edge of the Continental Shelf. Off all British coasts, ranges from Arctic waters to the Bay of Biscay.

22 GLOSSIDAE

Shell equivalve, inequilateral, umbones prominent, enrolled and coiled. Ligament external. Hinge line with cardinal and lateral teeth. Pallial line without sinus.

Glossus humanus (Linnaeus) FIG. 13.8

Shell thick and strong, very tumid, with prominent, rounded and tightly enrolled umbones, curving away from hinge line; umbones anterior to mid-line. Valves strongly convex, roughly circular, deeper posteriorly than anteriorly. Dull white or beige, periostracum thick, light brown or olive, darkening to deep brown or green in larger specimens. Up to 100 mm long. Lunule broad, ill-defined, hinge line pouting; escutcheon shallow, negligible. Three cardinal teeth in each valve, parallel to hinge, thick and prominent. In right valve both posterior cardinals are closely set, almost in continuous line, separated by deep groove from anterior tooth; in left valve, narrow ridged posterior cardinal is separated by groove from two closely set anterior teeth. Single ridge-like posterior lateral tooth in each valve. Sculpture of numerous concentric lines, growth stages distinct. Inside of shell dull white, adductor scars distinct, anterior smaller than posterior; pallial line broad, glossy, without sinus.

In soft substrata, offshore; mostly on southern and western coasts, from west Scotland and Ireland to the Bristol Channel. Distributed from Norway to the Mediterranean.

23 CARDIIDAE

Shell equivalve, tumid, both valves strongly convex; inequilateral, umbones just anterior to mid-line, prominent. Sculpture of bold radiating ribs, often with conspicuous spines, tubercles, or scales. Two peg-like cardinal teeth in each valve. Lateral teeth present. Adductor scars about equal, pallial line without sinus.

1. Radiating ribs with very low relief, most prominent ventrally, no knobs or spines. Prodissoconch smooth and glossy **_Laevicardium crassum_**
   Radiating ribs conspicuous, visible even on prodissoconch, typically with knobs or spines **2**

2. Ribs visible on the inner surface of the shell as grooves, extending practically to the region below the umbones **6**
   Ribs visible only on the ventral part of the inner surface, the grooves fading rapidly beyond the pallial line; area beneath umbones completely smooth **3**

3. Right valve with two posterior lateral teeth; left valve with single anterior and posterior lateral teeth equally developed **_Cerastoderma edule_**
   Right valve with one posterior lateral tooth; left valve with anterior lateral tooth more prominent than posterior **4**

4. Up to 20 ribs, each with regularly spaced spines, distinct ventrally but forming sharp spiny ridge dorsally, or on young specimens. Up to 90 mm long **_Acanthocardia tuberculata_**
   20 or more ribs, occasionally with small tubercles close to the margins but without spines. Not exceeding 20 mm **5**

5. Up to 28 ribs, each with transverse grooves, developing flat, quadrate tubercles about the margins. **_Parvicardium scabrum_**
   Up to 22 ribs, with sparsely developed, conical tubercles close to the margins **_Parvicardium exiguum_**

6. Each rib with sharp keel bearing numerous pointed spines (bases visible in worn specimens) **7**
   Ribs with or without blunt tubercles developed close to anterior and posterior margins, but not keeled **8**

7. Posterior margin convex; spines long and prominent anteriorly, usually shorter posteriorly. Cardinal teeth of left valve equal-sized. **_Acanthocardia echinata_**
   Posterior margin almost straight, spines short and thick anteriorly, longer and sharper posteriorly. In left valve, anterior cardinal tooth marginally longer than posterior **_Acanthocardia aculeata_**

8. Right valve with two posterior lateral teeth. Concentric sculpture extending over surface of ribs, occasionally raised to form flat scales but not tuberculate. Up to 60 mm **Cerastoderma glaucum**

   Right valve with one posterior lateral tooth. Concentric sculpture restricted to grooves between ribs. Ribs smooth, with few quadrate tubercles **9**

9. 28 or more ribs, each with short, flattened tubercles; concentric sculpture well-developed, forming series of lattices between ribs **Parvicardium minimum**

   Up to 26 ribs; tubercles sparsely developed, most prominent close to anterior and posterior margins **Parvicardium ovale**

Acanthocardia echinata (Linnaeus) FIG. 13.9

Shell brittle, obliquely oval, anterior hinge line sloping gently to convex anterior margin; posterior hinge line more steeply inclined, posterior margin only slightly convex. Up to 75 mm long. Light fawn to deeper brown, in concentric bands, often darker or bluish-tinged along posterior margin. Periostracum thin, typically stained greyish by fine sediment. 18–23 ribs, each with sharp central keel, regularly produced into erect, sharp spines, continuous basally with keel. On small specimens the spines are little more prominent than keel, in later growth stages keel is less obvious; old specimens may have both severely eroded; spines most prominent on anterior ribs. Concentric sculptures of numerous wavy ridges, pronounced between ribs, less so on them. Growth stages clear. Two cardinal teeth in each valve; right valve with two anterior and one posterior lateral teeth, left valve with single anterior and posterior laterals, the anterior being longer, thicker, and more prominent than posterior. Inner surfaces white, glossy, occasionally bluish beneath umbones. External sculpture visible as grooves which extend whole depth of shell. Adductor scars and pallial line indistinct.

In fine sand and gravel, usually with mud, offshore; off all British coasts, ranges from Norway to the Mediterranean and the Canary Isles.

Acanthocardia tuberculata (Linnaeus) FIG. 13.9

Shell thick and strong, approximately rhombic in shape: hinge line sloping gently from umbones on each side, posterior margin scarcely convex, almost straight, anterior margin more strongly convex. Up to 90 mm long. Off-white, yellow, or light brown, often in concentric bands of different shades, frequently darker about umbones. Periostracum thin, yellowish. Sculpture of 18–20 bold ribs, and fine concentric grooves and ridges. Dorsally, each rib has central keel, bearing short pointed spines; ventrally keel is obscured and spines appear separate. Growth stages clear. Two cardinal teeth in each valve; right valve with two anterior and one posterior lateral teeth, left valve with single anterior and posterior laterals. Inner surface white, glossy ventrally, becoming dull beneath the umbones; pallial line and adductor scars distinct. External sculpture visible as grooves extending from ventral margin to pallial line, fading rapidly beyond it and smooth beneath umbones.

On muddy sand and gravel, from the lower shore into the shallow sublittoral. Southern coasts only, ranging south to the Mediterranean and north-west Africa.

Acanthocardia aculeata (Linnaeus) FIG. 13.9

Shell markedly eccentric: anterior margin regularly convex, continuous with hinge line; posterior margin almost straight, with marked angle between it and posterior hinge line. Up to 100 mm long. Off-white, yellowish, or light chestnut, usually with concentric bands of colour, occasionally with pink tinge. Periostracum thin, yellowish; often stained grey with silt. About 20 bold ribs, with distinct concentric grooves and ridges between. Each rib with sharp keel, bearing regularly spaced spines; longest and sharpest spines towards posterio-ventral margin, anteriorly they are short, blunt, and recurved towards hinge line. Growth stages distinct. Each valve with two cardinal teeth, in left valve anterior (lower) tooth is larger than posterior; right valve with two anterior and one posterior lateral teeth, left with single anterior and posterior teeth. Inner surfaces white, with pink tints along the ribs; sculpture visible as grooves extending through whole of shell. Adductor scars clear, pallial line indistinct.

South and west coasts only, offshore. Distributed south to the Mediterranean and north-west Africa.

Cerastoderma edule (Linnaeus) FIG. 13.9

Shell broadly oval, up to 50 mm long. Cream, light yellowish, or pale brown; periostracum yellowish or greenish brown, thin. About 24 broad ribs, and closely spaced concentric ridges; ridges on ribs may be developed as flattened, scale-like spines. Growth stages distinct. Each valve with two small cardinal teeth; right valve with two anterior and two posterior laterals. Inner surfaces dull white, adductor scars and pallial line distinct; in larger specimens the posterior adductor, and areas around it, may be tinted chestnut or light purple. External sculpture visible as grooves extending from ventral margin, fading rapidly beyond pallial line.

In sandy mud, sand, or fine gravel, from mid-tidal level to just below ELWS. Common on all British coasts, tolerant of salinities as low as 10‰, and often abundant in sheltered bays and estuaries. Distributed from north-east Norway to West Africa.

Cerastoderma glaucum (Poiret) FIG. 13.9

Shell rather thin and brittle; markedly eccentric: anterior margin shallowly convex, forming sharp junction with hinge line, posterior margin continuous with hinge line, sloping steeply from umbones. Off-white, yellowish, or greenish, with darker brown areas, overlain by a greenish-brown periostracum. Up to 50 mm long. About 24 ribs, with closely spaced concentric corrugations, forming small, flattened scales on ribs. Growth stages distinct. Each valve with two small cardinal teeth; right valve with two anterior and two posterior laterals, left valve with single anterior and posterior laterals. Inner surfaces dull white or brown, ribs visible as distinct grooves throughout whole of shell, extending beneath umbones. Posterior margin not crenulate.

In brackish water, overlapping the distribution of *Cerastoderma edule*. known positively only from the southern parts of the British Isles, from East Anglia to South Wales, but probably present elsewhere. Recorded from many parts of western Europe, the Mediterranean, and the Black Sea.

Laevicardium crassum (Gmelin) FIG. 13.9

Shell thick, shallowly convex; obliquely oval, up to 70 mm long, as deep as its greatest length. Cream, light yellow to fawn, with blotches of pink, chestnut, or brown, particularly about the umbones. Periostracum greenish, thin. Sculpture of numerous (about 50) faint ribs, most conspicuous ventrally, concentrated on middle regions of shell and absent from areas adjoining anterior and posterior margins. Prodissoconch and first growth stages typically smooth and glossy. Growth lines clear. Each valve with two cardinal teeth, right posterior being considerably larger than right anterior. Right valve with two anterior and one posterior lateral; left valve with single anterior and posterior laterals, anterior being much larger than posterior. Inner surfaces smooth, glossy white tinted with pink, darkest close to margin. Adductor scars distinct, pallial line less so; external ribs visible as grooves extending from margin to pallial line but fading rapidly beyond it.

Offshore, to the edge of the Continental Shelf, in sand and shelly gravel. Off all British coasts; ranges from Norway to the Mediterranean and West Africa.

Parvicardium ovale (Sowerby) FIG. 13.9

Shell obliquely oval, small, not exceeding 13 mm; yellowish white, periostracum thin and indistinct. Up to 26 flat ribs; concentric ridges poorly developed, apparent between ribs close to shell margin. Tubercles sparsely developed on ribs, usually present only at anterior and posterior margins; anteriorly forming flat scales, posteriorly more spine-like. Growth stages clear. Each valve with two small cardinal teeth; right valve with two anterior and one posterior lateral teeth, left valve with single anterior and posterior laterals, the anterior being the larger. Inner surfaces glossy white; adductor scars clear, pallial line broad, ill-defined, external sculpture visible as grooves which extend beneath umbones.

In sand and gravel, offshore to about 100 m. Off all British coasts. Distributed from Iceland to the Mediterranean and Canary Isles.

Parvicardium minimum (Philippi) FIG. 13.9

Shell obliquely oval, thin and fragile, not exceeding 12 mm in length; yellowish white, periostracum thin and indistinct. Sculpture of 28 or more broad ribs, with indistinct concentric ridges between. Each rib with numerous small, closely spaced, flattened spines, present over whole of shell although sometimes worn on greatest areas of curvature. Growth stages indistinct. Each valve with two small cardinal teeth; right valve with two anterior and one posterior lateral teeth, left valve with single anterior and posterior teeth, the anterior being the largest. Inner surfaces white; adductor scars clear, pallial line broad and ill-defined; external sculpture visible as grooves extending throughout whole of shell.

In mud, sand, and fine gravel, offshore to about 160 m. Present off western coasts of Britain and Ireland; ranging from Iceland to the Mediterranean and north-west Africa.

Parvicardium exiguum (Gmelin) FIG. 13.9

Shell thick, obliquely oval, hinge line sloping strongly to anterior, umbones well in front of mid-line. Yellowish white, with a thick, patchy brown periostracum; up to 13 mm long. Up to 22 broad ribs, with well-marked concentric ridges between; small tubercle-like spines present on all ribs in young specimens, later persisting only at anterior and posterior margins, and sparingly ventrally. Growth stages distinct. Each valve with two small cardinal teeth; right valve with two anterior and one posterior lateral teeth, the upper anterior being very small. Left valve with single anterior and posterior laterals, the anterior being more prominent. Inner surfaces dull white, or tinted light brown or green. Ribs developed as deep marginal crenulations, not extending into the interior. Adductor scars and pallial line clear.

In sand, gravel, or mud; from lower shore into the shallow sublittoral, often extending into estuaries. Off all British coasts. Distributed from Norway to the Mediterranean and Black Sea.

P. papillosum (Poli), known only from the Channel Isles and the Isles of Scilly, is very similar to *P. exiguum*. It is distinguished by more numerous tubercles, and by the well-developed concentric sculpture, which forms alternating ridges and slots, indenting the edges of the ribs.

Parvicardium scabrum (Philippi) FIG. 13.9

Shell broadly oval, small, up to 12 mm; yellowish white with pale brown periostracum. Up to 28 ribs, interstices between almost smooth, with a few concentric lines or grooves. Broad, scale-like spines present on all ribs, over whole shell but most pronounced along posterior margin. Growth stages clear. Each valve with two small cardinal teeth; right valve with two anterior and one posterior lateral teeth, the upper anterior being very small. Left valve with single anterior and posterior laterals, the anterior more prominent than the posterior. Inner surfaces white, adductor scars and pallial line distinct. Marginal crenulations extending as grooves as far as pallial line, fading rapidly beyond it.

In sand, gravel, or shell, offshore, perhaps to the edge of the Continental Shelf. Off all British coasts; distributed from Norway to the Mediterranean and north-west Africa.

24 VENERIDAE

Shell typically thick and strong, equivalve; inequilateral, umbones anterior to mid-line, usually prominent. Sculpture principally of concentric grooves. Characteristically with three cardinal teeth in each valve, occasionally with anterior lateral teeth. Lunule distinct. Ligament external. Adductor scars about equal, pallial line with a sinus.

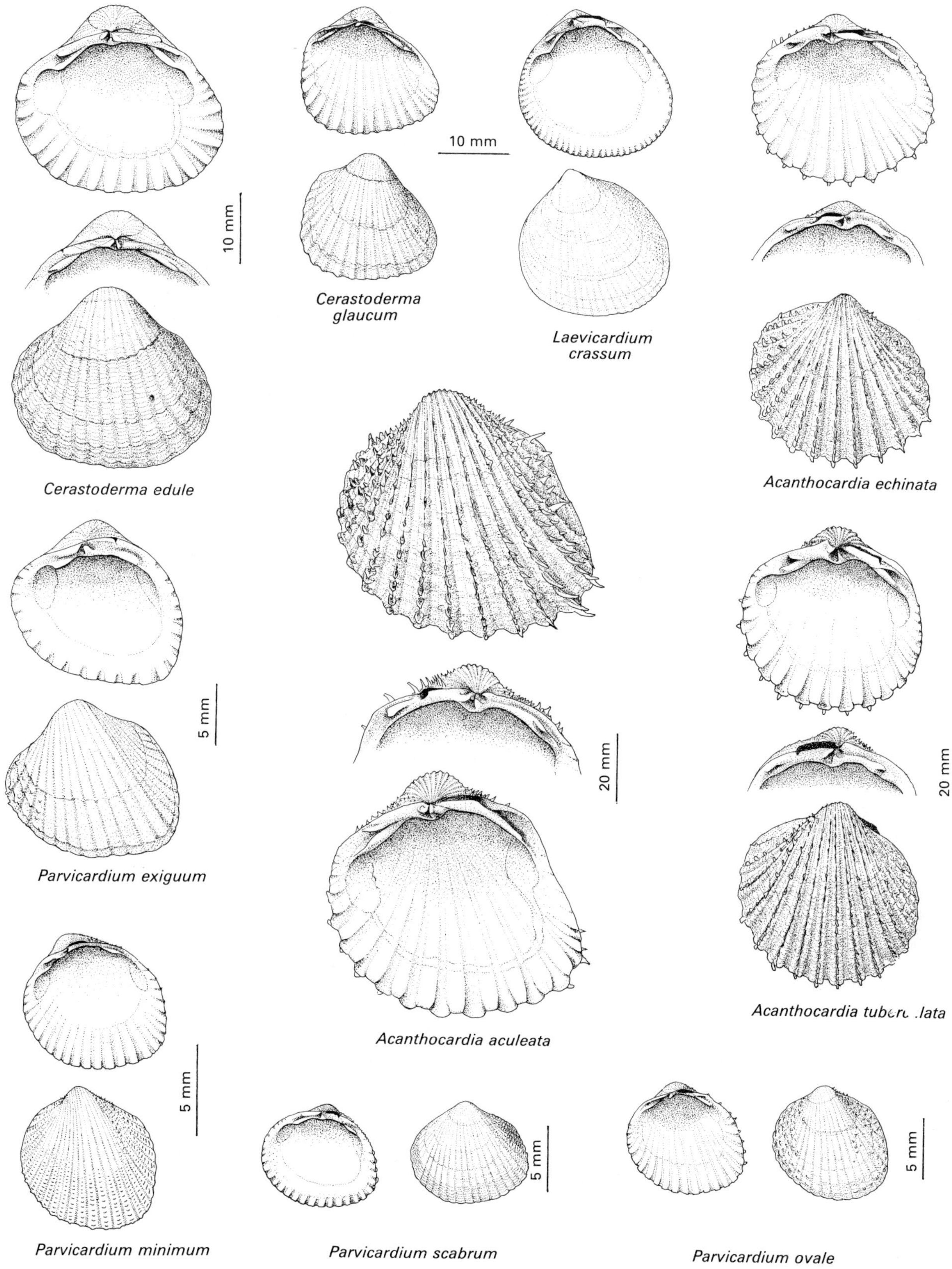
10 mm
10 mm
10 mm
Cerastoderma glaucum
Laevicardium crassum
Cerastoderma edule
Acanthocardia echinata
5 mm
20 mm
20 mm
Parvicardium exiguum
Acanthocardia tubercu.lata
Acanthocardia aculeata
5 mm
5 mm
5 mm
Parvicardium minimum
Parvicardium scabrum
Parvicardium ovale

Fig. 13.9

1. Shell oval, smooth, and plump; minute, less than 3 mm long. Inhabiting crevices between tidemarks, attached by byssus. Cardinal teeth poorly developed, three in each valve but only two in each distinctly visible **_Turtonia minuta_**
   Adult shell considerably larger. Cardinal teeth distinct. Distributed above and below ELWS **2**

2. Shell with bold concentric ridges, extending over whole surface of valve, not simply developed at posterior margin **3**
   Shell smooth, or with radiating ridges or striations, or with concentric grooves and lines, occasionally forming ridges posteriorly or around umbones, but without regular concentric ridges **8**

3. Shell oblong, elongate-oval, or irregular. concentric sculpture friable, tending to wear. Inhabiting holes and crevices **4**
   Shell oval or triangular, regular, often thick and solid. Concentric sculpture thick, not easily broken **5**

4. Concentric sculpture with raised and frilled edges. Shell often irregular. Pallial sinus short, triangular **_Irus irus_**
   Concentric sculpture more prominent posteriorly, but not frilled. Shell regularly oblong. Pallial sinus deep, U-shaped **_Venerupis saxatilis_**

5. Concentric ridges few in number—about 15—broad and flat, varying in width and imparting a stepped appearance to surface of valve **_Clausinella fasciata_**
   Concentric ridges numerous, closely spaced, with acute edges **6**

6. Concentric ridges of equal height and thickness over whole of valve, generally rather fine. Shell with three broad bands of red-brown or chestnut radiating from umbones **_Chamelea galina_**
   Concentric ridges coarse, prominent, and often varying in width and height. Shell uniform off-white or cream, with or without light brown patches about the umbones **7**

7. Concentric sculpture intersecting with radiating grooves or ridges, accentuated posteriorly where the concentric ridges are disrupted to form coarse raised knobs **_Venus verrucosa_**
   No radiating sculpture, concentric ridges entire, most prominent posteriorly **_Circomphalus casina_**

8. Inner margin of shell with distinct crenulations **9**
   Inner margin of shell smooth **10**

9. Sculpture of numerous radiating ribs intersected by concentric grooves, giving a regularly decussate effect. Shell up to 20 mm long, superficially resembling a small cockle **_Timoclea ovata_**
   Sculpture of concentric grooves and striations, developed as fine ridges anteriorly and posteriorly, and about the umbones. Lunule heart-shaped, very distinct. Up to 120 mm long **_Mercenaria mercenaria_**

10. Pallial sinus very shallow, represented only by a small indentation in the pallial line **_Gouldia minima_**
    Pallial sinus deep and distinct, U- or V-shaped **11**

11. Shell almost circular, with an arched anterior margin **12**
    Shell broadly oval or rhomboidal, the anterior margin sloping gently away from the umbones **13**

12. Lunule as broad as long; in front of it the anterior margin is strongly convex **_Dosinia exoleta_**
    Lunule longer than broad; in front of it the anterior margin is gently convex **_Dosinia lupinus_**

13. Left valve with a stout anterior lateral tooth, right valve with a corresponding elongate pit. Pallial sinus mamilliform **_Callista chione_**
    No lateral teeth. Pallial sinus regularly U-shaped **14**

14. Shell sculpture of concentric grooves only **_Paphia rhomboides_**
    Radiating ribs or lines present, as well as concentric grooves or lines **15**

15. Pallial sinus extending beyond mid-line of shell, lower edge confluent with pallial line for part of its length **16**
    Pallial sinus not extending beyond mid-line of shell, lower edge distinct from pallial line for whole of its length **17**

16. Sculpture of fine radiating striae and concentric grooves and ridges, the latter not prominent. A burrower in sand and gravel **_Venerupis pullastra_**
    Sculpture principally of concentric grooves and ridges, often prominent posteriorly. Radiating striae indistinct. A borer in rock **_Venerupis saxatilis_**

17. Sculpture principally of concentric grooves, radiating striae indistinct. Shell elongate-oval **_Paphia aurea_**
    Sculpture of concentric grooves and light ridges, intersecting with radiating ridges; more pronounced posteriorly, where the shell has a marked decussate appearance **_Tapes decussatus_**

Venus verrucosa Linnaeus FIG. 13.10

Shell oval, tumid, rugose, up to 60 mm long; off-white to light brown, darkest about umbones and anterio-dorsal margin, periostracum deeper brown. Sculpture of numerous (>20) prominent, concentric ridges; prodissoconch with distinct radiating ridges, early stages appearing decussate, but only posteriorly in later stages where radiating grooves disrupt concentric sculpture, giving rise to erect warty knobs. Growth stages indistinct.

10 mm
Dosinia exoleta
10 mm
Dosinia lupinus
10 mm
Callista chione
10 mm
Gouldia minima
10 mm
Circomphalus casina
10 mm
Venus verrucosa
10 mm
Timoclea ovata
10 mm
Clausinella fasciata
10 mm
Chamelea gallina
10 mm
Mercenaria mercenaria

Fig. 13.10

Lunule deeply impressed, oval, with fine striations, deep brown or chestnut; escutcheon elongate, unequally developed: prominent on left valve, broad, smooth, and crossed by irregular bands of deep brown, largely obscured on right valve by shell sculpture. Three cardinal teeth in each valve, with a small lateral tooth in front of anterior cardinal of left valve, and a corresponding pit in right. Inner surfaces white, very glossy. Posterior adductor scar often tinted pinkish brown, chestnut, or light purple. Pallial line faint, sinus small and triangular, extending no further forward than inner edge of posterior adductor scar. Inner shell margin very finely crenulate.

From lower shore to about 100 m, in sand and gravel. Predominantly southern and western in distribution, from the English Channel to western Scotland; ranges south to the Mediterranean, West and South Africa.

Circomphalus casina (Linnaeus) FIG. 13.10

Shell broadly oval, tumid, rugose, up to 50 mm long; dull white, often tinted light pinkish brown with a few darker blotches; periostracum deep brown. Sculpture of numerous (25–40) concentric ridges, of variable width; appearing flat but with acute dorsal edges, posteriorly more prominent and almost perpendicular to shell surface. Growth stages indistinct. Lunule deeply impressed, distinct, with fine radiating striations. Escutcheon elongate, shallow, unequally developed: prominent in left valve, smooth, with fine striations and transverse bands of colour, narrow in right valve, almost obscured by shell sculpture. Three cardinal teeth in each valve, with a small accessory knob (lateral tooth) in front of anterior tooth of left valve, and a corresponding pit in right. Inner surfaces dull white, adductor scars and pallial line distinct, pallial sinus short, triangular; inner shell margin finely crenulate.

In coarse sand and gravels, offshore, perhaps to the edge of the Continental Shelf. Widespread and common off all British coasts. Distributed from the North Sea to the Mediterranean and West Africa.

Timoclea ovata (Pennant) FIG. 13.10

Shell thin, subtriangular, tumid, up to 20 mm long; white, light yellow, or pale brown, often with streaks and blotches of pink, chestnut, or light purple. Periostracum thin, indistinct. Sculpture of numerous radiating ridges, pronounced near margins, with equally numerous concentric grooves, giving a decussate effect. Growth stages clear. Lunule elongate, lanceolate, not clearly defined but distinctly pigmented. Eschutcheon scarcely discernible. Three cardinal teeth in each valve. Inner surfaces white, glossy; adductor scars and pallial line distinct, pallial sinus small, U-shaped; inner shell margin finely crenulate.

In sand and gravel; offshore, perhaps to the edge of the Continental Shelf. Widespread off all British coasts. Distributed from Norway to the Mediterranean and Cape Verde Islands.

T. ovata superficially resembles a species of *Parvicardium*, but examinaton of the hinge teeth will prevent confusion.

Clausinella fasciata (da Costa) FIG. 13.10

Shell subtriangular, rather flat, umbones prominent, up to 25 mm long. Colour very variable, from white through shades of yellow and brown to pink and deep purple; typically with bands, rays, or streaks of colour on a paler ground. Periostracum thin, indistinct. Sculpture of up to 15 broad, bold, concentric ribs, round in section with fine concentric striations between. Each growth stage indicated by two closely conjoined ribs. Anterior margin distinctly concave in front of umbones; lunule shallow but distinct, usually deeply pigmented. Escutcheon elongate, narrow; clear on left valve but partly obscured on right. Three cardinal teeth in each valve. Inner surfaces white; adductor scars and pallial line distinct, pallial sinus small, U-shaped. Margin very finely crenulate.

In coarse gravels, usually with sand or shell; offshore to about 100 m. Widespread off all British coasts. From Norway to the Mediterranean and the Canary Isles.

Chamelea gallina (Linnaeus) FIG. 13.10

Shell broadly triangular, umbones prominent, dorsal margin sloping steeply posteriorly, concave anteriorly; up to 40 mm long. Off-white or cream, often tinted pale yellowish and frequently with numerous, very fine chestnut or pinkish streaks; typically with three broad bands of deep chestnut or reddish-brown radiating from umbones, fading in long dead material. Periostracum thin, indistinct. Sculpture of numerous fine, concentric ridges, growth stages clear. Lunule deeply impressed, heart-shaped, light brown or reddish brown, with fine striations. Escutcheon elongate, lanceolate, with fine ridges and bands of colour on left valve, more coarsely ridged and less pigmented on right. Three cardinal teeth in each valve. Inner surface glossy white, adductor scars and pallial line distinct; pallial sinus V-shaped. Margin crenulate.

In sand, from the lower shore into the shallow sublittoral. Common off all British coasts. Distributed from Norway to the Mediterranean and the Canary Isles.

Mercenaria mercenaria (Linnaeus) FIG. 13.10

Shell subtriangular, thick; umbones prominent, anterior to midline; up to 120 mm long. Light brown to grey, often in varying concentric bands; periostracum thin, greyish. Sculpture of numerous thin concentric ridges, sharp and raised on early growth stages, and close to anterior or posterior margins, but usually worn smooth on central areas of larger shells. Growth stages clear. Lunule deeply impressed, heart-shaped, finely striated; escutcheon elongate, narrow, poorly defined. Three conspicuous cardinal teeth in each valve, with a group of small corrugations behind posterior tooth. Inner surfaces glossy white, often with deep bluish-purple tints posteriorly or about the muscle scars. Adductor scars and pallial line distinct, pallial sinus short, triangular; inner margin very finely crenulate.

In mud, from the lower shore into the shallow sublittoral. The North American edible clam, or Quahog, has been introduced into British waters, deliberately or otherwise, on several occasions since the middle of the last century. Populations may still be found at several points on the south and south-east coasts, most notably in Southampton water and the Solent, and at Burnham-on-Crouch.

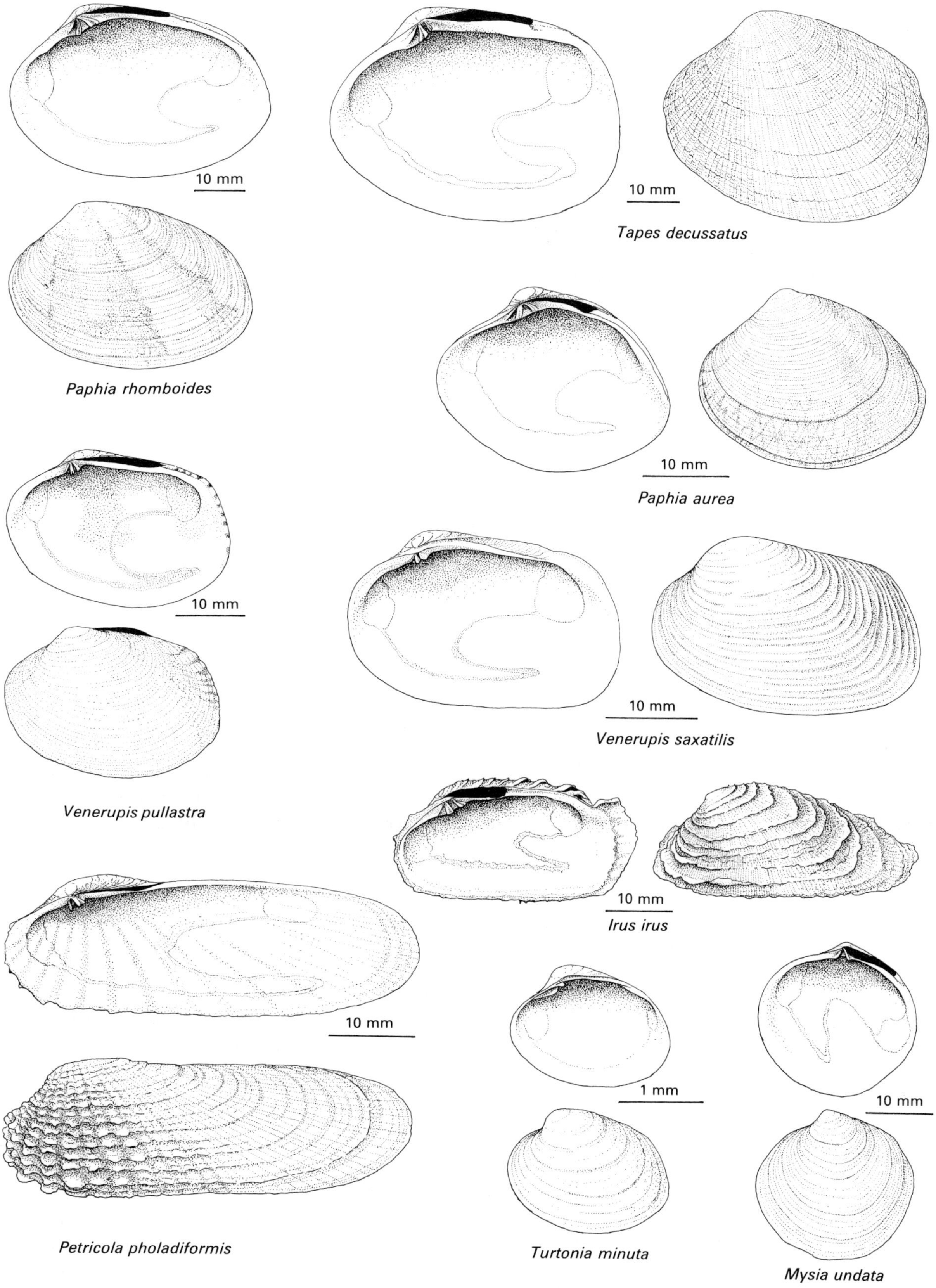
10 mm
10 mm
Tapes decussatus
Paphia rhomboides
10 mm
Paphia aurea
10 mm
10 mm
Venerupis saxatilis
Venerupis pullastra
10 mm
Irus irus
10 mm
1 mm
10 mm
Petricola pholadiformis
Turtonia minuta
Mysia undata

Fig. 13.11

Dosinia exoleta (Linnaeus) FIG. 13.10

Shell almost circular; inequilateral, umbones small but distinct, just anterior to mid-line. Up to 60 mm long. Dorsal margin gently convex posteriorly, deeply concave immediately anterior to umbones, with a highly arched junction with anterior margin. Lunule heart-shaped, as broad as long; escutcheon poorly defined. White, yellowish, or light brown, with irregular rays, streaks, or blotches of darker brown or pinkish brown. Sculpture of numerous fine concentric ridges, surface with a smooth feel. growth stages clear. Three cardinal teeth in each valve; a short, rounded knob in front of the interior teeth of the left valve engages in a shallow pit in the right. Inner surfaces glossy white; adductor scars and pallial line distinct, pallial sinus deep, narrow, U-shaped, extending anterio-dorsally well into the anterior half of the shell.

In muddy gravel or shell-gravel, from the lower shore to about 100 m. Widespread and often common off all British coasts. Distributed from Norway to the Mediterranean and West Africa.

Dosinia lupinus (Linnaeus) FIG. 13.10

Shell approximately circular, inequilateral, umbones just anterior to mid-line, small but prominent. Anterior hinge line shallowly concave below lunule; boundary with anterior margin distinct, but margin not highly arched beyond it. Sculpture of numerous fine concentric ridges, shell surface smooth to touch. Off-white, fawn, or light brown, umbones often tinted yellow or pink. Periostracum thin. Up to 40 mm long. Lunule deeply impressed, finely striated, heart-shaped, longer than wide. Escutcheon poorly defined. Growth stages clear. Each valve with three cardinal teeth; left valve with a single anterior lateral, right valve with a corresponding pit. Inner surfaces white; adductor scars and pallial line distinct, pallial sinus deep, narrow, and U-shaped, extending anterio-dorsally into the anterior half of the shelf.

In sandy mud, sand, and shell-gravel, from the lower shore to at least 120 m. Off all British coasts; distributed from Iceland to the Mediterranean and West Africa.

Gouldia minima (Montagu) FIG. 13.10

Shell subtriangular, almost equilateral, umbones on mid-line, up to 15 mm long; white to light brown, red, or purple, often with streaks, spots, or rays of darker pigment on a lighter ground. Periostracum thin, indistinct. Sculpture of numerous smooth, concentric ridges; growth stages clear. Lunule distinct, elongate, lanceolate; escutcheon ill-defined. Each valve with three cardinal teeth; two anterior laterals in the right valve, one in the left. Inner surfaces white, glossy; adductor scars and pallial line distinct; pallial sinus very small, represented by a shallow concavity immediately below posterior adductor scar.

In sandy mud, fine sand, or gravel; offshore to at least the edge of the Continental Shelf. Off all British coasts. Ranges from Norway to the Mediterranean and West Africa.

Callista chione (Linnaeus) FIG. 13.10

Shell oval, with prominent umbones, up to 90 mm long; light reddish brown to chestnut, in concentric bands, with darker streaks radiating from umbones; periostracum thin, glossy. Sculpture of numerous fine, concentric lines, growth stages clear. Lunule lanceolate, distinct, escutcheon absent. Three conspicuous cardinal teeth in each valve; left valve with a single, prominent anterior lateral tooth, with corresponding elongate pit in right valve. Inner surfaces dull white, with a glossy ventral border; adductor scars and pallial line distinct, pallial sinus broad, with concave borders, mamilliform.

In sand, offshore to at least 100 m. South-west only: from North Wales, south-west England, and the Channel Isles, distributed south to the Mediterranean, Canary isles, and the Azores.

Paphia rhomboides (Pennant) FIG. 13.11

Shell elongate-oval, umbones distinctly anterior; hinge line sloping anteriorly, almost straight posteriorly, posterior margin smoothly rounded. Cream, fawn, or light reddish brown, typically with streaks, blotches, or irregular rays of chestnut, light purple, or pinkish brown. Up to 60 mm long. Concentric sculpture of numerous fine grooves, no radiating sculpture. Growth stages clear. Lunule elongate, shallow, darkly tinted; escutcheon elongate, narrow, ill-defined. Three cardinal teeth in each valve, centre one of left valve, and centre and posterior of right, bifid. Inner surfaces glossy white, with a pinkish tinge beneath hinge line. Adductor scars and pallial line distinct; pallial sinus U-shaped, extending to a point below the posterior edge of ligament.

In coarse sands, gravel, and shell-gravel, from the lower shore, perhaps to the edge of the Continental Shelf. Off all British coasts; distributed from Norway to the Mediterranean and north-west Africa.

Paphia aurea (Gmelin) FIG. 13.11

Shell oval to subtriangular, hinge line sloping away from umbones on both sides; regularly convex anteriorly and posteriorly. Umbones just anterior to mid-line. Colour variable: off-white, cream, yellow, or light brown, usually with deeper reddish or purplish-brown markings, in a variety of patterns. Up to 45 mm long. Sculpture of numerous, fine, concentric ridges and grooves, and faint radiating striations. Lunule elongate, distinct; escutcheon poorly defined. Each valve with three cardinal teeth; centre one of left valve, and centre and posterior of right, bifid. Inner surfaces glossy, white or yellow, often with orange or purple markings. Adductor scars and pallial line distinct. Pallial sinus U-shaped, not extending beyond mid-line of shell, reaching a point below posterior part of the ligament; lower limb of sinus distinct from pallial line for the whole of its length.

A shallow burrower in mixed soft bottoms, from the lower shore into the shallow sublittoral. Off all British coasts, though apparently rarely reported from Scotland. Distributed from Norway to the Mediterranean and north-west Africa.

Tapes decussatus (Linnaeus) FIG. 13.11

Shell broadly oval to quadrate, umbones distinctly anterior; posterior hinge line straight, posterior margin truncate, anterior hinge line grading into the down-sloping anterior margin. Cream, yellowish, or light brown, often with darker markings. Up to 75 mm long. Sculpture of fine concentric striae and bolder radiating lines; prominent posteriorly, where the shell is conspicuously decussate. Growth stages clear. Lunule and escutcheon poorly defined. Each valve with three cardinal teeth: centre one in left valve, and centre and posterior in right are bifid. Inner surfaces glossy white, often with yellow or orange tints, and with a bluish tinge along dorsal edge. Pallial line and adductor scars distinct. Pallial sinus U-shaped, not extending beyond mid-line of shell, but reaching a point below the posterior part of the ligament. Lower limb of sinus distinct from pallial line for the whole of its length.

Lower shore and shallow sublittoral, in sand, muddy gravel, or clay. Mostly off southern and western coasts. Distributed south to the Mediterranean and West Africa.

Venerupis pullastra (Montagu) FIG. 13.11

Shell elongate, oval to quadrate, umbones distinctly anterior; hinge line sloping anteriorly, straight posteriorly, forming a sharp angle with posterior margin. Cream, light fawn to brown, in irregular patches, bands, and rays, usually darkest close to posterior margin. Up to 50 mm long. Sculpture of very fine concentric lines and radiating striations, equally developed, with the concentric element predominant posteriorly. Growth stages clear. Lunule elongate, shallow, escutcheon narrow. Each valve with three cardinal teeth: centre one of left valve, and centre and posterior of right are bifid. Inner surfaces glossy white, often tinted blue or purple beneath umbones and along posterior margin. Adductor scars and pallial line distinct. Pallial sinus deep, U-shaped, extending beyond mid-line of shell; posteriorly, the lower limb of the pallial sinus may be confluent with the pallial line for a short distance.

A shallow burrower in mixed sandy bottoms, usually attached by a byssus. From the lower shore into the shallow sublittoral. Off all British coasts, distributed from northern Norway to the Mediterranean and north-west Africa.

Venerupis saxatilis (Fleuriau) FIG. 13.11

Shell brittle, roughly oblong, umbones anterior to mid-line and incurved. Posterior hinge line straight, posterior margin truncate; anterior hinge line continuous with regularly convex anterior margin. Off-white, cream, or light brown. Up to 40 mm long. Sculpture of closely spaced, small corrugated ridges and grooves, most pronounced close to posterior margin, often worn about umbones; fine radiating striations also present. Lunule and escutcheon shallow, indistinct. Three cardinal teeth in each valve, centre one in left, and centre and posterior in right, bifid. Inner surface white or orange, often tinted blue posteriorly. Adductor scars and pallial line distinct; pallial sinus deep, U-shaped, extending well beyond mid-line of shell, to a point below the anterior half of the ligament. Lower limb of pallial sinus posteriorly confluent with the pallial line.

Intertidal. Attached by a byssus, in holes and crevices in soft rocks, including those bored by other molluscs. Of limited distribution in British waters: reported from the south coasts, and the west coast as far north as the Clyde Sea. Ranges south to north-west Africa.

Irus irus (Linnaeus) FIG. 13.11

Shell elongate, oblong, or irregular, umbones anterior to mid-line. Yellowish white or fawn, usually darker about the umbones. Up to 25 mm long. Sculpture of about ten thin, concentric ridges, raised and convoluted, conspicuous posteriorly where they form distinct frills. Fine radiating ridges visible between the concentric sculpture. Growth stages indistinct. Lunule obscured, escutcheon elongate, narrow, most prominent on left valve. Three cardinal teeth in each valve; centre one of left valve, centre and posterior of right, bifid. Inner surfaces white, with blue or chestnut patches, particularly posteriorly. Adductor scars and pallial line distinct. Pallial sinus short, U-shaped.

In holes and crevices in rock, or in *Laminaria* holdfasts, from the lower shore into the shallow sublitoral. Southern and western coasts only. Ranges south to the Mediterranean and West Africa.

Turtonia minuta (Fabricius) FIG. 13.11

Shell thin, tumid, oval, umbones anterior to mid-line. Off-white to light brown, sometimes pinkish dorsally. Very small, not exceeding 3 mm long. Sculpture of very fine concentric lines, growth stages clear; periostracum smooth and glossy. Lunule and escutcheon indistinct. Three cardinal teeth in each valve, anterior of right and posterior of left very reduced, inconspicuous. Inner surfaces dull white to pale brown, adductor scars and pallial line indistinct.

Intertidal only, on rocky shores. Attached by byssus, in holes and crevices, among or in barnacle shells and algae, often very common. On all British coasts. Distributed south as far as the Mediterranean, and north into boreal/arctic waters, where it is apparently circumpolar in distribution.

25 PETRICOLIDAE

Shell equivalve, inequilateral. Ligament external. Right valve with two cardinal teeth, left with three. No lateral teeth. Adductor scars about equal. Pallial line with sinus.

1. Shell elongate. Sculpture of concentric ridges and radiating lines, developed as pronounced ribs anteriorly
   Petricola pholadiformis

   Shell orbicular. Sculpture of fine concentric lines only
   Mysia undata

Petricola pholadiformis Lamarck FIG. 13.11

Shell thin and brittle, very elongate, oval, shaped rather like a date; umbones close to anterior margin. Dull white or fawn, with brown or greyish periostracum. Up to 60 mm long. Sculpture of numerous fine concentric ridges, prominent anteriorly where they may be raised and corrugated, forming flat scales; radiating ridges also present, developed anteriorly into about ten raised ribs. Left valve with three cardinal teeth, the centre one bifid;

right valve with two cardinal teeth, the posterior bifid. Inner surfaces glossy white; adductor scars and pallial line distinct, pallial sinus deep, U-shaped, extending to mid-line of shell, anterior ribs visible as light grooves, anterior margin crenulate.

Boring into clay, peat, or soft rocks, this species ranges from the middle shore into the shallow sublittoral. Reported from southern and south-eastern coasts only, from Lincolnshire to Dorset and Cornwall, but most common off Essex and the Thames estuary. Distributed from southern Norway to the Mediterranean and West Africa.

Mysia undata (Pennant) FIG. 13.11

Shell thin and brittle, almost circular, umbones just anterior to mid-line. White or light yellow, periostracum indistinct. Up to 38 mm long. Sculpture of fine, closely spaced concentric lines; growth stages clear. Lunule and escutcheon scarcely evident. Inner surfaces glossy white. Adductor scars and pallial line distinct; pallial sinus U-shaped, extending anterio-dorsally into anterior half of shell. Left valve with three cardinal teeth, the centre one bifid; right valve with two cardinal teeth, the posterior may be bifid.

Burrowing in muddy, coarse bottoms. Offshore, but probably extending little deeper than 50 m. Off all British coasts, distributed from Norway to the Mediterranean and the Canary Isles.

26 MACTRIDAE

Shell equivalve, equilateral or just inequilateral. Ligament external, thin, and internal, in a small chondrophore recessed within the hinge line. Two or three cardinal teeth in each valve; in the left, two cardinals fused to form a single forked structure. Lateral teeth present. Pallial line with a sinus.

1. Lateral teeth smooth, inner and outer portions of ligament separated by a distinct calcareous septum on dorsal side of chondrophore **2**

   Lateral teeth serrated along interlocking surfaces. Inner and outer portions of ligament not separated **3**

2. Up to 50 mm long. Right anterior cardinal tooth almost parallel to hinge line. Common on all coasts ***Mactra stultorum***

   Up to 110 mm long. Right anterior cardinal tooth at a marked angle to hinge line. South-western only ***Mactra glauca***

3. Forked cardinal tooth of left valve small, extending only half-way down the hinge plate. Shell subtriangular ***Spisula solida***

   Forked cardinal tooth of left valve larger, extending practically to the edge of the hinge plate **4**

4. Shell broadly elliptical, anterior and posterior regularly and equally convex ***Spisula elliptica***

   Shell subtriangular, asymmetrical; posterior hinge line sloping more steeply than anterior, posterior margin somewhat drawn out ***Spisula subtruncata***

Mactra stultorum (Linnaeus) FIG. 13.12

Shell thin and brittle, oval, umbones just anterior to mid-line. White, tinted purple about the umbones, with light brown rays of varying width radiating from umbones; periostracum light brown, thin. Up to 50 mm long. Sculpture of very fine concentric lines, growth stages clear. Shell margin prominent at hinge line: lunule and escutcheon absent. Right valve with two cardinal teeth, the anterior parallel to hinge line, and paired, elongate anterior and posterior laterals. Left valve with three cardinal teeth, anterior two joined to form a single, widely forked structure, the third poorly developed, and single elongate anterior and posterior laterals. Chondrophore triangular, posterior to the cardinal teeth in each valve, with a small dorsal septum isolating it from the external ligament. Inner surfaces glossy, white, tinted purple. Adductor scars and pallial line indistinct; pallial sinus broad and rounded, not extending far into shell.

In clean sand, from the lower shore into the shallow sublittoral. Widespread, and often abundant, off most British coasts. Distributed from Norway to the Mediterranean and West Africa.

Mactra glauca Born FIG. 13.12

Shell rather thin and brittle for its size, oval, umbones just anterior to mid-line. White or cream, often with pale reddish-brown or fawn rays; periostracum greenish-brown. Up to 110 mm long. Sculpture of very fine concentric lines, visible with hand lens, coarser near the posterior margin in large specimens; growth stages clear. Right valve with two cardinal teeth, anterior short and at an angle to margin, and paired, elongate anterior and posterior laterals. Left valve with three cardinal teeth; anterior two joined to form a single, forked structure, third small and fragile; with single elongate, anterior and posterior laterals. Chondrophore posterior to cardinal teeth, separated from external ligament by a slender septum. Inner surfaces glossy, adductor scars and pallial line not always distinct; pallial sinus broad and rounded, but not very deep.

Uncommon, reported only from extreme south-west and Channel Isles. Occurs offshore. Distributed south to the Mediterranean and north-west Africa.

Spisula solida (Linnaeus) FIG. 13.12

Shell thick and strong, subtriangular, anterior more regularly convex than posterior but not markedly so; umbones more or less on mid-line, anterior and posterior hinge with approximately equal slope. Dull white to light fawn; periostracum greyish brown, usually prominent around margin and along growth lines. Up to 50 mm long. Sculpture of fine concentric lines, developed as fine ridges in front of and behind umbones; growth stages distinct. Lunule and escutcheon poorly defined. Right valve with two short cardinal teeth and paired, elongate anterior and posterior laterals. Left valve with three cardinal teeth; anterior two forming a forked structure which extends only half-way down hinge plate, third very small; single, elongate anterior

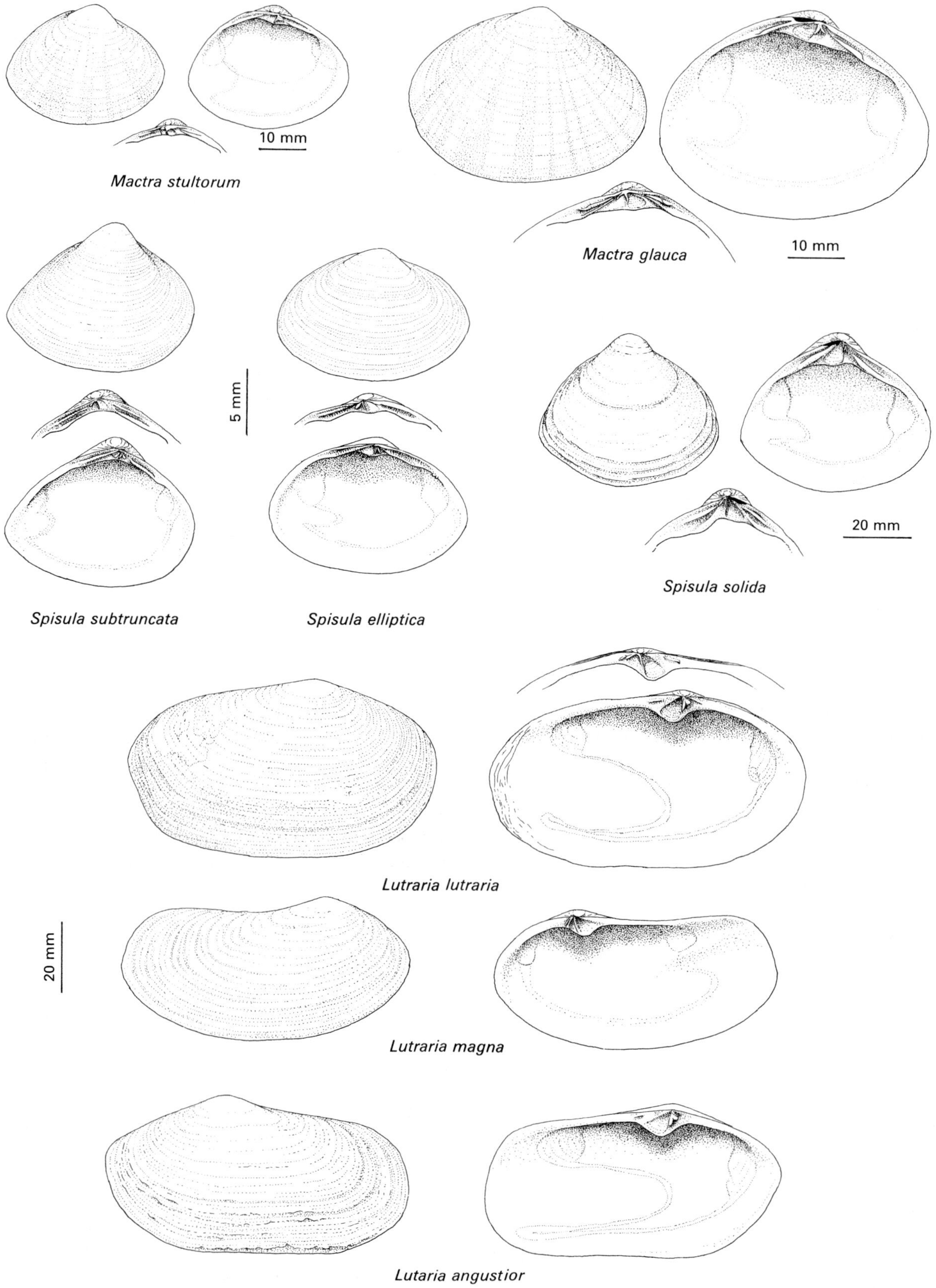
10 mm
Mactra stultorum
Mactra glauca
10 mm
5 mm
Spisula solida
20 mm
Spisula subtruncata
Spisula elliptica
Lutraria lutraria
20 mm
Lutraria magna
Lutaria angustior

Fig. 13.12

and posterior laterals. Interlocking surfaces of lateral teeth serrated. Chondrophore posterior to cardinal teeth. Inner surfaces glossy, white; adductor scars and pallial line distinct, pallial sinus linguiform, extending to a point below and beyond the mid-line of the posterior laterals.

Burrows in sand, from the lower shore down. Widespread in British waters; distributed from south Iceland and Norway to Spain and Morocco.

Spisula subtruncata (da Costa) FIG. 13.12

Shell thick and strong, subtriangular but distinctly asymmetrical; umbones close to mid-line, posterior hinge line and margin sloping more steeply than anterior, posterior end appearing slightly drawn out. Dull white to cream, periostracum greyish brown. Up to 30 mm long. Sculpture of fine concentric lines and grooves, growth stages clear. Lunule and escutcheon broad and elongate, latter bounded by low ridges extending from umbones. Right valve with two short cardinal teeth and paired, elongate anterior and posterior laterals. Left valve with three cardinal teeth; anterior two forming a single forked structure extending almost to edge of the hinge plate, third very small, indistinct; with single elongate, anterior and posterior laterals. Interlocking surfaces of lateral teeth serrated. Chondrophore posterior to cardinal teeth. Inner surfaces glossy, white; adductor scars and pallial line distinct, pallial sinus broad and rounded, extending to a point below and behind the mid-line of the posterior lateral teeth.

Burrowing in muddy or silty sand, from the lower shore into the shallow sublittoral. Widespread and common off most British coasts; distributed from Norway to the Mediterranean and Canary Isles.

Spisula elliptica (Brown) FIG. 13.12

Shell thin, elongate oval, umbones close to mid-line. Dull white with greenish or greyish brown periostracum. Up to 30 mm long. Sculpture of fine concentric lines and grooves, growth stages clear. Lunule and escutcheon poorly defined, the former perhaps most distinct. Right valve with two separate, but closely spaced cardinal teeth and paired, elongate anterior and posterior laterals. Left valve with three cardinal teeth: the anterior two forming a single forked structure extending almost to edge of the hinge plate, the third small and indistinct; with single anterior and posterior laterals. Interlocking surfaces of lateral teeth serrated. Chondrophore posterior to cardinal teeth. Inner surfaces glossy, white; adductor scars and pallial line distinct, pallial sinus oval, extending to a point below and beyond mid-line of posterior lateral teeth.

In mixed soft substrata, offshore to about 100 m. Off all British coasts. Distributed northwards to the Barents Sea.

27 LUTRARIIDAE

Shell elongate, gaping anteriorly and posteriorly. Equivalve, inequilateral. Ligament external and internal, latter in a chondrophore recessed into the hinge line. Cardinal and lateral teeth present, two of left cardinal teeth partly fused, forming a forked structure. Adductor scars about equal. Pallial line with a sinus.

1. Anterior cardinal tooth of right valve bifid. Dorsal margin distinctly concave posterior to umbones **_Lutraria magna_**
   Anterior cardinal tooth of right valve entire. Dorsal margin straight or convex **2**
2. Shell elongate-oval, dorsal hinge line gently convex. Pallial line distinct from lower edge of pallial sinus **_Lutraria lutraria_**
   Shell somewhat quadrate, dorsal hinge line more or less straight. Pallial line largely fused with lower edge of pallial sinus **_Lutraria angustior_**

Lutraria lutraria (Linnaeus) FIG. 13.12

Shell elongate, elliptical, gaping at each end; umbones anterior to mid-line. White or yellowish, often tinted pink or purple, glossy; periostracum olive, thick, brittle, and peeling readily in dead specimens. Up to 130 mm long. Sculpture of fine concentric lines, developing as fine ridges anteriorly and posteriorly; growth stages clear. Right valve with two thin cardinal teeth anterior to chondrophore, and a single, poorly developed posterior lateral tooth. Left valve with two cardinals forming a solid, forked structure; a third, very fine posterior cardinal, and very thin, single anterior and posterior laterals. Inner surfaces dull white. Adductor scars and pallial line distinct, anterior adductor larger and more elongate than posterior; lower edge of pallial sinus distinct from pallial line.

In mixed soft substrata, from the lower shore to about 100 m. Off all British coasts, distributed from Norway south to the Mediterranean and West Africa.

Lutraria angustior (Philippi) FIG. 13.12

Shell elongate, somewhat quadrate, anterior hinge line sloping more steeply than posterior; gaping at both ends. Umbones anterior to mid-line. Dull white or yellowish, periostracum pale yellowish brown. Up to 100 mm long. Sculpture of numerous fine grooves, developing as fine ridges close to the margin; growth stages clear. Right valve with two cardinal teeth and a single, poorly developed, posterior lateral. Left valve with anterior two cardinal teeth forming a solid, forked structure, third thin and indistinct; and single, thin anterior and posterior laterals. Inner surfaces white, glossy, adductor scars and pallial line distinct; anterior adductor scar elongate, irregular, lower edge of pallial sinus largely fused with pallial line.

In mixed soft substrata, offshore to about 50 m. On southern coasts only, ranging south to the Mediterranean and West Africa.

Lutraria magna (da Costa) FIG. 13.12

Shell elongate, oblong, umbones anterior to mid-line; dorsal margin distinctly concave posterior to umbones, with almost equal curvature to ventral margin. Gaping at both ends. Posterior margin somewhat truncate. Light yellow or fawn, with brown periostracum. Up to 120 mm long. Sculpture of fine concentric lines, growth stages clear. Right valve with two rather prominent cardinal teeth anterior to chondrophore, the anterior distinctly bifid; left valve with anterior two cardinals forming a single forked structure, the third very thin. Inner surfaces white, pallial line and adductor scars distinct; anterior adductor scar larger, elongate, and irregular; pallial sinus deep, lower edge largely fused with pallial line.

In shell-gravels, offshore. South-western coasts only, ranging south to the Mediterranean and North Africa.

28 DONACIDAE

Shell equivalve, inequilateral. Ligament external. Hinge line with two cardinal teeth in each valve, anterior and posterior teeth also present. Adductor scars about equal. Pallial line with a sinus. Cruciform muscles present extending between valves, involved in siphon retraction: scars often visible close to lower edge of pallial sinus. Shell margin crenulate.

Donax vittatus (da Costa) FIG. 13.13

Shell roughly wedge-shaped, umbones posterior to mid-line. dorsal margin sloping gently anteriorly from umbones, continuous with rounded anterior margin; posteriorly sloping more steeply, posterior margin bluntly pointed. Ventral margin distinctly crenulate, visible from both sides of each valve. White, yellowish, light brown, or purple, frequently lighter about the umbones, with pale radiating rays and often with pigmented bands along growth lines. Rarely larger than 35 mm. Periostracum light brown to olive-brown, glossy. Sculpture of fine concentric grooves and numerous fine radiating striations. Growth stages clear. Right valve with two cardinal teeth, the anterior small, the posterior broad and bifid; left valve with two slender cardinal teeth, the anterior weakly bifid. One anterior and two posterior lateral teeth in right valve; single small, anterior and posterior laterals in left valve. Inner surfaces glossy, white, often tinted purple, yellow, or orange, with growth stages showing as pale bands. Adductor scars, pallial line, and cruciform muscle scars only lightly impressed. Pallial sinus broadly oval, extending to mid-line of shell.

In sand, from the lower shore into the shallow sublittoral. On most British coasts, distributed from Norway south to the Mediterranean and north-west Africa.

Donax variegatus (Gmelin) occurs in similar environments on the southern and south-west coasts of Britain. More regularly oval than *D. vittatus*, it is distinguished principally by the marginal crenulations which are so fine as to feel smooth to the touch.

29 TELLINIDAE

Shell equivalve, inequilateral, posterior ventral border frequently with a dent or twist. Ligament external. Hinge line with two cardinal teeth in each valve, lateral teeth usually present. Adductor scars about equal. Pallial line with sinus. Cruciform muscles present, linking valves, and leaving scars close to lower edge of pallial sinus.

---

1. Hinge line with cardinal teeth only. Shell almost circular, colour variable but often in concentric bands. Intertidal, often abundant in soft estuarine deposits **_Macoma balthica_**

   Hinge line with cardinal and lateral teeth, the latter usually conspicuous in the right valve, but in the left appearing simply as slender ridges close to the margin **2**

2. Shell broadly oval, little longer than deep. Lower edge of pallial sinus distinctly separated from the pallial line **_Arcopagia crassa_**

   Shell elongate oval, at least 1.5 times as long as deep. Lower edge of pallial sinus confluent with pallial line **3**

3. Umbones close to mid-line of shell. No anterior lateral tooth in left valve **4**

   Umbones clearly situated in posterior half of the shell. A small anterior lateral tooth in left valve **6**

4. Dorsal and ventral margins concave posteriorly, imparting a characteristic hooked outline to that part of the shell **_Angulus squalidus_**

   Posterior part of shell not hooked **5**

5. Posterior margins convex, shell subtriangular **_Angulus tenuis_**

   Posterior margin convex ventrally, slightly concave dorsally; shell somewhat attenuated posteriorly **_Fabulina fabula_**

6. Posterior dorsal margin of shell sloping gently from umbones, posterior part of shell regularly rounded **_Moerella pygmaea_**

   Posterior dorsal margin of shell sloping steeply from umbones, posterior part of shell irregularly truncate **_Moerella donacina_**

Angulus tenuis (da Costa) FIG. 13.13

Shell thin and brittle, oval, valves shallowly convex, umbones just posterior to mid-line; anterior dorsal margin sloping gently, forming a continuous curve with anterior margin, posteriorly more steeply inclined, posterior margin slightly truncate. Ligament very conspicuous, a fusiform, brown prominence posterior to umbones. Colour very variable, white, through shades of pink, yellow, and orange, often in distinct bands. Periostracum glossy, thin. Up to 28 mm long. Sculpture of fine concentric lines, growth stages clear. All hinge teeth very small, delicate and liable to be broken: right valve with two cardinal teeth, anterior small, posterior bifid, and single anterior and posterior laterals; left valve with bifid anterior and very small posterior cardinals, and a single small, posterior lateral. Inner surfaces glossy, unicolorous with exterior; adductor scars and pallial line distinct, pallial sinus broadly oval, its lower edge largely fused with pallial line. Cruciform muscle scars usually visible.

In sand, from the middle shore to the shallow sublittoral, often very abundant. Widespread and common, off all British coasts. Distributed from Norway to the Mediterranean and the north-west coast of Africa.

Angulus squalidus (Pulteney) FIG. 13.13

Shell thin and brittle, flat, broadly oval with the posterior end somewhat attenuated. Umbones just posterior to mid-line. Anterior hinge line downward sloping, continuous with anterior margin; posterior margin attenuated, concave dorsally and ventrally, appearing hooked. Posteriorly, ventral margin has a lateral dent, concave on right and convex on left. Fawn, yellow, orange, or pink, often banded; periostracum light brown, usually conspicuous around margins. Up to 45 mm long. Sculpture of numerous, fine concentric grooves, with a straight keel extending posteriorly from umbones on each valve. Growth stages distinct. Right valve with two cardinal teeth, the posterior bifid, and single anterior and posterior laterals; left valve with two cardinals, the anterior bifid, and a single, poorly developed posterior lateral. Inner surfaces yellow or orange, paler towards margins; adductor scars and pallial line distinct, cruciform muscle scars visible with hand lens. Pallial sinus deep, lower edge fused with pallial line.

In muddy sand, offshore, though probably not beyond 40 m. Not common, reported mostly from southern and western coasts of Britain, including the Isles of Scilly and the Channel Isles. Its distribution elsewhere appears little known.

Fabulina fabula (Gmelin) FIG. 13.13

Shell thin and brittle, elongate-oval, umbones close to mid-line; anterior margin broadly rounded, posterior tapering and curved to right. White, pale yellow, or orange, often in concentric bands; up to 20 mm long. Sculpture of very fine concentric lines, right valve, additionally, with wavy striations extending from dorsal anterior to ventral posterior margins. Growth stages clear. Right valve with two cardinal teeth, the posterior bifid, and single anterior lateral tooth; left valve with two cardinal teeth, the anterior bifid, and small posterior lateral tooth. Adductor scars and pallial line distinct; pallial sinus very deep, lower edge completely fused with pallial line. Cruciform muscle scars often clear. Inner surface white, or tinted with yellow and orange.

In mixed sandy deposits, from the lower shore into the shallow sublittoral. Off all British coasts, distributed from Norway to the Mediterranean and north-west Africa.

Moerella donacina (Linnaeus) FIG. 13.13

Shell thin and brittle, elongate-oval, with posterior somewhat attenuated. Umbones in posterior half of shell. Anterior hinge line sloping gently; posterior hinge line sloping more steeply, forming abrupt junction with posterior margin. Yellowish white to cream, with bands of pink or red, and discontinuous rays of cream or red. Up to 25 mm long. Sculpture of numerous, fine, concentric ridges and grooves, becoming coarser posteriorly; growth stages distinct. Right valve with two cardinal teeth, the posterior bifid, and single anterior and posterior laterals; left valve with two cardinals, the anterior bifid, and single, indistinct anterior and posterior laterals. Inner surfaces glossy, white, yellow, or deep orange, with the external rays visible. Adductor scars and pallial line distinct, cruciform muscle scars visible with hand lens. Pallial sinus very broad and deep, extending almost as far as anterior adductor scar, lower edge fused with pallial line.

In coarse sand and shell-gravel, in shallow coastal waters. Southern and western coasts only, distributed south to the Mediterranean and West Africa.

Moerella pygmaea (Lovén) FIG. 13.13

Shell thin, fragile, elongate-oval, umbones posterior to mid-line; hinge line sloping gently from umbones anteriorly, more steeply posteriorly. Colour variable, white to light orange, pink, or reddish brown, often in patches; up to 10 mm long. Sculpture of very fine, closely spaced concentric grooves; growth stages clear. Right valve with two cardinal teeth, the posterior bifid, and single anterior and posterior laterals; left valve with two cardinal teeth, the anterior bifid, and single, poorly developed, anterior and posterior laterals. Inner surface white or pink. Adductor scars and pallial line indistinct; pallial sinus irregular, extending to just beyond mid-line of shell.

In sand and shell-gravel, offshore, perhaps to the outer edge of the Continental Shelf. Reported from south-west and west coasts, and from the east and north-east of Scotland. Distributed from northern Norway to the Mediterranean and West Africa.

Arcopagia crassa (Pennant) FIG. 13.13

Shell oval, little longer than deep; umbones posterior to mid-line, conical and prominent. Posterio-ventral margin with a shallow dent: concave on left and convex on right. Off-white or fawn; frequently pale orange or light red about the umbones, with a few irregular rays of red. Up to 60 mm long. Sculpture of numerous, fine concentric ridges; growth stages clear. Right valve with two cardinal teeth: anterior small, posterior larger and bifid, with single anterior and posterior laterals; left valve with bifid anterior and slender posterior cardinals, and single, small anterior and posterior lateral teeth. Inner surfaces glossy, white, tinted with patches of yellow, orange, pink, or purple. Adductor scars and pallial line clear, pallial sinus broadly oval, extending deep into interior of shell. Cruciform muscle scars visible close to junction of pallial line and pallial sinus.

In mixed sandy deposits, from the lower shore, extending far out into shelf waters. Not common; most frequent on south-western coasts, but apparently occurring around Scotland. Distributed from Norway southwards to Senegal.

Macoma balthica (Linnaeus) FIG. 13.13

Shell broadly oval, umbones more or less on mid-line; anterior hinge line and margin regularly convex, posterior hinge line and margin slightly attenuated. Colour very variable: white, yellow, pink, or purple, in various shades, unicolorous or banded. Periostracum light brown, most conspicuous at margins. Up to 25 mm long. Sculpture of numerous, fine, concentric lines; growth stages clear, usually marked by bands of colour. Two small cardinal teeth in each valve, the right posterior and left anterior being bifid. No lateral teeth. Inner surfaces glossy; white, pink, or purplish. Adductor scars and pallial line distinct,

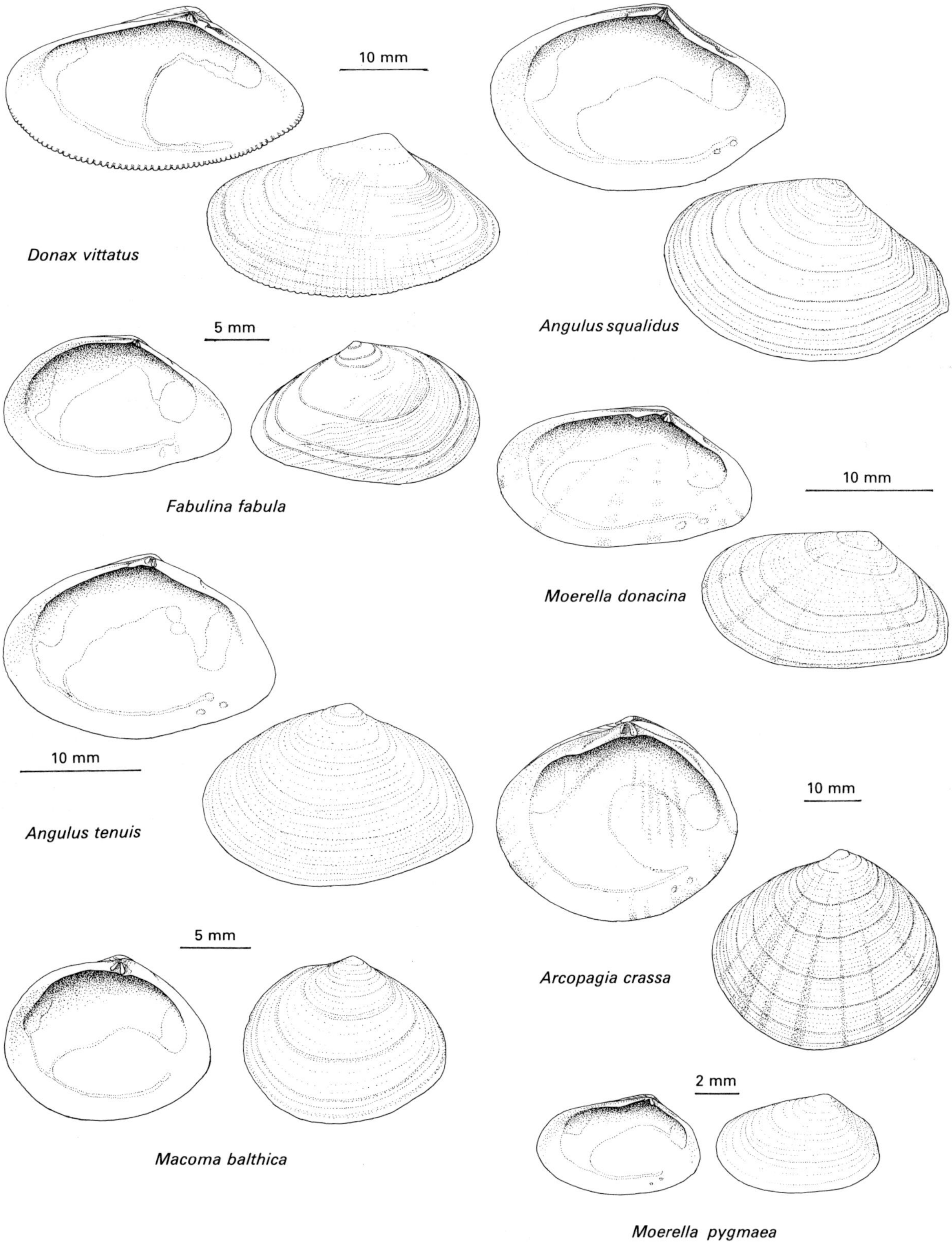
10 mm
Donax vittatus
Angulus squalidus
5 mm
Fabulina fabula
10 mm
Moerella donacina
10 mm
Angulus tenuis
10 mm
Arcopagia crassa
5 mm
Macoma balthica
2 mm
Moerella pygmaea

Fig. 13.13

cruciform scars unclear; pallial sinus iregular, deep, lower edge largely fused with pallial line.

In soft substrata, particularly in estuaries and on tidal flats, where it may be abundant; intertidal only. Present on most British coasts, distributed from the White Sea south to Spain and Portugal.

30 SCROBICULARIIDAE

Shell equivalve, equilateral or just inequilateral. Ligament with small external element, and larger internal element in a triangular chondrophore recessed within the hinge line. Right valve with two cardinal teeth, left with one. Lateral teeth variously developed. Pallial line with a sinus. Cruciform muscles present, scars indistinct.

---

1. Without lateral teeth. Shell yellow-white to grey, with coarse concentric grooves; up to 60 mm long, intertidal only, often abundant in estuarine muds **_Scrobicularia plana_**

   Lateral teeth present, usually prominent in right valve. Shell white, small (<40 mm); mostly offshore, one intertidal species **2**

2. Shell elongate oval or fusiform, about twice as long as deep **3**

   Shell broadly oval or subtriangular, at the most only 1.5 times as long as deep **4**

3. Anterior and posterior dorsal margins with approximately equivalent slope from the umbones. Beaks almost on the mid-line; proportionately less slender **_Abra nitida_**

   Posterior dorsal margin sloping more steeply from the umbones than the anterior. Beaks just posterior to mid-line; proportionately more slender **_Abra prismatica_**

4. Shell subtriangular, equilateral, umbones prominent. Growth stages indistinct **_Abra tenuis_**

   Shell oval, inequilateral, umbones in posterior half. growth stages clear **_Abra alba_**

Scrobicularia plana (da Costa) FIG. 13.14

Shell thin and brittle, oval, umbones just anterior to mid-line. Dull white, yellowish, or grey, usually darkest about the growth lines. Up to 65 mm long. Periostracum thin, conspicuous only at margins. Sculpture of numerous, fine concentric lines and grooves. Internal ligament supported in each valve by a triangular, backwardly directed chondrophore; two small cardinal teeth anterior to chondrophore in right valve, a single one in left. Inner surfaces glossy, white or yellowish; adductor scars and pallial line not readily apparent in fresh material. Pallial sinus broad, lower edge largely fused with pallial line; posteriorly, cruciform muscle scars may be apparent.

Intertidal only, burrowing in soft substrata in estuaries and tidal flats, in conditions of fluctuating salinity. Often abundant. Widely distributed around the British Isles, ranges from Norway to the Mediterranean and West Africa.

Abra alba (Wood) FIG. 13.14

Shell thin and brittle, broadly oval, umbones posterior to mid-line. White and glossy; periostracum thin, apparent only at margins. Up to 25 mm long. Sculpture of very fine concentric lines, growth stages clear. Chondrophore elliptical, posteriorly directed. Two small, peg-like cardinal teeth anterior to chondrophore in right valve, a single one in left; single, thin anterior and posterior lateral teeth in right valve, present but less prominent in left. Inner surfaces white; adductor scars and pallial line clear, cruciform muscle scars visible with hand lens. Pallial sinus deep, rather iregular.

In soft substrata; occasionally on the lower shore but most abundant in shallow, offshore waters (to about 60 m) where it may be the dominant member of the benthic infauna. Widespread around the British Isles; distributed from Norway to the Mediterranean and west Africa.

Abra nitida (Müller) FIG. 13.14

Shell thin and brittle, elongate-oval, umbones just posterior to mid-line; anterior and posterior hinge line sloping equally from umbones. Glossy white. Rarely longer than 12 mm. Sculpture of very fine concentric lines, scarcely visible with hand lens; growth stages unclear. Chondrophore slender, posteriorly directed. Right valve with two short cardinal teeth anterior to chondrophore and single, elongate, anterior and posterior laterals; left valve with one cardinal tooth and poorly developed, single anterior and posterior laterals. Inner surfaces white, glossy; adductor scars and pallial line visible with hand lens, cruciform muscle scars indistinct. Pallial sinus deep, lower edge partly confluent with pallial line.

Burrowing in mixed soft substrata, offshore to the edge of the Continental Shelf. Widespread, and often common, around the British Isles. Distributed from Norway south to the Mediterranean and north-west Africa.

Abra prismatica (Montagu) FIG. 13.14

Shell thin and brittle, elongate-oval to fusiform, approximately twice as long as deep. Glossy white. Up to 13 mm long. Sculpture of very fine concentric lines, visible with hand lens. Growth stages indistinct. Chondrophore slender, posteriorly directed. Right valve with two small cardinal teeth anterior to chondrophore, and single anterior and posterior lateral teeth; left valve with one small cardinal tooth, and single, poorly developed anterior and posterior laterals. Inner surfaces white, glossy, rendering adductor scars and pallial line indistinct. Pallial sinus with lower edge partly fused with pallial line.

In mixed sandy bottoms, from the lower shore to about 60 m. Widespread around the British Isles, distributed from Norway to the Mediterranean and north-west Africa.

Abra tenuis (Montagu) FIG. 13.14

Shell thin and brittle, subtriangular; umbones prominent, with hinge line sloping away evenly on each side. Dull white, with

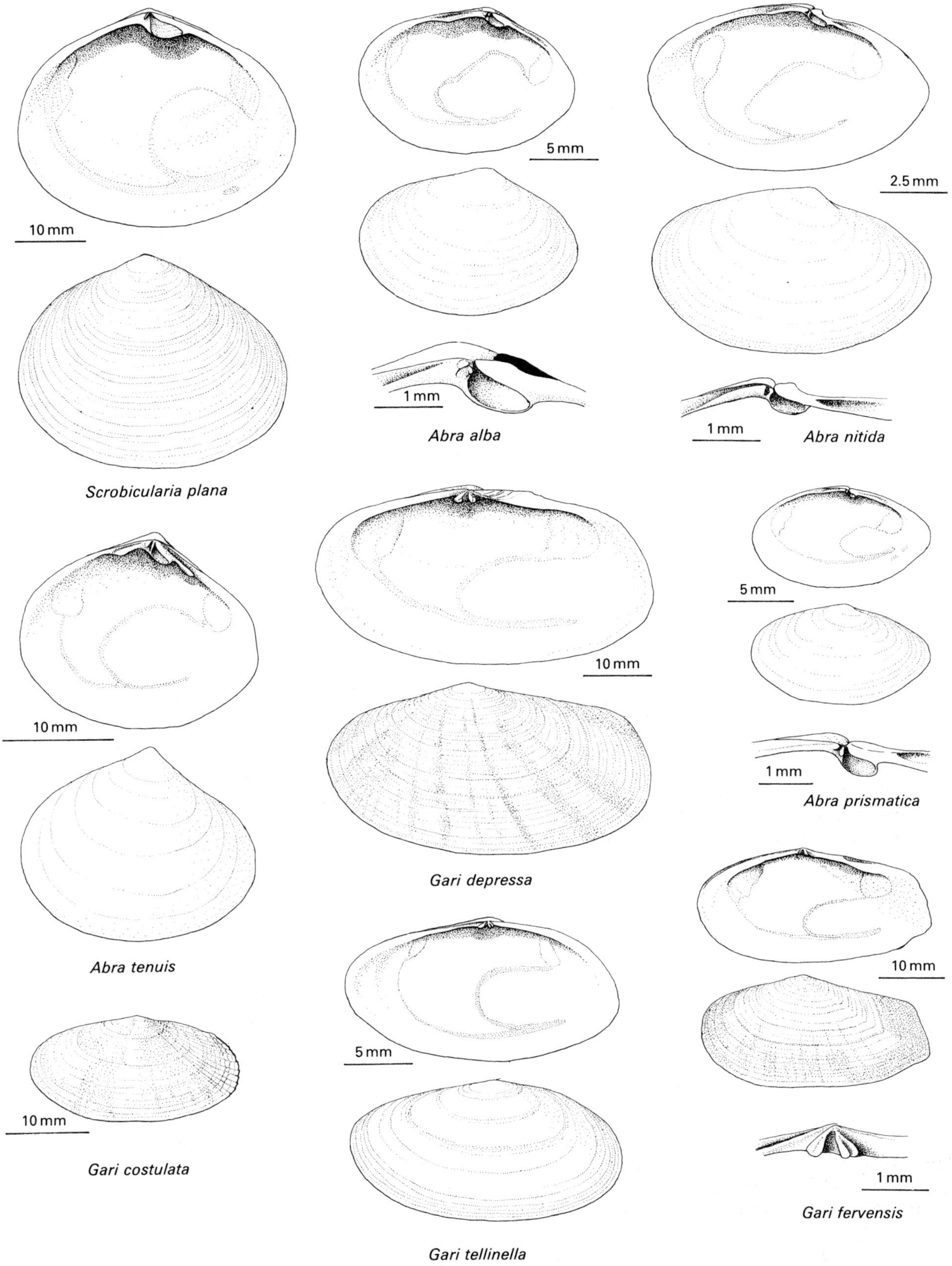

10 mm
Scrobicularia plana
5 mm
1 mm
Abra alba
2.5 mm
1 mm
Abra nitida
10 mm
Abra tenuis
10 mm
Gari depressa
5 mm
1 mm
Abra prismatica
10 mm
Gari costulata
5 mm
Gari tellinella
10 mm
1 mm
Gari fervensis

Fig. 13.14

pale grey or brown periostracum. Up to 13 mm long. Sculpture of numerous fine concentric lines, growth stages visible with hand lens. Chondrophore slender, posteriorly directed. Right valve with two short cardinal teeth anterior to chondrophore, and single, elongate anterior and posterior laterals; left valve with one small cardinal tooth, and single, poorly developed, anterior and posterior laterals. Inner surfaces white, adductor scars and pallial line usually distinct; pallial sinus deep, its lower edge partly fused with pallial line. Cruciform muscle scars distinct.

In soft substrata, in estuaries and tidal flats subject to fluctuating salinity. Intertidal only. Reported from east coasts, as far north as Northumberland, the south coast, and the west as far north as Scotland. Distributed south to the Mediterranean and north-west Africa.

A. longicallus (Scacchi), a marine species reported from the west coasts of Scotland and Ireland, is similar in shape to *A. tenuis*. It is distinguished principally by fine radiating striations on the shell.

31 PSAMMOBIIDAE

Shell equivalve, or slightly inequivalve; almost equilateral, with beaks just in front of mid-line. Ligament external, supported by a distinct plate posterior to beaks. Two cardinal teeth in each valve, usually unequal, one or both typically bifid. No lateral teeth. Pallial line with sinus.

---

1. Posterior end of shell abruptly truncate, with a raised keel extending from the umbone to the posterior ventral corner of each valve — ***Gari fervensis***
   Posterior end of shell not abruptly truncate, regularly or irregularly rounded — **2**
2. Posterior end of shell with a group of fine ribs extending from the ligament to the posterior margin, which is finely crenulate where they meet it — ***Gari costulata***
   Without a posterior group of ribs — **3**
3. Posterior dorsal margin almost horizontal, the posterior margin sloping steeply from it. Valves with a distinct posterior gape — ***Gari depressa***
   Posterior dorsal margin sloping away from the umbones, continuous with the gently rounded posterior margin. Without a posterior gape — ***Gari tellinella***

Gari fervensis (Gmelin) FIG. 13.14

Shell thin, elongate, more than twice as long as broad. Umbones anterior to mid-line; anterior hinge line forming a smooth continuous curve with anterior margin, posterior hinge line sloping more steeply, forming a sharp angle with truncate posterior margin. A distinct keel extends posteriorly from the umbone to postero-ventral corner of each valve. Dull white, yellowish, fawn, or shades of red, pink, or purple; usually in concentric bands of different intensity, with a few white or cream rays radiating from umbones. Periostracum light or dark brown, persisting as a coarse band around margins. Up to 50 mm long. Sculpture of numerous, fine concentric grooves, growth stages clear. Right valve with two cardinal teeth, the anterior grooved, posterior distinctly bifid; left valve with a large, bifid anterior cardinal, and a slender posterior cardinal. Inner surfaces glossy, white tinted with pink or purple. Adductor scars and pallial line distinct, cruciform muscle scars unclear. Pallial sinus deep, U-shaped, its lower edge largely fused with the pallial line.

In mixed sand or gravelly bottoms, from the lower shore downwards. Widespread and common off all British coasts. Distributed from Norway to the Mediterranean and West Africa.

Gari depressa (Pennant) FIG. 13.14

Shell elongate-oval, umbones anterior to mid-line. Anterior hinge line sloping gently, continuous with convex anterior margin; posteriorly almost straight, forming an abrupt junction with the truncate, oblique, posterior margin. Valves with a distinct posterior gape. Glossy white, cream, or yellow, often with radiating streaks of reddish brown or lilac. Periostracum light or dark brown, thick and prominent around margins. Up to 60 mm long or more. Sculpture of numerous fine concentric lines, forming low ridges anteriorly and posteriorly; fine radiating striations may also be present. Growth stages clear. Right valve with two bifid teeth, anterior broader than posterior; left valve with a bifid anterior cardinal and thin posterior cardinal. Inner surfaces glossy, white, or yellowish, often tinted purple or blue, with external rays showing through. Adductor scars and pallial line distinct, cruciform muscle scars unclear. Pallial sinus deep, U-shaped, its lower edge largely fused with the pallial line.

In sandy substrata, from the lower shore into the shallow sublittoral. Most frequent on southern and western coasts, ranging south to the Mediterranean and West Africa.

Gari costulata (Turton) FIG. 13.14

Shell thin, elongate-oval, umbones just posterior to mid-line. Anterior hinge line sloping very gently from umbones, anterior margin somewhat upcurved; posterior hinge line sloping more steeply, continuous with the regularly convex posterior margin. White, yellowish, or pink, often with patches or rays of red or purple. Up to 25 mm long. Sculpture of numerous, fine concentric lines, with about 20 flat ridges extending from umbones to posterior margin. Growth stages clear. Right valve with two bifid cardinal teeth, anterior larger than posterior; left valve with bifid anterior and thin posterior cardinals. Inside of shell showing external colour patterns. Adductor scars and pallial line distinct, cruciform scars unclear; pallial sinus deep, U-shaped, lower edge largely fused with pallial line.

In muddy sand, offshore though not to great depths. Widespread, though not common, around the British Isles. Distributed south to the Mediterranean and West Africa.

Gari tellinella (Lamarck) FIG. 13.14

Shell thin, elongate-oval, umbones posterior to mid-line. Anterior hinge line sloping very gently, posterior more steeply. Colour variable: yellowish white, orange, pink, chestnut, or dull red, often in contrasting bands, frequently with rays of red or chestnut. Up to 25 mm long. Sculpture of numerous fine, concentric lines, and finer radiating striae; growth stages clear, often marked by colour bands. Right valve with two, indistinctly bifid, cardinal teeth, anterior larger than posterior; left valve with broad, bifid anterior and thin, indistinct posterior cardinals. Inner surfaces glossy; white, yellow, orange, or reddish brown. Adductor scars and pallial lines distinct, cruciform scars unclear; pallial sinus broad and deep, lower border largely fused with pallial line.

In coarse sand and gravel, offshore to about 60 m. Present off most British coasts, but apparently absent from much of the North Sea. Distributed from southern Iceland to the Mediterranean.

32. SOLECURTIDAE

Shell elongate, equivalve, inequilateral; gaping anteriorly and posteriorly. Ligament external. Hinge line with peg-like cardinal teeth, no more than two in each valve; lateral teeth present or absent. Pallial line with sinus. Cruciform muscles present, scars indistinct.

---

1. Shell 4–5 times as long as deep; dorsal and ventral margins almost parallel, anterior and posterior margins rounded: resembling a pea-pod. Ligament elongate, situated medially on the dorsal margin **_Pharus legumen_**

2. Shell with concentric sculpture only **_Azorinus chamasolen_**

   Shell with concentric sculpture, and distinct radiating striations on the posterior two-thirds **_Solecurtus scopula_**

Solecurtus scopula (Turton) FIG. 13.15

Shell elongate, anterior and posterior margins rounded, shaped rather like a date; umbones close to mid-line. Dull white, periostracum thin, olive or yellow-brown. Up to 60 mm long. Sculpture of fine concentric lines and two groups of radiating ridges, one in middle region of shell, the other extending posteriorly from umbones. Growth stages clear. Right valve with two cardinal teeth, posterior larger than anterior, perpendicular to hinge line and curving outwards from beneath umbo; left valve with one identical tooth. Inner surfaces white, adductor scars and pallial line distinct. Cruciform muscle scars unclear; pallial sinus broad, irregular, extending into anterior half of shell.

In sandy or shelly gravel, from the lower shore into the shallow sublittoral. South and west coasts only, as far north as Orkney. Distributed south to the Mediterranean and West Africa.

Azorinus chamasolen (da Costa) FIG. 13.15

Shell elongate, anterior and posterior margins rounded and gaping; shaped rather like a date, umbones just anterior to mid-line. Dull white or fawn; periostracum dark brown to olive, coarse and corrugated. Up to 60 mm long. Sculpture of fine concentric lines only, growth stages indistinct. Right valve with two cardinal teeth, posterior larger than anterior, perpendicular to hinge line, curving outwards from beneath umbo; left valve with one identical tooth. Inner surfaces white, adductor scars and pallial line distinct. Cruciform muscle scars unclear; pallial sinus broad, extending up to but not beyond mid-line of shell.

In mud or mixed muddy bottoms, from the shallow sublittoral to well offshore. South and west coasts of the British Isles. Distributed from Norway to the Mediterranean and North Africa.

Pharus legumen (Linnaeus) FIG. 13.15

Shell thin and brittle, elongate, about four times as long as deep; anterior and posterior margins rounded, gaping at both ends. Anterior end distinctly tapered. Umbones low and indistinct, sited about one-third of length from anterior margin; ligament immediately posterior to umbones, situated in middle third of shell. White or light brown, with a glossy, light olive or yellow periostracum, postero-dorsal section of shell fawn. Sculpture of numerous fine concentric lines, with a group of very fine striae radiating from umbones to middle third of ventral margin. Growth stages visible but not conspicuous. Right valve with a single cardinal tooth projecting perpendicularly from hinge line, a single, elongate anterior lateral and a short, peg-like, posterior lateral tooth. Left valve with two elongate, closely spaced, cardinal teeth; a single, elongate, anterior lateral tooth and a single, projecting, posterior lateral tooth, almost as long as cardinals. Inner surfaces white, glossy. Adductor scars and pallial line distinct, pallial sinus short, quadrate; cruciform scars unclear.

Burrowing in sand, from the lower shore into the shallow sublittoral. Off south-west coasts of England, Wales, and Ireland; distributed south to the Mediterranean and north-west Africa.

33 SOLENIDAE

Shell elongate, equivalve, inequilateral: umbones at anterior end. Gaping anteriorly and posteriorly. Ligament external. One or two peg-like cardinal teeth in each valve; lateral teeth present or absent. Adductor scars unequal, anterior typically elongate, posterior small. Pallial sinus present.

---

1. Ligament in middle third of shell, close to mid-line; anterior end of shell noticeably tapered **_Pharus legumen (p. 777)_**

   Ligament close to anterior margin of shell; anterior end not noticeably tapered **2**

2. Each valve with a deeply impressed dorsoventral groove externally, close to anterior margin **_Solen marginatus_**

   Without such a groove **3**

3. Lateral tooth in each valve projecting, similar in size to cardinal teeth and situated close to them. Anterior margin rounded, posterior margin gently truncate **_Phaxas pellucidus_**

Lateral teeth elongate, ridge-like, clearly differentiated from cardinal teeth. At least one margin sharply truncate **4**

4. Dorsal and ventral margins of shell practically straight, and parallel ***Ensis siliqua***

   Either or both dorsal and ventral margins curved, not parallel **5**

5. Anterior end of shell rounded, dorsal and ventral margins both equally curved ***Ensis ensis***

   Anterior end of shell truncate. Ventral margin curved, dorsal margin almost straight ***Ensis arcuatus***

Solen marginatus Montagu FIG. 13.15

Shell thin and brittle, elongate; dorsal and ventral margins straight, parallel, anterior end of shell obliquely truncate, with a distinct dorsoventral groove close to margin. White or yellow, periostracum glossy, light olive or brown. Up to 120 mm long. Sculpture of fine concentric lines, distinct along growth stages. A single cardinal tooth in each valve, no laterals. Inner surfaces white or yellowish; anterior adductor scar almost as long as ligament, posterior small and round, often close to the dorsal edge of the pallial sinus.

Burrows in sand, from the lower shore into the shallow sublittoral; south eastern, southern, and western coasts only. Distributed from Norway to the Mediterranean and north Africa.

Ensis ensis (Linnaeus) FIG. 13.15

Shell thin and brittle, dorsal and ventral margins distinctly and symmetrically curved about mid-line of shell; anterior margin rounded. Dull white or cream, with pale reddish or purplish brown streaks and spots; periostracum glossy, light to dark olive or green. Up to 130 mm long. Sculpture of fine lines following growth stages. Left valve with two, projecting peg-like cardinal teeth, and two elongate, posterior laterals, situated one above the other; right valve with one short cardinal and a single, elongate posterior lateral. Inner surfaces white, often tinted blue or purple. Adductor scars and pallial line clear; anterior adductor elongate, dorsal, extending to below ligament, posterior small and round, well-spaced from pallial sinus.

Burrows in fine sand on the lower shore and in the shallow sublittoral. Off all British coasts, distributed from Norway south to the Mediterranean and north-west Africa.

Ensis siliqua (Linnaeus) FIG. 13.15

Shell thin and brittle, elongate; dorsal and ventral margins parallel, practically straight, anterior and posterior margins obliquely truncate, with rounded corners. White or cream, with pale purple, pink, or reddish streaks; periostracum glossy, light olive or yellow-brown, grey-green along posterior dorsal region. Up to 200 mm long. Sculpture of fine concentric lines, following the growth increments. Left valve with two projecting, peg-like cardinal teeth and two elongate, posterior laterals, situated one above the other; right valve with a single cardinal and a single, elongate posterior lateral. Inner surfaces white, often tinted blue or purplish. Adductor scars and pallial line clear; anterior adductor scar elongate, dorsal, extending to behind ligament; posterior scar small and round, separated from pallial sinus.

Burrowing in sand, from the lower shore into the shallow sublittoral. Off all British coasts, distributed from Norway to the Mediterranean and northwest Africa.

Ensis arcuatus (Jeffreys) FIG. 13.15

Shell thin and brittle, elongate; dorsal margin almost straight, ventral margin distinctly curved, shell deepest at mid-line. Anterior and posterior margins obliquely truncate. White or cream, with reddish or purplish brown streaks and blotches; periostracum glossy, light olive or yellow-brown, darker grey-green along posterior dorsal region. Up to 150 mm long. Sculpture of fine lines following the growth stages. Left valve with two, projecting, peg-like cardinal teeth and two, elongate, posterior laterals, one above the other; right valve with one cardinal and a single, elongate, posterior lateral. Inner surfaces white, tinted with pink or pale purple; adductor scars and pallial line distinct. Anterior adductor elongate, dorsal, extending to behind ligament, posterior small, round, separated from pallial sinus.

Burrowing in sand and gravel, from the lower shore into the shallow sublittoral. Off all British coasts, distributed from Norway to Spain.

Phaxas pellucidus (Pennant) FIG. 13.15

Shell thin and brittle, elongate; dorsal margin practically straight, ventral margin curved. Anterior end rounded and upturned, posterior slightly truncate. White or cream, sometimes with dark markings; periostracum glossy, light yellow-brown or olive. Up to 40 mm long. Sculpture of fine lines, following the distinct growth stages. Left valve with a group of three projecting teeth: two cardinals, the posterior being broad and bifid, and posteriorly a single lateral. Right valve with two teeth, a single cardinal and a backwardly directed posterior lateral. Inner surfaces white. Adductor scars and pallial line distinct; anterior adductor as long as ligament, posterior small and largely fused with the dorsal edge of the irregular pallial sinus.

In mixed fine substrata, offshore to about 100 m. Widespread, and often abundant, around Britain; distributed from Norway to north-west Africa.

34 MYACIDAE

Shell equivalve, inequilateral, typically with a pronounced posterior gape. Ligament largely internal, supported by a chondrophore—usually projecting and prominent in the left valve. Hinge teeth absent. Pallial sinus present.

---

1. Shell thin and brittle, not exceeding 20 mm. Chondrophore small. Umbones anterior to mid-line; posterior hinge line sloping from umbones, gently truncate posteriorly ***Sphenia binghami***

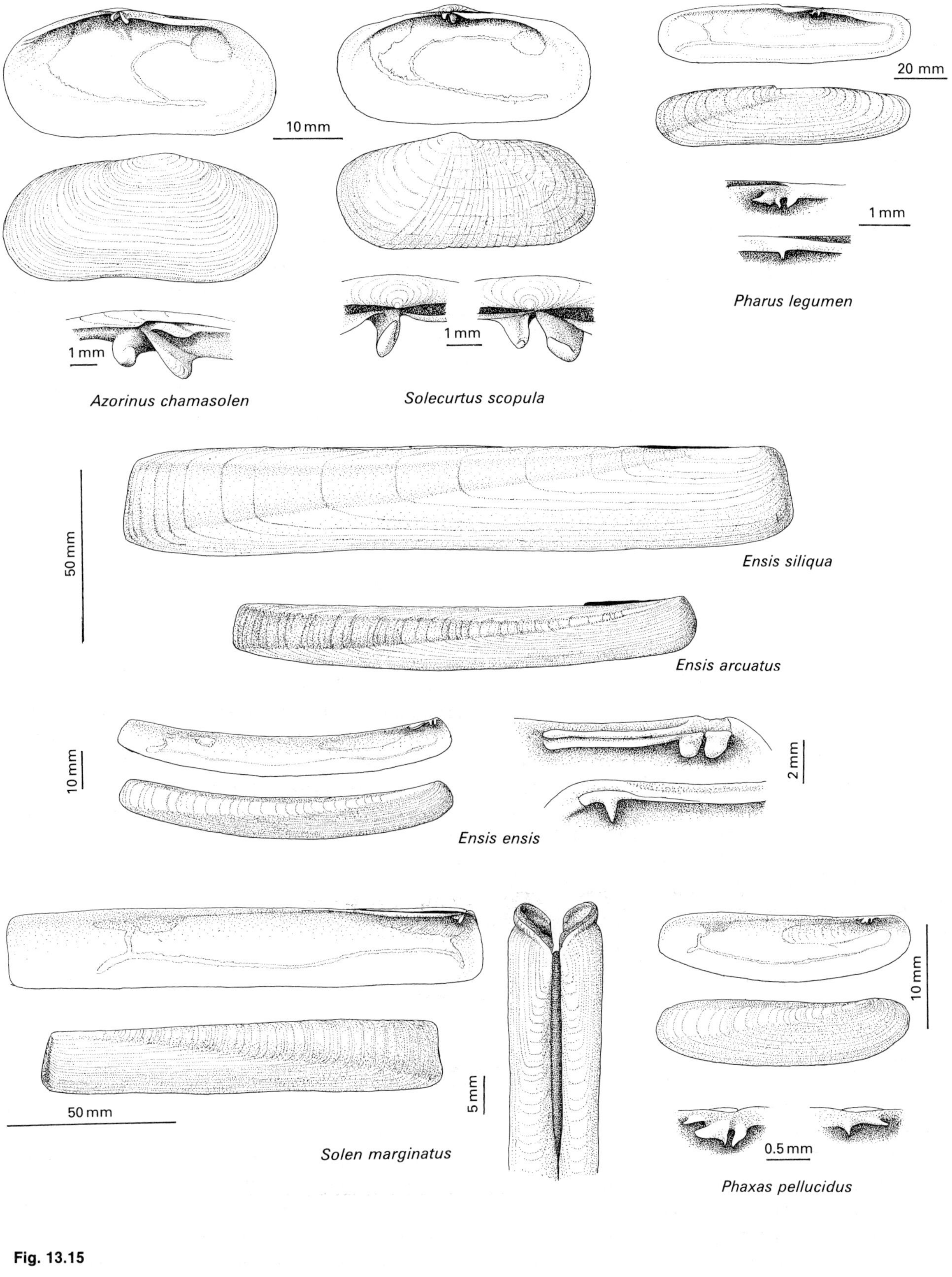
10 mm
1 mm
Azorinus chamasolen
1 mm
Solecurtus scopula
20 mm
1 mm
Pharus legumen
50 mm
Ensis siliqua
Ensis arcuatus
10 mm
2 mm
Ensis ensis
50 mm
5 mm
Solen marginatus
10 mm
0.5 mm
Phaxas pellucidus

Fig. 13.15

Shell thick and strong, up to 150 mm long. Chondrophore prominent, projecting in left valve, concealed beneath umbo in right. Umbones close to or posterior to mid-line **2**

2. Posterior half of shell semi-elliptical, margin rounded ***Mya arenaria***

Posterior half of shell quadrate, abruptly truncate ***Mya truncata***

Mya arenaria Linnaeus FIG. 13.16

Shell thick and strong, oval, umbones just posterior to mid-line; anterior end regularly rounded, posterior somewhat tapered. Off-white, yellowish, or fawn, dark greyish brown about the umbones; periostracum light brown, often stained by iron deposits. Up to 150 mm long. Sculpture of fine concentric lines, becoming coarser and more corrugated close to margins, also a few, faint, radiating striae. Growth stages clear. Left valve with a prominent spatulate chondrophore, projecting at a right angle to hinge line, with a distinct tooth-like ridge along its posterior edge. Right valve with a concave, spatulate chondrophore recessed beneath umbo. Inner surfaces white. Adductor scars and pallial line distinct; anterior scar elongate, posterior rounded.

In sand, often mixed with mud or gravel, on the lower shore and offshore to about 20 m. Often very common in estuaries, where it may occur in extensive beds. Off all British coasts; circumboreal, not reaching the Mediterranean.

Mya truncata Linnaeus FIG. 13.16

Shell thick and strong, quadrate, umbones on mid-line; anterior end regularly rounded, posterior abruptly truncate. Dull white to light brown, periostracum light olive to brown. Up to 70 mm long. Sculpture of coarse concentric grooves, growth stages clear. Left valve with a prominent spatulate chondrophore, projecting at a right angle to hinge, bounded anteriorly and posteriorly by a distinct ridge. Right valve with a concave, spatulate chondrophore recessed beneath umbo, a blunt ridge bounding it anteriorly. Inner surfaces white. Pallial line and adductor scars distinct, anterior scar elongate, posterior rounded.

In mixed sandy substrata, from the lower shore to about 70 m. Off all British coasts; circumboreal, in the north-east Atlantic extending south to Biscay.

Sphenia binghami Turton FIG. 13.16

Shell thin and fragile, irregularly oval, umbones anterior to mid-line; anterior end rounded, posterior sloping away from umbones, somewhat attenuated, irregularly rounded or gently truncate. White, with pale green or brown periostracum. Not exceeding 20 mm. Sculpture of fine concentric lines, growth stages clear. Left valve with a short, broad chondrophore projecting horizontally from beneath umbo, continuous posteriorly with the thickened dorsal margin. Left valve with a recessed chondrophore, completely obscured beneath umbo. Adductor scars and pallial line distinct; anterior scar elongate, posterior rounded.

Attaches by a byssus, in holes and crevices or algal holdfasts, in shallow sublittoral waters. Off most British coasts; distributed southwards to the Mediterranean and north-west Africa.

35 CORBULIDAE

Shell inequivalve, more or less equilateral. Ligament internal, in a chondrophore recessed within the hinge line. True cardinal teeth absent, lateral teeth present. Pallial line with a very small sinus.

Corbula gibba (Olivi) FIG. 13.16

Shell thick, broadly oval to subtriangular, umbones close to mid-line; right valve convex, enclosing and overlapping left, its ventral margin covered only by a fringe of periostracum from the left valve. Dull white to cream; periostracum coarse, grey-brown, usually worn at the umbones, most conspicuous on left valve. Up to 15 mm long. Sculpture of coarse concentric grooves and ridges, and a few radiating striae on left valve. Right valve with a single, massive tooth, anterior to a recessed, triangular chondrophore, and single elongate anterior and posterior lateral teeth. Left valve with a shallow chondrophore posterior to a deep socket, no lateral teeth. Inner surfaces white or yellowish, sometimes tinted blue; adductor scars and pallial line usually clear, pallial sinus very shallow.

In muddy sand and gravel, occasionally on the lower shore, most abundant offshore. Common and widespread off all British coasts; distributed from Norway south to the Mediterranean and west Africa.

36 HIATELLIDAE

Shell equivalve, inequilateral. Ligament external. Right valve with one cardinal tooth, left with two or none. No lateral teeth. Pallial line comprising a series of disjunct scars, with a small sinus.

---

1. Pallial sinus distinct, discontinuous. Shell elongate, quadrate; umbones close to anterior margin. In holes and crevices ***Hiatella arctica***

Pallial sinus indistinct. Shell oval or subtriangular, umbones anterior to mid-line. Among bottom deposits ***Saxicavella jeffreysi***

The third representative of this family in British waters, *Panomya arctica* (Lamarck) (Fig. 13.16), is restricted to Scotland and the North Sea, where it occurs offshore in mixed soft substrata. Oblong in shape and often exceeding 70 mm in length, it is easily distinguished from both the described species.

Hiatella arctica (Linnaeus) FIG. 13.16

Shell thick, elongate, rectangular or rather irregular; umbones close to anterior end, anterior margin rounded, posterior truncate. Dull white; periostracum coarse, yellow-brown. Up to 40 mm long. Sculpture of coarse, concentric ridges, less evident

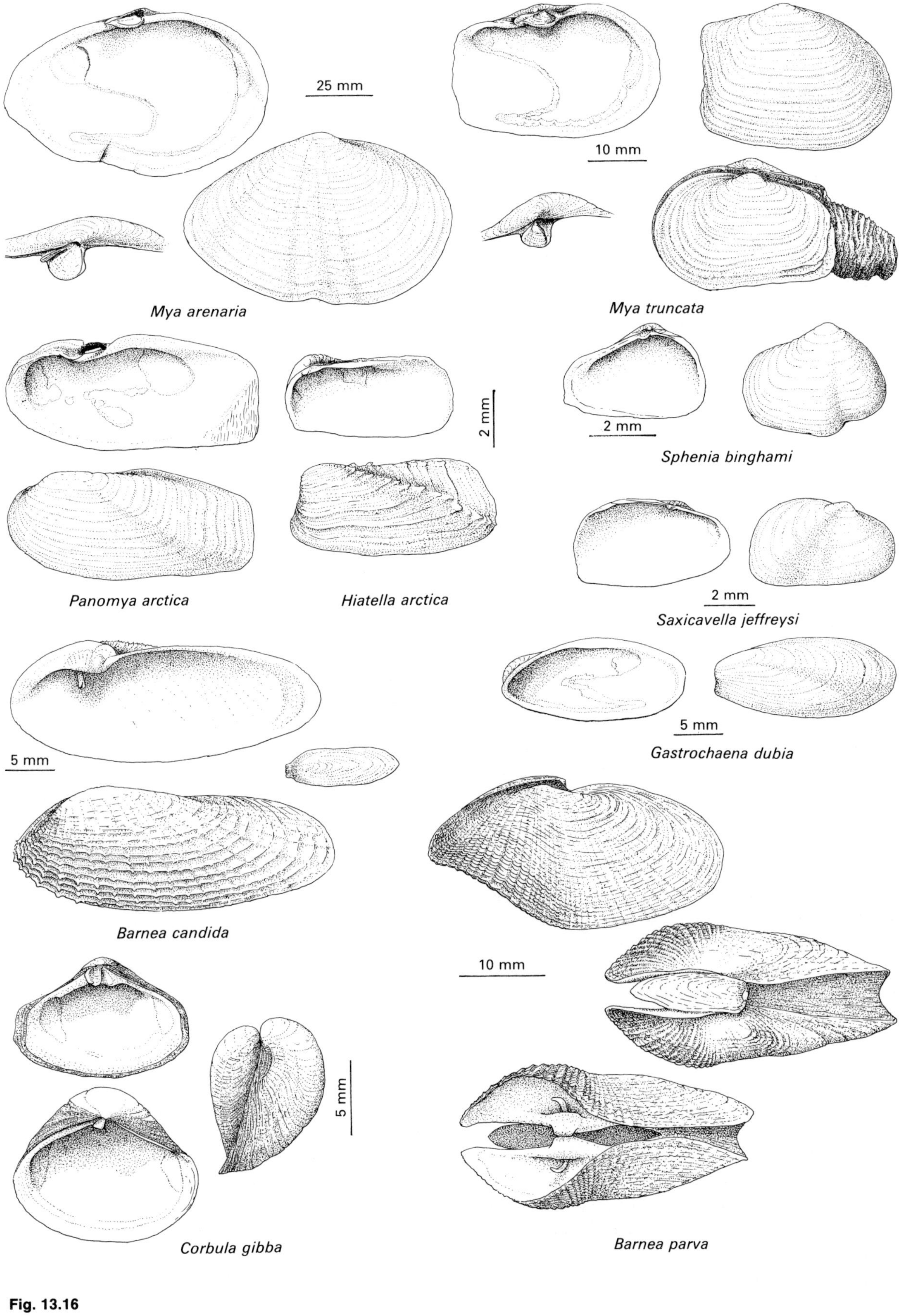
25 mm
10 mm
Mya arenaria
Mya truncata
5 mm
2 mm
2 mm
Sphenia binghami
Panomya arctica
Hiatella arctica
2 mm
Saxicavella jeffreysi
5 mm
Gastrochaena dubia
5 mm
Barnea candida
10 mm
5 mm
Corbula gibba
Barnea parva

Fig. 13.16

about the umbones, where the shell is usually abraded. Two distinct ridges extend posteriorly from umbo on each valve: in juveniles these bear a few short, coarse tubercles, worn away in older specimens. Right valve with a single cardinal tooth, left valve with two; visible in juveniles only, in older shells these are worn completely away, the hinge line below the umbo and ligament being undulating and thickened. Inner surfaces white, adductor scars large and distinct, pallial line discontinuous, comprising a series of separate scars.

Attaches by a byssus, in holes, crevices, or algal holdfasts; also bores into soft calcareous rocks and shells. On the lower shore and well out on to the Continental Shelf; widespread, and common, off all British coasts. Distributed from arctic waters south to, at least, the Mediterranean and north-west Africa.

Saxicavella jeffreysi Winckworth — FIG. 13.16

Shell thin and fragile, oval to subtriangular; umbones anterior to mid-line, hinge sloping steeply to anterior. White, up to 10 mm long. Sculpture of very fine concentric lines, becoming coarser and somewhat corrugated towards margins. Growth stages clear. Right valve with a single small cardinal tooth, adjacent to a triangular plate supporting the external ligament; left valve without teeth, but with hinge line posterior to umbones broad and thickened to support the ligament. Inner surfaces white, adductor scars and pallial line indistinct, pallial sinus very small.

In mixed soft substrata, offshore. Reported from various localities around the British Isles, uncommon; distributed from southern Iceland and Norway to the Mediterranean and Canary Isles.

37 GASTROCHAENIDAE

Shell equivalve, inequilateral, gaping widely postero-ventrally. Ligament external, hinge line without teeth. Pallial sinus present.

Gastrochaena dubia (Pennant) — FIG. 13.16

Shell thin and brittle, elongate; umbones close to anterior end, both valves strongly convex, gaping widely along the whole ventral margin. Dull white or yellowish, periostracum coarse, light brown; up to 25 mm long. Sculpture of coarse concentric lines, growth stages indistinct. Inner surfaces dull white. Adductor scars and pallial line clear: anterior adductor scar close to anterior margin, thickened, pallial line broad; pallial sinus triangular, extending beyond mid-line of shell.

A borer in sandstone, limestone, and organic carbonates, lower shore and shallow sublittoral. South and south-west coasts only. Distributed southwards to the Mediterranean.

38 PHOLADIDAE

Shell equivalve, inequilateral, often elongate or irregular; sculpture often pronounced anteriorly, forming a rasping file which assists the animal in boring. Typically gaping along part or all of the anterior, ventral and posterior margins. Ligament internal, reduced; hinge teeth absent. Antero-dorsal margin reflected, curving back towards umbones. Accessory plates present along the dorsal margin: a *protoplax* anterior to umbones, a *mesoplax* above or immediately posterior to the umbones and a *metaplax* posteriorly; not always present together. Additional elements may be developed to close an impermanent pedal gape (*callum*), or to protect posteriorly projecting siphons (*siphonoplax*). Foot muscles supported, in each valve, by a long, curved process (*apophysis*) extending ventrally from beneath the umbo. Pallial line with a sinus.

---

1. Shell with a marked anterior gape, anterior margin oblique to hinge line — **2**
   Shell not gaping anteriorly, margin rounded — **4**

2. Umbonal reflections supported by distinct vertical septa — ***Pholas dactylus***
   Umbonal reflections without septa — **3**

3. Shell twice as long as deep; coarse anterior sculpture grading into finer posterior sculpture. With an elongate, lanceolate protoplax anteriorly, but no mesoplax — ***Barnea parva***
   Shell length less than twice depth; coarse anterior sculpture separated from finer posterior sculpture by a vertical groove. With a small triangular mesoplax, but no protoplax (do not confuse with juvenile *P. loscombiana*, see 4) — ***Zirfaea crispata***

4. Shell elongate-oval, with a posterior gape; sculpture continuous, accentuated anteriorly. Protoplax elongate-oval, with a longitudinal groove. Posterior margins rounded — ***Barnea candida***
   Shell quadrate, anterior gape present in juveniles, closed by the development of a smooth flap, or callum, on each valve. Sculpture posterior to callum discontinuous, separated by a ventral groove. Posterior margin truncate, not gaping, developing on each valve an accessory flap, or siphonoplax — ***Pholadidea loscombiana***

Pholas dactylus Linnaeus — FIG. 13.17

Shell thin and brittle, elongate oval, umbones anterior to mid-line; anterio-ventral margin deeply concave about a large, elliptical pedal gape, posterior margin regularly rounded, not gaping. Both valves strongly convex anteriorly. Dull white or grey, periostracum yellowish, often discoloured; up to 150 mm long. Sculpture of concentric ridges and radiating lines, produced as prominent tubercles where they intersect, most pronounced anteriorly; growth stages distinct in larger specimens. Umbonal reflection pronounced, with ten or more vertical septa linking it with umbone. Dorsal margin with a broad, oval protoplax, a small mesoplax and an asymmetric metaplax overlying the right side. Inner surfaces white. Apophysis broadening towards its free edge, extending from beneath umbo, in each valve, halfway towards the ventral margin. Adductor scars and pallial line clear, pallial sinus deep, U-shaped.

Bores into a wide range of substrata including wood, peat,

compacted sand, and various soft rocks. Lower shore and shallow sublittoral. South and south-west coasts only, ranging to the Mediterranean and north-west Africa.

Barnea candida (Linnaeus) FIG. 13.16

Shell thin and brittle, elongate-oval, shaped rather like a date; anterior and posterior margins rounded, gaping posteriorly. White, periostracum yellowish or light brown; up to 65 mm long. Sculpture of corrugated concentric ridges and radiating lines, developed as sharp tubercles where they intersect, most pronounced anteriorly. Umbonal reflections prominent anteriorly, closely applied to umbones posteriorly and without septa. An elongate, oval protoplax dorsally. Inner surfaces white. Apophysis slender, extending one-third of distance to ventral margin. Adductor scars and pallial line indistinct, pallial sinus deep, U-shaped.

Bores into wood, peat, and various soft rocks. Lower shore and sublittoral. Off all British coasts; distributed from Norway to the Mediterranean and West Africa.

Barnea parva (Pennant) FIG. 13.16

Shell thin and brittle, obliquely oval, umbones anterior to mid-line; anterior margin deeply concave about a broad, oval pedal gape. Both valves strongly convex. Dull white, periostracum yellowish, often discoloured; up to 50 mm long. Sculpture of numerous corrugated concentrated lines and radiating ribs, forming short, triangular tubercles where they intersect, most numerous and pronounced anteriorly. Umbonal reflections without septa. An elongate, lanceolate protoplax dorsally. Inner surfaces white. Apophysis delicate, flat, extending half-way towards ventral margin. Adductor scars and pallial line distinct, pallial sinus rather irregular.

Bores into soft limestones, sandstones, and argillaceous rocks; lower shore and shallow sublittoral. South and south-west coasts only, distributed southwards to the Mediterranean.

Zirfaea crispata (Linnaeus) FIG. 13.17

Shell thin and brittle, obliquely oval, both valves strongly convex; umbones anterior to mid-line, anterior margins obliquely truncate above a broad pedal gape; posterior gaping widely. Dull white, periostracum light yellowish brown; up to 90 mm long. Sculpture of fine concentric lines, raised as coarse ridges anteriorly, where they intersect with fine radiating ribs to form sharp tubercles. A median groove extending from umbones to ventral margin separates the two areas. Umbonal reflections broad, closely applied to, and extending just posterior to, the umbones; a small mesoplax present. Inner surfaces white. Pallial line and adductor scars clear, pallial sinus deep, extending to beyond mid-line of shell. External median groove visible as a distinct ridge. Apophysis slender, slightly broadened distally, extending no more than one-third of distance to ventral margin.

Bores into peat, clay, and soft rocks, rarely into waterlogged wood, from the lower shore into the shallow sublittoral. Widespread around the British Isles, distributed from Norway to the Bay of Biscay.

Pholadidea loscombiana Turton FIG. 13.17

Shell thin, brittle and very fragile, elongate-oval, truncate posteriorly, umbones anterior to mid-line. Juveniles with an obliquely truncate anterior margin about a large pedal gape, closed in adults by a thin additional element on each valve, the callum. Posteriorly, adults develop a tubular siphonoplax, two lightly calcified plates enveloping the siphons. White, with a pale yellow periostracum; up to 40 mm long. Sculpture of concentric ridges, flat and smooth posteriorly, raised and frilled anteriorly, the two components separated by a deep groove extending from umbones to ventral margin. Callum smooth. Umbonal reflection low and narrow; accessory plates limited to a small, bipartite mesoplax posterior to umbones. Inner surfaces white, glossy; junction of callum distinct, external groove marked by a median ridge. Apophysis short, directed anteriorly; left valve with a small chondrophore just projecting from beneath umbo, right valve with larger chondrophore. Adductor scars and pallial line distinct, pallial sinus extending into anterior half of shell.

Bores into compacted mud or clay, peat, and soft sandstones, from the lower shore down. South and south-west coasts only, distributed south to the coast of Spain.

39. TEREDINIDAE

Shell thin, equivalve, inequilateral and highly modified, gaping anteriorly and posteriorly, appearing trilobed and composed of several distinct elements: a triangular *anterior lobe*, typically with fine ridges parallel to the hinge; a deeper *anterior disc*, flanked by slender ridges; a broader *posterior disc*, and finally a roughly triangular *auricle*. Umbones prominent, ligament and hinge teeth absent, but a dorsal articulating condyle present below umbo, and a ventral condyle on the ventral shell margin. Animal elongate, enclosed within its burrow, which is lined with calcareous deposits. A pair of accessory plates (*pallets*) present posteriorly at the tips of the siphons.

1. No apophyses present. Shell gaping anteriorly, closed posteriorly. Dorsal mesoplax present **_Xylophaga dorsalis_**

   Long, slender apophyses present. Shell small, trilobed, gaping anteriorly and posteriorly; body elongate, enclosed by calcareous deposits, with paired accessory plates (pallets) posteriorly **2**

2. Posterior lobe of shell (auricle) much smaller than anterior. Broad end of pallet truncate, slender handle extending as a ridge along the mid-line of the blade **_Nototeredo norvegica_**

   Posterior lobe of shell almost as large as anterior. Broad end of pallet with concave edge, slender handle not extending along mid-line of blade **_Teredo navalis_**

Many other species of Teredinidae may occur, frequently or sporadically, on British shores as a consequence of their unique habitat, in drifting waterlogged wood. Complete coverage is not possible here, but all species likely to occur are described by Tebble (1976).

Teredo navalis Linnaeus FIG. 13.17

Shell thin and fragile, up to 10 mm long, white, periostracum light brown. Anterior lobe triangular, with fine horizontal ridges; anterior disc with fine vertical to oblique ridges; median disc narrow, concave, with concentric sculpture; posterior disc with fine concentric lines; auricle semi-elliptical, almost as large as anterior lobe. Both valves strongly convex. Umbonal reflection closely applied to umbo; posterior hinge line broad, concave, with a bounding ridge. Inner surfaces white; junction of auricle with posterior disc visible as a ridge, median disc generally indistinct. Apophysis long, curved, slender, and rod-like, extending two-thirds of the depth of the shell. A knoblike condyle below the umbone and another on the opposite, ventral, border. Pallet blade elliptical with a concave edge, handle short, cylindrical, not extending on to blade.

The borings of *T. navalis* are frequently found in drifted, waterlogged wood, each lined with a thin tubular, calcareous deposit. Widespread around the world, it has been collected on all British coasts.

Nototeredo norvegica (Spengler) FIG. 13.17

Shell thin and fragile, up to 20 mm long. White, periostracum brown, darkening to almost black in larger specimens. Anterior lobe triangular with fine horizontal ridges; anterior disc with even finer oblique ridges; median disc narrow, concave, with concentric sculpture; posterior disc with concentric lines; auricle smooth, crescent-shaped, much smaller than anterior lobe. Both valves strongly convex. Umbonal reflections small, closely applied to anterior face of umbones; hinge line posterior to umbo broad and concave, delimited by a raised ridge. Inner surfaces white. Median disc, and junction of auricle with posterior disc, evident as distinct ridges. Apophysis broad, bladelike, extending half-way to opposite ventral border. A knoblike condyle below the umbo, and another ventrally. Pallet concave, blade elliptical with a straight edge, handle cylindrical, extending along mid-line of blade as a distinct ridge.

Bores into fixed and floating wood; widespread in the north-east Atlantic, from Iceland to the Mediterranean, reported off most British coasts.

Xylophaga dorsalis Turton FIG. 13.17

Shell thin and brittle, globular, each valve strongly convex, rounded posteriorly, with a right-angled identation anteriorly, exposing a pedal gape. Shell constructed of several distinct elements: a triangular *anterior lobe* dorsal to the gape, with regular horizontal ridges; an elongate *anterior disc* extending from the umbo to the ventral margin, forming the posterior edge of the gape, with regular, oblique ridges; the *median disc*, a concave band with raised margins, sculptured with very fine concentric lines; the *posterior disc*, comprising more than half the shell, with fine concentric lines. Umbonal reflection small, closely applied to umbo posteriorly, obscured by a bipartite mesoplax. Inner surfaces white, anterior edge of median disc developed as a prominent ridge. No apophyses, although left valve has a prominent chondrophore.

Boring into wood, present off all British coasts. Distributed from Iceland and Norway south to the Mediterranean.

Xylophaga properly belongs to the Pholadidae but so closely resembles species of Teredinidae that it is most usefully keyed out here. The rare *Xylophaga praestans* (Smith), at present known only from wood borings off the north-east coast of England, differs principally from *X. dorsalis* in gaping both anteriorly and posteriorly.

40 PERIPLOMATIDAE

Shell inequivalve, right valve more convex than left; inequilateral. Ligament external and internal, latter component in an oval chondrophore. No hinge teeth. Umbones typically with a discontinuity in the calcification. Pallial line with a sinus.

Cochlodesma praetenue (Pulteney) FIG. 13.17

Shell thin and brittle, oval, umbones just posterior to mid-line; anterior margin rounded, posterior somewhat truncate, hinge line posterior to umbones gently sloping. Dull white, periostracum light yellowish brown; up to 40 mm long. Sculpture of fine concentric lines, growth stages discernible with hand lens; on each valve a distinct crack extends for a short distance posteriorly from the umbones—a lens is necessary for small specimens. Posterio-dorsal surface of shell finely granular, with coarser periostracum, and often encrusted detritus. Chondrophore elliptical, projecting, arising posterior to the umbone and directed obliquely anteriorly. Inner surfaces white. Adductor scars and pallial line clear, pallial sinus regular, U-shaped, extending almost as far as the chondrophore.

In mixed fine deposits, from the lower shore to about 100 m. Reported from various localities around the British Isles; distributed from Norway and southern Iceland to the Mediterranean.

41 THRACIIDAE

Shell inequivalve, right valve slightly larger or more convex than left, inequilateral. Ligament external and internal, the latter in a triangular chondrophore; supported by a *lithodesma*, a short calcified peg fitting into a notch anterior to chondrophore. No hinge teeth. Pallial line with a sinus.

1. Pallial sinus shallow, extending to just beyond front edge of posterior adductor scar **_Thracia pubescens_**
   Pallial sinus deep, extending to below chondrophore **2**
2. Shell granulation fine, not visible with × 10 hand lens **_Thracia phaseolina_**
   Shell granulations coarser, visible with a × 10 hand lens **_Thracia villosiuscula_**

Thracia phaseolina (Lamarck) FIG. 13.17

Shell thin, brittle, oblong-oval, umbones just posterior to mid-line; anterior margin rounded, posterior abruptly truncate. Dull

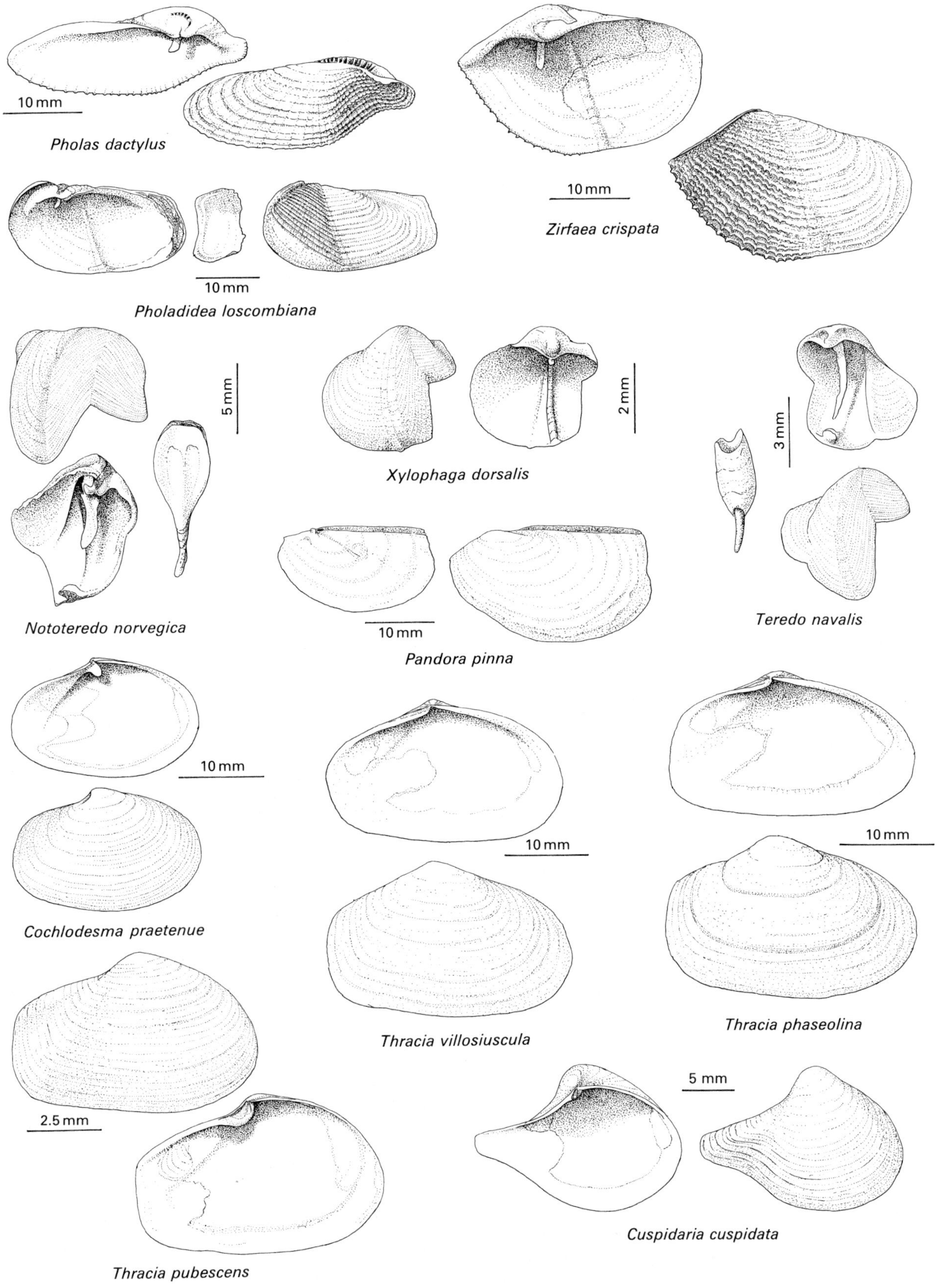
10 mm
Pholas dactylus
10 mm
Zirfaea crispata
10 mm
Pholadidea loscombiana
5 mm
2 mm
3 mm
Xylophaga dorsalis
Nototeredo norvegica
10 mm
Pandora pinna
Teredo navalis
10 mm
10 mm
10 mm
Cochlodesma praetenue
Thracia villosiuscula
Thracia phaseolina
2.5 mm
5 mm
Thracia pubescens
Cuspidaria cuspidata

Fig. 13.17

white, periostracum thin, light yellowish brown; up to 38 mm long. Sculpture of very fine concentric lines, growth stages clear; surface very finely granulated, the structure not visible under a × 10 hand lens. Chondrophore developed in each valve as a slender, triangular concavity in the hinge line, posterior to umbo; anterior to the chondrophore a small notch receives the short, peg-like lithodesma. Inner surface dull white. Adductor scars and pallial line clear, pallial sinus U-shaped, extending dorsally and anteriorly to a point below chondrophore.

In mixed sand, from the lower shore to about 50 m. Reported from numerous localities around Britain. Geographical range uncertain, through confusion with *T. villosiuscula* (q.v.), but is present from Norway to the Mediterranean.

Thracia villosiuscula (Macgillivray) FIG. 13.17

Shell thin, brittle, oblong-oval, umbones just posterior to mid-line; anterior margin rounded, posterior abruptly truncate. Dull white, periostracum thin, yellowish brown; up to 30 mm long. Sculpture of very fine concentric lines, growth stages clear; surface granular, the structure visible with a × 10 lens. Chondrophore developed in each valve as a slender, triangular concavity in the hinge line, posterior to the umbo; lithodesma short, peg-like, fitting into a notch in front of the chondrophore in each valve. Inner surface white. Adductor scars and pallial line distinct, pallial sinus U-shaped, extending anterio-dorsally to a point just below the chondrophore.

This species is difficult to distinguish from *T. phaseolina*. The posterior margin tends to be obliquely truncate, rather than vertical as in *T. phaseolina*, but according to Tebble (1976) this is a variable feature. The relative coarseness of the shell granulations is perhaps the most useful character.

In coarse sand and gravel, from the lower shore to about 60 m. Reported from various localities around Britain but distribution elsewhere uncertain.

Thracia pubescens (Montagu) FIG. 13.17

Shell thin, brittle, oblong-oval, umbones on the mid-line; anterior margin rounded, posterior abruptly truncate. Dorsal margin of each valve inflected, posterior to umbo, forming a narrow shelf. Dull white, periostracum thin, light yellowish brown; up to 90 mm long. Sculpture of fine concentric lines and ridges, growth stages clear. Chondrophore broadly triangular, thickened, and concave; lithodesma in a small notch immediately anterior to umbones. Inner surfaces white. Adductor scars and pallial line distinct, pallial sinus irregular, shallow, extending to just beyond front edge of posterior adductor scar.

In sand and mud, offshore. South-western coasts only, distributed south to the Mediterranean and west Africa.

Two other species of *Thracia* are known from British waters, *T. convexa* (Wood) and *T. distorta* (Montagu). The former has the same shallow pallial sinus as *T. pubescens* but is broadly triangular in shape, without the truncate posterior margin of the three species described here. *T. distorta*, as suggested by the name, is often irregular in outline, being an inhabitant of crevices; unlike other species, the umbones are in the anterior half of the shell.

42 PANDORIDAE

Shell inequivalve, left valve convex, overlapping the thin, flat right valve. Ligament internal, in a prominent ridge-like chondrophore. Right valve with a single, ridged, tooth-like structure, with a corresponding pit in the left, but true teeth absent. Pallial line comprising a disjunct series of small round scars, without a sinus.

Pandora pinna (Montagu) FIG. 13.17

Shell thin and brittle, elongate, elliptical, umbones close to anterior margin; dorsal margin straight posteriorly, sloping anteriorly, ventral margin strongly convex. Left valve convex, right valve flat, with a distinct dorsal flange where it joins the left. Dull white, periostracum light brown. Up to 20 mm long. Sculpture of fine concentric lines, conspicuous only on right valve; left valve with a single, fine ridge extending posteriorly from umbo. Right valve with a single tooth-like ridge anterior to a slender, raised chondrophore; left valve with a corresponding socket, a ridge anterior to it and a slender, raised chondrophore posteriorly. Inner surfaces white; adductor scars distinct, pallial line a series of small, round scars.

Occurs offshore, reported from various localities, off all British coasts, but generally uncommon. Distributed south to the Mediterranean and north-west Africa. A similar species, *P. inaequivalvis* (Linnaeus), has been reported off southern and western coasts from the English Channel to the Isle of Man; it is distinguished from *P. pinna* by its strongly concave dorsal margin.

43 CUSPIDARIIDAE

Shell equivalve; inequilateral, with posterior end of shell drawn out as a slender spout. Ligament internal, in a small chondrophore; lithodesma present. Cardinal and lateral teeth present or absent, usually small. Pallial line with a small sinus.

Cuspidaria cuspidata (Olivi) FIG. 13.17

Shell thin and brittle, umbones close to mid-line, prominent and bulbous. Anterior half of shell semi-elliptical, posterior drawn out as a tapered spout, truncate at its end. Both valves strongly convex. White; periostracum brown, coarse, often corrugate, with adhering sediment. Up to 20 mm long. Sculpture of numerous fine lines and ridges, somewhat coarser posteriorly and towards margins. Ligament supported in each valve by an elliptical chondrophore, prominent in right valve but recessed below hinge line in left. Right valve with one elongate posterior lateral tooth, fitting beneath the left hinge line. Inner surfaces white, adductor scars and pallial line distinct, pallial sinus shallow.

In muddy sand and gravel, offshore. Reported from north-east England and the east and west coasts of Scotland. Distributed from Norway to the Mediterranean and West Africa.

This is one of four species of *Cuspidaria* found in British waters, none of which is common. All have the same kind of hinge line, and have the posterior spout variously developed. Descriptions of the other species are given by Tebble (1976).

G. Class **Cephalopoda**

In the Cephalopoda the molluscan body displays its most highly modified form, and molluscan evolution achieves its most sophisticated expression. The Class Cephalopoda includes the largest invertebrate animals. The cephalopod body is elongated along a dorsoventral axis, so that the defined anterior end, bearing the head, is morphologically ventral, while the rounded or pointed posterior end is in fact dorsal. A chambered shell which completely encloses the body persists in *Nautilus*, the only living genus of the ancient Nautiloidea, but in most cephalopods the shell is reduced and internal, or absent. Two significant features of cephalopod morphology derive from the modification of the head and foot, viz. the tentacles or *arms* which surround the mouth, and the *funnel* through which water is expelled from the mantle cavity.

The cephalopod body is rounded, elongate, or distinctly torpedo-shaped. The *mantle* is thickened and muscularized, completely enveloping the *visceral mass*. The *mantle cavity* is ventral in position, containing the *ctenidia*, anus, and reproductory and excretory orifices. Water is actively pumped into the mantle cavity and expelled through the anteriorly situated *funnel*. In many cephalopods this is flexible, and while rapid water expulsion is utilized in jet-propulsion, the funnel may be directed to allow more deliberate, slower movement. In all squids and cuttlefishes, however, lateral or terminal fins are employed in normal swimming. A substantial internal shell, the familiar cuttlebone, is present in cuttlefish. A light, laminar calcareous structure, it imparts a firmness and rigidity to the body, and functions as a very finely balanced device for buoyancy control. In the squids, in which neutral buoyancy is achieved through regulation of ionic balances, the shell is represented by a slender, quill-like structure of keratin. Cephalization is marked; the enlarged cerebral ganglia are enclosed in a cartilaginous capsule, and the head bears large, complex eyes. The mouth is armed with tough jaws, resembling a parrot's beak, and most cephalopods have modified salivary glands which secrete a venom used to immobilize prey. The tentacles, or *arms*, surround the mouth; they are muscular, prehensile, and bear numerous cup-shaped suckers, which may be armed with chitinous hooks and spines. Octopods have eight equal-sized arms, while cuttlefish and squid have additionally a pair of longer, retractile *tentacles*, typically with the suckers grouped on terminal pads. In most cephalopods the mantle epidermis is richly supplied with chromatophores, and octopods and squids are often strikingly coloured, with a potential for rapid change of both pattern and shade in response to environmental or other cues. A final characteristic of cephalopods is the ink sac; this is discharged when the animal is threatened, the dense ink cloud serving to mask its escape. Sexes are separate, and breeding involves the transfer of packaged sperm by the male to the female, usually following a specific behavioural display. In most species one arm of the male, the *hectocotylous* arm, is specially modified for sperm transfer.

All cephalopods are predatory. The Octopoda are for the most part benthic or bentho-pelagic, living in holes or crevices, beneath large rocks or in caves, and tend to capture passing prey. The bentho-pelagic cuttlefish are active hunters, as are the fast swimming, pelagic squids.

REFERENCES

Clarke, M. R. (1966). A review of the systematics and ecology of oceanic squids. *Advances in Marine Biology*, **4**, 91–300.

Clarke, M. R. ed. (1986). *A handbook for the identification of Cephalopod beaks*. Clarendon Press, Oxford.

Cohen, A. C. (1976). The systematics and distribution of *Loligo* in the western North Atlantic with descriptions of two new species. *Malacologia*, **15**, 299–367.

Jaeckel, S. G. A. (1968). *Die Tierwelt der Nord-und Ostsee*, **37**. Cephalopoden. Geest and Partig, Leipzig.

Nixon, N. and Messenger, J. B. (1977). Biology of Cephalopods. *Symposium of the Zoological Society of London*, **38**.

Robson, G. (1931). *A monograph of the Recent Cephalopoda*, Parts I, II. British Museum, London.

Roper, C. F. E., Sweeny, M. J., and Nauen, C. E. (1984). FAO species catalogue. Vol. 3, Cephalopods of the world. *FAO Fisheries Synopsis*, no. 125, Vol. 3.

Roper, C. F. E., Young, R. E., and Voss, G. L. (1969). An illustrated key to the families of the Order Teuthoidea (Cephalopods). *Smithsonian Contributions to Zoology*, **13**, 1–32.

KEY TO FAMILIES

1. Head with eight arms and two tentacles. Suckers stalked **2**
   Head with eight arms only. Suckers sessile **5. Octopodidae**

2. Body broadly oval and rather flattened, or short and rounded. Internal shell substantial, calcified; or smaller, chitinous. Fins always paired, lateral **3**
   Body typically torpedo-shaped. Fins paired, terminal or subterminal **4**

3. Body flattened. Fins extending along entire length of body, on each side, but not joining posteriorly. Internal shell calcified. Male with left ventral hectocotylized arm **1. Sepiidae**

Body short, ovoid. Fins forming rounded, projecting lateral lobes, not reaching posterior end of body. Internal shell chitinous, reduced. Male with left, or left and right, dorsal hectocotylized arm **2. Sepiolidae**

---

4. Mantle edge projecting forwards dorsally, between eyes. Arms little longer than head. Fins elongate, extending more than half length of mantle **3. Loliginidae**
   Mantle edge straight dorsally. Arms stout, twice length of head. Fins broadly rhombic, extending less than half-way along mantle length from posterior **4. Ommastrephidae**

I. SEPIIDAE

Inner shell substantial, extending whole length of mantle, white, calcified. Body elongate-oval, flattened, with lateral fins extending whole length of body, not joined posteriorly. Male with left ventral arm hectocotylized.

---

1. Body broadly oval. Dorsal anterior mantle edge forming a rounded lobe, extending to just posterior of the eyes. Shell parallel-sided, evenly rounded posteriorly ***Sepia officinalis***
   Body slender oval. Dorsal anterior mantle edge forming an acute lobe extending between eyes. Shell tapered anteriorly and posteriorly ***Sepia elegans***

Sepia officinalis Linnaeus FIG. 13.18

Body broadly oval, maximum mantle length 400 mm. Dorsal anterior edge of mantle developed as a blunt, rounded lobe. Arms with four rows of suckers; relatively short, comprising less than half combined length of head and arms; central pair of arms broadly flattened dorso-ventrally, all arms with acute outer edges. Shell elongate-oval, almost parallel-sided, smoothly rounded posteriorly.

Shallow sublittoral, and offshore to 250 m; common inshore during summer months. Widely distributed, abundant on all British coasts.

Sepia elegans d'Orbigny FIG. 13.18

Body elongate-oval, maximum mantle length 80 mm. Dorsal anterior mantle edge produced as an acute lobe extending between eyes. Arms with two rows of suckers; short, comprising less than half combined length of head and arms. Shell with lanceolate outline, distinctly narrowed close to rounded posterior end.

Southern and western coasts only, in similar habitats to *S. officinalis*. A third species, *S. orbignyana* Ferussac occurs rarely on extreme south-western shores. It may be distinguished by its slender oval body, with an acute anterior dorsal mantle lobe, and its relatively long arms bearing four rows of suckers. The shell of *S. orbignyana* is slender, elongate-oval, with a conspicuous posterior spine.

2 SEPIOLIDAE

Inner shell reduced, chitinous. Body rounded; fins paired, lateral, not extending whole length of body, not meeting posteriorly. Male with left, or both left and right, dorsal hectocotylous arm.

---

1. Dorsal edge of mantle free, not fused to head. No membrane between arms ***Rossia macrosoma***
   Dorsal edge of mantle fused with dorsal surface of head. With or without proximal membranous connexion between dorsalmost pair of arms **2**
2. Dorsal pair of arms not connected by a membrane, the cleft between them as deep as between other arms. Tentacle club with eight rows of suckers **3**
   Dorsal pair of arms with a short, proximal connecting membrane, the cleft between them thus shorter than between other arms. Tentacle club with 16 or more rows of suckers **4**
3. Ventral mantle edge straight, showing short, broad siphon ***Sepiola atlantica***
   Ventral mantle edge deeply cleft, showing elongate, slender siphon ***Sepiola aurantiaca***
4. Body broad and squat ***Sepietta oweniana***
   Body slender, elongate ***Sepietta neglecta***

Rossia macrosoma (Delle Chiaje) FIG. 13.18

Body smooth, dorsal surface more densely and finely pigmented than ventral surface. Maximum mantle length 60 mm, comprising about one-third total length of animal; dorsal anterior edge of mantle not fused to head, with well-marked lip continuous with rest of mantle opening. Arms about twice length of head, more or less rounded in section; each with two rows of suckers proximally, four or more rows distally. Fins short and rounded, rather fleshy.

Shallow sublittoral and offshore. All British coasts, more common in north and west.

Sepiola atlantica d'Orbigny FIG. 13.18

Body smooth, translucent, with variably sized, blotchy patches of pigment, particularly dense on dorsal surface. Fat, saccular, with rounded posterior end; maximum mantle length about 20 mm, comprising about half total body-length. Dorsal anterior mantle edge fused to dorsal surface of head, between eyes; ventral edge of mantle with straight, thickened edge. Arms more

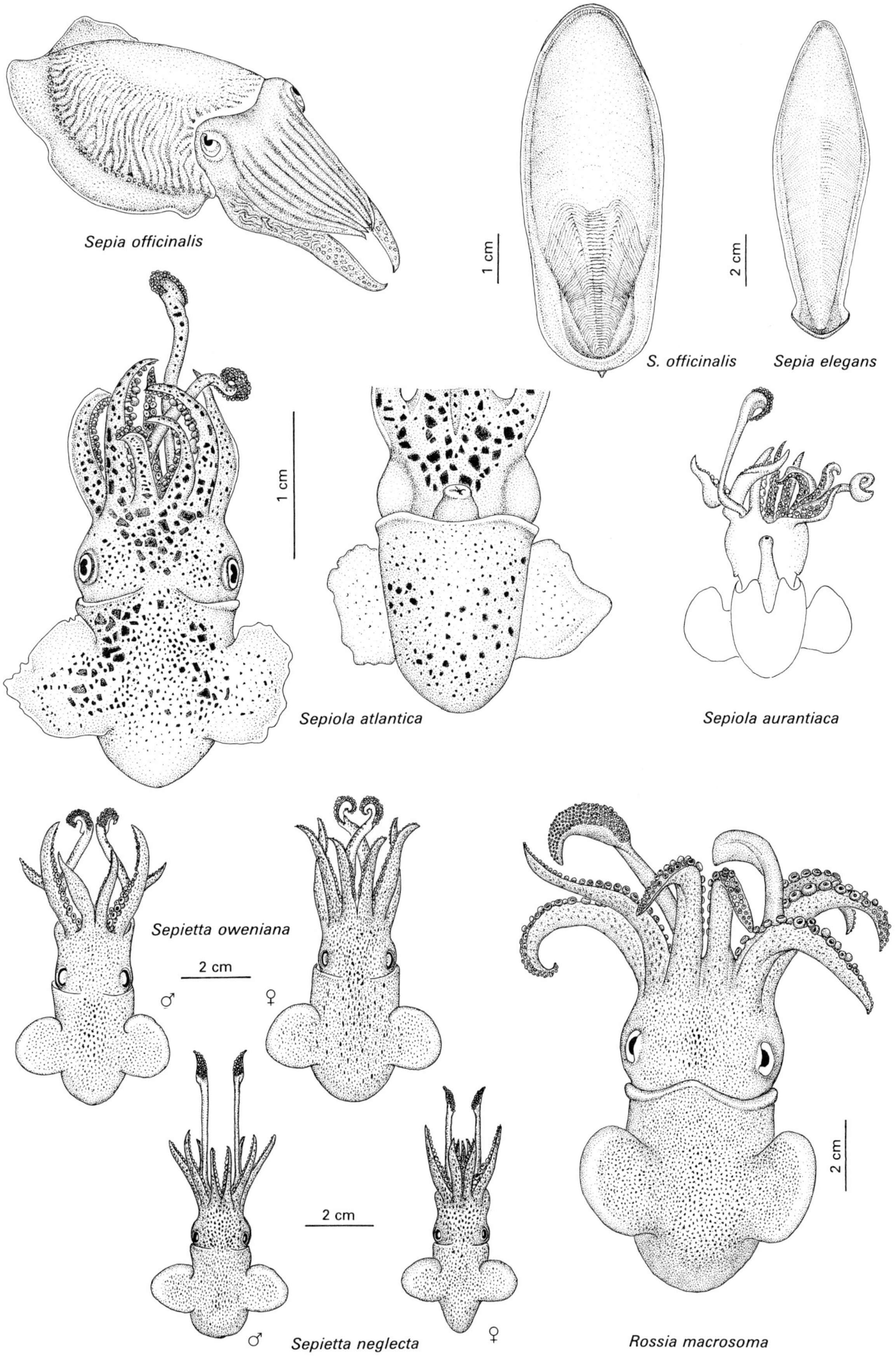
Sepia officinalis
1 cm
2 cm
S. officinalis
Sepia elegans
1 cm
Sepiola atlantica
Sepiola aurantiaca
Sepietta oweniana
2 cm
♂
♀
2 cm
♂
Sepietta neglecta
♀
2 cm
Rossia macrosoma

Fig. 13.18

or less rounded in section; with two longitudinal rows of suckers, increasing to four to eight rows distally on dorsal arms. Tentacle club with eight longitudinal rows of suckers. Fins thin and delicate, each as broad as body at point of attachment, irregularly rounded.

Shallow subtidal and offshore; frequently in intertidal rock pools, often taken on sandy shores in shrimp nets. Widespread and common on all British coasts.

Sepiola aurantiaca Jatta FIG. 13.18

Similar to *S. atlantica*. Distinguished immediately by the ventral surface of the mantle, which bears a deep cleft in its anterior edge. Maximum mantle length only 13 mm. All arms with just two rows of suckers.

A northern species reported infrequently from Scottish coasts.

Sepietta oweniana (d'Orbigny) FIG. 13.18

Body ovoid, stout; maximum total body-length 65 mm. Dorsal anterior edge of mantle fused to dorsal surface of head between eyes. Dorsal pair of arms linked proximally by a membrane, the cleft between them being noticeably shorter than between other arms. Tentacle club with up to 30 rows of very small suckers. Fins thin and delicate, oval, as broad as body at point of attachment.

Offshore, uncommon. Reported mostly from north-western shores.

Sepietta neglecta Naef FIG. 13.18

Similar to *S. oweniana*, but with a proportionately more slender, elongate body. Total length up to 50 mm. Tentacle club with less than 20 rows of suckers.

Offshore, uncommon. Reported from the northern North sea, and off the west coasts of Scotland.

3 LOLIGINIDAE

Internal shell reduced to a slender, horny pen. Body slender, with well-developed fins equivalent to more than half length of mantle. Arms relatively short. Dorsal anterior mantle edge produced, extending between eyes. Male with left ventral arm hectocotylized.

---

1. Posterior end of body blunt. Fins angular, forming a diamond shape in dorsal view **2**

   Posterior end of body produced and pointed. Fins with rounded outline, forming a heart shape in dorsal view ***Alloteuthis subulata***

2. Tentacle club with very large median suckers and much smaller marginal suckers ***Loligo vulgaris***

   Tentacle club with median suckers not, or only slightly larger than marginal suckers ***Loligo forbesii***

Loligo forbesii Steenstrup FIG. 13.19

Body elongate, solid and fleshy, bluntly rounded posteriorly; up to 600 mm long, maximum mantle length about 350 mm. Dorsal surfaces densely pigmented reddish brown, ventral surfaces lighter. Fins angular, forming an elongate diamond shape in dorsal view, comprising about two-thirds of total body length. Anterior dorsal edge of mantle produced as a blunt lobe extending to a point just posterior to eyes. Arms about twice length of body, angular in section, ventral pair particularly broad and flat; bearing two longitudinal rows of suckers. Tentacle club with median suckers of the middle region less than twice the diameter of adjacent marginal suckers.

The common squid of commercial fisheris, *L. forbesii* is widespread and abundant, in coastal waters and offshore, around the whole of the British Isles.

Loligo vulgaris Lamarck FIG. 13.19

Very similar to *L. forbesii*, reaching 750 mm total length. Distinguished by the tentacle club, on which the median suckers of the middle region are three or four times the diameter of adjacent marginal suckers.

Southern and western coasts only; eastwards as far as the Low Countries, and northwards as far as the Irish Sea.

Alloteuthis subulata (Lamarck) FIG. 13.20

Body slender, up to 210 mm, with maximum mantle length 140 mm; produced posteriorly to a finely tapered point; in males the portion posterior to the fins constitutes one-third to one-half total mantle length, in females less than one-quarter. Fins narrow, rounded, together with a heart shape outline in dorsal view. Dorsal anterior edge of mantle produced as a quadrate projection extending between eyes. Arms up to twice length of head, with flattened, angular section, dorsal pair half length of ventral pair; all with two longitudinal series of suckers. Tentacle club with median suckers of middle region up to four times diameter of adjacent marginal suckers.

Widespread and abundant, in coastal waters and offshore, on all British coasts.

Alloteuthis media (Linnaeus) FIG. 13.20

Very similar to the preceding species, *A. media* is difficult to distinguish from *A. subulata*. Arms and tentacles tend to be proportionately more slender, but the chief characteristic by which the two species are separated is the lateral mantle edge, which in *A. media* extends obliquely posteriorly between the dorsal and ventral margins.

Uncommon, but apparently present on all British coasts.

4 OMMASTREPHIDAE

Body cylindrical, with rhombic terminal fin on posterior third of mantle. Anterior dorsal edge of mantle straight, not produced as a lobe. Funnel in a deep pit. Arms stout, twice length of head, each with two rows of suckers. Tentacle club with four rows of suckers, those of the central rows larger than the marginal series.

Todaropsis eblanae (Ball) FIG. 13.20

Body rather stout, up to 287 mm, maximum mantle length 105 mm. Terminal fin broadly rhombic, wider than long, rounded posteriorly. Arms stout, more than twice length of

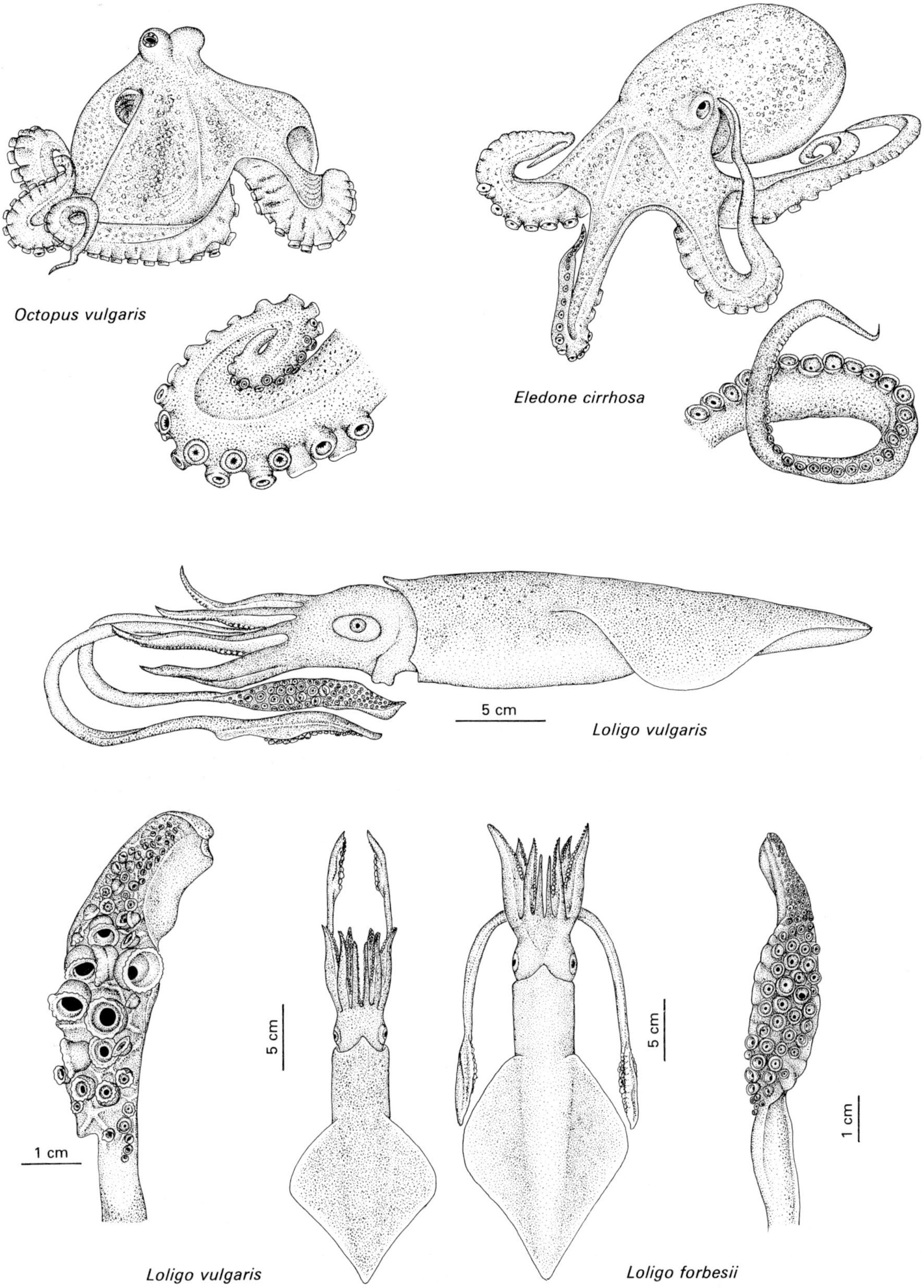
Octopus vulgaris
Eledone cirrhosa
5 cm
Loligo vulgaris
1 cm
5 cm
5 cm
1 cm
Loligo vulgaris
Loligo forbesii

Fig. 13.19

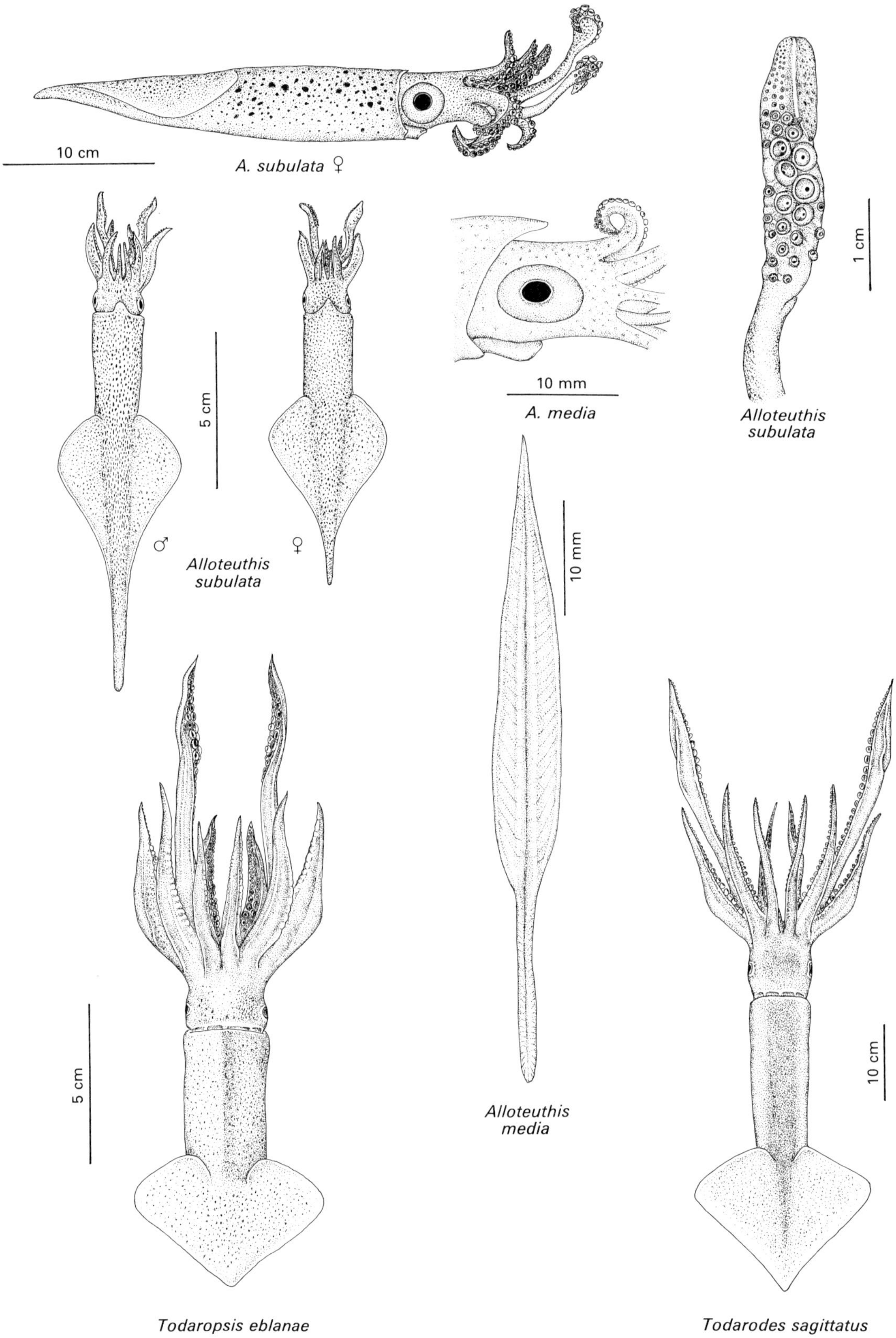
10 cm
A. subulata ♀
1 cm
10 mm
A. media
Alloteuthis subulata
5 cm
♂
♀
Alloteuthis subulata
10 mm
5 cm
Alloteuthis media
10 cm
Todaropsis eblanae
Todarodes sagittatus

Fig. 13.20

head, with angular section. Tentacle club slender, with four rows of suckers, not extending proximally on to stalk.

Offshore. Present on western and south-western coasts only.

Todarodes sagittatus (Lamarck) FIG. 13.20

Body slender, up to 1500 mm, maximum mantle length 730 mm. Terminal fin scarcely wider than long, tapered to a point posteriorly. Arms slender, more than twice length of head, angular in section. Tentacle club slender, with four rows of suckers which extend proximally down whole of the tentacle stalk.

Offshore. Predominantly northern distribution, ranging south to the Wash in the east, and the Isle of Man in the west.

5 OCTOPODIDAE

Body saccular, without an internal shell. Head and mantle fused dorsally, with only a restricted ventral mantle aperture. Eight arms only, bearing sessile suckers without chitinous teeth.

Octopus vulgaris Lamarck FIG. 13.19

Body distinctly warty, up to 1000 mm total length, with maximum arm spread of 3000 mm, but commonly much smaller. Arms thick and stout, bearing two longitudinal rows of suckers.

On rocky coasts, from the shallow sublittoral to 100 m. Widely distributed in the warm temperate Atlantic, *O. vulgaris* reaches its north-eastern limit in south-western Britain and the western English Channel, where it varies in abundance from year to year.

Eledone cirrhosa (Lamarck) FIG. 13.19

Body smooth or finely tuberculate, up to 500 mm total length, with maximum arm spread of 700 mm. Arms slender, finely tapered distally and curled when at rest, with a single longitudinal row of suckers.

On rocky coasts, often on the lower shore, and subtidally to 120 m. Widely distributed in the north-east Atlantic, present on all British coasts.

14 THE LOPHOPHORATE PHYLA PHORONIDA, BRYOZOA, AND BRACHIOPODA

THE Phoronida, Bryozoa, and Brachiopoda are smallish phyla of quite different outward appearance; yet they are linked by a common body plan. In all three the body is coelomate and divided into three regions. The mouth is situated within the *lophophore*, which is a crescentic or annular ridge bearing slender ciliated tentacles and morphologically an extension of the *mesosome* (middle region). The alimentary canal is deeply recurved, with the anus opening near but outside the lophophore. The three may conveniently be grouped as the lophophorate phyla. Despite their fundamental similarities, members of the three phyla would never be confused. Phoronids are small, tubicolous worms bearing the lophophore anteriorly; bryozoans form colonies consisting of great numbers of microscopic zooids; and brachiopods are superficially lamellibranch-like, living within a bivalved shell. The phyla are considered in the sequence Phoronida, Bryozoa (p. 797), and Brachiopoda (p. 833).

PHYLUM PHORONIDA

Phoronids are sedentary worm-like animals that occupy tubes buried vertically in muddy sand or borings in shells and limestone (Fig. 14.1). The tube is membranous and secreted by the animal but, if free, is largely covered by incorporated sand-grains and shell-fragments; it lines the borings made in shell and rock. The body is cylindrical and somewhat widened posteriorly (the *ampulla*). The anterior end bears the conspicuous lophophore: it is not (except in *P. ovalis*) introversible, but can be spread for feeding or folded when the animal retracts into its tube. The lophophore is crescentic (except in *P. ovalis*), comprising two tentaculated ridges. The convexly curved *outer ridge* passes ventral to the elongated mouth and the concave *inner ridge* dorsal to the mouth; the horns of the crescent thus curl towards the dorsal side, in the concavity of which is situated the anus. Tentacles are numerous, slender, and ciliated; new ones arise, and are therefore shorter, on either side of a median break on the inner ridge (on both ridges in *P. muelleri*). The mouth is covered by a small flap, the *epistome*.

The lophophoral coelom (*mesocoel*) is separated from the trunk coelom (*metacoel*) by an oblique septum. The metacoel is divided longitudinally by *oral* and *anal mesenteries*, so that there are usually four compartments, oral left (OL), oral right (OR), anal left (AL), and anal right (AR). The number of longitudinal muscle bundles in each compartment (quoted with the numerals disposed in quadrants, thus: $\frac{\text{OL}\quad\text{OR}}{\text{AL}\quad\text{AR}}$) is taxonomically useful.

There are two *metanephridia* associated with the lateral mesenteries, opening one on each side of the anus. The nephridia are J- or U-shaped, with the variously developed descending lobe bearing one or two *nephrostomes* opening from the metacoel; in the latter case, one opens from the oral side and one from the anal side of the mesentery. The ascending lobe terminates with the nephridiopore.

Eggs may be shed via the nephridia and fertilized externally to develop into distinctive planktonic (*actinotrocha*) larvae, or they may be brooded in the dorsal concavity of the lophophore. Ear-like *lophophoral organs* may be present in the concavity of mature males, and *nidimentary glands* occur in incubating females. Asexual propagation occurs in some species, which encourages the development of dense aggregations.

British species belong to the genus *Phoronis*. Three species are known to occur in British waters and two more seem likely to be present. The most recent and comprehensive descriptions of all the species have been given by Emig (1979).

LITERATURE

Cori, C. I (1932). Phoronidea. *Tierwelt der Nord- und Ostsee*. Teil 7, c. 2: 101–32.

Cori, C. I. (1939). Phoronidea. *Bronn's Klassen und Ordnungen des Tierreichs*, Bd. 4, Abt. 4, Buch 1, Teil 1.

Emig, C. C. (1979). British and other phoronids. *Synopses of the British Fauna*, N. S., **13**, 1–57.

Selys-Longchamps, M. de (1907). Phoronis. *Fauna Flora Golfo di Napoli*, **30**.

Silén, L. (1953). Researches on Phoronidea of the Gullmar Fiord area (west coast of Sweden). *Arkiv för Zoologi, Stockholm*, **4**, 95–140.

KEY TO SPECIES

1.	Tubes closely aggregated, either as borings in rock or shell or attached to a firm surface	**2**
	Tubes not closely aggregated, situated in muddy sand or soft substrata	**4**

2. Very small (to 6 mm); lophophore oval, with about 20 tentacles **P. ovalis**

   Larger (to 50 mm or more); lophophore horseshoe-shaped, with up to 100 or more tentacles **3**

---

3. Boring or on firm surfaces. Hermaphrodites brooding eggs in two masses, one each side of the lophophoral concavity; lophophoral organs small. Nephridia G-shaped, with two unequal nephrostomes (see Fig. 14.1) **P. hippocrepia**

   Not boring. Dioecious. Females brooding eggs in a single mass in the lophophoral concavity. Breeding males with large lophophoral organs. Nephridia J-shaped, with a single nephrostome terminating the descending lobe (see *P. muelleri*, Fig. 14.1) **P. psammophila**

---

4. Tentacles decreasing in length medially in both the inner and outer series **P. muelleri**

   Tentacles decreasing in length medially in the inner series only **5**

---

5. Tubes 60 mm or more in length. Nephridia J-shaped, with a single nephrostome terminating the descending lobe. About 12 (7–16) longitudinal muscle bundles in each of the oral metacoelic compartments **P. psammophila**

   Tubes less than 40 mm long. Nephridia with two elongated nephrostomes. Five longitudinal muscle bundles in each of the oral metacoelic compartments **P. pallida**

Phoronis hippocrepia Wright FIG. 14.1

[EMIG 1979, 32]

Boring in calcareous substrata or forming aggregations of tubes attached to firm surfaces. Borings extend for up to 100 mm, the emergent part of the tube being about 1 mm wide and covered with sand-grains. Animal 30–40 (–100) mm long. Lophophore horseshoe-shaped, with 50–150 tentacles of up to 2–3 mm in length; new tentacles forming medially on the inner ridge only. Longitudinal muscles $\frac{9\text{–}12 \quad 9\text{–}13}{4\text{–}6 \quad 5\text{–}7}$. Nephridia G-shaped, with two nephrostomes, the anal larger than the oral. Hermaphrodite. Embryos brooded in two masses, one in each horn of the lophophoral concavity; associated with nidimentary glands. Small lophophoral organs present during the reproductive season. Also propagates by transverse fission.

Bores in limestone, mollusc (e.g. *Ostrea*) shells, and thick *Lithothamnion* crusts, or attaches its tubes to firm surfaces; may be present in huge numbers. From low in the intertidal zone to 50 m; widely distributed.

Phoronis muelleri Selys-Longchamps FIG. 14.1

[EMIG 1979, 36]

Sand dwelling and non-aggregated. Tube long, straight, and slender, 50 – 80 (–120) × 1 mm, constricted to a small pore basally; membranous part delicate, uncoloured; immersed part thickly covered with sand-grains, stiff but fragile. Lophophore horseshoe-shaped, with 40–100 tentacles of 1 mm maximum length; new tentacles forming medially on both ridges. Longitudinal muscles $\frac{9\text{–}10 \quad 9\text{–}10}{3 \quad 3}$, the left lateral mesentery incompletely formed. Nephridia J-shaped, with a single nephrostome terminating the descending lobe. Probably hermaphrodite. Embryos not brooded. Lophophoral organs well-developed. No fission.

In muddy sand from shallow water (7–140 m); widely distributed.

Phoronis ovalis Wright FIG. 14.1

[EMIG 1979, 28]

Forms aggregations in dead shells. Tubes parallel the shell surface but open to the exterior by a sharp bend. Animal small, up to 6 (–15) mm long. Distal end of body introversible. Lophophore oval, hardly incurved anally. About 20 (15–28) tentacles 0.3–1.0 mm in length. No lateral mesenteries, longitudinal muscles 13–17 13–17. Nephridia simple, I-shaped, with a single nephrostome opening directly into the ascending lobe. No lophophoral organs. Embryos not brooded. Propagates by transverse fission.

Bores into gastropod and bivalve shells (e.g. of *Arctica, Modiolus, Ostrea*), and may be present in huge numbers. Intertidal and from shallow water (to 50 m); widely distributed.

Phoronis pallida Schneider FIG. 14.1

[EMIG 1979, 42]

Sand dwelling and non-aggregated. Tube vertically placed, curved, about 15–30 (–140) mm long, 2–4 mm wide, narrowing distally; basal end wide open; membranous part tinged yellow or brown; immersed part thinly covered with sand-grains, compressible and tough. Lophophore horseshoe-shaped, with 50–140 tentacles about 2.5 mm in length; new tentacles forming medially on the inner ridge only. Longitudinal muscles $\frac{5 \quad 5}{4 \quad 4}$. Nephridia U-shaped, with two elongate nephrostomes occu-

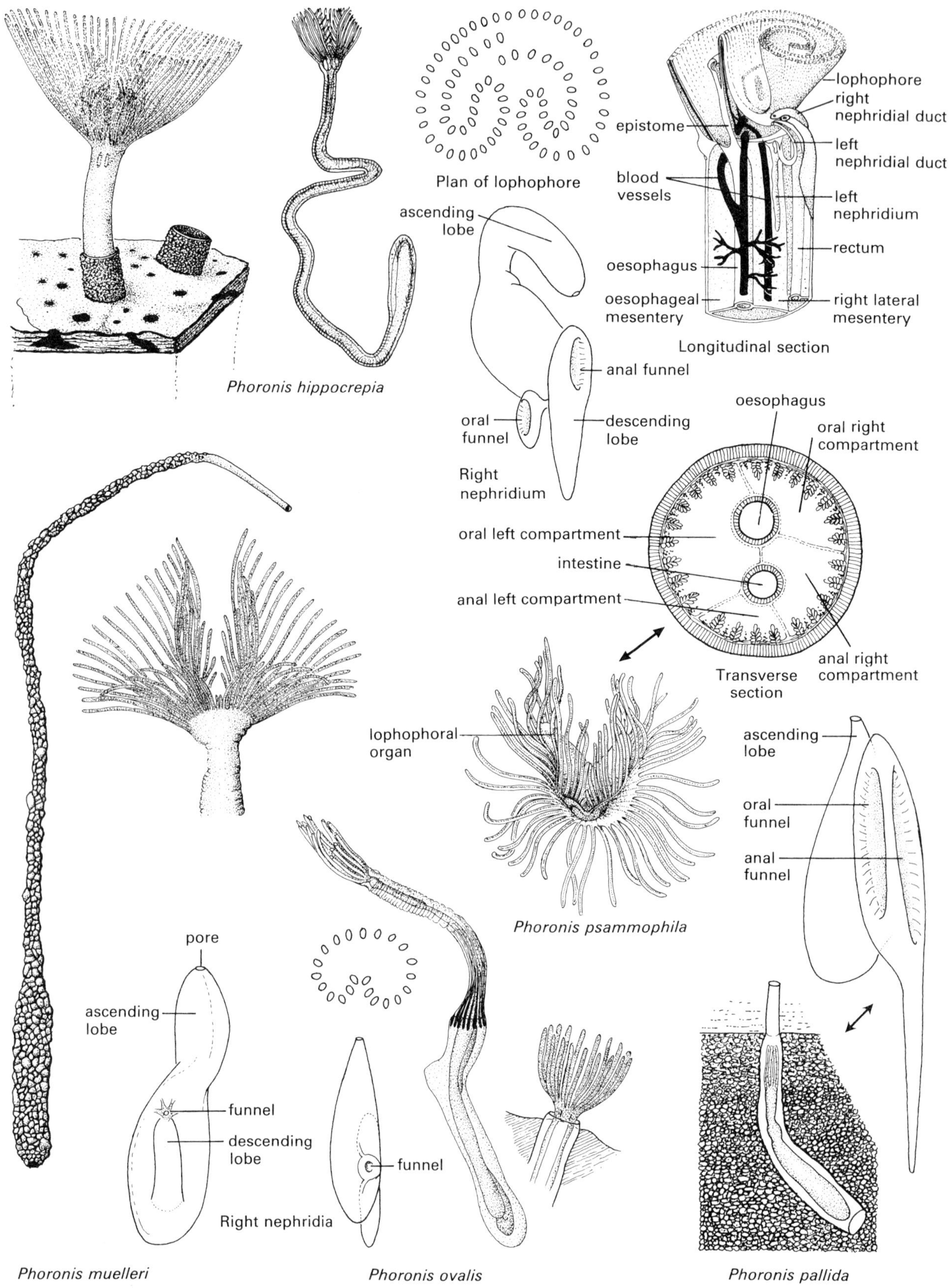
Plan of lophophore
lophophore
right nephridial duct
epistome
left nephridial duct
blood vessels
left nephridium
oesophagus
rectum
oesophageal mesentery
right lateral mesentery
Longitudinal section
ascending lobe
anal funnel
oral funnel
descending lobe
Right nephridium
Phoronis hippocrepia
oesophagus
oral right compartment
oral left compartment
intestine
anal left compartment
anal right compartment
Transverse section
lophophoral organ
ascending lobe
oral funnel
anal funnel
Phoronis psammophila
pore
ascending lobe
funnel
descending lobe
funnel
Right nephridia
Phoronis muelleri
Phoronis ovalis
Phoronis pallida

Fig. 14.1

pying the length of the descending lobe. Hermaphrodite. Lophophoral organs large. Embryos not brooded. No fission.

In muddy sand from shallow water (1–12 m). Adults in Europe recorded only from west coast of Sweden, but larvae have also been found in the southern North Sea area.

Phoronis psammophila Cori FIG. 14.1
[EMIG 1979, 40]
Sand dwelling and sometimes aggregated, or attached to shells. Tube long and slender, 50–100 (–190) × 1–2 mm. Lophophore horseshoe-shaped with 60–130 tentacles 1.5–2.5 mm in length; new tentacles forming medially on the inner ridge only. Longitudinal muscles $\frac{7\text{–}16}{4\text{–}9}\,\frac{7\text{–}16}{4\text{–}9}$. Nephridia J-shaped, with a single nephrostome terminating the descending lobe. Dioecious. Lophophoral organs large in males. Embryos brooded in a single mass within the lophophoral concavity.

In fine sand or amongst sea-grasses, but also among *Sabellaria* reefs, *Merceriella* tubes, and on shells, often to very high densities. From the intertidal zone to 18 m. Not yet recorded from British coasts but known from several localities in Brittany (France).

PHYLUM BRYOZOA
(Polyzoa, Ectoprocta)

Bryozoans are sessile, colony-forming coelomates. Each colony arises by asexual budding from an *ancestrula* formed by the metamorphosis of a sexually produced larva. The individual units of the colony, or *zooids*, remain in communication with each other through pores. Zooid polymorphism is developed to a varying degree: the normal feeding zooids are termed *autozooids* and specialized morphs are called *heterozooids*. Each autozooid has a *lophophore*, circular in marine bryozoans, bearing a series of slender, ciliated, post-oral tentacles. The funnel of tentacles, with the mouth at its vertex, constitutes a ciliary feeding apparatus. The anterior part of the body forms an introvert, called the *tentacle sheath*, within which the lophophore and tentacles can be withdrawn. The alimentary canal is deeply looped, so that the anus opens near the mouth, but just outside the lophophore. Each zooid secretes a rigid or gelatinous wall, so providing support for the colony. The skeletal component of the zooid is sometimes termed the *zooecium*, in contrast to the viscera which constitute the *polypide*. Colonies, but not necessarily zooids, are hermaphrodite; embryos are generally brooded, either in the mother zooid or in special brood chambers.

The most primitive zooid shape appears to be cylindrical, and the common intertidal species *Bowerbanka imbricata* shows the basic structure clearly (Fig. 14.2). In this species the autozooids are borne along slender stolons. In undisturbed zooids the bell of ciliated tentacles will be visible at the free end. The processes of withdrawal and eversion of the tentacles are dependent on muscles, the incompressibility of coelomic fluid and the flexibility of the body wall. The terminal opening, apparent in the wall of the retracted zooid, is termed the *orifice*.

More commonly bryozoan zooids are squat and adjacent, so that the colony forms a continuous encrusting layer. Such zooids have a lower or *basal* surface applied to the substratum, and an upper or *frontal* surface; in free-rising colonies the zooids retain their distinctive basal and frontal surfaces. In the most fundamental type of zooid the frontal surface, or most of it, remains flexible as a *frontal membrane*, but the lateral and basal walls are calcified (Fig. 14.2). The end of the zooid nearest to the origin of the colony is termed *proximal*, while the end furthest away is *distal*. The orifice is situated near to the distal end of the frontal membrane, and is closed by a muscular sphincter, paired lips, or, most typically, a hinged flap, the *operculum*.

There are two classes of marine Bryozoa.

Class I Stenolaemata

Zooids long, slender, cylindrical, tapering proximally, with well-calcified walls. A single order.

Order 1 Cyclostomata

Orifice of zooid circular, inoperculate. Embryos developing in special reproductive *gonozooids*.

Class II Gymnolaemata

Zooids cylindrical or squat, the walls calcified, cuticular, or gelatinous. Two orders.

Order 1 Cheilostomata

Zooids generally shaped like a flat box; some or all the walls calcified. Orifice frontal, closed by a hinged operculum. Specialized heterozooids, such as *avicularia*, commonly present. Embryos often develop inside special brood chambers known as *ovicells* or *ooecia*.

Avicularia are generally small, non-feeding zooids, often attached to an autozooid; their main characteristic is a jaw-like *mandible*, homologous with the operculum of an autozooid. Another specialized heterozooid is the *vibraculum* in which the operculum or mandible has evolved into a whip-like *seta*. Ovicells are more or less globular chambers, which hold a single embryo, produced at the distal end of fertile zooids. Their position and appearance vary a good deal. Often they rest on the next distal zooid, and may be immersed in it. They may open directly to the exterior or be closed by the operculum of the confluent zooid. Two suborders may be recognized.

Suborder Anasca

Frontal surface of zooid wholly or partly membranous, the gut and retracted tentacles visible inside. Operculum simply a hinged semicircle of the frontal membrane (Fig. 14.2). Anascans are generally recognizable by having a visible frontal membrane, but occasionally this is covered by ribs or by a porous shield derived from fused ribs. The flexible frontal membrane is essential for everting the tentacles.

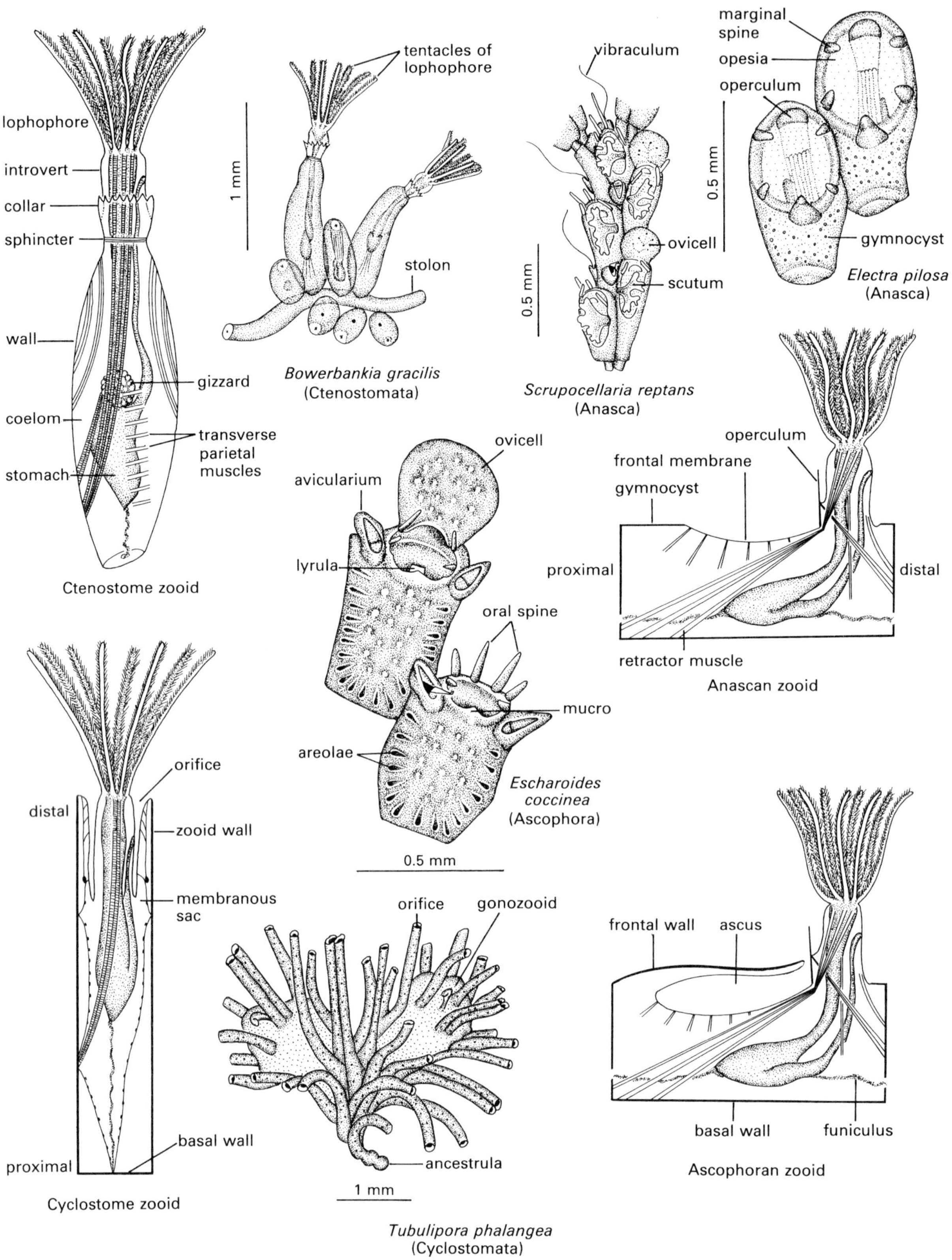
lophophore
introvert
collar
sphincter
wall
gizzard
coelom
transverse parietal muscles
stomach
Ctenostome zooid
tentacles of lophophore
1 mm
stolon
Bowerbankia gracilis
(Ctenostomata)
vibraculum
ovicell
scutum
0.5 mm
Scrupocellaria reptans
(Anasca)
marginal spine
opesia
operculum
0.5 mm
gymnocyst
Electra pilosa
(Anasca)
ovicell
avicularium
lyrula
oral spine
mucro
areolae
Escharoides coccinea
(Ascophora)
0.5 mm
operculum
frontal membrane
gymnocyst
proximal
distal
retractor muscle
Anascan zooid
orifice
distal
zooid wall
membranous sac
basal wall
proximal
Cyclostome zooid
orifice
gonozooid
ancestrula
1 mm
Tubulipora phalangea
(Cyclostomata)
frontal wall
ascus
basal wall
funiculus
Ascophoran zooid

Fig. 14.2

Suborder Ascophora

Frontal surface of the zooid calcified. Calcification thin or thick, smooth or rough, plain or apparently porous, but always opaque. Orifice a definite hole at the distal end of the frontal wall, and the operculum a discrete structure within the orifice. Hydraulic system for everting the tentacles preserved by the presence of a sac, the *ascus*, beneath the frontal wall (Fig. 14.2).

The pores of ascophorans, which may be confined to the margin or extend all over the frontal surface, are not holes leading to the ascus but merely gaps in the calcification: they are best called *pseudopores*. The frontal wall often rises to a median point (*mucro*) or prominence (*umbo*) proximally to the orifice, this position being referred to as *suboral*. In some families (e.g. Smittinidae) there is commonly a suboral avicularium, but in others (e.g. Schizoporellidae) avicularia are placed beside the orifice. The orifice varies in shape from family to family: in the Hippoporinidae it is oval or cordate with a pair of lateral teeth (*condyles*) on which the operculum pivots. In Schizoporellidae it is suborbicular with a notch (*sinus*) in the proximal margin. The operculum again pivots on condyles, though these are not denticulate, and the sinus represents the exterior opening of the ascus. A development from this is the D-shaped orifice with separate, suboral *ascopore* found, for example, in the Microporellidae.

Order 2 Ctenostomata

Zooids cylindrical or squat; walls cuticular or gelatinous, not calcified. Orifice terminal or nearly so, often closed by a pleated collar (Fig. 14.2). No ovicells. No avicularia or vibracula. Two suborders, probably not natural entities, have often been recognized.

Suborder Carnosa

Colonies encrusting or diffuse; zooids squat or cylindrical, not borne on stolons. The colonies sometimes appear stolonate, as in the Nolellidae and Victorellidae, but the apparent stolons are, in fact, slender extensions from the base of an autozooid.

Suborder Stolonifera

Colonies diffuse or tufted; zooids cylindrical, borne on stolons. True *stolons* are greatly attenuated, empty zooids (*kenozooids*), which bud autozooids singly or in groups.

With few exceptions, such as the very obvious *Membranipora membranacea* colonies found on *Laminaria* fronds, bryozoans are inconspicuous and often hidden from view. A few, such as *Alcyonidium hirsutum, A. gelatinosum, Bowerbankia imbricata, Electra pilosa,* and *Flustrellidra hispida*, are commonly found on intertidal fucoids, where they form tufts, gelatinous crusts, or lacy patches. More species occur on the lower shore, as hard, calcified, white or coloured encrustations in *Laminaria* holdfasts or on stones and below rocks; or as hanging tufts, perhaps among anemones and ascidians, under large boulders or in caves. In general they prefer dark places and protection from desiccation. As with other filter feeders, they are favoured by non-destructive movement of water and may reach spectacular abundance in straits and rapids systems. *Laminaria* fronds may here be smothered with bushy *Scrupocellaria* colonies and innumerable scabby patches of encrusting bryozoans. They will be dredged on rocks and shells.

The collection of bryozoans generally necessitates the removal of stones or pieces of rock split off with a hammer and cold chisel. Some may profitably be examined alive: *Flustrellidra hispida*, for example, has very large tentacle bells, and *Membranipora membranacea* is another species that will readily evert its tentacles when left undisturbed in a bowl of water. Note that in a few species a count of tentacles may be necessary to confirm specific identification.

All calcified bryozoans, particularly delicate, tufted forms like *Bugula* and *Crisia*, are best preserved in 70% ethanol, but the fleshy or membranous colonies of ctenostomes should be fixed in sea-water formalin. They may then be stored in 1% aqueous propylene phenoxetol, especially if the colonies are on algae. Strongly calcified species on stones, particularly ascophorans, are often best simply washed in fresh water and allowed to dry.

For identification a high-powered stereomicroscope (up to × 50) must be employed in conjunction with a powerful, focusing light-source. Preparations of cleaned zooid skeletons, to display details, can be made by immersing the specimen in hypochlorite solution (diluted domestic bleach will do). In delicate, branching genera *Bugula, Crisia, Scruparia*, and *Scrupocellaria* identification may be easiest in specimens lightly stained in borax carmine and mounted in balsam (or a synthetic substitute). Ctenostomes, such as *Bowerbankia* and *Walkeria*, however, do not make satisfactory preparations when mounted in standard media.

LITERATURE

Bryozoans by Ryland (1970) is a general biological account of the phylum. The standard monograph for the identification of British marine species for many years was Hincks' *British marine Polyzoa* (1880), used with the key devised by Harmer (1896) and, latterly, with the nomenclatural index of Ryland (1969). Prenant and Bobin have comprehensively monographed the Ctenostomata (1956) and Cheilostomata Anasca (1966) for western Europe but for users in the British Isles complete coverage is available in a series of four Linnean Society synopses (Ryland and Hayward 1977; Hayward and Ryland 1979; Hayward 1985; Hayward and Ryland 1985). These and some other useful publications are listed below.

Eggleston, D. (1969). Marine fauna of the Isle of Man: revised lists of phylum Entoprocta (=Kamptozoa) and phylum Ectoprocta (=Bryozoa). *Report of Marine Biological Station, Port Erin,* **81**: 57–80.

Eggleston, D. 1975. The marine fauna of the Cullercoats district. Ectoprocta. *Report of Dove Marine Laboratory*, Series 3, **18**, 1–30.

Hamond, R. (1973). The marine and brackish-water Bryozoa of Norfolk. *Transactions of Norfolk and Norwich Naturalists' Society,* **22,** 406–23.

Harmer, S. F. (1896). Polyzoa. *The Cambridge Natural History,* **2**: 463–533.

Hayward, P. J. (1971). The marine fauna and flora of the Isles of Scilly. Bryozoa and Entoprocta. *Journal of Natural History*, **5**, 481–9.

Hayward, P. J. (1976). The marine fauna and flora of the Isles of Scilly. Bryozoa II. *Journal of Natural History*, **10**, 319–30.

Hayward, P. J. (1977). The marine fauna of Lundy. Bryozoa. *Report of Lundy Field Society*, **27**, 16–34.

Hayward, P. J. (1985). Ctenostome bryozoans. *Synopses of the British Fauna*, N.S., **33**, 1–169.

Hayward, P. J. and Ryland, J. S. (1979). British ascophoran bryozoans. *Synopses of the British Fauna*, N.S., **14**, 1–312.

Hayward, P. J. and Ryland, J. S. (1985). Cyclostome bryozoans. *Synopses of the British Fauna*, N.S., **34**, 1–147.

Hincks, T. (1880). *A history of the British marine Polyzoa*. Van Voorst, London.

Kluge, G. A. (1975). *Bryozoa of the northern seas of the USSR*. Amerind, New Delhi.

Marcus, E. (1940). Mosdyr (Bryozoa eller Polyzoa). *Danmarks Fauna*, **46**, 1–401.

Prenant, M. and Bobin, G. (1956). Bryozoaires, 1re partie. Entoproctes, Phylactolèmes, Cténostomes. *Faune de France*, **60**, 1–598.

Prenant, M. and Bobin, G. (1966). Bryozoaires, 2e partie. Chilostomes Anasca. *Faune de France*, **68**, 1–647.

Roper, R. E. (1915). The marine Polyzoa of Northumberland. *Report of Dove Marine Laboratory*, Series 2, **2**: 36–57.

Ryland, J. S. (1965). Polyzoa. *Catalogue of main marine fouling organisms*, **2**. O.E.C.D., Paris.

Ryland, J. S. (1969). A nomenclatural index to "A history of the British marine Polyzoa" by T. Hincks (1880). *Bulletin of the British Museum (Natural History), Zoology*, **17**: 207–60.

Ryland, J. S. (1970). *Bryozoans*. Hutchinson University Library, London.

Ryland, J. S. (1982). Bryozoa. In: S. P. Parker (ed.) *Synopsis and classification of living organisms*, 2. McGraw-Hill, New York, pp. 743–63.

Ryland, J. S. and Hayward, P. J. (1977). British anascan bryozoans. *Synopses of the British Fauna*, N. S., **10**, 1–188.

KEY TO FAMILIES

1. Zooids cylindrical or elongated, arising singly, in clusters, or in fused groups from long, creeping or free stolons **2**

   Zooids variously shaped, but not rising upwards from long stolons **4**

2. Zooids calcified, whitish, either rising freely in branching chains or borne singly; long, tubular, and with a lateral membranous area at the free end. Orifice not terminal **3**

   Zooids uncalcified, generally more or less buff coloured, not rising freely in branching chains; if long and cylindrical, then externally symmetrical. Orifice terminal. **(Ctenostomata) 33**

3. Zooids arising singly from stolons, long and cylindrical **6. Aeteidae**

   Zooids arising in chains, roughly diamond shaped **7. Scrupariidae**

4. Colony firm, gelatinous, encrusting or lobate; or erect and muddy. Non-calcified. Operculate orifices, ovicells, avicularia, and vibracula always absent **32**

   Colony calcified (though not necessarily hard and brittle). Operculate orifices, ovicells or other brood chambers, avicularia, and vibracula sometimes present **5**

5. Zooids cylindrical, perhaps united for the greater part of their length but often ending in a free distal portion which bears the terminal orifice. Individual zooids often partly obscured by calcification surrounding their basal parts. Chitinous operculum never present. Avicularia and vibracula absent. **(Cyclostomata) 40**

   Zooids more or less box- or boat-shaped, not long and tubular. Orifice not at the end of a free cylindrical portion. A more or less distinct operculum usually present. Avicularia and vibracula may be present **6**

6. Upper or frontal surface wholly or largely membranous; the polypide visible; operculum semicircular, not greatly differentiated from the membrane. The membrane often surrounded by spines borne on the side wall of the zooid; these spines may overarch and obscure the membrane like a series of ribs. The membrane sometimes underlain

by a calcified lamina (and therefore not very apparent), in which case colony bushy and composed of jointed, dichotomously divided branches. Colony often erect and tufted **(Cheilostomata Anasca) 7**

Frontal surface, except for the operculum, covered by a calcified wall, which may be smooth or granular, porous or imperforate; the polypide always completely obscured. Orifice a distally situated opening in the frontal wall, closed by a distinct operculum, and often partly surrounded by spines. Colony sometimes erect but never bushy or tufted **(Cheilostomata Ascophora) 16**

---

7. Colony incrusting, forming continuous expansions **8**

Colony not incrusting **11**

---

8. Ovicells and avicularia never present. True spines on the rim of the membranous area absent (but spinules, small protuberances, or pointed tooth-like projections from the rim of the membranous area, or a single bristle or thorn-like process proximal to the frontal membrane, may be present). Often on algae **9. Membraniporidae**

Ovicells and avicularia usually present. Rim of membranous area bearing one to many spines, which may arch over the frontal membrane and be more or less fused together **9**

---

9. Marginal spines modified to form a cage over the frontal membrane, by fusing with one another, the sutures being marked by lines of pores **12. Cribrilinidae**

Marginal spines erect or flattened and overarching the frontal membrane, but never fused with adjacent spines on the same side **10**

---

10. Spines erect or sloping slightly over the frontal membrane, or with about five pairs of flattened spines arching over the membrane and meeting each other in the mid-line. The ovicell not closeable by the zooidal operculum **11. Calloporidae**

Generally eight or more pairs of flattened spines arching over the frontal membrane and meeting each other in the mid-line. The operculum of the zooid also closing the ovicell **12. Cribrilinidae**

---

11. Colony in the form of creeping chains of zooids in single series which give rise to similar but erect chains, each like a string of beads **7. Scrupariidae**

Colony not as above **12**

---

12. Branches of colony composed of two single series of zooids joined basally (i.e. back to back) **8. Eucrateidae**

Branches composed of two or more adjacent series of zooids arranged as a single layer, joined back to back with others, or disposed around an axis **13**

---

13. Colony strongly calcified (branches can be snapped), made up of cylindrical, dichotomously dividing rami, in which the segments between bifurcations are separated by non-calcified joints. Zooids arranged all around the branch axis **13. Cellariidae**

Colony often weakly calcified; in the form of wide foliaceous lobes or narrow and much divided branches, in which two or more series of zooids are arranged side by side **14**

---

14. Colony lobate, with wide unjointed branches. Zooids in two layers, joined back to back **10. Flustridae**

Colony tufted, with narrow branches, generally either jointed or with pedunculate (bird's head) avicularia. Zooids in a single layer **15**

15. Branches composed of short series of zooids separated at bifurcations by distinct joints. Pedunculate avicularia absent. Vibracula often present **14. Scrupocellariidae**

Branches not jointed at bifurcations. Pedunculate avicularia usually present. Vibracula absent **15. Bugulidae**

16. Zooids heaped and jumbled together, lacking any common orientation **17**

Zooids clearly orientated, arranged in regular series; or separated from each other by short stolons **18**

17. Orifice with a distinct notch or sinus in its proximal margin. Either one short avicularian process suborally or a pair of lateral avicularia raised on processes **25. Celleporidae**

Orifice circular. A single tall, acuminate process beside the orifice **25. Celleporidae (*C. pumicosa*)**

18. A conspicuous median pore (usually the ascopore) present proximally to the orifice **19**

No conspicuous median pore **21**

19. Frontal wall below the pore clearly marked with radiating ribs separated by rows of punctures **12. Cribrilinidae**

Pores in the frontal wall not arranged in radiating lines **20**

20. Ascopore separated from the orifice by a median, distally directed avicularium **21. Adeonidae**

Ascopore situated directly below the orifice **20. Microporellidae**

21. Orifice with a distinct notch or sinus in its proximal margin **22**

Orifice without a proximal sinus **25**

22. Avicularia absent. Zooids distant, not forming a continuous crust; or zooids contiguous but with the ovicells borne by differentiated, shorter zooids **18. Hippothoidae**

Avicularia present suborally or elsewhere on the frontal wall **23**

23. A lyrula (median tooth) present in the orifice **23. Smittinidae**

Lyrula absent **24**

24. Frontal wall with marginal pores only; suboral avicularium present. Ovicell non-porous **23. Smittinidae**

Frontal wall generally covered with pores; avicularium (-a) suboral or lateral. Ovicell porous or imperforate **24. Schizoporellidae**

25. A lyrula (median tooth) present in the orifice, or colony erect and branched **26**
Lyrula absent, colony incrusting or foliaceous **28**

26. Avicularia absent **22. Escharellidae**
Suboral or lateral avicularia present **27**

27. Frontal wall proximal to the orifice produced into a raised tapering point (mucro); flanked each side by an avicularium **17. Exochellidae**
Proximal margin of orifice not produced into a mucro, a single suboral or lateral avicularium **23. Smittinidae**

28. Orifice flanked each side by an avicularium, and bordered distally (in non-ovicellate zooids at least) by a series of spines. (Frontal wall perforated by large pores) **12. Cribrilinidae**
Flanking avicularia and spines absent **29**

29. Frontal wall with marginal and central pores. Orifice with condyles on which operculum pivots **26. Hippoporinidae**
Frontal wall imperforate or with marginal pores only. Orifice without condyles **30**

30. Avicularium (if present) suboral. Marginal pores large, separated by buttresses. Ovicells absent **16. Umbonulidae**
Avicularium distal to the orifice, borne on the ovicell of ovicellate zooids. Frontal wall entire. Spaces in the calcification present between adjacent zooids **19. Chorizoporidae**

31. Colony a continuous adherent crust, or forming thick, fleshy, free lobes; or erect, branching and muddy **32**
Colony composed of creeping or free stolons bearing or merging with solitary, clustered, or fused groups of (usually cylindrical) zooids **33**

32. Colony erect with spirally disposed branches; zooids cylindrical and covered with muddy particles **30. Nolellidae**
Colony erect with fleshy, free lobes and immersed zooids, or forming a continuous adherent crust **35**

33. Colony boring in shells or calcified tubes, or in the lining of polychaete tubes, or on the carapace or appendages of decapod crustaceans **34**
Colony not found as above **38**

34. Colony boring in mollusc shells, cirripede plates, or calcified worm tubes; openings round or kidney-shaped **35. Penetrantiidae**
Colony not inhabiting calcified structures **36**

35. Zoological bearing spines. Orifice two lipped. On algae **28. Flustrellidridae**

Zooids without spines or, if apparently spiny, encrusted with particles and growing on hydroids. Orifice not two-lipped **27. Alcyonidiidae**

36. Colony living in the inner layers of polychaete tubes (especially in those of *Chaetopterus*) **32. Hypophorellidae**

Colony adhering to the carapace or appendages of decapod crustaceans **33. Triticellidae**

37. Stolons radiating from zooid bases **38**

Colony composed of a network of stolons bearing zooids in clusters (the clusters often coinciding with the sites of stolonal branching) **39**

38. Zooid walls opaque with fine incrusting particles. New zooids not budded off from the sides of existing zooids. Marine **30. Nolellidae**

Zooid walls translucent. New zooids frequently arising from the sides of existing zooids. Brackish water **29. Victorellidae**

39. Zooids very small, less than 0.5 mm in length. Gizzard absent. Expanded tentacles not forming a perfect circle but the dorsalmost pair curving strongly outwards. Branch stolons frequently arising in opposite pairs and diverging from the main axis at right angles. Often found on *Corallina* or *Halidrys* **31. Walkeriidae**

Zooids generally larger, exceeding 0.5 mm in length. Gizzard present. Expanded tentacles forming a perfect circle. Stolonal branches not normally arising in opposite pairs as above. Various substrata **34. Vesiculariidae**

40. Colony erect. Branches composed of a single or double series of zooids, divided at intervals by non-calcified joints. Brood chambers pear-shaped **1. Crisiidae**

Colony adherent; linear, lobed, or circular **41**

41. Colony more or less circular or discoidal, sometimes forming secondary colonies by marginal budding **42**

Colony linear or lobed, not circular **43**

42. Colony flat. Zooids contiguous, not separated by spaces. Calcified walls (especially of brood chambers) punctate. Margin of orifice not spiny **4. Diastoporidae**

Colony convex. Zooids separated by a series of walls and spaces. Calcified walls not punctate. Margin of orifice spiny **5. Lichenoporidae**

43. Colony adnate, its branches (lobes) as broad as long; zooids generally in transverse rows; gonozooid much divided between the autozooids, its opening on a tube decumbent on an autozooid **3. Tubuliporidae**

Colony adnate or erect, its branches longer than broad; zooids in linear series or in short transverse rows; gonozooid elongate or divided between the autozooids, its opening on a free-standing tube **2. Annectocymidae**

Class **Stenolaemata**

Order **Cyclostomata**

I CRISIIDAE

Colonies erect, composed of slender branches divided by *joints* into *internodes* made up of one, two, or several zooids, the number generally increasing towards the distal part of the colony. One or more branches may arise from each internode. The lowest zooid in each branch is traversed by a joint, that part below the joint, which attaches the branch to the main axis, being termed the *branch base*. In fertile internodes, one or more zooids become transformed into *gonozooids*. Colonies are attached to the substratum by *rhizoids*.

1. Sterile internodes consisting of a single zooid — 2
   Sterile internodes, apart from the basal ones, consisting of a number of zooids — 3
2. Each zooid bearing a filiform spine; gonozooid free for most of its length — ***Crisidia cornuta***
   Zooids without spines; gonozooid adherent to an autozooid for most of its length — ***Filicrisia geniculata***
3. Internodes short, in the main consisting of nine zooids or fewer (fertile zooids in the distal parts of the colony may be a little longer) — 4
   Internodes long, in the main consisting of 11 zooids or more — 5
4. Internodes predominantly of five to seven zooids, the branches strongly incurved; zooids never bearing spines — ***Crisia eburnea***
   Internodes often composed of more than seven zooids in the distal parts of the colony, branches not incurved; many zooids bearing spines — ***Crisia aculeata***
5. Joints black (except near the growing points); branch base situated low on the zooid bearing it, appearing wedged-in between successive zooids on the same side; opening of gonozooid transversely elongate, forwards facing — ***Crisia denticulata***
   Joints pale; branch base not wedged-in between successive zooids; opening of gonozooid on an upwards directed, trumpet-like spout — ***Crisia ramosa***

Crisia aculeata Hassall — FIG. 14.3
[HAYWARD AND RYLAND 1985, 50]
Colonies forming sparse tufts up to 20 mm high; internodes mainly short, commonly of five to seven zooids, but longer distally; branches usually arising from the first or second zooid, straight; branch base short; at least some zooids usually bearing long, jointed spines; gonozooid situated fairly high in its internode, adnate, inflated distally, then abruptly flattened to the inconspicuous, sessile opening.

On algae and rocks; LWST and shallow water; all coasts; quite common.

Crisia denticulata (Lamarck) — FIG. 14.3
[HAYWARD AND RYLAND 1985, 54]
Colonies forming coarse tufts up to 25 mm high; branches usually arising alternately from the main axes, one from each internode; internodes long, most often of 11 zooids; joints jet black (except near the growing points); branch base situated low on the bearing zooid, wedged-in between it and the next zooid below; gonozooid pyriform, positioned high in its internode, its opening facing forwards, not on a long tube.

On hydroids, algae and rocks; LWST and shallow water; not found north of the British Isles; common.

Crisia eburnea (Linnaeus) — FIG. 14.3
[HAYWARD AND RYLAND 1985, 49]
Colonies forming dense tufts up to 20 mm high, the branches incurved; internodes short, of an odd number of zooids, commonly five or seven; branches usually arising from the lowest zooid in each odd numbered internode, coming off from alternate sides of the main axes but characteristically all from the same side in distal parts of the colony; joints yellowish; branch base short; spines absent; gonozooid replacing the second or third zooid, adnate, its opening subterminal, transversely elongated, facing forwards.

On various substrata, especially red algae; LWST and shallow water; all coasts; common.

Crisia ramosa Harmer — FIG. 14.3
[HAYWARD AND RYLAND 1985, 52]
Colonies forming straggling tufts up to 20 mm high, much branched; internodes of variable length, but often very long and consisting of numerous zooids, giving rise to up to five branches; joints yellowish; branch base long, reaching but not wedged inside the next zooid below it; gonozooid pyriform, most often in the 6th–8th position, its opening circular, on a long funnel-shaped tube.

On algae, bryozoans, and rocks; shallow water; mainly southern, but reaching the Skagerrak; frequent.

Crisidia cornuta (Linnaeus) — FIG. 14.3
[HAYWARD AND RYLAND 1985, 42]
Colonies forming tufts 10–15 mm high; internodes consisting of a single zooid, separated by yellowish joints; zooids curved, giving rise to the next internode plus a long slender spine, or two internodes; gonozooid pyriform, the only member of its internode, free for most of its length, situated at a bifurcation so that a pair of internodes arises from its base; opening of gonozooid terminal, spout-like.

On hydroids, bryozoans, algae, and rocks; LWST and shallow water; all coasts; common.

Filicrisia geniculata (Milne Edwards) — FIG. 14.3
[HAYWARD AND RYLAND 1985, 40]
Colonies forming slender tufts up to 20 mm high; sterile inter-

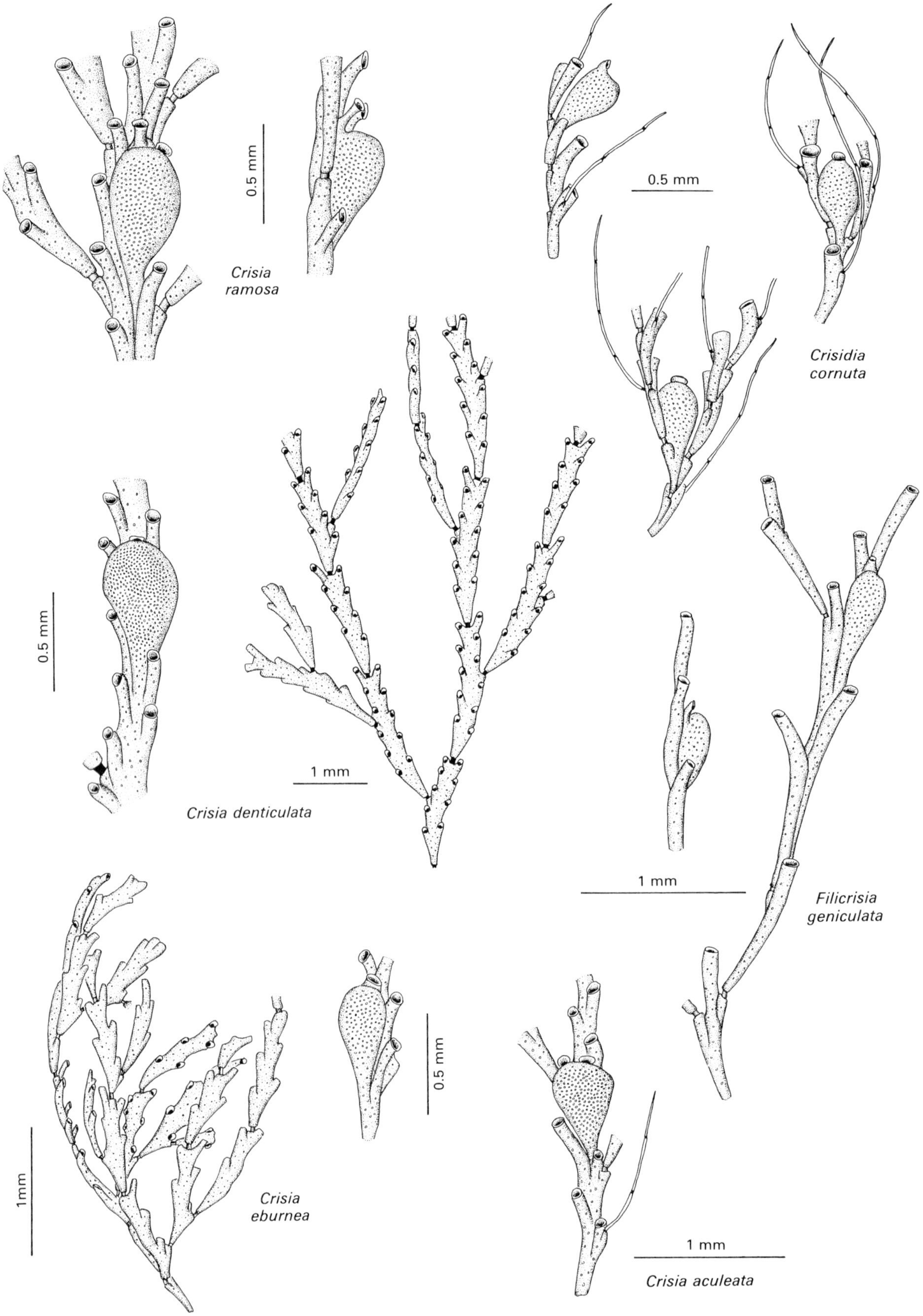
Crisia ramosa
0.5 mm
0.5 mm
Crisidia cornuta
0.5 mm
1 mm
Crisia denticulata
1 mm
Filicrisia geniculata
1mm
Crisia eburnea
0.5 mm
1 mm
Crisia aculeata

Fig. 14.3

nodes consisting of a single zooid, separated by yellowish joints; zooids long and slender, curved, without spines; fertile internodes of three to five zooids, the gonozooid in the form of a slender club, adherent to an autozooid, its opening on a short, curved, subterminal tube.

On hydroids, bryozoans, rocks, and algae; LWST and shallow water; all coasts; quite common.

2 ANNECTOCYMIDAE

Colonies adnate or erect; autozooids in alternating longitudinal series or in irregular transverse rows. Gonozooid elongate, pyriform to irregular, its cavity sometimes traversed by autozooids; its opening tubular.

---

1. Ancestrula budding a single zooid which gives rise to the axial branch, but also a wider, recurving 'adventitious' branch; colony adnate at first, the zooids in two to five (commonly three) series, erect branches few; opening of gonozooid compressed, reflected, on a short tube ***Annectocyma major***

   Ancestrula budding a single zooid, from which arise a pair of zooids, each the first in its branch (no 'adventitious' branches); colony adnate at first, the zooids in a single series (two at bifurcations), erect growths then frequent; opening of gonozooid circular, on a long slender tube ***Entalophoroecia deflexa***

Annectocyma major (Johnston) FIG. 14.4

[HAYWARD AND RYLAND 1885, 109]

Colony encrusting, frequently branching, with occasional erect growths in some colonies. Ancestrula budding a single distal zooid from which continues the gradually widening axial branch of the colony; but also arising from near the origin of the colony is a recurving 'adventitious' branch in which one to three buds quickly establish normal branch width. Other adventitious branches may occur as the colony grows. Erect parts of zooids in two or more (commonly three) series; their openings 0.15–0.20 mm diameter. Gonozooid variable in shape, only slightly more punctate than the autozooids; its opening in the distal third of the gonozooid, compressed, on a short reflected tube.

On stones and shells; offshore; common off southern and western coasts.

Entalophoroecia deflexa (Couch) FIG. 14.4

[HAYWARD AND RYLAND 1985, 113]

Colonies at first adnate, linear, dichotomously branching, with the zooids arranged in single series except at bifurcations and in the vicinity of a gonozooid; later erect, broader. Two diverging branches arising from the first daughter zooid. Zooids slender, curving upwards, free distally, the orifice more or less elevated. Gonozooid located in the distal part of an adnate or erect branch, forming an elongated inflation surrounding the bases of several contiguous zooids and traversed by others; its opening circular, on a slender free-standing tube, facing upwards.

In *Laminaria* holdfasts, on hydroids, shells, and stones; shallow water; all coasts; quite common but small colonies inconspicuous.

3 TUBULIPORIDAE

Colonies adherent or nearly so; fan-shaped or variously lobate with zooids free or in transverse rows, the zooid ends then rising upwards like miniature organ pipes. No joints. Gonozooid an irregular inflation on the surface of the colony, more strongly punctate than the autozooids and extending between them; its opening at the end of a short tube.

---

1. Opening of gonozooid tubular, flaring, much larger than an orifice, upwards or sidewards facing ***Tubulipora plumosa***

   Opening of gonozooid tubular, but not flaring, smaller or not much larger than an orifice, sideways or downwards facing (the opening itself often invisible from above) **2**

2. Opening of gonozooid facing sideways, a little larger than an orifice ***Tubulipora liliacea***

   Opening of gonozooid facing downwards, completely invisible from above, not larger than an orifice ***Tubulipora phalangea***

Tubulipora liliacea (Pallas) FIG. 14.4

[HAYWARD AND RYLAND 1985, 74]

Colonies adherent or partly erect, commonly divided into several slender, radiating lobes; the living colony strongly flushed with purple. In each lobe the zooids are contiguous in transverse rows; these diverging alternately to each side of the mid-line; the inner zooids tallest, so that the height diminishes towards the margin of the lobe. Orifice diameter about 165 μm. Gonozooid rather zigzag or star-shaped, fitting in between rows of zooids; its opening, about 260 μm in diameter, facing horizontally at the end of a curved tube decumbent on an autozooid.

Found particularly on hydroids in shallow water; all coasts; frequent.

Tubulipora phalangea Couch FIG. 14.4

[HAYWARD AND RYLAND 1985, 78]

Colonies adherent, almost circular or variously lobed, up to about 15 mm across, white or lightly flushed with purple. Zooids long and slender, contiguous in rows; transverse and diverging alternately to each side of the mid-line in a narrow lobe, radial in the presence of a gonozooid. Orifice diameter about 160 μm. Gonozooid spreading between rows of zooids; its opening about 160 μm in diameter, facing downwards at the end of a curved tube, decumbent on an autozooid.

On stones, shells, and algae; shallow water; all coasts; frequent.

Tubulipora plumosa Thompson in Harmer FIG. 14.4

[HAYWARD AND RYLAND 1985, 80]

Colonies adherent, variously lobed, up to about 20 mm across,

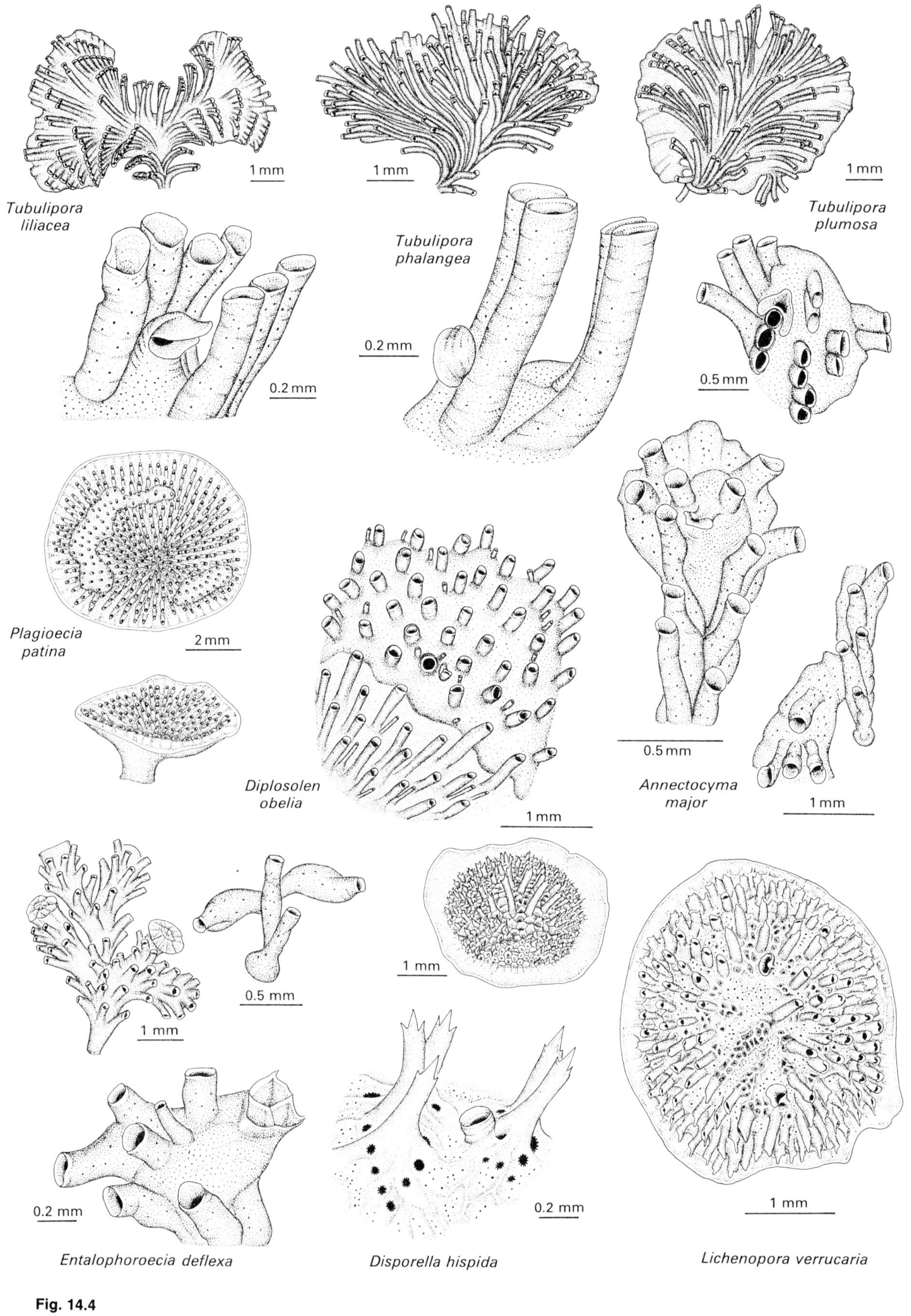
1 mm
1 mm
1 mm
Tubulipora liliacea
Tubulipora plumosa
Tubulipora phalangea
0.2 mm
0.2 mm
0.5 mm
Plagioecia patina
2 mm
Diplosolen obelia
1 mm
0.5 mm
Annectocyma major
1 mm
1 mm
0.5 mm
1 mm
1 mm
0.2 mm
0.2 mm
1 mm
Entalophoroecia deflexa
Disporella hispida
Lichenopora verrucaria

Fig. 14.4

white or lightly flushed with purple. Zooids contiguous in rows, transverse and diverging alternately to each side of the mid-line in a narrow lobe, or radiating from a flat central area. Orifice diameter about 185 μm. Gonozooid spreading between rows of zooids; its opening about 350 μm in its largest diameter, facing directly or obliquely upwards, at the end of a short flaring tube, decumbent on an autozooid.

On algae, particularly on *Saccorhiza* and *Laminaria* bases in shallow water; all coasts; frequent.

4 DIASTOPORIDAE

Colonies adherent, usually discoid; zooids largely or wholly adnate, with a subcircular orifice. No joints. Gonozooid an irregular inflation on the surface of the colony, its opening at the end of a short tube.

---

1. Dwarf zooids present among the autozooids — ***Diplosolen obelia***
   Dwarf zooids absent — ***Plagioecia patina***

Diplosolen obelia (Johnston) FIG. 14.4
[HAYWARD AND RYLAND 1985, 102]

Colonies flat and adherent, variable in outline. Zooids wholly adnate or with the distal ends slightly raised, interspersed with dwarf zooids (*nanozooids*). Gonozooid rather small, spreading between several zooids, its opening borne on a short, free-standing tube.

On stones, shells, and algae; shallow water; all coasts; fairly common.

Plagioecia patina (Lamarck) FIG. 14.4
[HAYWARD AND RYLAND 1985, 97]

Colonies discoid, flat, and entirely adherent or bowl-shaped; surrounded by a white border; about 5 mm across. Zooids arranged in radiating lines, wholly adnate or with the distal ends slightly raised; orifice more or less elliptical. Gonozooid roughly crescentic, spreading among the surrounding several zooids, its opening borne on a short free-standing tube.

On stones, shells, and algae; shallow water; all coasts; common.

5 LICHENOPORIDAE

Colonies discoid, adnate, convex; zooids erect or suberect, radiating from a zooid-free porous (alveolar) central area in which a colonial brood chamber develops by fusion of *alveoli*.

---

1. Spaces (alveoli) between the zooids thin-walled, sometimes roofed over; margin of orifices generally produced above as a pointed process; opening of brood chamber a flaring tube — ***Lichenopora verrucaria***
   Alveoli thick-walled; margin of orifices often trifid; opening of brood chamber circular, lacking any tube — ***Disporella hispida***

Disporella hispida (Fleming) FIG. 14.4
[HAYWARD AND RYLAND 1985, 128]

Colonies discoid, somewhat convex, sometimes surrounded by a border; or composed of several confluent disks. Zooids disposed around the central area, the orifices raised, frequently terminating in one to three slender points. Central alveoli with thick walls, the openings often reduced to small central pores. Opening of brood chamber round, tubeless.

On algae, stones and shells; shallow water; all coasts; common.

Lichenopora verrucaria (Fabricius) FIG. 14.4
[HAYWARD AND RYLAND 1985, 126]

Colonies discoid, somewhat convex, sometimes surrounded by a border. Zooids disposed around a small central area, the orifices raised, terminating in a slender process. Central alveolae with thin walls, sometimes roofed over. Opening of brood chamber carried on an upright flaring tube.

On algae, hydroids, shells, and stones; LWST to shallow water; northern coasts only.

Class **Gymnolaemata**

Order **Cheilostomata**

Suborder **Anasca**

6 AETEIDAE

Colonies made up from a series of basically cylindrical zooids, the proximal part of each being recumbent, the distal part being erect and free. The adherent portions adjoin in series and thus appear as stolons supporting uprights at intervals. Each upright consists of a stem and a distal region in which a basal surface and frontal membrane may be distinguished; the operculum is terminal. No avicularia or vibracula. Embryos develop in delicate, evanescent *ovisacs*, not in ovicells.

---

1. Terminal portion of zooid wider than the stem, usually downwards facing — ***Aetea anguina***
   Terminal portion not wider than the stem; zooid straight — **2**
2. Stem region of zooid appearing striated when viewed by transmitted light under a microscope — ***Aetea sica***
   Stem region of zooid not striated when viewed as above — ***Aetea truncata***

Aetea anguina (Linnaeus) FIG. 14.5
[RYLAND AND HAYWARD 1977, 45]

Colony white, adnate portion ramifying, with uprights 0.6–0.8 mm high arising from pronounced dilatations. Stem of upright faintly but clearly annulate by transmitted light; distal region one-quarter to one-third of the free length, wider than the stem, finely punctate, usually facing downwards.

In *Laminaria* holdfasts and on hydroids, bryozoans, and shells; LWST to shallow water; all coasts; common.

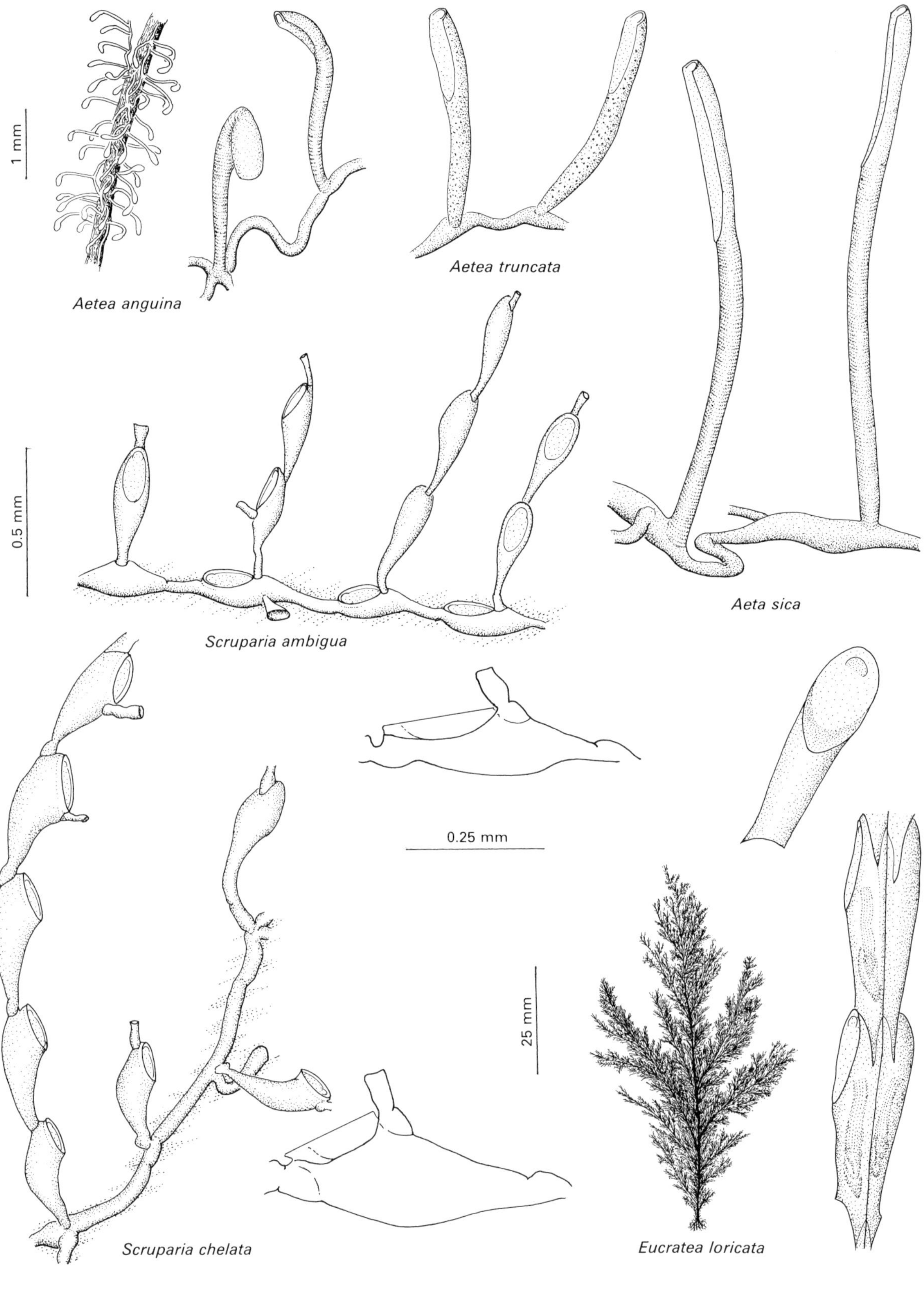

1 mm
Aetea anguina
Aetea truncata
Aeta sica
0.5 mm
Scruparia ambigua
0.25 mm
25 mm
Scruparia chelata
Eucratea loricata

Fig. 14.5

Aetea sica (Couch) FIG. 14.5

[RYLAND AND HAYWARD 1977, 46]

Colony white, adnate portion ramifying, with uprights 0.8–1.8 mm high arising from pronounced dilatations. Stem of upright faintly but clearly annulate when viewed by transmitted light; distal region about one-third of the free length, scarcely wider than the stem, finely punctate, erect.

Various substrata; LWST to shallow water; all coasts; frequent.

Aetea truncata (Landsborough) FIG. 14.5

[RYLAND AND HAYWARD 1977, 48]

Colony white, adnate portion ramifying, with uprights 0.6–0.8 mm high rising from weak dilatations. Stem of upright punctate, not striated when viewed by transmitted light; distal region one-third to one-half of the free length, scarcely wider than the stem, finely punctate, erect.

Various substrata; LWST to shallow water; all coasts; frequent.

7 SCRUPARIIDAE

Colonies erect, with the zooids arranged uniserially. Zooids elongated, with an oval frontal membrane situated in the distal part of the zooid. No avicularia or vibracula. Embryos develop in bivalved chambers borne on special reproductive zooids.

---

1. Creeping part of colony a line of zooids. Zooids slender, with the frontal membrane almost parallel to the zooidal axis **_Scruparia ambigua_**

   Creeping part of the colony a stolon. The frontal membrane at a marked angle to the axis of the zooid **_Scruparia chelata_**

Scruparia ambigua (d'Orbigny) FIG. 14.5

[RYLAND AND HAYWARD 1977, 50]

Colony small, white, with free branching chains of zooids arising from similar but adnate chains. Zooids long and slender, the frontal membrane almost parallel to the basal wall. Brood chamber subglobular, bivalved, about 200 μm across, scarcely wider than its supporting zooid.

On algae, hydroids, bryozoans, and shells; LWST to shallow water; all coasts; common.

Scruparia chelata (Linnaeus) FIG. 14.5

[RYLAND AND HAYWARD 1977, 52]

Colony small, white, with free branching chains of zooids arising from weak dilatations in a creeping stolon. Zooids horn-shaped, the frontal membrane and basal wall diverging at an angle of about 25°. Brood chamber subglobular, bivalved, about 350 μm across, wider than its supporting zooid.

Various substrata; LWST to shallow water; uncommon in Scotland, frequent elsewhere.

8 EUCRATEIDAE

Colonies erect, tufted. Zooids elongate, paired back to back in branching series. No spines, avicularia, or ovicells. Embryos developing in evanescent *ovisacs*.

Eucratea loricata (Linnaeus) FIG. 14.5

[RYLAND AND HAYWARD 1977, 54]

Colonies forming tall, dense, slender clumps 10 cm or more high; whitish brown. Zooids elongate, broader distally, with an oval frontal membrane; joined back to back in pairs, all the pairs in the branch having the same orientation. Embryos develop in membranous ovisacs.

Sublittoral, but often washed up; all coasts; local.

9 MEMBRANIPORIDAE

Colonies encrusting. Zooids squat, longer than broad, with all or most of the frontal surface membranous. No avicularia or vibracula. The lophophore intermittently bears a small, internally ciliated funnel, the *intertentacular organ*, between the dorso-medial pair of tentacles. Embryos not brooded, larva a *cyphonautes*.

---

1. Membranous area oval, occupying about one-half of the total length of the zooid. Proximal part of the zooid covered by a punctate wall, and bearing a median tooth or bristle. One or more thorn-like denticles present on each lateral wall **_Electra pilosa_**

   Membranous area occupying all or most of the frontal surface **2**

2. Membranous area more or less rectangular. Zooid walls bearing a tubercle at each angle. On *Laminaria* (rarely other algae) **_Membranipora membranacea_**

   Membranous area more or less oval. Zooid walls without tubercles. Not on *Laminaria* (often in brackish water) **3**

3. Usually a thorn-like spine proximal to the membranous area. Operculum calcified and opaque **_Electra crustulenta_**

   No median spine proximal to the membranous area. Operculum transparent **4**

4. Small triangular spaces usually present in the angles between zooids. Sometimes a few delicate spinules on the margin of the membranous area. Marine or brackish **_Conopeum reticulum_**

   Triangular spaces never present between zooids. Marginal spinules generally absent except for a distal pair. Brackish water only **_Conopeum seurati_**

Conopeum reticulum (Linnaeus) FIG. 14.6

[RYLAND AND HAYWARD 1977, 60]

Colonies forming gauze-like incrustations, often quite extensive. Zooids oval, about 0.5 × 0.3 mm. Membranous area occupying the whole of the frontal surface, the margin somewhat thickened, granular, sometimes bearing a few delicate spines. Triangular dwarf zooids usually present in the angles between normal zooids.

On stones, shells, and submerged structures; MLWN to shallow water; open coasts, also estuaries and brackish water (to less than 20‰); widespread and common.

Conopeum seurati (Canu) FIG. 14.6
[RYLAND AND HAYWARD 1977, 62]
Colonies forming gauze-like incrustations. Zooids oval, about 0.5×0.35 mm. Membranous area occupying the whole of the frontal surface, its margin usually bearing a pair of distal spines, and sometimes a few delicate lateral spines. Triangular dwarf zooids absent.

In brackish water (as low as 1‰), on aquatic plants, shells, and submerged structures; locally common, particularly in brackish lagoons, in England southwards to the Mediterranean; northern limit unknown.

Electra crustulenta (Pallas) FIG. 14.6
[RYLAND AND HAYWARD 1977, 68]
Colonies usually forming encrusting patches but in enclosed lagoons sometimes producing erect growths and large clumps. Zooids ovate hexagonal 0.5×0.3 mm. Membranous area occupying most of frontal surface; the proximal part non-porous. Operculum calcified, markedly opaque compared to the rest of the frontal membrane. Margin crenulate, bearing a single, short, proximal spine.

In brackish water; on available substrata; lower shore and shallow water; estuaries; locally common.

Electra pilosa (Linnaeus) FIG. 14.6
[RYLAND AND HAYWARD 1977, 64]
Colonies forming irregular or stellate, greyish patches. Zooids ovate-oblong, about 0.45×0.3 mm; the proximal part porous, the distal half or two-thirds occupied by the oval frontal area; operculum not calcified, inconspicuous. The margin bearing three to 13 short tooth-like spines, the median one at the proximal end usually the longest, sometimes bristle-like.

Marine; various substrata, particularly *Fucus serratus* and other algae; lower shore and shallow water; all coasts; very common.

Membranipora membranacea (Linnaeus) FIG. 14.6
[RYLAND AND HAYWARD 1977, 56]
Colonies forming extensive, lacy coverings on algae. Zooids rectangular, about 0.4×0.15 mm, with tubercles or short spines on the corners. Membrane covering the whole frontal surface. Lateral walls incorporating vertical uncalcified bands to provide flexibility.

On algae, particularly *Laminaria*; LWST to shallow water; all coasts; common and often conspicuous.

10 FLUSTRIDAE

Colonies erect, foliaceous; fronds flexible and made up of two layers of zooids placed back to back; arising from an encrusting base. Zooids ovate-quadrangular, with all the frontal surface membranous. Avicularia usually present, in series with the autozooids. Ovicells present, completely immersed in the next distal zooid.

---

1. Colony generally with broad lobes. Two pairs of spines on most non-ovicellate zooids ***Flustra foliacea***

   Colony with rather narrow branches. At most a single pair of spines present on non-ovicellate zooids **2**

2. One pair of spines at the distal end of the zooids. Avicularia absent. Fronds not noticeably wedge-shaped ***Chartella papyracea***

   Spines absent. Avicularia present. Fronds wedge-shaped ***Securiflustra securifrons***

Chartella papyracea (Ellis and Solander) FIG. 14.6
[RYLAND AND HAYWARD 1977, 80]
Colonies forming delicate, whitish, bushy tufts of much divided, rather narrow fronds; about 30 mm high. Zooids oblong, with a single short spine on each of the distal angles. Avicularia absent. Ovicell underlying the frontal membrane of the distal zooid, somewhat pointed from above; separated from the maternal zooid by a closing membrane which opens distally to the operculum. Embryos orange.

Below boulders at LWST, and on stones in shallow water; southern Britain only; frequent.

Flustra foliacea (Linnaeus). Hornwrack FIG. 14.6
[RYLAND AND HAYWARD 1977, 76]
Colony a coarse greyish tuft of broadly foliaceous or narrower strap-shaped fronds; commonly about 100 mm high; often smelling of lemon. Zooids narrow proximally, broader and rounded distally, with a pair of spines on each side. Avicularia situated at the bifurcation of zooid rows, the mandible rounded. Ovicell egg-shaped, its calcified chamber bulging into both the distal and maternal zooids; opening distally to the operculum. Embryos pale orange.

On stones and shells in shallow water; all coasts; common. Dead colonies often washed up.

Securiflustra securifrons (Pallas) FIG. 14.6
[RYLAND AND HAYWARD 1977, 84]
Colony a straw-coloured tuft, divided and subdivided into narrowly based, wedge-shaped segments; about 100 mm high. Zooids oblong, without spines. Avicularia situated in the lines of zooids, the mandible rounded. Ovicell underlying the frontal membrane of the distal zooid, rather rounded from above; separated from the maternal zooid by a closing membrane which opens distally to the operculum; between the closing membrane and the operculum are two juxtaposed calcified plates.

On stones and shells in shallow water; northern Irish Sea, Scotland, and north-east England; locally common.

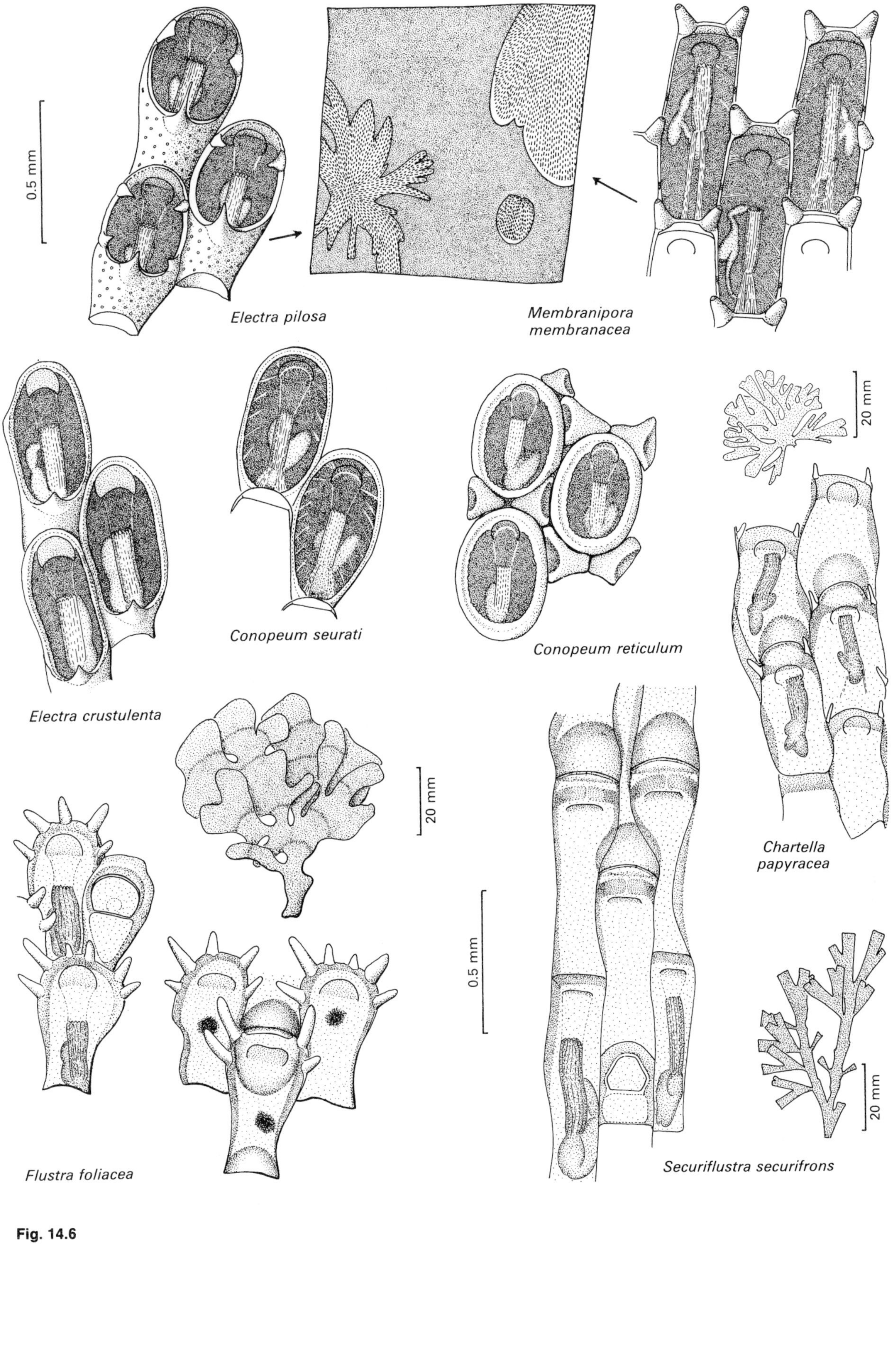
0.5 mm
Electra pilosa
Membranipora membranacea
Electra crustulenta
Conopeum seurati
Conopeum reticulum
20 mm
Chartella papyracea
20 mm
Flustra foliacea
0.5 mm
Securiflustra securifrons
20 mm

Fig. 14.6

II CALLOPORIDAE

Colonies encrusting. Zooids with most of the frontal surface membranous; sometimes underlain proximally by a calcified shelf (*cryptocyst*). Avicularia generally present. Ovicells globular, prominent, opening above the zooidal operculum.

---

1. A single large spine beside the membranous area, or one to two pairs of distal spines **2**
   Membranous area surrounded by spines (five pairs or more) **4**
2. Surface of ovicell uniformly granular. One or two pairs of spines present at the distal end of the membranous area ***Callopora dumerilii***
   Surface of ovicell including a distinct and strongly demarcated rounded or triangular area of different texture. Except on marginal zooids, one or two spines present on the side of the membranous area, one of them often greatly developed, or spines absent **3**
3. Differentiated area of ovicell more or less circular or rectangular. Developing embryos orange. Frontal membrane underlain by a conspicuous calcified lamina surrounding a trilobate opening ***Amphiblestrum flemingii***
   Differentiated area of ovicell triangular. Developing embryos white. No calcified lamina under the frontal membrane ***Callopora aurita***
4. Avicularia stalked, borne on the rim of the membranous area, and replacing marginal spines which they closely resemble ***Cauloramphus spiniferum***
   Some or more zooids bearing sessile, triangular avicularia proximally to the membranous area **5**
5. Spines, apart from the distalmost pair(s), flattened, arched over the frontal membrane and meeting in the mid-line, like ribs. Developing embryos orange ***Callopora rylandi***
   Spines not rib-like, more or less erect. Developing embryos pink **6**
6. Spines rounded in section, up to 12 in total number but usually fewer, almost perfectly erect ***Callopora lineata***
   Spines, except for the distalmost pair(s), flattish in section, 12 or more in number, closely set and sloping inwards across the membranous area ***Callopora craticula***

Amphiblestrum flemingii (Busk) FIG. 14.7

[RYLAND AND HAYWARD 1977, 104]

Colonies white, forming smallish patches. Zooids pear-shaped, widest proximally; the frontal membrane underlain in its proximal third by a cryptocyst. Often four to six spines situated near the orifice, sometimes one only: this, or one of the lower spines, greatly enlarged, tall, and curved. Generally two avicularia on the proximal part of the zooid, variously orientated. Ovicells globular, the upper surface often bearing a crescentic ridge enclosing a finely granulated area. Embryos orange.

On stones and shells; LWST and offshore; all coasts; common in shallow water.

Callopora aurita (Hincks) FIG. 14.7

[RYLAND AND HAYWARD 1977, 90]

Colonies white, forming smallish patches. Zooids ovate, the membranous area occupying most of the frontal surface; no underlying cryptocyst. Up to four spines situated near the orifice, one of them often enlarged. Generally two avicularia on the proximal part of the zooid, variously orientated. Ovicells globular, the upper surface bearing a rib which rises to an apex distally and encloses a small triangular space. Embryos white.

Stones, shell, and *Laminaria* holdfasts; shore and shallow sublittoral (possibly deeper); all coasts but especially in north.

Callopora craticula (Alder) FIG. 14.7

[RYLAND AND HAYWARD 1977, 92]

Colonies white, forming small subcircular patches. Zooids hexagonal, 0.3–0.4 × 0.25 mm, the oval membranous area occupying most of the frontal surface. 12–15 slender spines surrounding the area, the distalmost pairs erect, the remainder slanted inwards towards the mid-line of the zooid. A single avicularium on the proximal part of the zooid, variously orientated. Ovicells rather small, smooth, with a transverse rib. Embryos pink.

On shells and seaweeds; offshore; Isle of Man, Scotland, north-east England; locally common.

Callopora dumerilii (Audouin) FIG. 14.7

[RYLAND AND HAYWARD 1977, 89]

Colonies white, forming small patches. Zooids hexagonal, the oval membranous area occupying most of the frontal surface. Generally one or two pairs of spines situated near the orifice. One or two avicularia on the proximal part of the zooid. Ovicells with a granular surface; not ribbed. Embryos orange.

Chiefly on stones and shells, below LWST; all coasts; quite common.

Callopora lineata (Linnaeus) FIG. 14.7

[RYLAND AND HAYWARD 1977, 88]

Colonies white, forming small subcircular patches. Zooids hexagonal, about 0.5 × 0.25 mm, the membranous area oval, occupying most of the frontal surface; usually eight to ten more or less erect spines around the area. A single avicularium on the proximal part of the zooid, variously orientated. Ovicells fairly large, smooth, with a transverse rib. Embryos pink.

On shells, stones, and algae from the shore and shallow water, sometimes abundant on *Laminaria* fronds in sheltered situations; all coasts; common.

Callopora rylandi Bobin and Prenant FIG. 14.7

[RYLAND AND HAYWARD 1977, 91]

Colonies white, forming small patches. Zooids oval-hexagonal, about 0.5 × 0.35 mm, the membranous area oval, occupying most of the frontal surface. seven to 11 flattened spines, like ribs,

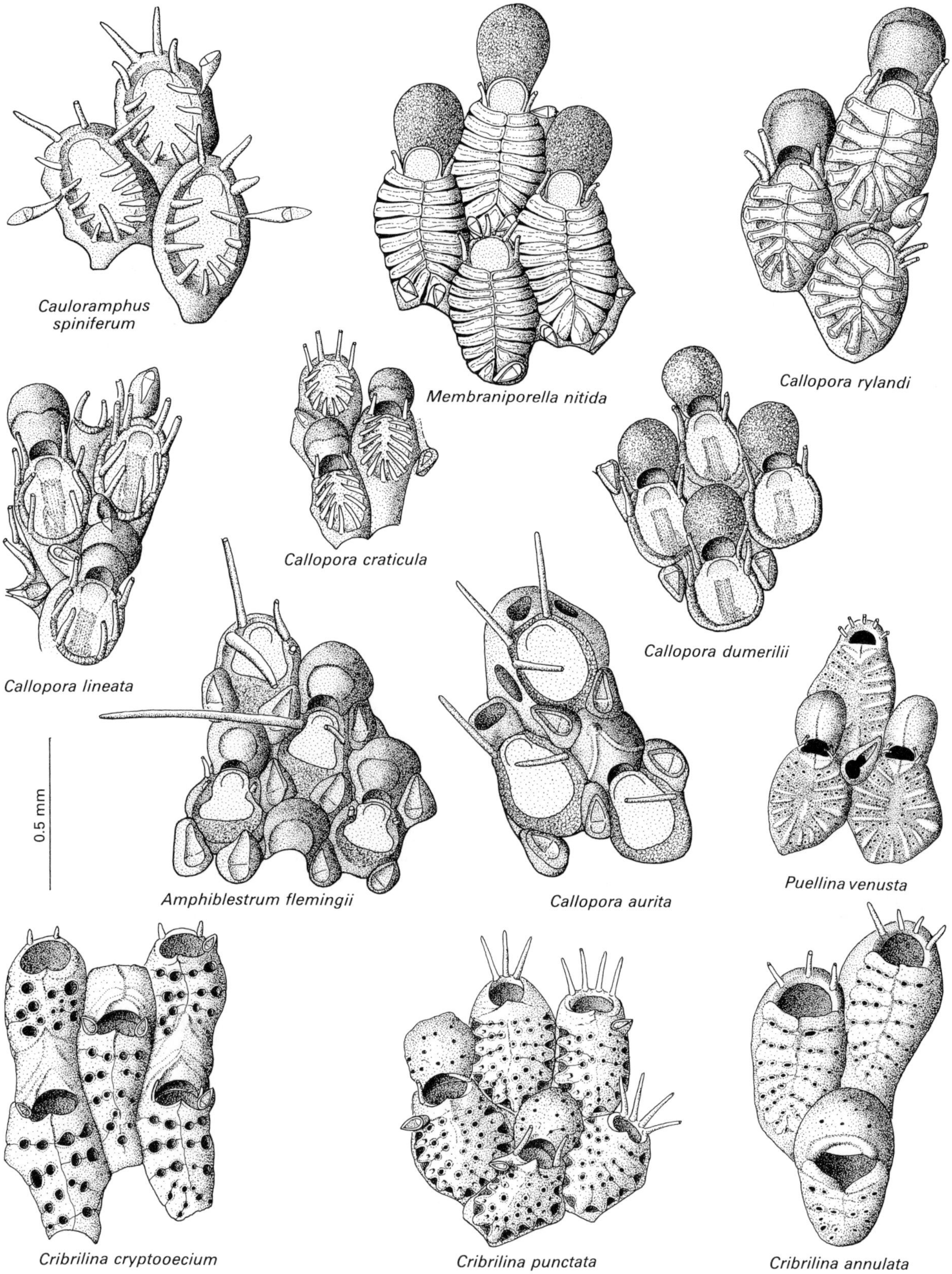
Cauloramphus spiniferum
Membraniporella nitida
Callopora rylandi
Callopora craticula
Callopora lineata
Callopora dumerilii
0.5 mm
Amphiblestrum flemingii
Callopora aurita
Puellina venusta
Cribrilina cryptooecium
Cribrilina punctata
Cribrilina annulata

Fig. 14.7

overarch the area and meet in the mid-line. A single avicularium sometimes present on the proximal part of the zooid. Ovicell globular, finely granular, with an inconspicuous rim distally. Embryos orange. This species is rather similar in appearance to *Membraniporella nitida* (next family), but has fewer ribs.

On rocks, stones, and around *Himanthalia* holdfasts; lower shore; south and west of the British Isles; frequent.

Cauloramphus spiniferum (Johnston) FIG. 14.7
[RYLAND AND HAYWARD 1977, 98]
Colonies forming small brownish patches. Zooids elongate-oval, the membranous area occupying most of the frontal surface. 14–16 close-set, slightly inbent spines surrounding the area; interspersed among them one or more slender, pedicellate, club-shaped avicularia; mandible subterminal. Ovicells shallow, with a frontal rib. Embryos orange.

On stones and shells; LWST and offshore; all coasts; quite common, particularly in the south and west.

12 CRIBRILINIDAE

Colonies incrusting. Zooids with the frontal membrane spanned by flattened spines, with or without lateral fusions. Avicularia generally present. Ovicells globular, opening below (and closed by) the zooidal operculum.

---

1. Frontal membrane covered by a rib-like cage of flattened spines **Membraniporella nitida**
   Frontal membrane covered by a shield of fused spines separated from each other only by rows of pores **2**
2. Large, elongate avicularia scattered among the zooids; small, setiform avicularia often present beside the orifice. Openings in frontal shield small, in radiating rows which clearly delimit the spines comprising the shield **Puellina venusta**
   No large scattered avicularia; usually a small triangular avicularium on each side of the orifice. Openings in frontal shield rather large, not very clearly in rows **3**
3. Ovicell imperforate, subimmersed, partly obscured by a thick arched rib. Developing embryos orange **Cribrilina cryptooecium**
   Ovicell punctate, not immersed or obscured by a rib. Developing embryos pink **Cribrilina punctata**

Cribrilina cryptooecium Norman FIG. 14.7
[HAYWARD AND RYLAND 1979, 58]
Colonies encrusting, up to 25 mm across. Zooids elongate-oblong, 0.4–0.55 × 0.3 mm. Membranous area covered and obscured by a frontal shield of seven to 11 fused ribs; generally two large pores in the suture line between adjacent ribs. The distalmost ribs thickened, sometimes produced into a median point (mucro), delimiting the orifice. Small lanceolate avicularia may be present on one or both sides of the orifice. Ovicell not prominent, smooth, imperforate, becoming overgrown by an A-shaped bar. Embryos orange.

On rocks and algae on the shore and in the shallow sublittoral; all coasts; common.

Cribrilina punctata (Hassall) FIG. 14.7
[KLUGE 1975, 473; HAYWARD AND RYLAND 1979, 56, in part]
Colonies encrusting, up to 15 mm across. Zooids ovate-hexagonal, 0.4–0.55 × 0.3–0.45 mm. Membranous area covered and obscured by a frontal shield of 11–13 fused ribs; generally three smallish pores in the suture line between adjacent ribs. The distalmost ribs thickened, raised into a median point (mucro), delimiting the orifice. Small lanceolate avicularia, directed outwards, present on one or both sides of the orifice, somewhat proximally to it. Ovicells prominent, punctured; not overgrown by a bar. Embryos pink.

On rocks and algae; LWST and offshore; all coasts; uncommon intertidally, frequent sublittorally.

The form (known as var. α), from western coasts as far north as Shetland, in which the ovicell bears small tubercles and is surmounted by an avicularium must be regarded as an unnamed species of *Collarina* (Bishop, J. D. D. 1986, *Bull. Zool. Nomencl.* **43**, 288.)

Membraniporella nitida (Johnston) FIG. 14.7
[RYLAND AND HAYWARD 1977, 110]
Colonies encrusting. Zooids elongate-oval, about 0.5 × 0.3 mm. Membranous area covered and obscured by 17–21 flattened, close-set ribs, which meet and fuse in the mid-line. Orifice deeply D-shaped, flanked by two to four spines. Generally a pair of avicularia at the proximal end of each zooid, variously orientated. Ovicell longer than broad, prominent; its surface sculptured with tiny spherules. Embryos orange. Liable to be confused with *Callopora rylandi*.

On stones, shells, and algae; LWST and shallow water; all coasts; not very common.

Puellina venusta Canu and Bassler FIG. 14.7
[HAYWARD AND RYLAND 1979, 62, as *Cribrilaria radiata* (Moll); Bishop, J. D. D. and Househam, B. C. 1987 *Bull. Br. Mus. Nat. Hist.* (Zool.), **53**, 28]
Colonies white or silvery, forming patches. Zooids oval, with six to nine ridges on each side, radiating from the mid-line; the furrow between each pair occupied by a row of small pores. Orifice D-shaped, with a distinct operculum, and a large pore just proximally to it. Four to six spines placed around the orifice. Avicularia dimorphic: large and elongate, triangular, scattered among the autozooids; small and setiform, beside the orifice. Ovicells globular, prominent, with a keel down the mid-line. Embryos orange-red.

On stones, shells, hydroids, etc.; offshore; all coasts; generally fairly common.

A somewhat similar species, in which, however, conspicuous avicularia are lacking, is *Cribrilina annulata* (Fabricius) which may occur in abundance on *Laminaria* fronds from Scotland and Denmark northwards.

13 CELLARIIDAE

Colonies erect, composed of cylindrical, dichotomously dividing, jointed branches, in which the zooids are disposed around an imaginary axis; well-calcified. Zooids hexagonal or rhomboidal, separated by raised margins, most of the frontal membrane underlain by a calcified shelf (*cryptocyst*). Avicularia present. Ovicells immersed, entirely below the colony surface.

1. Avicularia replacing autozooids in the regular alternating arrangement. Zooids appearing as ovals inside hexagons — ***Cellaria salicornioides***

   Avicularia wedged between two zooids in a linear series. Zooids not appearing as ovals inside hexagons — **2**

2. Mandible of avicularium directed distally. Branches not widening distally, their ends narrow (about 0.6 mm diameter) — ***Cellaria fistulosa***

   Mandible of avicularium directed proximally. Branches widening distally (to about 1.0 mm diameter) — ***Cellaria sinuosa***

Cellaria fistulosa (Linnaeus) FIG. 14.8

[RYLAND AND HAYWARD 1977, 122]

Colonies forming diffuse, rigid tufts up to about 50 mm high; branches slender, not widening distally, about 0.6 mm diameter. Zooids hexangular, orifice D-shaped. Avicularia placed in the zooid rows, between two zooids; mandible semicircular, wider than long, facing distally. Opening of ovicell reniform or rounded, distal to the orifice. Embryos yellow.

On stones and shells; offshore; all coasts; common.

Cellaria sinuosa (Hassall) FIG. 14.8

[RYLAND AND HAYWARD 1977, 121]

Colonies forming coarse rigid tufts; branches widening distally to about 1 mm diameter. Zooids hexangular, orifice D-shaped with a projecting lower lip. Avicularia placed in the zooid rows, between two zooids; mandible obtusely triangular, as long as wide, facing proximally. Opening of ovicell reniform, distal to the orifice. Embryos yellow.

On stones and shells; offshore; all coasts; frequent.

Cellaria salicornioides Lamouroux FIG. 14.8

[RYLAND AND HAYWARD 1977, 124]

Colonies forming rigid tufts about 30 mm high; branches slender, not widening distally. Zooids with the appearance of ellipses within hexagons, alternating; orifice D-shaped or reniform. Avicularia replacing autozooids in the zooid rows; mandible semicircular, facing distally. Opening of ovicell circular, distal to the orifice.

A southern species most likely to be encountered in the English Channel; offshore; probably uncommon

14 SCRUPOCELLARIIDAE

Colonies erect or suberect, with narrow unilamellar, dichotomously dividing, jointed branches; attached to the substratum by rhizoids. Zooids in two series, alternately arranged. Basal and lateral walls calcified; frontal wall calcified apart from the oval membranous area, which may be partly covered by a modified spine (*scutum*); generally with spines on the distal angles. Avicularia usually present. Ovicells present, opening above the orifice, recumbent on the distal zooid.

1. A vibraculum present on the basal surface of each autozooid. — **2**

   Vibracula absent — **4**

2. A single vibraculum situated in the axil of each bifurcation (view from the basal side). Scutum divided like an antler — ***Scrupocellaria reptans***

   A pair of vibracula in each axil. Scutum more or less undivided, or absent — **3**

3. Scutum absent — ***Scrupocellaria scruposa***

   Scutum present — ***Scrupocellaria scrupea***

4. Scutum, frontal and lateral avicularia present; two or more spines present on the zooid margin — ***Tricellaria ternata***

   Scutum, frontal and lateral avicularia absent. A single short spine present on the outer distal angle of each zooid — ***Tricellaria peachii***

Scrupocellaria reptans (Linnaeus) FIG. 14.8

[RYLAND AND HAYWARD 1977, 134]

Colonies forming spreading straggling tufts, the branches frequently anchored by rhizoids; mostly with five to seven zooids between bifurcations. Zooids narrow proximally, broader distally, membranous area overarched by a branched scutum; two to three spines on the outer angle, one on the inner. Vibracula present on the lower surface, with one in the axil of each bifurcation. A rather small lateral avicularium present, and often a larger frontal avicularium situated just proximally to the membranous area. Ovicells globular, with small punctures. Embryos red.

On *Flustra*, algae and submerged structures, and below boulders; LWST and shallow water; all coasts; common. Especially abundant on *Laminaria* fronds in sheltered channels.

Scrupocellaria scrupea Busk FIG. 14.8

[RYLAND AND HAYWARD 1977, 138]

Colonies forming erect tufts about 20 mm high; up to 20 zooids between bifurcations. Zooids narrow proximally, broader distally, membranous area overarched by an oval or reniform scutum; three spines on the outer angle, one to two on the inner. Vibracula present on the basal surface, with two in the axil of each bifurcation. A large lateral avicularium present, and sometimes a small frontal avicularium situated just proximal to the membranous area. Ovicells globular, imperforate.

On algae and stones; offshore; southern Britain; frequent.

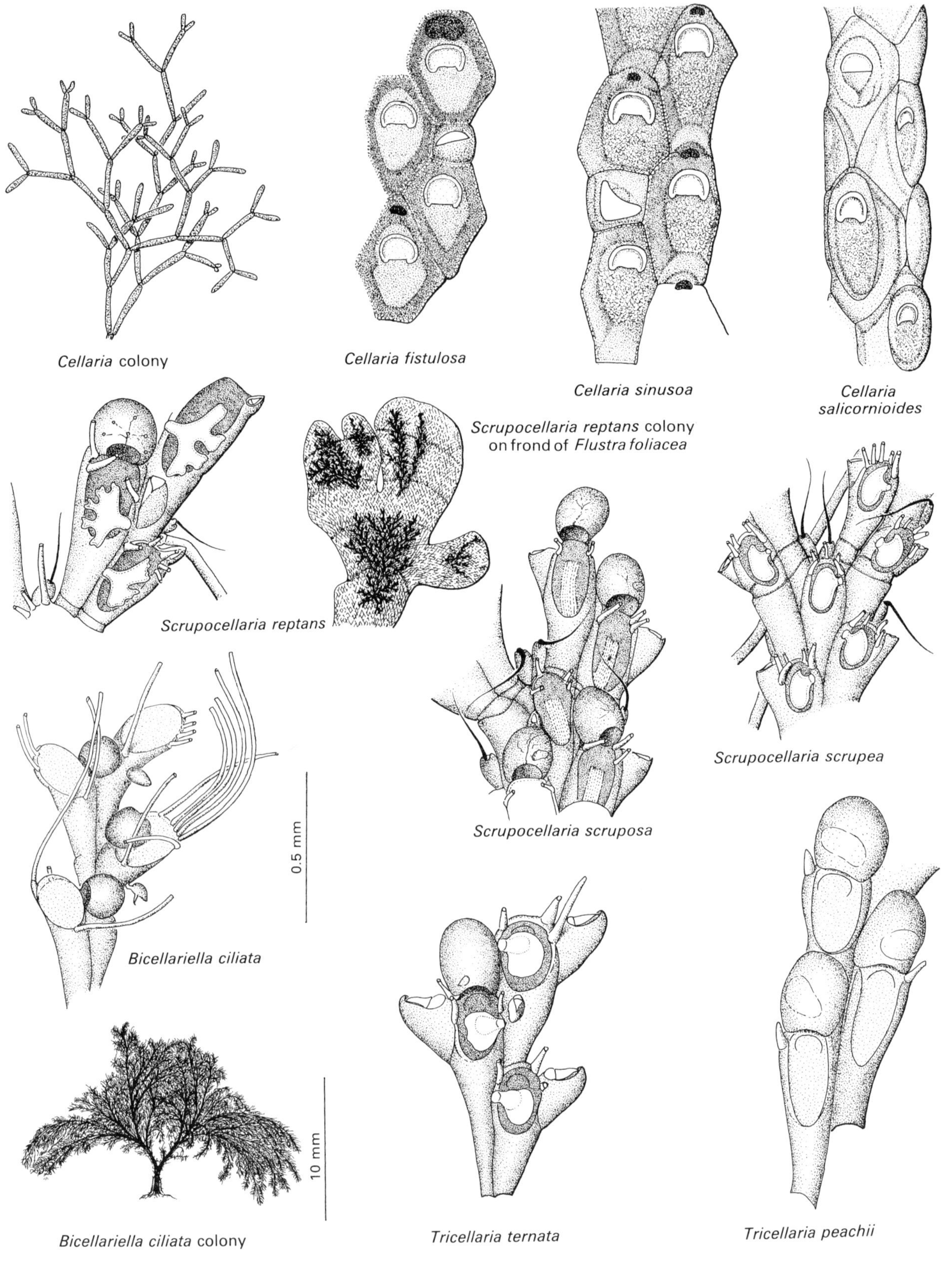
Cellaria colony
Cellaria fistulosa
Cellaria sinusoa
Cellaria salicornioides
Scrupocellaria reptans colony on frond of Flustra foliacea
Scrupocellaria reptans
Scrupocellaria scrupea
Scrupocellaria scruposa
0.5 mm
Bicellariella ciliata
10 mm
Bicellariella ciliata colony
Tricellaria ternata
Tricellaria peachii

Fig. 14.8

Scrupocellaria scruposa (Linnaeus) FIG. 14.8
[RYLAND AND HAYWARD 1977, 140]
Colonies forming erect tufts; seven to 11 zooids between bifurcations. Zooids slightly narrower proximally, no scutum over the membranous area; two to three spines on the outer angle, two on the inner. Vibracula present on the basal surface, with two in the axil of each bifurcation. A large lateral avicularium present, small frontal avicularia found on top of ovicells. Ovicells subglobular, imperforate. Embryos orange-red.

On stones, shells, *Flustra*, and submerged structures; LWST and shallow water, all coasts; common, though rarely as prolific as *S. reptans*.

Tricellaria peachii (Busk) FIG. 14.8
[RYLAND AND HAYWARD 1977, 145]
Colonies white, forming slender tufts about 25 mm high; seven–nine zooids between bifurcations. Zooids more slender proximally; no scutum over the membranous area; at most a small spine on the outer angle. Vibracula and avicularia absent. Ovicells globular.

On stones, shells, hydroids, and bryozoans; offshore only; absent from southern Britain.

Tricellaria ternata (Ellis and Solander) FIG. 14.8
[RYLAND AND HAYWARD 1977, 144]
Colonies forming delicate, white, straggling tufts, attached by rhizoids and tendrils; most commonly with three zooids between joints. Zooids more slender proximally, of varying length; a small spatulate scutum overarching the membranous area; two to three spines on the outer angle. No vibracula. Large lateral avicularia present on the two lower zooids in the internode; a small frontal avicularium situated proximally to the area on all zooids or just on the uppermost one in the internode. Ovicells elongated.

On algae, hydroids, shells, etc,; LWST and offshore; absent from southern Britain.

15 BUGULIDAE

Colonies erect and tufted; developing from an upright ancestrula; attached to the substratum with rhizoids. Branches unilamellar; dividing dichotomously; without joints. Zooids in two or more series, alternating. Basal and lateral walls lightly calcified, the frontal surface mainly or wholly membranous. Orifice closed by a sphincter muscle; operculum absent. Importance is attached to the arrangement of spines on the distal angles of the zooids. A stalked avicularium, shaped like a bird's head, frequently present on the side of the membranous area. Ovicells present, opening above the orifice, not recumbent on the distal zooid.

---

1. Branches made up from zooids attached to each other only by their proximal part. Free distal part occupied mainly by an oval membranous area, and bearing several very long spines. Colony white **_Bicellariella ciliata_**

   Zooids wholly attached to one another by their inner sides. Colony buff, brown, or yellow **2**

2. Spines present on the sides of the membranous area as well as on the upper distal angles of the zooid. Avicularium attached to the frontal wall of the zooid below the membranous area **_Dendrobeania murrayana_**

   Spines restricted to the distal angles of the zooid. Avicularium attached to the side of the membranous area, or absent **3**

3. Avicularia absent. Ovicell attached to the inner distal angle of the zooid, orientated transversely. Colony a deep purplish-brown **_Bugula neritina_**

   Avicularia present. Ovicell attached across the distal end of the zooid, facing proximally. Colony yellowish- to purplish-buff in colour **4**

4. Branches comprising two series of zooids (occasionally four series at the tips of branches) **5**

   Branches comprising several series of zooids **10**

5. Inner distal angle of zooid unarmed, outer angle with one projection or spine. Branch bifurcations in basal view never involving one or more completely enclosed zooids (check basal view) **6**

   Both distal angles bearing at least one spine. A majority of bifurcations usually involving one or more completely enclosed zooids **7**

6. Branches of colony disposed in spirals. Ovicell appearing globular. Avicularia very small, their length less than the width of a zooid. Outer distal angle of zooid produced as a fingerlike spine **_Bugula plumosa_**

   Colony tufted, branches not in spirals. Ovicell hemispherical. Avicularia as long as the width of a zooid. Outer distal angle bearing a jointed spine **_Bugula purpurotincta_**

7. Colony made up of branches clearly disposed in spirals **8**

   Colony tufted, branches not in spirals **9**

8. Outer distal angle of zooid bearing two spines. Beak of avicularium downcurved in profile **_Bugula avicularia_**

   Outer distal angle bearing one spine only. Beak of avicularium abruptly hooked **_Bugula turbinata_**

9. Outer distal angle of zooid bearing two spines, inner angle bearing one **_Bugula stolonifera_**

   Both distal angles bearing two to three spines **_Bugula fulva_**

10. Ovicells hemispherical. One spine on each distal angle of the zooid (sometimes two on the outer angles of marginal zooids) **_Bugula simplex_**

    Ovicells nearly spherical. At least two spines present on each distal angle of all zooids **_Bugula flabellata_**

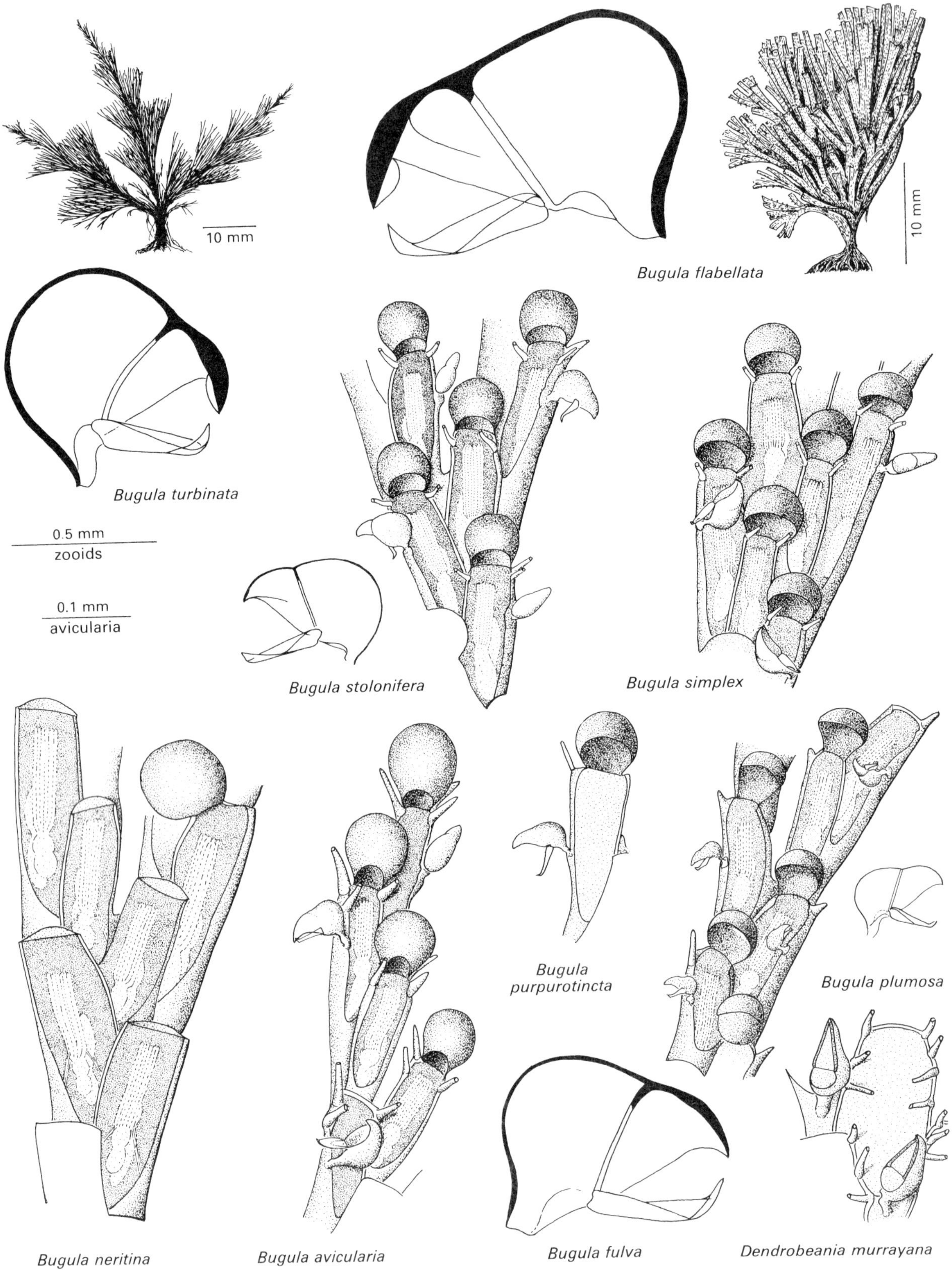
10 mm
10 mm
Bugula flabellata
Bugula turbinata
0.5 mm
zooids
0.1 mm
avicularia
Bugula stolonifera
Bugula simplex
Bugula purpurotincta
Bugula plumosa
Bugula neritina
Bugula avicularia
Bugula fulva
Dendrobeania murrayana

Fig. 14.9

Bicellariella ciliata (Linnaeus) FIG. 14.8
[RYLAND AND HAYWARD 1977, 146]
Colonies forming white, feathery tufts, 10–20 cm high. Zooids narrow below, much enlarged distally; membranous area oval, occupying about one-third of the frontal surface; four to seven very long, slender, curving spines present on the outer margin. Avicularia small, attached below the membranous area. Ovicells subglobular, helmet-shaped, contracting to a stem-like base. Embryos white.

On algae, hydroids, and bryozoans; LWST and shallow water; all coasts; frequent.

Bugula avicularia (Linnaeus) FIG. 14.9
[RYLAND AND HAYWARD 1977, 154]
Colonies bushy, tawny, 20–30 mm high, branches spirally disposed; zooids about 0.60 × 0.18 mm, arranged biserially. Membranous area occupying most of the frontal surface; two spines on the outer distal angle, one on the inner. Avicularia lateral, large, longer than the width of the bearing zooid, with downcurved beak. Ovicells globular; embryos yellow.

On shells, hydroids, and bryozoans; LWST to shallow water; all coasts; widespread but not very common.

Bugula flabellata (Thompson in Gray) FIG. 14.9
[RYLAND AND HAYWARD 1977, 158]
Colonies forming low dense tufts, 20–50 mm high, buff when living, greyish when dried or preserved in alcohol. Branches broad, wedge-shaped, about 0.80 × 0.17 mm, arranged in three to eight series. Membranous area occupying most of the frontal surface, three spines on the marginal distal angle, two on all inner angles. Large lateral avicularia present on marginal zooids, their length exceeding the width of the bearing zooid; small avicularia present on the inner zooids; the beak hooked at a right angle. Ovicell globular; embryos yellow.

In various habitats, but almost invariably attached to other bryozoans; LWST and shallow water; all coasts, but not found north of the British Isles; common.

Bugula fulva Ryland FIG. 14.9
[RYLAND AND HAYWARD 1977, 160]
Colonies forming dense bushy tufts, 20–30 mm high, tawny when living, greyish when dry. Branches not spirally disposed; zooids about 0.60 × 0.16 mm, arranged biserially. Membranous area occupying all of the frontal surface, three spines on the outer distal angle, two on the inner, and sometimes a sixth spine placed between the two groups. Avicularia lateral, somewhat longer than the width of the bearing zooid, with downcurved beak. Ovicells globular; embryos yellow.

In *Laminaria* holdfasts, below boulders, and in marine fouling; LWST and shallow water; coasts north to Northumberland; locally common.

Bugula neritina (Linnaeus) FIG. 14.9
[RYLAND AND HAYWARD 1977, 162]
Colonies forming luxuriant, purplish-brown tufts, up to 80 mm high. (Fixed material is a distinctive translucent brown.) Zooids large, about 0.75 × 0.25 mm, biserially arranged. Frontal area occupying the whole of the frontal surface; spines absent. Avicularia absent. Ovicell globular, attached to the inner distal angle of the zooid, its opening concealed; embryos pale brown.

On pier piles and in marine fouling; introduced on shipping to a few ports in southern Britain; locally common.

Bugula plumosa (Pallas) FIG. 14.9
[RYLAND AND HAYWARD 1977, 164]
Colonies forming buffish, feathery tufts, up to 80 mm high, with branches spirally disposed. Zooids about 0.50 × 0.15 mm, biserially arranged. Membranous area occupying three-quarters or more of the frontal surface; a single unjointed spine on the outer distal angle; inner angle unarmed. Avicularia small, lateral, shorter than the width of the bearing zooid; the beak slightly downcurved. Ovicells appearing globular when containing embryos, deciduous; embryos yellow.

On stones, shells, and submerged structures; LWST and shallow water; widespread and common along coasts of southern Britain.

Bugula purpurotincta Norman FIG. 14.9
[RYLAND AND HAYWARD 1977, 166]
Colonies forming bushy tufts, up to 80 mm high, purplish red when dry. Zooids about 0.56 × 0.17 mm, arranged biserially. Membranous area occupying most of the frontal surface; a single jointed spine on the outer distal angle; inner angle unarmed. Avicularia lateral, in length about equal to the width of the bearing zooid; the beak slightly downcurved. Ovicells more or less hemispherical.

On shells and hydroids; LWST and shallow water; Scotland and north-east England; locally common.

Bugula simplex Hincks FIG. 14.9
[RYLAND AND HAYWARD 1977, 168]
Colonies shortly tufted, 15–30 mm high, straw-coloured. Branches broad, zooids about 0.75 × 0.14 mm, arranged in three to six series. Membranous area occupying all of the frontal surface, generally one spine on each of the distal angles. Avicularia lateral, present only on the marginal zooids, their length slightly exceeding the width of the bearing zooid; beak downcurved. Ovicells more or less hemispherical; embryos yellow.

Recorded from several localities in southern Britain, chiefly in marine fouling.

Bugula stolonifera Ryland FIG. 14.9
[RYLAND AND HAYWARD 1977, 170]
Colonies forming dense tufts, 30–40 mm high, greyish-buff. Branches not spirally disposed. Zooids about 0.72 × 0.17 mm, arranged biserially. Membranous area occupying one-half to three-quarters of the frontal surface; two spines on the outer distal angle, one on the inner. Avicularia lateral, in length about equal to the width of the bearing zooid; the beak downcurved. Ovicells subglobular; embryos yellow.

Mainly a species of ports and harbours, occurring on pilings and other submerged structures; locally common in southern Britain.

Bugula turbinata Alder FIG. 14.9

[RYLAND AND HAYWARD 1977, 172]

Colonies forming buff-orange bushy tufts 30–60 mm high, with branches spirally disposed. Zooids about 0.55 × 0.18 mm, mainly arranged in two series. Membranous area occupying practically the whole of the frontal surface; each distal angle bearing one spine. Large lateral avicularia borne by the marginal zooids, small avicularia by inner zooids; the head very rounded and broad, the beak hooked at a right angle. Ovicells globular; embryos yellow.

Below boulders and in gullies from MLWN into shallow water; the southern part of the British Isles; common.

Dendrobeania murrayana (Johnston) FIG. 14.9

[RYLAND AND HAYWARD 1977, 174]

Colonies bushy, straw-coloured, about 30 mm high; branches broad, strap-shaped, with up to 12 rows of zooids. Membranous area occupying most of the frontal surface, each lateral margin bearing two or more spines. Avicularia situated proximal to the membranous area, those on the marginal zooids much larger than the others. Ovicells subglobular.

On hydroids, *Flustra*, shells, and stones; offshore only; absent from southern Britain.

Suborder **Ascophora**

16 UMBONULIDAE

Colonies encrusting. Zooids with a calcified frontal wall formed by overarching the frontal membrane; a row of pseudopores around the margin. Buttress-like ridges overgrow the frontal wall leaving spaces (*areolae*) above the pseudopores. Orifice large, without condyles.

Umbonula littoralis Hastings FIG. 14.10

[HAYWARD AND RYLAND 1979, 74]

Colonies encrusting, orange. Zooids large, about 1.0 × 0.7 mm; ovate. Frontal wall convex, rising to a central prominence from which radiate about 18 ridges. Orifice large, cordate. A small avicularium usually located on the distal face of the central prominence. Ovicells absent; embryos brooded within the parent zooids, red.

Below rocks and on *Laminaria* holdfasts; lower shore; all coasts; usually common.

A related species, *U. ovicellata* (Hastings), with fewer radiating ridges and incubating embryos in ovicells, is found sublittorally; not common.

17 EXOCHELLIDAE

Colonies encrusting. Zooids with a calcified frontal wall formed by overarching the frontal membrane; a row of pseudopores around the margin. Orifice without condyles. Spines distal to the orifice and avicularia both present.

Escharoides coccinea (Abildgaard) FIG. 14.10

[HAYWARD AND RYLAND 1979, 78]

Colonies encrusting, reddish orange. Zooids ovate; frontal wall convex, rough or granular, produced below the orifice into a median point (mucro) and two smaller, lateral denticles. Orifice subcircular, bearing six distal spines in non-ovicellate zooids; flanked by rather large, pointed, outwardly directed avicularia (sometimes only one is present). Ovicells globular, granular; embryos red.

Below rocks and on *Laminaria* holdfasts; lower shore and in shallow water; all coasts; common.

18 HIPPOTHOIDAE

Colonies adnate, forming a loose network or a crust. Frontal wall imperforate, generally showing transverse growth lines in the calcification. Ovicells present.

1. Colony forming more or less circular patches; zooids contiguous	***Celleporella hyalina***
Zooids distant, connected to each other by a slender extension of the proximal region	***Hippothoa divaricata***

Celleporella hyalina (Linnaeus) FIG. 14.10

[HAYWARD AND RYLAND 1979, 252]

Colonies encrusting, forming small circular, glistening patches. Zooids contiguous; frontal wall smooth, silvery or white; sometimes with a prominence just below the orifice. Orifice almost circular, with a shallow proximal sinus. No avicularia. Ovicells large, globular, and prominent, borne on special dwarf zooids, punctate. Embryos pale yellow.

Most characteristically found on *Laminaria* and other algae, often in great abundance; LWST and in shallow water; all coasts; common.

Hippothoa divaricata Lamouroux FIG. 14.10

[HAYWARD AND RYLAND 1979, 246]

Colonies adnate. Zooids pyriform, remote, each connected by a slender stolon-like region (cauda), forming branched chains. Frontal wall smooth, silvery, imperforate, keeled; orifice rounded distally, with a sinus proximally. No avicularia. Ovicells subglobular, imperforate, borne on special reproductive zooids.

On shells and stones from below LWST; all coasts; frequent but very inconspicuous.

H. flagellum Manzoni (Hayward and Ryland 1979, 248), which has much shorter female zooids, is a similar sublittoral species.

19 CHORIZOPORIDAE

Colony encrusting. Frontal wall of zooids thin, imperforate; orifice D-shaped, without spines. Ovicells imperforate, closed by the zooidal operculum. Avicularia present. *Pore-chambers* tubular, separating the zooids.

Chorizopora brongniartii (Audouin) FIG. 14.10

[HAYWARD AND RYLAND 1979, 240]

Colonies encrusting, white. Zooids ovate-elongate, distinct, separated from each other by a line of pores; frontal wall smooth and silvery or transversely furrowed, imperforate. Orifice D-shaped, with a small pointed avicularium placed distally to it. Ovicells elongate-globular, imperforate, sometimes keeled and surmounted by an avicularium, the opening closed by the zooidal operculum. Embryos pinkish-red.

Found particularly on shell; offshore, down to at least 100 m; all coasts; common.

20 MICROPORELLIDAE

Colonies encrusting. Zooids with the frontal wall punctate. Orifice D-shaped, the *ascopore* proximal to it; with or without distal spines. Avicularia often present. Ovicells present.

---

1. Orifice without associated spines; ascopore oval or round **2**

   Orifice with spines on the distal margin (except in presence of ovicell); ascopore crescentic **3**

2. Frontal surface bearing a very few scattered, simple pores. Usually on red algae ***Haplopoma impressum***

   Frontal surface bearing numerous pores, each one containing small, inwards-projecting teeth. Usually on stones or shells ***Haplopoma graniferum***

3. Avicularia (often setiform) present. Developing embryos red ***Microporella ciliata***

   Avicularia absent. Developing embryos yellow ***Fenestrulina malusii***

Fenestrulina malusii (Audouin) FIG. 14.11
[HAYWARD AND RYLAND 1979, 224]
Colonies forming white patches. Zooids hexagonal. Frontal wall convex, smooth, punctate (especally round the sides and near the orifice). Orifice deeply D-shaped, with three to four distal spines; a lunate ascopore near the centre of the zooid. No avicularia. Ovicells subglobular, with a distal series of alternating ribs and spaces. Embryos deep yellow.

Mainly on stones and shells; offshore; all coasts; frequent.

Haplopoma graniferum (Johnston) FIG. 14.10
[HAYWARD AND RYLAND 1979, 236]
Colonies encrusting, white. Zooids elongate-hexagonal; frontal wall convex, rather thick, with or without a prominence (umbo) proximal to the orifice, punctured by numerous pseudopores (under high magnification the pore lumina can be seen to be partly occluded by centripetally directed teeth). Orifice D-shaped, with ovate ascopore just proximal to it (sometimes obscured by the umbo). No avicularia. Ovicells more or less globular, with or without keel, punctate. Embryos pink.

On stones and shells; between tidemarks or in shallow water; all coasts; quite common.

Haplopoma impressum (Audouin) FIG. 14.10
[HAYWARD AND RYLAND 1979, 230]
Colonies encrusting. Zooids ovate-hexagonal; frontal wall vitreous, translucent in life, with or without a prominence (umbo) proximal to the orifice, punctured by a few scattered pseudopores (simple under high magnification). Orifice D-shaped, with circular ascopore just proximal to it (sometimes obscured by the umbo). Ovicells subglobular, with or without a keel, sparsely punctate. Embryos pink.

On algae, particularly Rhodophyta; LWST and shallow water; frequent in south and west Britain but unrecorded from the north-east.

Microporella ciliata (Pallas) FIG. 14.11
[HAYWARD AND RYLAND 1979, 222]
Colonies forming small white patches. Zooids hexagonal. Frontal wall silvery, frosted, punctate. Orifice D-shaped, with five to seven spines distally; a lunate ascopore close to its proximal margin. An avicularium, with acute or setiform mandible, frequently located to one side of the orifice. Ovicell globular, punctate, or granular. Embryos red.

On stones, shells, and algae, sometimes abundant on *Laminaria* fronds; LWST and offshore; all coasts; common.

21 ADEONIDAE

Colonies foliaceous or encrusting. Frontal wall of zooids with a row of marginal pseudopores. Orifice with a sinus, or D-shaped with a separate *ascopore*. Avicularia present. No ovicells; embryos develop in special reproductive zooids (*gonozooids*).

Reptadeonella violacea (Johnston) FIG. 14.11
[HAYWARD AND RYLAND 1979, 82]
Colonies encrusting, often extensive, generally of a dark purplish colour. Zooids ovate-hexagonal. Frontal wall flat, with pseudopores around the margin and a central circular ascopore. Orifice transversely oval. A single avicularium, with pointed, distally directed mandible, situated between the ascopore and the orifice. Gonozooids larger than infertile autozooids, and with a wider orifice. Embryos brown.

On stones and shells; LWST and shallow water; reaches its northern limit in south-west Britain.

22 ESCHARELLIDAE

Colonies encrusting. Zooids with the frontal wall imperforate except for the marginal pseudopores; generally raised into a point (mucro) at the orifice. Orifice suborbicular, with spines distally and an internal tooth proximally. Avicularia absent. Ovicells present.

---

1. Zooids small (about 0.5 mm in length), the sutures between them usually obscured by additional calcification. Orifice with six distal spines (four in presence of ovicell). Developing embryos pink ***Escharella immersa***

   Zooids larger (0.6–0.8 mm in length), the sutures

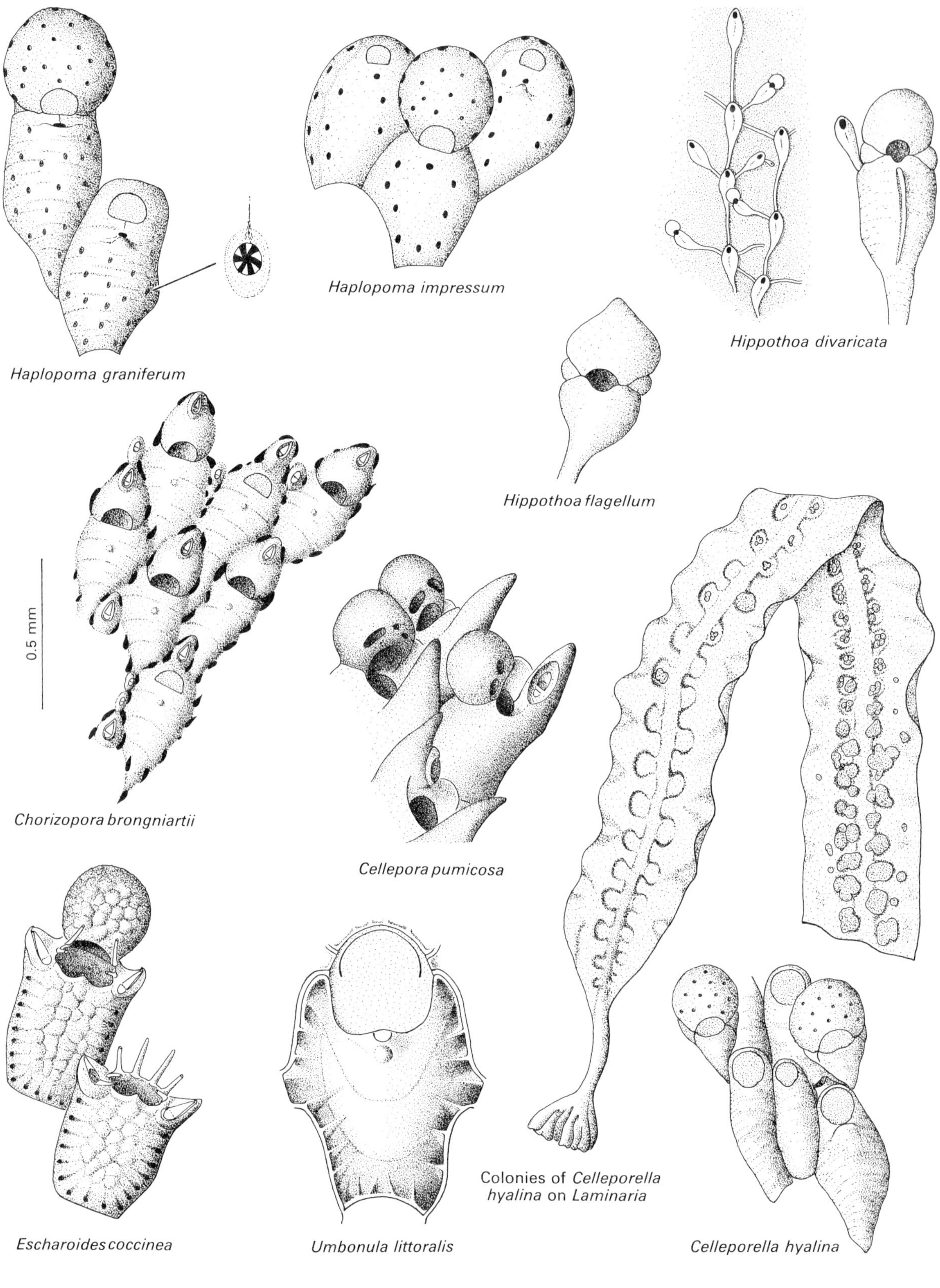
Haplopoma graniferum
Haplopoma impressum
Hippothoa divaricata
Hippothoa flagellum
0.5 mm
Chorizopora brongniartii
Cellepora pumicosa
Escharoides coccinea
Umbonula littoralis
Colonies of Celleporella hyalina on Laminaria
Celleporella hyalina

Fig. 14.10

remaining distinct. Orifice bearing four spines or fewer. Embryos not pink **2**

2. Orifice with four distal spines (sometimes two in presence of ovicell). A tall truncate mucro generally present. Embryos white ***Escharella ventricosa***

   Orifice with three spines or fewer. Mucro small, often bifid. Embryos orange ***Escharella variolosa***

Escharella immersa (Fleming) FIG. 14.11
[HAYWARD AND RYLAND 1979, 138]
Colonies forming small or quite extensive patches. Zooids hexagonal, varying in appearance with age. Frontal wall smooth and shining in young zooids; later covered in heavy calcification except for drop-shaped areolae marking the site of the marginal pseudopores; raised into a small pointed mucro at the orifice. Orifice suborbicular, longer than broad, with six spines distally (four in fertile zooids), and a small concave or bifid internal tooth. Ovicell globular; embryos pink.

On stones and shells; LWST and offshore; all coasts; common.

E. labiosa (Busk) is similar in size but has only four oral spines.

Escharella variolosa (Johnston) FIG. 14.11
[HAYWARD AND RYLAND 1979, 148]
Colonies forming white or yellowish patches. Zooids elongate-hexagonal, flat. Frontal wall granular, somewhat covered by additional calcification in older zooids; raised as a pointed or bifid mucro at the orifice. Orifice suborbicular, with two to five spines distally (two in fertile zooids), and a small square internal tooth proximally. Ovicells globular, with marginal pseudopores; embryos orange.

On stones and shells; offshore; all coasts; frequent.

Escharella ventricosa (Hassall) FIG. 14.11
[HAYWARD AND RYLAND 1979, 144]
Colonies silvery or white, forming patches. Zooids hexagonal. Frontal wall granular, convex, raised into a tall broad mucro. Orifice suborbicular, with four spines distally, and a forked internal tooth proximally. Ovicell globular; embryos white.

On stones and shells; offshore; all coasts; frequent.

E. labiosa (Busk) is smaller (zooids 0.4–0.5 mm long) and the peristomial mucro tapers to a point.

23 SMITTINIDAE

Colonies encrusting or erect and coralline. Zooids with frontal wall punctate or with marginal pseudopores only. Orifice often with spines distally and usually with condyles and a proximal internal tooth (*lyrula*). Avicularia present, often small and situated just above the tooth.

---

1. Colony erect, with flattened branches ***Porella compressa***

   Colony encrusting, rarely rising into foliaceous expansions **2**

2. Frontal wall perforate all over ***Smittina landsborovii***

   Frontal wall with marginal pores only **3**

3. Avicularium (occasionally two) placed to the side of the orifice ***Parasmittina trispinosa***

   Avicularium suboral (i.e. proximal to the orifice) **4**

4. Avicularian mandible pointed. Ovicell punctate ***Smittoidea reticulata***

   Avicularian mandible rounded. Ovicell imperforate ***Porella concinna***

Parasmittina trispinosa (Johnston) FIG. 14.11
[HAYWARD AND RYLAND 1979, 114]
Colonies forming orange-yellow crusts. Zooids ovate or elongate. Frontal surface often rather flat, granular, with marginal pseudopores. Orifice suborbicular, with two to four spines distally; an internal tooth and a notched raised lip proximally. Usually a large, distally directed avicularium to one side of the orifice, with acute mandible. Ovicell globular, its surface with a few large punctures; embryos orange.

On stones and shells; offshore; all coasts; common.

Porella compressa (Sowerby) FIG. 14.11
[HAYWARD AND RYLAND 1979, 116]
Colonies erect and branched, up to 70–80 mm, the branches flat, with jumbled zooids on both sides; buff-pink. Zooids ovate, convex, with marginal pseudopores. Orifice pyriform, wider distally, with a broad, shallow, median tooth proximally; a small suboral avicularium above the tooth. Ovicells globular, more or less immersed, smooth and imperforate; embryos pink-orange.

On stones and rocks; offshore; all coasts; locally frequent.

Porella concinna (Busk) FIG. 14.11
[HAYWARD AND RYLAND 1979, 118]
Colonies encrusting, forming reddish patches. Zooids ovate or rhomboid, varying in appearance with age. Frontal surface smooth, with marginal pseudopores, becoming covered by heavy calcification. Orifice wider distally, with two spines in young zooids; a variably developed, broad, shallow, median tooth proximally, and a small avicularium above it. Ovicells globular; embryos reddish-orange.

On stones and shells; offshore; all coasts; quite common.

Smittina landsborovii (Johnston) FIG. 14.11
[HAYWARD AND RYLAND 1979, 98]
Colonies encrusting, forming large reddish patches, often with a silvery sheen; occasionally erect and foliaceous. Zooids hexagonal or elongate, rather flat; the entire surface punctate. Orifice suborbicular, an internal tooth and an erect, notched lip proximally; a small median avicularium, with rounded mandible, in the notch. Ovicell globular, punctured; embryos red.

On stones and shells; LWST and offshore; all coasts; common.

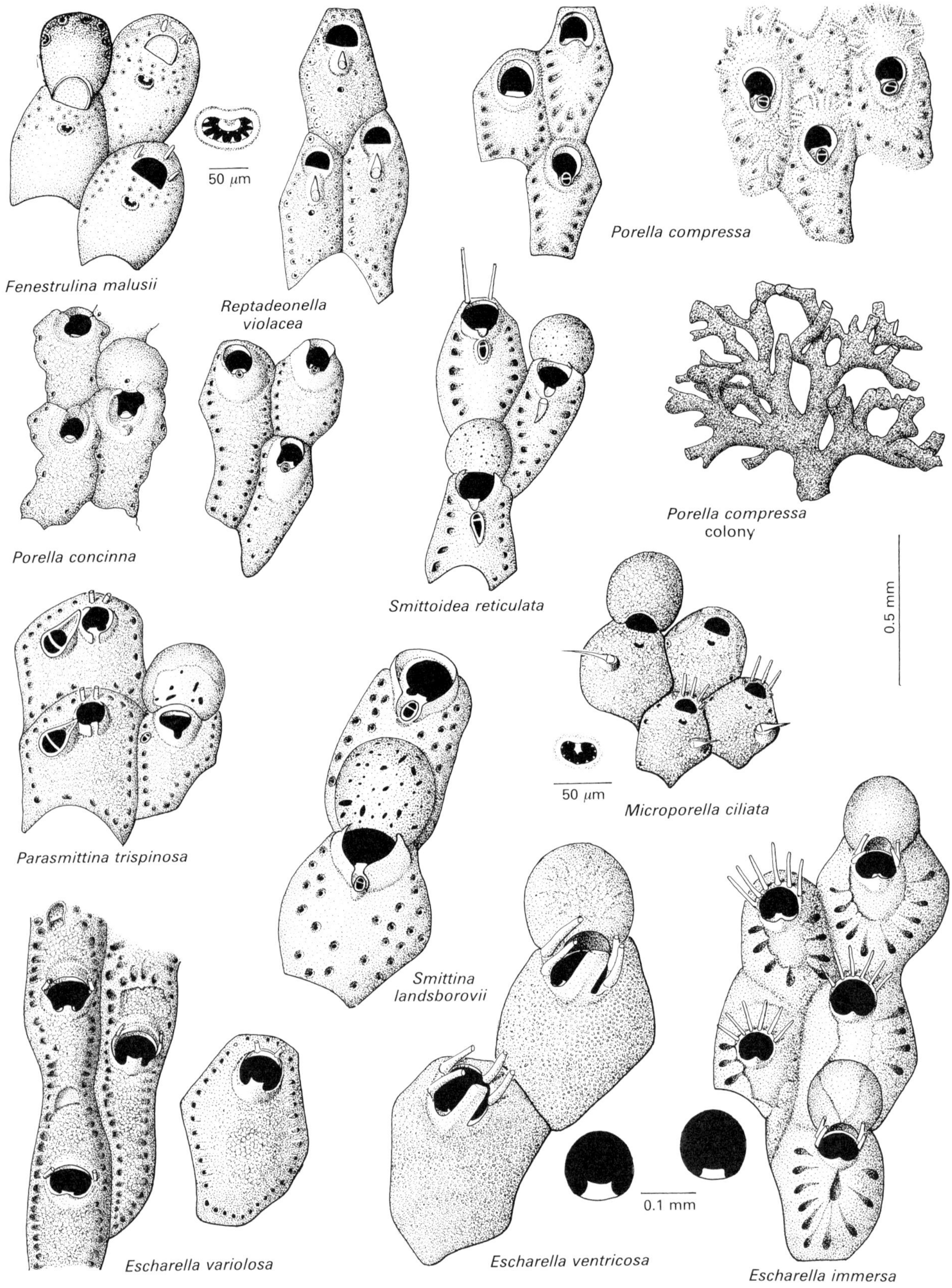
50 μm
Fenestrulina malusii
Reptadeonella violacea
Porella compressa
Porella compressa colony
Porella concinna
Smittoidea reticulata
0.5 mm
Parasmittina trispinosa
50 μm
Microporella ciliata
Smittina landsborovii
0.1 mm
Escharella variolosa
Escharella ventricosa
Escharella immersa

Fig. 14.11

Smittoidea reticulata (Macgillivray) FIG. 14.11
[HAYWARD AND RYLAND 1979, 108]
Colonies encrusting, silvery. Zooids ovate-elongate, slightly convex; frontal surface granular, with conspicuous marginal pseudopores. Orifice suborbicular, with three to four spines distally; an internal tooth and an erect, notched lip proximally. Avicularium suboral, large, with pointed, proximally directed mandible. Ovicells globular, punctured; embryos orange.

On stones and shells; offshore; all coasts; frequent.

24 SCHIZOPORELLIDAE

Colonies encrusting. Frontal wall of zooids usually punctate over the whole area. Orifice suborbicular, with a clearly defined notch (*sinus*) in its proximal border, delimited from the main part of the orifice by hinge teeth (*condyles*) for the operculum. Avicularia usually present. Ovicells opening above the orifice, often closed by the operculum.

---

1. Many zooids at least bearing a mamilla, below the orifice, to which is articulated a long, pointed spine **_Phaeostachys spinifera_**

   Frontal wall flat or umbonate, sometimes produced as a fingerlike projection upon which rests an avicularium, but never bearing a long, articulated spine **2**

2. Each zooid bearing one or two avicularia beside the orifice, each directed obliquely outwards. Ovicell imperforate **_Schizoporella unicornis_**

   Avicularia suboral or, if present in pairs, facing inwards. Ovicell punctured **3**

3. Avicularium small, suboral, with a rounded mandible **_Schizomavella auriculata_**

   Either a single avicularium with triangular mandible resting on a central prominence or directed up a fingerlike projection, or a pair of small lateral avicularia with inwardly directed mandibles **_Schizomavella linearis_**

Phaeostachys spinifera (Johnston) FIG. 14.12
[HAYWARD AND RYLAND 1979, 202]
Colonies forming small silvery or greyish patches. Zooids ovate to rhomboid, convex; the frontal wall lightly punctured. Orifice deeply D-shaped, with a well-developed sinus in the straight proximal margin. Five to seven distal spines, and a single, articulated spine proximal to the orifice. Frequently an avicularium, with slender pointed mandible directed outwards, situated beside the spine. Ovicell prominent, subglobular, imperforate, with fluted margin; embryos pinkish-red.

On *Laminaria* holdfasts, algal stems, and stones; lower shore; all coasts; fairly common.

Schizomavella auriculata (Hassall) FIG. 14.12
[HAYWARD AND RYLAND 1979, 176]
Colonies forming glistening, reddish patches. Zooids rhomboid, short; arranged in regular series radiating from the point of origin of the colony. Frontal surface flat, with pseudopores all over. Orifice suborbicular, the sinus shallow; two to four spines distal to the orifice, except in fertile zooids. A small avicularium, with rounded mandible, situated just proximal to the orifice, sometimes raised on a prominence. Ovicell subglobular, resting on the distal zooid, punctate, closed by the operculum. Embryos orange-red.

On stones and shells; offshore; all coasts; quite common.

Schizomavella linearis (Hassall) FIG. 14.12
[HAYWARD AND RYLAND 1979, 178]
Colonies forming glistening, reddish patches. Zooids rhomboid, short; arranged in regular series radiating from the point of origin of the colony. Frontal surface flat, with pseudopores all over. Orifice suborbicular, the sinus wedge-shaped; two to four spines distal to the orifice, except in fertile zooids. Either a single avicularium decumbent on a central prominence, with an upwards-directed, pointed mandible, or a pair of small avicularia, one each side of the orifice, pointing inwards. Ovicell subglobular resting on the distal zooid, punctate, closed by the operculum. Embryos orange-red.

On stones and shells; LWST and offshore; all coasts; common.

Schizoporella unicornis (Johnston in Wood) FIG. 14.12
[HAYWARD AND RYLAND 1979, 168]
Colonies forming whitish or pink patches. Zooids generally rectangular, arranged in regular series. Frontal wall slightly convex, with pseudopores all over; often a knob (umbo) proximal to the orifice. Orifice D-shaped, with a well-marked median sinus in the straight proximal edge. A raised avicularium on one or both sides of the orifice, tapering mandible directed obliquely outwards. Ovicell globular, prominent, resting on the distal zooid, imperforate, with a fluted margin, closed by the zooidal operculum. Embryos orange-red.

Usually on stones and under boulders, but on algae locally; lower shore; all coasts; common.

S. dunkeri (Reuss) is a similar but sublittoral species.

25 CELLEPORINIDAE

Colonies encrusting, nodular or erect. Zooids usually erect, superposed in disorderly manner. Orifice suborbicular, with or without a clearly defined notch (*sinus*) in its proximal edge; spines absent. Avicularia dimorphic, with large, scattered, interzooidal avicularia and small frontal avicularia mounted on one or two spiky prominences. Ovicells free-standing, punctate, opening above the orifice.

---

1. Orifice without a sinus. Ovicell with two or three pores close to aperture **_Cellepora pumicosa_**

   Orifice with U- or V-shaped sinus. Ovicell with scattered frontal pores **2**

2. A single, more or less suboral, avicularium on each zooid. Ovicells with a few large punctures **_Turbicellepora avicularis_**

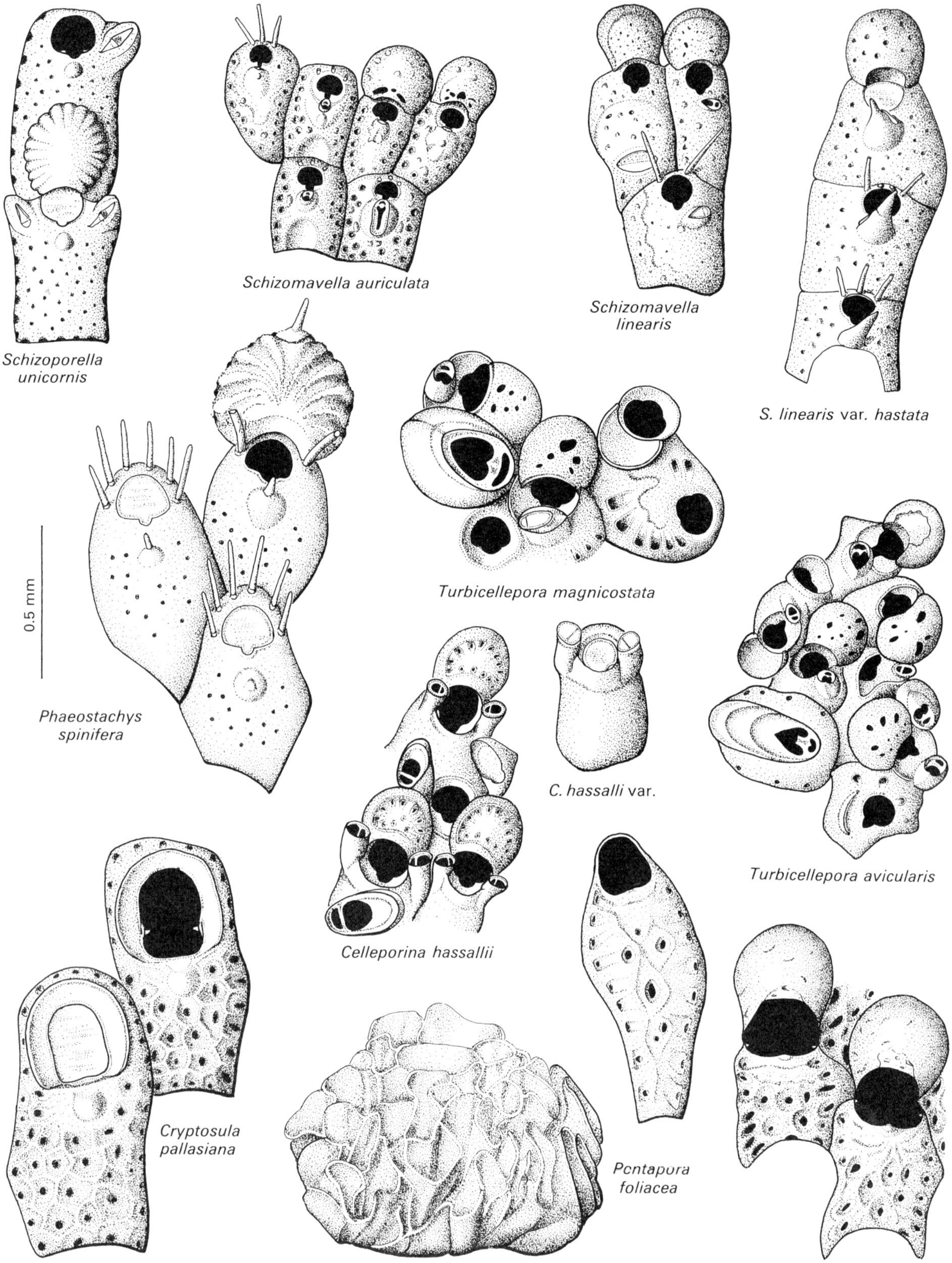

Schizoporella unicornis
Schizomavella auriculata
Schizomavella linearis
S. linearis var. hastata
Phaeostachys spinifera
0.5 mm
Turbicellepora magnicostata
C. hassalli var.
Turbicellepora avicularis
Celleporina hassallii
Cryptosula pallasiana
Pentapora foliacea

Fig. 14.12

Two avicularia, one each side of the orifice. Ovicell with a crescent of fine pores ***Celleporina hassallii***

Cellepora pumicosa (Pallas) FIG. 14.10

[HAYWARD AND RYLAND 1979, 274]

Colonies nodular or massive, of jumbled zooids in several layers. Zooids subcylindrical or ovoid, erect (except near the colony margin). Frontal wall convex, smooth, with marginal pseudopores; raised below the orifice as a slender, acuminate process. Orifice circular. Avicularia dimorphic: a small avicularium with pointed mandible on, and directed up, each suboral spike; and large avicularia, with rounded spatulate mandibles, scattered among the autozooids. Ovicells subglobular, smooth, imperforate or with a few large pores. Embryos orange-red.

Below rocks and on *Laminaria* holdfasts; lower shore and in shallow water; all coasts; quite common.

Celleporina hassallii (Johnston) FIG. 14.12

[HAYWARD AND RYLAND 1979, 276]

Colonies nodular or forming thick multilaminar crusts. Zooids erect, except at the growing edge of the colony. Frontal surface convex, smooth. Orifice suborbicular, with rounded sinus; flanked by a pair of columnar processes, each bearing a small terminal avicularium with a rounded mandible. Spatulate avicularia scattered among the autozooids. Ovicells subglobular, with a porous crescent on the upper surface. Embryos red.

A variety exists in which the columnar avicularia are linked by a tall rim along the proximal border of the orifice.

On stones, *Laminaria* holdfasts, larger hydroids, etc.; LWST and shallow water; all coasts; common.

Turbicellepora avicularis (Hincks) FIG. 14.12

[HAYWARD AND RYLAND 1979, 124]

Colonies nodular or forming thick multilaminar crusts. Zooids erect, except at the growing edge of the colony. Frontal surface convex, smooth. Orifice deeply D-shaped, with a wedge-shaped sinus. Proximal to the orifice, and slightly to one side of the sinus, a conical prominence bearing an avicularium with a pointed, upwards directed mandible. Spatulate avicularia scattered among the autozooids. Ovicells subglobular, with a few scattered punctures. Embryos deep yellow.

On stones, hydroids, etc.; LWST and shallow water; all coasts.

T. magnicostata (Barroso), found intertidally in the Isles of Scilly, has fewer, larger pores in the ovicell, a wider sinus to the orifice, and distinct marginal areolar ribs (Fig. 14.12).

26 HIPPOPORINIDAE

Colonies encrusting or foliaceous. Frontal wall of zooid with pseudopores all over or confined to a marginal series. Orifice usually longer than wide, arched distally, straighter proximally, with the operculum pivoting on pointed, lateral condyles. Avicularia usually present. Ovicells present or absent.

---

1. Colony encrusting. Ovicells never present ***Cryptosula pallasiana***

   Colony erect and foliaceous. Ovicells present ***Pentapora foliacea***

Cryptosula pallasiana (Moll) FIG. 14.12

[HAYWARD AND RYLAND 1979, 90]

Colonies encrusting, forming white or pinkish, more or less circular patches, up to about 30 mm across. Zooids hexagonal, with a thick frontal wall, porous over the whole area, each pseudopore sunk in a small pit. Orifice large, longer than broad, curved distally, narrowed a little below the middle; no spines, but usually a raised rim. Sometimes a small avicularium, with rounded mandible, situated just proximal to the orifice. No ovicells; the orange embryos develop inside the zooid.

Usually on stones, less often on algae and inert submerged structures; LWST and shallow water; all coasts; common.

Pentapora foliacea (Ellis and Solander) FIG. 14.12

[HAYWARD AND RYLAND 1979, 94]

Colonies erect, of bilamellar, strongly calcified, foliaceous, anastomosing fronds, forming orange-buff clumps generally up to about 150 mm across, but occasionally much larger. Young zooids elongate-hexagonal, with scattered pores in the frontal wall; becoming covered with heavy additional calcification, the outline becoming rectangular or broadly hexagonal. Orifice cordate. Sometimes a small avicularium, with rounded mandible, just proximal to the orifice. Ovicells at first globular, resting on the distal zooid, later immersed in additional calcification.

On shells and stones; offshore; south-west Britain; locally quite common.

Order **Ctenostomata**

27 ALCYONIDIIDAE

Colonies adnate, forming a network or a crust, or erect and lobate; fleshy or gelatinous. Zooids squat, contiguous or in series. Orifice circular, closed by a sphincter, often raised on a papilla. Some species intermittently possess a small, internally ciliated funnel, or *intertentacular organ*, between the dorso-medial pair of tentacles. Embryos developing in clusters in *gonozooids*.

---

1. Colony smothered in muddy particles; on hydroids ***Alcyonidium parasiticum***

   Colony not smothered in muddy particles; not on hydroids **2**

2. Colony in the form of free lobes. Surface not tuberculate ***Alcyonidium diaphanum***

   Colonies encrusting; rarely lobate, the surface then tuberculate **3**

3. Surface smooth	***Alcyonidium gelatinosum***
Surface tuberculate	***Alcyonidium hirsutum***

Alcyonidium diaphanum Lamouroux [*A. gelatinosum* of authors] FIG. 14.13
[HAYWARD 1985, 44]
Colonies lobate, digitiform, usually smooth, occasionally knobbly; firmly gelatinous; pale brown; commonly about 150 mm long, but attaining 500 mm. Orifices slightly raised; 15–17 tentacles; an intertentacular organ present at times.

Attached to stones and boulders, or detached; LWST and shallow water, may be washed up; all coasts; generally common, occasionally present sublittorally in great quantity. Apparently responsible for allergic dermatitis 'Dogger Bank Itch' in the North Sea.

Alcyonidium gelatinosum (Linnaeus) [Syn. *A. polyoum* (Hassall)] FIG. 14.13
[HAYWARD 1985, 52]
Colonies encrusting, forming glossy gelatinous patches on algae; rarely lobate; pale brown; the surface smooth. Interzooidal boundaries not visible. Orifices scarcely raised; 19–20 tentacles. Gonozooids containing small clusters of embryos, appearing yellowish-fawn, during autumn and winter.

On algae, especially *Fucus serratus*, and sometimes on stones; middle and lower shore; all coasts; common unless very exposed.

A. mytili Dalyell differs from *A. gelatinosum* by having visible interzooidal boundaries, 15–18 tentacles, and pink embryos; at certain times of year an intertentacular organ (never found in *gelatinosum*) is present. On *Mytilus* shells and perhaps on stones; LWST; widely distributed but far less common than *A. gelatinosum*.

Alcyonidium hirsutum (Fleming) FIG. 14.13
[HAYWARD 1985, 55]
Colonies encrusting or lobate, firmly gelatinous; lobes reaching about 100 mm. Surface matt, covered with minute, conical papillae; greyish brown. Orifices not raised; 16–20 tentacles. Gonozooids with conspicuous rings of white embryos from autumn to spring.

On algae, particularly *Fucus serratus* and *Gigartina stellata*; lower shore; all coasts; common.

Alcyonidium parasiticum (Fleming) FIG. 14.13
[HAYWARD 1985, 62]
Colonies encrusting hydroids, smothered in muddy particles. Zooids small, elongate, erect, hirsute; orifice terminal.

On hydroids such as *Hydrallmania*; offshore; all coasts; locally quite common.

28 FLUSTRELLIDRIDAE

Colonies encrusting or erect; gristly. Zooids squat; contiguous. Orifice closed by chitinous lips.

Flustrellidra hispida (Fabricius) FIG. 14.13
[HAYWARD 1985, 70]
Colonies encrusting, purplish-brown, hispid. Zooids large, with horny spines around the orifice, and sometimes around the zooid margin. Orifice bilabiate; polypides large, the expanded tentacle bells about 1 mm across. Embryos white, developing in summer.

On algae, especially *Fucus serratus* and *Gigartina stellata*; middle and lower shore; all coasts; common.

29 VICTORELLIDAE

Colonies creeping or loosely tufted. Zooids upright cylinders, not contiguous, linked by slender stolonate extensions from their bases. Daughter zooids arising from basal and upright portions.

Victorella pavida Kent FIG. 14.13
[HAYWARD 1985, 96]
Colonies consisting of chains of zooids connected by stolonic extensions from their bases; becoming very dense, like a sward. Zooids cylindrical above their bases, short at the edge of the colony but tall centrally, reaching 2–3 mm in height; budding additional zooids along their length; walls translucent; eight tentacles.

On reeds *Phragmites communis*, other plants, and submerged structures, in brackish water (1–35‰); very local but often abundant where it occurs.

V. pavida and the closely similar *V. muelleri* (Kraepelin) are distinguished by the position of the cardiac sphincter muscle. In the former it is situated midway between the oesophageo-cardiac valve and the point of merger with the caecum; in *V. muelleri* it demarcates the cardia from the caecum.

30 NOLELLIDAE

Colonies erect or creeping. Zooids upright cylinders, not contiguous; branching from their bases, with or without slender stolonic extensions. Zooid walls particle encrusted.

1. Colony erect, branching, forming a dense, grey-brown muddy tuft up to 100 mm high. Zooids cylindrical, budded directly from each other	***Anguinella palmata***
Colony repent, forming a cluster or sward of erect, cylindrical, sandy-coloured zooids about 2 mm in height attached to the substratum and spreading by stolonic extensions	***Nolella dilatata***

Anguinella palmata van Beneden FIG. 14.13
[HAYWARD 1985, 92]
Colonies large, to 100 mm or more, in form of erect stems with spirally arranged branches from which curving, cylindrical zooids are budded. Zooids to 1.5 mm, with 10–11 tentacles; the whole colony surface muddy with accreted particles. Unmistakable but easily overlooked on account of its muddy appearance.

On stones, shells, pier piles, and other solid structures; inter-

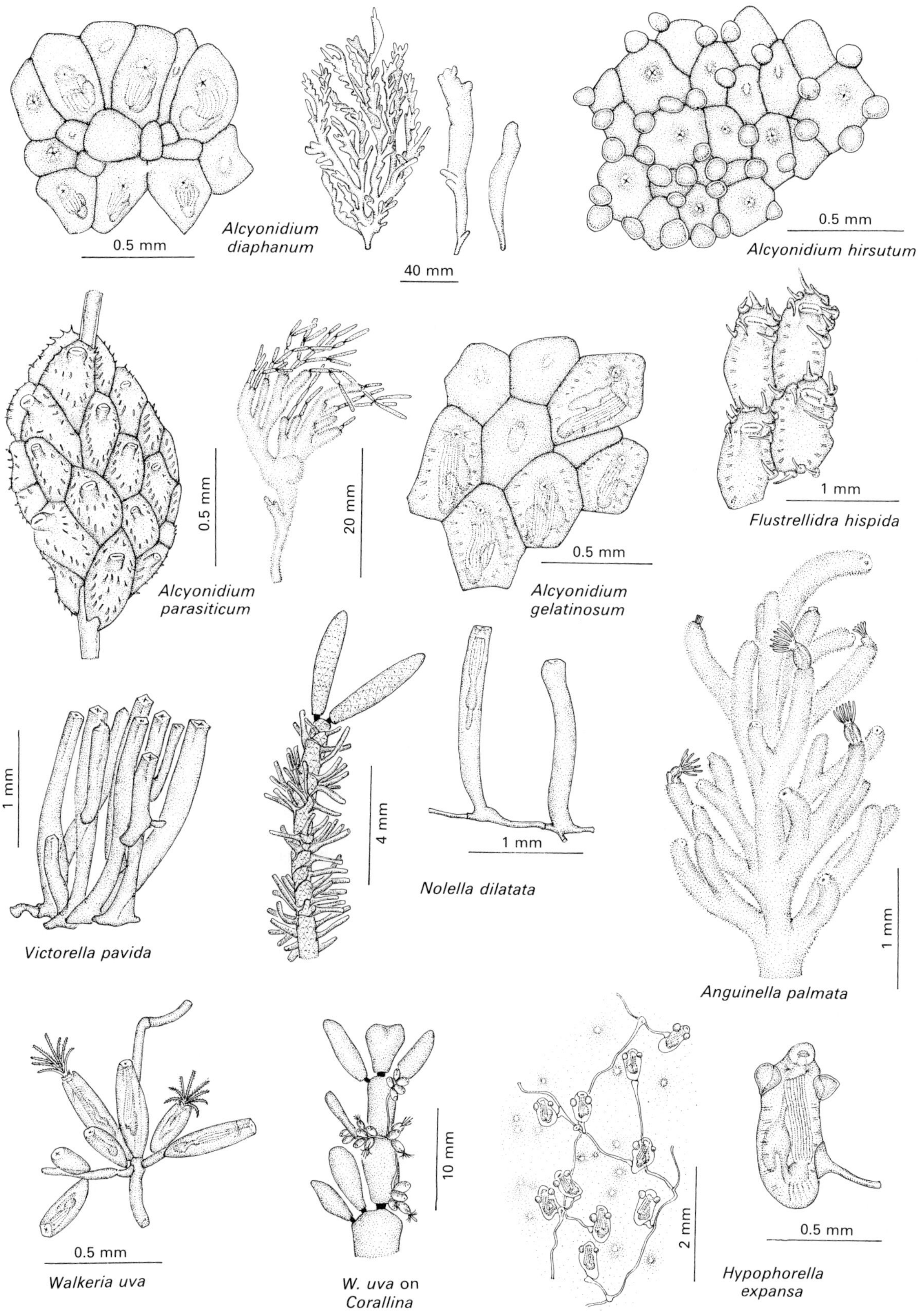
0.5 mm
Alcyonidium diaphanum
40 mm
0.5 mm
Alcyonidium hirsutum
0.5 mm
20 mm
0.5 mm
1 mm
Flustrellidra hispida
Alcyonidium parasiticum
Alcyonidium gelatinosum
1 mm
4 mm
1 mm
Nolella dilatata
1 mm
Victorella pavida
Anguinella palmata
10 mm
0.5 mm
Walkeria uva
W. uva on Corallina
2 mm
0.5 mm
Hypophorella expansa

Fig. 14.13

tidal in shelter or sublittoral, tolerating fluctuating salinities; southern Britain; probably quite common but biologically almost unknown.

Nolella dilatata (Hincks) FIG. 14.13
[HAYWARD 1985, 90]
Colonies consisting of a network of zooids connected by stolonic extensions from their bases. Zooids cylindrical above their bases, reaching about 1.5 mm in height; walls opaque, having the appearance of being sanded with minute particles. 16–20 tentacles.

On other bryozoans, hydroids, ascidians, shells, etc.; LWST and shallow water; all coasts; frequent.

31 WALKERIIDAE

Colonies stolonate, creeping or tufted. Stolons thin; autozooids not contiguous, grouped at the distal end of the stolonic segments (kenozooids). Autozooids cylindrical, the orifice terminal; eight tentacles, when everted not forming a perfect circle, two of them outcurved; no gizzard.

Walkeria uva (Linnaeus) FIG. 14.13
[HAYWARD 1985, 108]
Colonies creeping or tufted. Stolons very slender, branches typically arising in pairs at the nodes, in a small cluster of zooids. Zooids small, less than 0.5 mm in height. Embryos unpigmented.

Mainly on algae (creeping colonies are often found on *Corallina officinalis*; loose tufts may develop on *Halidrys siliquosa*); LWST and shallow water; all coasts; frequent.

32 HYPOPHORELLIDAE

Colonies stolonate, creeping; stolons thin; autozooids occurring singly, but each one bearing a pair of spherical abortive zooids.

Hypophorella expansa Ehlers FIG. 14.13
[HAYWARD 1985, 124]
Stolons slender, ramifying; zooids shortly cylindrical, adnate; orifice terminal, bilabiate, the upper lip toothed; 12–14 tentacles; no gizzard.

Commensal, living in the lining layers of tubes of *Chaetopterus variopedatus* and other polychaetes; usually present wherever there are substantial populations of the worms, often in a majority of tubes. The presence of *Hypophorella* can be detected if the *Chaetopterus* tubes are slit open and the lining layers stripped off in thin sheets; these can be examined under a microscope using transmitted light. Widely distributed.

33 TRITICELLIDAE

Colonies stolonate, creeping; zooids cylindrical, articulated on a slender peduncle (stolonic prolongation).

Triticella flava Dalyell FIG. 14.14
[HAYWARD 1985, 117]
Stolons adnate; autozooids fusiform, 0.5–1.0 mm in height, on peduncles one to four times their own height; often very dense.

Commensal on crustaceans, usually on the antennae, mouthparts, and pereopods, particularly on *Calocaris*, *Geryon*, *Goneplax*, *Nephrops*, and *Upogebia*. All coasts.

34 VESICULARIIDAE

Colonies stolonate, creeping or tufted; stolons bifurcating. Zooids cylindrical, contiguous or not, usually grouped; arising from a budding zone of variable length in each stolonic segment (kenozooid). Zooids with terminal orifice; eight to ten tentacles, which form a circle when everted; gizzard present.

---

1. Zooids fused together along their length, forming small groups **Amathia lendigera**
   Zooids remote or clustered, but not fused together **2**
2. Colony tufted, up to 300 mm in length, attached basally by rhizoids. Zooids in single series, generally more or less remote from each other **Vesicularia spinosa**
   Colony tufted or repent. Zooids in double series, generally clustered **3**
3. Zooid clusters disposed around the stolons in distinct helices. Colony a free-hanging, basally attached tuft **4**
   Zooid clusters not helical. Colony repent or forming dense clumps **5**
4. Living polypide pigmented a bright lemon-yellow **Bowerbankia citrina**
   Living polypides buff-coloured **Bowerbankia pustulosa**
5. Polypide with eight tentacles. Developing embryos pink or yellow. Stolon considerably narrower than a zooid. Zooids solitary, in pairs, or in small clusters; each often provided with a small proximal 'tail'. On various substrata **Bowerbankia gracilis**
   Polypide with ten tentacles. Developing embryos yellow. Stolon, at least when free from the substratum, about as wide as a zooid. Zooids often in dense clusters. Particularly on intertidal fucoids **Bowerbankia imbricata**

Amathia lendigera (Linnaeus) FIG. 14.14
[HAYWARD 1985, 134]
Colonies forming wiry tufts; the stolons about as wide as the zooids. Eight to 16 contiguous zooids in straight, biserial groups below bifurcations; the height in each group decreasing distally. Eight tentacles.

On algae, especially *Halidrys siliquosa*, and other substrata; LWST and shallow water; reaches its northern limit in Scotland; frequent.

Bowerbankia citrina (Hincks) FIG. 14.14
[HAYWARD 1985, 142]
Colonies forming flaccid tufts; the stolons about as wide as the zooids; zooids disjunct, in biserial, spirally disposed clusters

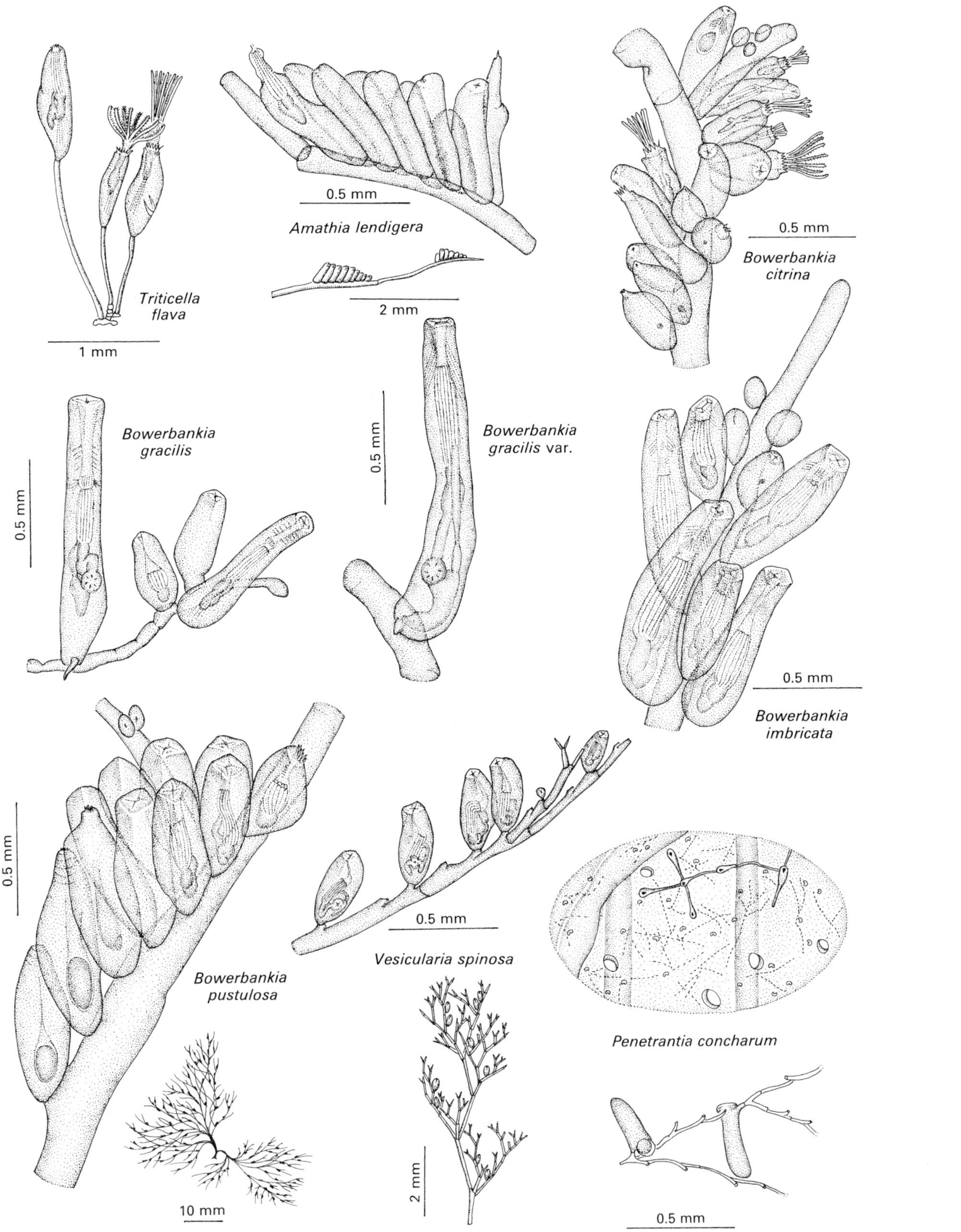
Triticella flava
1 mm
0.5 mm
Amathia lendigera
2 mm
0.5 mm
Bowerbankia citrina
Bowerbankia gracilis
0.5 mm
0.5 mm
Bowerbankia gracilis var.
0.5 mm
Bowerbankia imbricata
0.5 mm
Bowerbankia pustulosa
10 mm
0.5 mm
Vesicularia spinosa
2 mm
Penetrantia concharum
0.5 mm

Fig. 14.14

below bifurcations. Polypides (in life) bright lemon-yellow; eight tentacles.

On rocks and boulders; LWST and shallow water; southern coasts; local.

Bowerbankia gracilis Leidy FIG. 14.14
[HAYWARD 1985, 144]
Colonies adnate; the stolons less wide than the zooids; zooids disjunct, solitary, in pairs or in small clusters, 0.6–1.5 mm in height, often with a caudate process beside the point of origin from the stolon. The colony sometimes very dense, like a pile or sward. Eight tentacles. Two species seem currently included under this name: one has pink embryos, the other yellow.

On any submerged substratum, in fully saline or brackish water (down to 15‰); LWST and offshore; all coasts; common, but rarely conspicuous.

Bowerbankia imbricata (Adams) FIG. 14.14
[HAYWARD 1985, 139]
Colonies adnate or tufted; the stolons slender when adnate, but wider than the zooids when free; zooids disjunct, in straight, biserial clusters; 0.8–1.0 mm (occasionally to 1.5 mm) in height; ten tentacles; embryos yellow.

On intertidal fucoids, especially *Ascophyllum nodosum* and *Fucus vesiculosus*; mainly intertidal; all coasts; common.

Bowerbankia pustulosa (Ellis and Solander) FIG. 14.14
[HAYWARD 1985, 150]
Colonies forming flaccid tufts; the stolons about as wide as the zooids; zooids disjunct, in biserial, spirally disposed clusters below bifurcations. The polypides unpigmented; eight tentacles; embryos yellow.

On rocks and boulders; LWST and shallow water; all coasts; frequent.

Vesicularia spinosa (Linnaeus) FIG. 14.14
[HAYWARD 1985, 130]
Colonies tufted; the stolons much branched, tapering to spinelike points terminally; zooids ovoid to fusiform, disjunct, in single series, more or less remote from each other, irregularly placed on the internodes; eight tentacles.

On stones and shells; in shallow waters; all coasts; locally common.

35 PENETRANTIIDAE
Colonies stolonate, boring (by means of phosphoric acid) in shells. Autozooids cylindrical, never clustered, borne on short, lateral to distally subterminal peduncles; the orifice terminal, closed by an operculum; 10–14 tentacles; muscular cardia, sometimes termed a gizzard, present; embryos developing in gonozooids.

Despite the presence of stolons it is far from certain that the Penetrantiidae are correctly placed in the Ctenostomata; the so-called gizzard lacks teeth and is not like that of Vesiculariidae. They may be a highly aberrant family of Cheilostomata.

Penetrantia concharum Silén FIG. 14.14
[HAYWARD 1985, 128]
Stolons slender, branching, bearing zooids on both sides; autozooids cylindrical, attached to the stolon by a short subterminal stalk, suberect (at right angles to the shell surface). Perforations in the shell reniform.

In dead gastropod and lamellibranch shells from shallow water; widespread; locally abundant (up to 90 per cent of shells being bored).

PHYLUM BRACHIOPODA

Brachiopods (Fig. 14.15), like other lophophorates, are sessile and benthic, but they are readily recognized by their bivalved shell and distinctive internal structure. The two *valves* of the shell are morphologically dorsal and ventral, not lateral as in the Lamellibranchia. In most (class Articulata) the valves are hinged together, though not held by a ligament; the ventral valve projects behind the hinge as a projection or *umbo*, which is deeply indented or perforated by an aperture, the *foramen*. A stalk or *pedicle* projects between the valves or through the foramen and attaches the animal to the substratum: *Crania* has no pedicle. The living brachiopod is not necessarily orientated dorso-ventrally; in articulates the ventral valve often faces upwards. The shell valves are lined and secreted by the *mantle lobes*; they often contain internally originating channels or *punctae*; the valves open somewhat for feeding, the more widely gaping end being anterior. The median, exhalant region of the dorsal valve may be indented (*sulcate*). The lophophore, which occupies much of the mantle cavity, is basically horseshoe-shaped, though the arms (*brachia*) may be long and variously coiled. The tentacles (or *filaments*) are long and ciliated. There are two classes.

Class I **Inarticulata (Ecardines)**

The shell valves are not hinged but held together by muscles. The lophophore lacks skeletal support, and the alimentary canal is U-formed, with an anus. *Lingula* (not British) and *Crania* belong to this class.

Class II **Articulata (Testicardines)**

The shell valves are dissimilar and hinged together by a tooth and socket arrangement. The foramen may be open and bounded anteriorly by the margin of the dorsal valve, or it may be closed by a pair of small *deltidial plates*. Within the hinge axis, the inner surface of the ventral (or pedicle) valve is produced into a pair of *hinge teeth*. The dorsal (or brachial) valve has no pronounced umbo, but is produced internally at its posterior midpoint into a *cardinal process*, flanked by sockets, which fits between the hinge teeth. The inner surface of the dorsal valve may bear a ridge or *median septum*. The lophophore is supported internally by skeletal projections, the *crura* (sing. *crus*), arising from the dorsal valve near and on either side of the cardinal process. This support may extend anteriorly as a *brachial loop* of varying length and complexity. The alimentary canal is looped, but lacks an anus. Most extant brachiopods belong to this class.

Brachiopods are most usually found on stones dredged from deep water, e.g. on the continental slope. Rather few species have been reported from shallow water, and one of these (*Gwynia capsula*) is also found living in shell-gravel; *Terebratulina retusa* is probably the commonest inshore. Cori (1933) has given an account of the commoner European species, and almost all are described and well-illustrated in the older monograph by Davidson (1886–8). A recent but inconsistent key is by Brunton and Curry (1979); there are descriptions of all British species but disappointing illustrations.

LITERATURE

Brunton, C. H. C. and Curry, G. B. (1979). British Brachiopods. *Synopses of the British Fauna*, N.S., **17**, 1–64.

Cori, C. I. (1933). Brachiopoda. *Tierwelt der Nord- und Ostsee*, **7** (3), 133–50.

Davidson, T. (1886–8). A monograph of the Recent Brachiopods. *Transactions of the Linnean Society of London*, Ser. 2, **4**, 1–248.

KEY TO SPECIES

The shell valves of articulates must be forced apart, with a scalpel blade, to break the tooth and socket articulation. To see the brachial loop and punctae place the dorsal valve in dilute bleach, *after noting the form of the lophophore*, for the shortest effective time.

1. Pedicle absent; animal attached to the substratum by the flatter, ventral valve (INARTICULATA) — ***Crania anomala***

   Pedicle, by which the animal is attached to the substratum, present (ARTICULATA) — **2**

2. Articulating margin of dorsal valve straight or slightly sinuate — **3**

   Articulating margin of dorsal valve convex or tapered — **6**

3. Brachial loop well-developed, attached to the dorsal valve at three to five submarginal points; lophophore with four lobes — ***Megathiris detruncata*** (p. 838)

   Brachial loop rudimentary or, if well-developed, attached only to the median septum; lophophore variously shaped — **4**

4. Shell longitudinally oval; brachial loop rudimentary, no median septum — ***Gwynia capsula***

   Shell transversely oval; brachial loop attached to the median septum — **5**

5. Small, not exceeding 3 mm in length; valves smooth — ***Argyrotheca cistellula***

   Adults reaching 20 mm; valves with fine radial ribs — ***Megerlia truncata*** (p. 838)

6. Shell without punctae; lophophoral support (crura) without a loop; lophophore with two lateral spirals — ***Hemithiris psittacea***

   Shell with punctae (if these are not immediately visible externally, view the shell under a dissecting microscope by transmitted light); brachial loop present; lophophore with long lateral lobes and a median coil — **7**

7. Brachial loop less than half the length of the dorsal valve — ***Terebratulina retusa***

   Brachial loop exceeding half the length of the dorsal valve — **8**

8. Shell outline almost triangular; anterior of shell sulcate; median septum present — ***Dallina septigera*** (p. 838)

   Shell outline ovate, non-sulcate; no median septum — ***Macandrevia cranium*** (p. 838)

Class Inarticulata

Order Acrotretida

Dorsal shell valve larger than ventral; lophophore spirolophous (two lateral spirals); epifaunal.

1 CRANIIDAE

Shell hingeless; dorsal valve conical, ventral valve convex or conforming to the shape of the surface to which it is attached; no pedicle. Could easily be mistaken at first sight for a limpet or saddle-oyster, but separation of the valves will display the lophophore.

Crania anomala (Müller) FIG. 14.15
[DAVIDSON 1888, 183]
Dorsal valve of rounded outline or squarely orbicular, the posterior margin almost straight, the remainder convex. In profile conical; the apex subcentral or near the posterior margin, obtuse or acute; with concentric growth lines. Ventral valve usually obscured by the dorsal; convex and in part attached to another object, or flat, thin, with its external face moulded and cemented to the substratum. Dorsal valve reddish-brown; to 14 mm in length. Valves gaping in life, with approximately 60 non-retractile setae projecting in each anterio-lateral crescent. Brachia curled.

On stones and shells (e.g. the concave side of scallop valves), often in clusters; hard grounds in shallow (15 m) to moderately deep water where currents may be strong; coasts of Ireland, Scotland and Isle of Man; also Brittany; Arctic to the Canary Is.

Class Articulata

Order Rhynchonellida

Shell without punctae. Lophophore spirolophous (lateral spirals and no median coil), partly supported by the crura.

2 HEMITHIRIDIDAE

Dorsal valve with low median ridge; pedicle short and broad. The spelling Hemithyrididae is not permissible [Article 35 (d) (i) of the 1985 *Code of Zoological Nomenclature*].

Hemithiris psittacea (Gmelin) FIG. 14.15
[DAVIDSON 1887, 163, as *Rhynchonella*]
Shell ovate, with fine ribs and concentric growth lines; grey, brown, or blackish; to about 25 mm. Ventral umbo produced as a curved beak. Shell without punctae; no brachial loop. Lophophore spirolophous, the two spirals projecting dorsally. Pedicle short.

Deeper than 15 m in northern waters; mainly Arctic in distribution.

Order Terebratulida

Shell light-coloured; punctate. Pedicle variably developed.

3 CANCELLOTHYRIDIDAE

Both valves convex, capillate (striated). Brachial loop simple; median septum usually absent. Lophophore plectolophous (i.e. with lateral lobes and a dorsally directed median coil).

Terebratulina retusa (Linnaeus) FIG. 14.15
[DAVIDSON 1886, 17, as *T. caput-serpentis*]
Shell ovate, almost pear-shaped, longer than wide, broadest about the middle; rounded, straight or slightly indented anteriorly. Dorsal valve uniformly convex, tapering posteriorly; ventral valve somewhat deeper, convex, the beak rather short; foramen incomplete, margined laterally by two small deltidial plates and anteriorly by the umbo of the dorsal valve; surface covered with fine, radiating striations (approx. 12 in a 5 mm width at the midpoint of the dorsal valve). Shell whitish, sometimes tinged with yellow or rust; to 33 × 25 mm. No median septum. Brachial loop at first a simple U, becoming annular through the union of the oral processes. Lophophore horseshoe-shaped, with a medial coil.

On stones, *Modiolus*, and hydroids; often from fairly shallow (but not less than 15 m) inshore waters, especially vertical faces below 30 m; may be abundant at 80–200 m; absent from east coast; Scandinavia to Mediterranean.

4 MEGATHIRIDIDAE

Shell biconvex; posterior margin of dorsal valve straight or nearly so (*strophic*). Brachial loop composed of two descending branches converging anteriorly to join the end of the median septum, or absent. Lophophore lacking a median coil. Spelling Megathyrididae is not permissible [Article 35 (d) (i) of the 1985 *Code of Zoological Nomenclature*].

Argyrotheca cistellula (S. Wood) FIG. 14.15
[DAVIDSON 1887, 139]
Shell small, squarish in outline or tapering towards the pedicle, as wide or wider than long; to 2 × 3 mm, tawny or yellowish. Dorsal valve D-shaped, with the posterior margin straight; moderately convex. Ventral valve deeper, with a straight umbo; foramen incomplete, margined laterally by small deltidial plates and anteriorly by the umbo of the dorsal valve; surface with concentric growth lines and rather large punctae. Median septum present; the brachial loop comprising two semicircles from the hinge to the anterior end of the septum. Lophophore bilobed.

On stones and shells, sometimes in clusters; in shallow (20 m) to moderately deep water; all coasts except eastern England but infrequently recorded; Norway to the Mediterranean.

Gwynia capsula (Jeffreys) FIG. 14.15
[DAVIDSON 1887, 150]
Shell minute, pouch-shaped, rounded anteriorly, longer than wide, broadest anteriorly; to 1 × 0.5 mm; white, glossy, with large and well-spaced punctae. Dorsal valve convex, its posterior margin nearly straight. Ventral valve as deep, convex, with a slight umbo; foramen incomplete. Brachial support rudi-

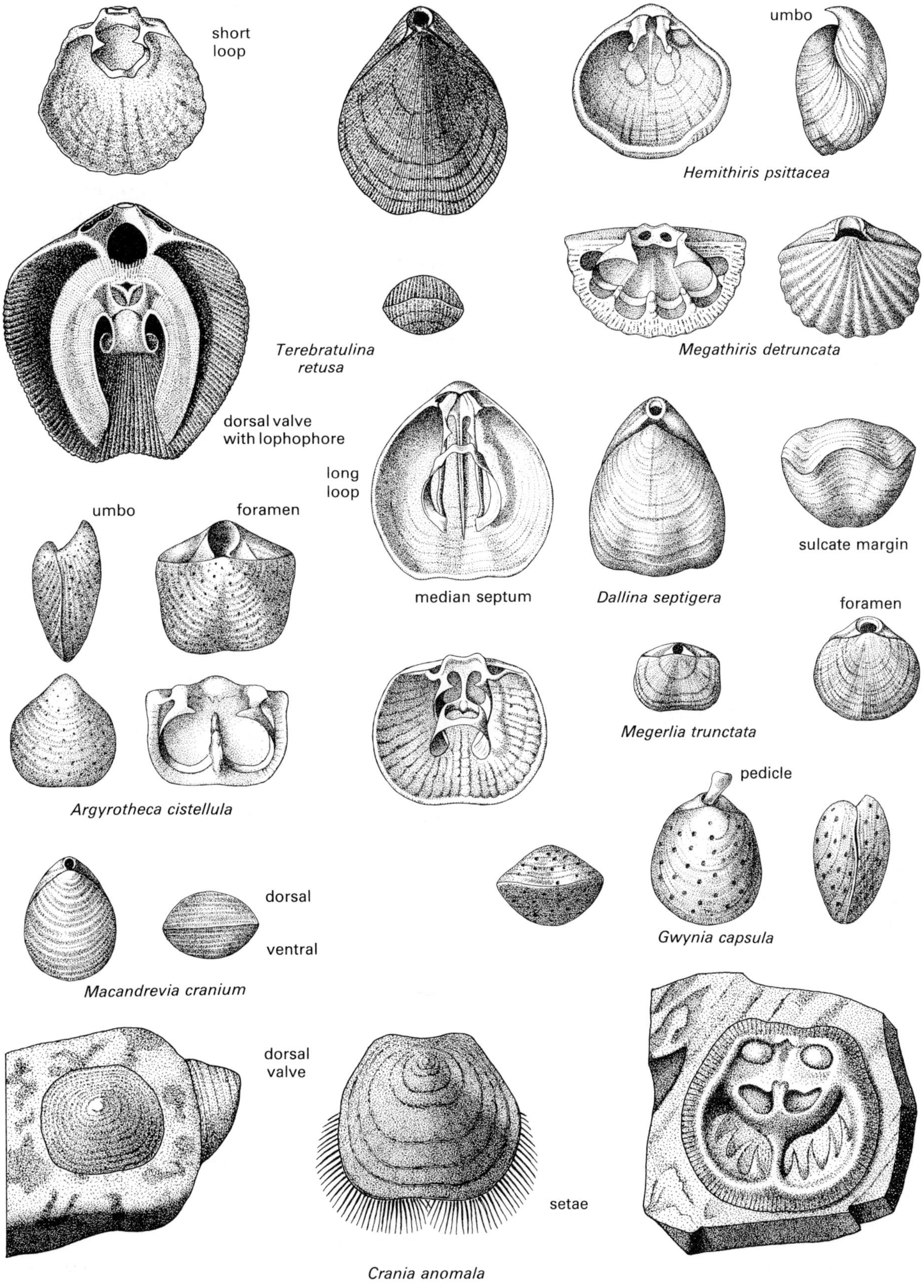
short
loop
umbo
Hemithiris psittacea
Terebratulina
retusa
Megathiris detruncata
dorsal valve
with lophophore
long
loop
median septum
Dallina septigera
sulcate margin
umbo
foramen
foramen
Megerlia truncata
Argyrotheca cistellula
pedicle
dorsal
ventral
Gwynia capsula
Macandrevia cranium
dorsal
valve
setae
Crania anomala

Fig. 14.15

mentary; no median septum. Lophophore circular (with an anterio-median break), attached to the dorsal mantle.

On stones and shells (often with *Argyrotheca cistellula*) on the bottom, or attached to shell-fragments and inside broken serpulid tubes in shell-gravel; intertidal (on rocks embedded in sand, Jersey) and from shallow water; widely distributed.

Megathiris detruncata (Gmelin) FIG. 14.15
[DAVIDSON 1887, 128, as *Argiope decollata*]
Shell transversely oval or diamond-shaped; to about 5 × 7 mm; ventral umbo broad and prominent; posterior margin of dorsal valve straight; broadly ribbed. Foramen large. Brachial loop lobed, attached submarginally to three to five septa on the dorsal valve. Lophophore with four lobes (no coil or spirals).

From depths in excess of 37 m: Channel Is., Brittany, Western Approaches, and Mediterranean.

5 KRAUSSINIDAE

Shell form variable. Brachial loop with a central (septal) ring and descending branches; dorsal median septum present.

Megerlia truncata (Linnaeus) FIG. 14.15
[DAVIDSON 1887, 103]
Shell outline rounded when small, becoming wider than long; white to yellowish or reddish, up to about 20 mm long. Dorsal valve margin straight. Foramen large and circular. Valves with four radial ribs and concentric growth lines. Brachial loop attached to dorsal valve in three places: base of the hinge plate, to the short median septum, and by the lateral branches curving to the anterior extremity of the septum. Lophophore with large lateral lobes and a median coil.

Found with *Terebratulina retusa* on vertical walls, 27–55 m off Brittany; also Bay of Biscay, Western Approaches, and the Mediterranean; 80–200 m.

6 DALLINIDAE

Shell biconvex, smooth. Posterior margin of dorsal valve tapered (*non-strophic*). Brachial loop long. Median septum distinct, as long or longer than the loop; cardinal process small or absent.

Dallina septigera (Lovén) FIG. 14.15
[DAVIDSON 1886, 56, as *Waldheimia*]
Shell distinctively triangular, widest anteriorly, to about 35 mm; externally smooth apart from the growth lines, white or yellowish; broadly and deeply sulcate. Pedicle foramen circular, at the apex of the shell; deltidial plates meeting in the mid-line. Median septum long and prominent; loop long, simple, reflected ventrally. Lophophore with long lateral lobes and a median spiral.

A deep water (below 200 m) species; Scotland, North Sea, and down the western continental edge; northern Norway to Canary Is.

7 LAQUEIDAE

Shell biconvex, smooth or ribbed. Posterior margin of dorsal valve convex. Brachial loop long; median septum weak, shorter than the adult loop. Cardinal process small or absent

Macandrevia cranium (Müller) FIG. 14.15
[DAVIDSON 1886, 61, as *Waldheimia*)
Shell outline oval; to about 30 mm long; white, yellowish, or grey; smooth except for concentric growth lines. Valves not sulcate anteriorly. Pedicle foramen with small deltidial plates that do not join in the mid-line. Loop long, simple, reflected ventrally; no median septum. Lophophore with long lateral lobes and a median spiral.

Sublittoral; from 10 m in parts of its range; Shetlands, along western coasts, Western Approaches; Arctic to Mediterranean.

15 ECHINODERMATA

PHYLUM ECHINODERMATA

MOST echinoderms have a distinctive radial pattern which may take the form of five *arms* radiating from a central *disk* (Fig. 15.1) or a more globular or cylindrical shape, again with structures arranged in five rows (or multiples of five) (Fig. 15.2). A few exceptional types have more than five *radii* and in others a bilateral shape has been superimposed on the basic radial plan. Compared to many other groups of marine invertebrates, echinoderms are large animals and, with rare exceptions, they are benthic (bottom living) and very slow moving. They are characterized by a number of unique internal organ systems and external appendages, the detailed form of the latter being used in precise identification of species. Up-to-date accounts of the general biology of the group are given in several good textbooks, some of which are listed in the bibliography. For the present purposes it is necessary to mention only selected characters.

Typically, echinoderms have a superficial layer of calcareous *skeletal plates*. These may be joined together to form a complete shell or '*test*' as in the sea urchins; more usually they are separated by intervening flexible integument as in the starfishes. Such calcareous plates or *ossicles* have a lattice structure with a very efficient strength to weight ratio and form the rigid elements of *spines* and other structures in addition to the skeleton proper. The detailed shapes and naming of these plates and ossicles vary from class to class and are described in appropriate places below.

Running along the length of each of the five rays of an echinoderm there are usually two rows of mobile little structures called *tube feet*. These usually terminate in a circular *sucker-disc*, about a millimetre or so across, used for attaching to a surface. There are often hundreds of tube feet to one ray, and in the starfish and some other groups they carry out stepping actions, all out-of-step but stepping in the same direction at a given time, to produce the locomotion of the whole animal. For this reason the ray with its complement of tube feet is referred to as an *ambulacrum* or walking surface. Associated skeletal plates are referred to as *ambulacral plates*; the spaces and associated plates between the rays are referred to as *interambulacral*. It should be noted that in the urchins and sea cucumbers the ambulacra and interambulacra run all the way from the oral to the anal end of the animal. In the stars however they are confined to one surface of the animal—the *adoral* or *adambulacral* surface; the other surface, lacking ambulacra, is referred to as the *aboral* surface. In addition to the spines and tube feet already mentioned, there is a variety of other surface structures of which we need be concerned with only two. The first are *papulae* or gills. These are most obvious in live specimens and take the form of thin-walled protuberances, often very numerous, over the aboral surface of starfish for instance. Secondly there are the unique structures called *pedicellariae* found in asteroids and echinoids. These are minute jawed structures used for removing trash from the surface of the animal. Sometimes they are set on long mobile stalks and are three-jawed, others are sessile and scissor-like. Some species have four or five different types. One type has poison glands for dealing with small intruding crustacea etc.

It is important to realize that internally the echinoderms have another unique system—the hydraulic *water vascular system* which controls the action of the tube feet. It consists of a ring-main around the throat and a radiating vessel internal to each ambulacrum. There is also a replenishment vessel (*stone canal*) running to the ring canal from the surface of the animal where a special structure, the sieve plate or *madreporite*, serves to monitor and control the intake of sea-water.

The echinoderm alimentary canal varies considerably in different classes. Most echinoderms take in only small particles; some partially digest their prey externally. The brittle stars have no anus.

It is impossible to give general instructions for the collection of echinoderms since they occupy such a diversity of habitats in which they are frequently quite obvious. A few points can be made. They are fully marine and barely penetrate estuaries. Intertidally they avoid desiccation by being confined to rock-pools and under boulders. Some small species of brittle stars and asteroids only become evident when algal clumps from rock-pools are sorted through. Collecting burrowing forms in intertidal sands requires careful digging as for other infauna. Diving, when possible, is a better method than dredging or trawling for sublittoral specimens since it is much less likely to result in damage. Nevetheless most echinoderms are quite tough and only a few of the brittle star species spontaneously shed their arms. Care needs to be taken when narcotizing. Gradual introduction of fresh water is effective but can lead to osmotic distortion. Otherwise any of the common agents can be used. Many of the more highly ossified species are best preserved dry for most purposes. It is sometimes possible to dry out quite large starfish specimens in the sun without the body wall collapsing. Urchins need to be drained of internal fluids; with care their spines can be preserved in life-like position. Wet preservation of echinoderms should always be in an excess of 70% ethanol.

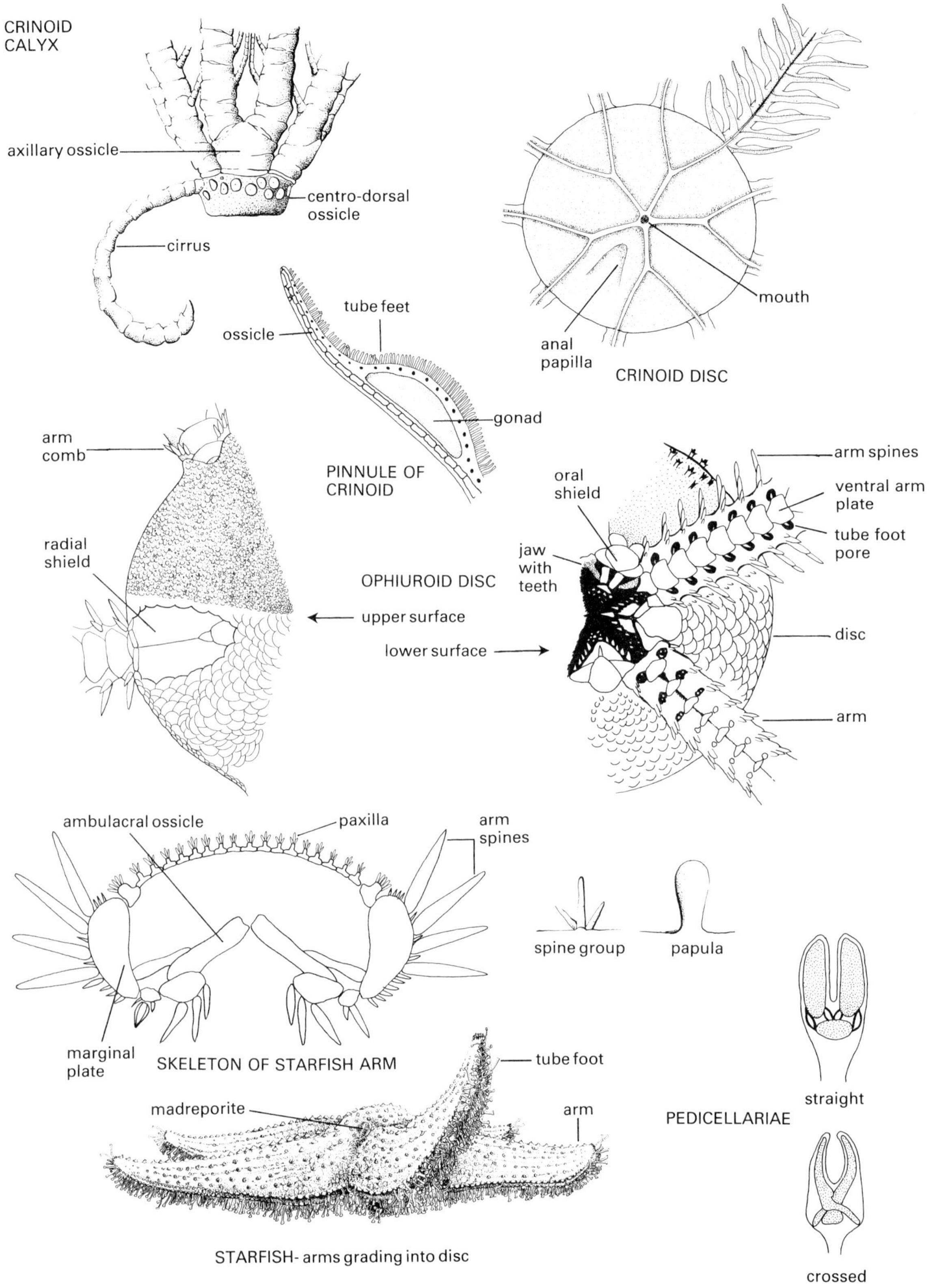
CRINOID
CALYX
axillary ossicle
centro-dorsal
ossicle
cirrus
tube feet
ossicle
gonad
PINNULE OF
CRINOID
mouth
anal
papilla
CRINOID DISC
arm
comb
radial
shield
OPHIUROID DISC
upper surface
lower surface
oral
shield
jaw
with
teeth
arm spines
ventral arm
plate
tube foot
pore
disc
arm
ambulacral ossicle
paxilla
arm
spines
spine group
papula
marginal
plate
SKELETON OF STARFISH ARM
tube foot
madreporite
arm
straight
PEDICELLARIAE
crossed
STARFISH- arms grading into disc

Fig. 15.1

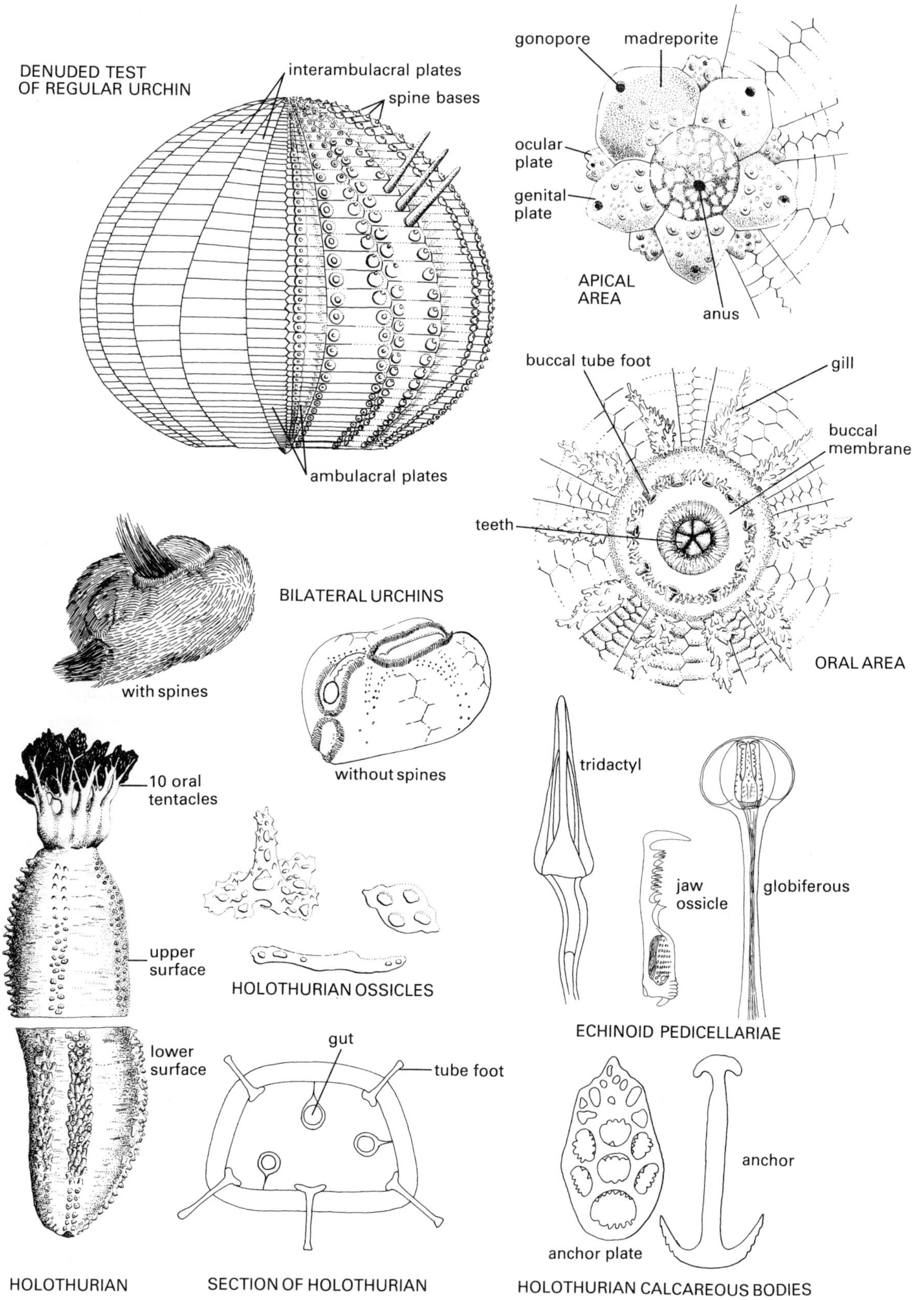
DENUDED TEST
OF REGULAR URCHIN
interambulacral plates
spine bases
ambulacral plates
gonopore
madreporite
ocular plate
genital plate
APICAL AREA
anus
buccal tube foot
gill
buccal membrane
teeth
ORAL AREA
BILATERAL URCHINS
with spines
without spines
10 oral tentacles
upper surface
lower surface
HOLOTHURIAN OSSICLES
tridactyl
jaw ossicle
globiferous
ECHINOID PEDICELLARIAE
gut
tube foot
anchor
anchor plate
HOLOTHURIAN
SECTION OF HOLOTHURIAN
HOLOTHURIAN CALCAREOUS BODIES

Fig. 15.2

This should be replaced after the first few days because of dilution by body fluids. Alcohol is preferred because formalin so readily turns acid and dissolves the calcareous ossicles and plates. This is disastrous in the case of those holothurians whose accurate identification depends on examination of the small body wall ossicles.

CLASSIFICATION

Class CRINOIDEA: 'Feather Stars'. Distinct central disc with apparently ten feathery arms curled upwards and numerous hooked cirri below for attachment.

Class ASTEROIDEA: 'Starfishes'. Five (or rarely more) heavy arms radiating from a centre, but no distinct central disc.

Class OPHIUROIDEA: 'Brittle Stars'. Five narrow arms radiating from a distinct central disc; usually small.

Class ECHINOIDEA: 'Sea Urchins'. Globular or more or less flattened and disc-shaped; usually covered in spines.

Class HOLOTHURIOIDEA: 'Sea Cucumbers'. Elongated and nonrigid; ten large, often branched, oral tentacles.

REFERENCES

Barnes, R. D. (1982). *Invertebrate Zoology*, 4th edn. Holt-Saunders.

Boolootian, R. A. (1966). *Physiology of Echinodermata*. Wiley.

Binyon, J. (1972). *Physiology of Echinoderms*. Pergamon.

Clark, A. M. (1977). *Starfishes and related Echinoderms*. British Museum (Natural History).

Emson, R. H. and Crump, R. G. (1979). Description of a new species of *Asterina* (Asteroidea), with an account of its ecology. *Journal of the Marine Biological Association of the U.K.*, **59**, 1, 77–94.

Hyman, L.H. 1955. *The Invertebrates*. Vol. 4, *Echinodermeta*. McGraw-Hill.

Lawrence, J. M. (ed.) (1981). Echinoderms. *Proceedings of the International Conference, Tampa Bay*. Balkema.

Mortensen, Th. (1927). *Handbook of the Echinoderms of the British Isles*. Oxford University Press (also reprint edition available).

Nichols, D. (1969). *Echinoderms*. Hutchinson University Library.

A. Class **Crinoidea**

Echinoderms with a cup-shaped body and ten feathery arms (Fig. 15.1). The mouth and *ambulacral grooves* are on the upper surfaces of these. Some crinoids, the 'sea-lilies', live permanently attached to the substratum by a stalk and rootlets; other forms, the comatulids, attach temporarily by prehensile *cirri* which arise from their lower surfaces. The stalked crinoids do not occur in shallow British waters.

The outer part of the crinoid body, the *calyx*, is formed by symmetrically arranged plates of importance in identification. In the comatulids (stalkless), a large *ossicle*, the *centrodorsal* (corresponding to the top of the stalk) is surrounded by series of other large ossicles. These consist of five *basals* followed by five *radials* alternating with them. The radials are succeeded in the same radius by *first* and *second brachials*. The second brachial is also known as the *axillary* since it has two articular faces at its distal end and is the point at which the ray divides into two, so resulting in ten arms.

Each arm is supported throughout its length by a series of articulating *brachial ossicles*; successive ossicles may be united by both ligaments and muscles, or by ligaments alone. Side branches or *pinnules* arise alternately along the arms. These too, are supported by a row of progressively smaller articulating ossicles. The few lowermost, or *oral*, pinnules are usually distinctive and useful in identification; they are numbered P_1, P_2, P_3, etc. All pinnules are furnished with a ciliated ambulacral food groove with papillate but suckerless tube feet arranged along each side in groups of three. Gonads are present in the proximal pinnules.

The cirri arise from sockets on the centrodorsal ossicle in a pattern characteristic of the taxon. The cirri themselves are unbranched structures greatly varying in length between species; the terminal ossicle is usually claw-shaped.

Only one family, the Antedonidae, is represented in the area.

1 ANTEDONIDAE

Crinoids lacking stalks. Free-moving, attaching temporarily by clawed cirri. Ten arms.

---

1. Cirri over half length of arms; cirri arise in vertical rows on large columnar centrodorsal ossicle; oral pinnules 1 and 2 similar and longer than 3
   Leptometra celtica

2. Cirri much less than half arm length; cirri arise in transverse rows on low conical centrodorsal ossicle. First oral pinnules with elongated segments
   Antedon bifida

Antedon bifida (Pennant) FIG. 15.3

Proximal (oral) pinnules long and modified; about 35 segments. Distal edges of arm ossicles (brachials) prominent. Genital pinnules with slight spine at outer corner of the first six joints. Gonad confined to proximal half of pinnule. Cirri short, up to 20 (rarely 30). Arm length up to 100 mm, usually less. Colour variable, dark reddish to orange and yellow, often mottled or banded.

On hard substrata amongst colonial hydroids, bryozoans, etc; pentacrinoid stage stalked and attached to hydroids, algae, etc. Immediate sublittoral down to 200 m. British Isles, sometimes very abundant, all coasts except east England; elsewhere Shetland to Portugal.

Frequently parasitized by *Myzostoma*, a hermaphrodite polychaete.

Leptometra celtica (Barrett and McAndrew) FIG. 15.3

First two oral pinnules similarly elongated with about 20 segments; brachial ossicles lack distal spines. About 20–30 very long cirri with proximal segments longer than broad. Arm length up to 100 mm. Green when alive, fading to brown when preserved.

British Isles; only locally off north and west coasts 70–100 m; elsewhere south to Morocco.

B. Class Asteroidea

The starfishes. Echinoderms with a star-shaped body-plan, the five (or more) arms grading into the central disc, in contrast to the condition in the Ophiuroidea (brittle stars) in which the disc is sharply delimited from the arms. The short account given here is confined to features important in identification.

The normally lower surface bears the central mouth and is also referred to as the *oral, ambulacral,* or *actinal* surface, the upper being the *aboral* or *abactinal* (Fig. 15.1). The arms are also referred to as *rays* and their axes as *radii*, areas of the disc between the radii are referred to as *interradial.* The taxonomically important relationship between arm-length and disc-diameter is expressed as R/r, where R, is the length from disc centre to arm tip, and r is the length from disc centre to the disc edge between two arms. Running from the mouth to the end of each arm is a deep open furrow, the *ambulacrum* or *ambulacral groove*, in which the two or four rows of *tube feet* are situated. The tube feet may or may not have a terminal disc-shaped attachment sucker; internally each is connected through a pore in the ossicles to a small water-filled *ampulla*, which may be single or double. Paired *ambulacral ossicles* bound the groove internally and are flanked by *adambulacral ossicles* (otherwise referred to as *adambulacral plates*) which form the sills of the furrow. From these may arise *furrow spines*, singly, in rows or groups, *pedicellariae* and *ventrolateral spines*. The angle between two ambulacra adjacent to the mouth is sometimes marked by a special ossicle, the *ventral mouth plate*. The edges of the arms may be sharply marked by the presence of distinctive or large ossicles known as *marginal plates*. The general outer surface is usually furnished with spines of various types which may be prominent and arranged in rows. *Paxillae* are spines which are tipped with a bunch of smaller elongated ossicles like a little brush. Pedicellariae of the Asteroidea are always only two-jawed but come in two basic types: *crossed,* in which the jaw ossicles have a scissor-like action, and *straight* with a forceps action. *Papulae* are thin-walled outpushings of the body surface that act as gills.

The Asteroidea are divided into three orders:

1. Phanerozonia—with large marginal plates;
2. Spinulosa—characterized by groups of spines on the upper surface;
3. Forcipulata—armed with crossed pedicellariae.

KEY TO FAMILIES

1.	Starfish with large conspicuous marginal plates. Papulae (gills) mostly on the upper surface; no crossed pedicellariae	**3**
	Starfish with inconspicuous marginals. Papulae on both upper and lower surfaces	**2**
2.	No crossed pedicellariae, straight ones rare. Spines of upper surface in groups. Normally only two rows of tube feet	**6**
	Crossed pedicellariae present. Spines of upper surface usually arranged singly rather than grouped. Tube feet in four rows	**8**
3.	Tube feet without a terminal sucker disc, pedicellariae simple	**4**
	Tube feet with a terminal attachment sucker disc, pedicellariae valvate	**5**
4.	Well-developed upper marginal plates; simple papulae; one pair of gonads at base of each arm	**1. Astropectinidae**
	Upper marginals replaced by paxillae; papulae bush-shaped; series of paired gonads down length of arms	**2. Luidiidae**
5.	Large thin-skinned marginal plates; upper surface covered with flat spines or nodules; edge of disc blunt	**3. Goniasteridae**
	Small indistinct marginal plates covered by naked skin or a few spines; upper surface covered by naked skin or minute spines; edge of disc sharp	**4. Poraniidae**

6. Animal more or less flattened with sharp edge between upper and lower surfaces	**5. Asterinidae**
No sharp edge between upper and lower surfaces	**7**
7. Very small disc; five long cylindrical arms, small spines irregularly arranged singly or in small groups like paxillae	**7. Echinasteridae**
Large disc; more than five arms, long, well-formed paxillae	**6. Solasteridae**
8. Ossicles of body wall arranged in reticulate pattern; spines occur singly	**9. Asteriidae**
Ossicles of upper surface arranged in discernible transverse and longitudinal rows; these plates covered with groups of small spines	**8. Stichasteridae**

1 ASTROPECTINIDAE

Phanerozoan starfish with well-developed upper and lower marginal plates covered with fine spines. Upper surface covered with paxillae. Tube feet pointed and suckerless; provided with double ampullae. Six genera in British seas but only *Astropecten* within the depths included here.

Astropecten irregularis (Pennant) FIG. 15.3

Diameter often 100 mm but occasionally up to 200 mm, pale violet or yellowish. Five, rather short, stiff, tapering arms meeting at distinct angles. Upper and lower rows of marginal plates well-marked. The whole outline has a continuous fringe of spines, which arise four or five to each lower marginal plate, the second or third spine of each group being the longest. A row of four to seven ventrolateral plates in each lower (oral) interradius. Adambulacral ossicles each bear three furrow spines.

Partly buried in sandy substrata; sublittoral, 10–1000 m. Common on all British coasts; elsewhere Norway to Morocco.

2 LUIDIIDAE

Phanerozoan starfish of the suborder Paxillosa in which the upper marginal plates are represented by marginal paxillae; lower marginals well-formed and spiny. Upper surface covered with paxillae. Tube feet lack terminal sucker disc but used for burrowing and are powered by double ampullae. Gut lacks pyloric caeca and anus. Gonads serially arranged along sides of each arm. Only one genus in the area.

Luidia sarsi Düben and Koren FIG. 15.3

Diameter up to 340 mm (usually smaller); reddish brown, with darker bands. Five gently tapering medium length arms, (R/r 6–10). Lower row only of marginal ossicles, well-defined, each bearing (three or four) fringing spines. Upperside clothed with paxillae. Each two marginal paxillae are followed by four lateral paxillae, of which there are several rows, then 15–20 median paxillae. Bivalved pedicellariae occur on underside.

Feeds on other echinoderms, in addition to polychaetes, molluscs, and crustaceans. Sublittoral on muddy substrata 10–600 (1300) m. Common on all British coasts except south. Elsewhere, distributed from Trondheim to Cape Verde and Mediterranean.

Luidia ciliaris (Philippi) FIG. 15.3

Diameter up to 600 mm but usually smaller; brick-red. A fine starfish with seven arms (R/r 7–8), only the distal parts tapering. Lower marginal ossicles each bear four or five spines. Abundant paxillae of upper surface consist of two marginals to every three laterals, the transverse rows of which are each followed by about ten median ossicles. Three-valved pedicellariae sometimes occur on the ventrolateral plates.

Feeds on other echinoderms, more exclusively than *L. sarsi*, including *Ophiothrix*, *Echinocardium*, and *Asterias*. Sandy substrata sublittorally, in shallower water than *L. sarsi*, 4–400 m. Common on all British coasts except southern North Sea; elsewhere distributed southwards from Faeroes to Cape Verde and Mediterranean.

3 GONIASTERIDAE

Phanerozoan asteroids of the suborder Valvata with large distinct marginal plates. Upper surface with short flat spines arranged like paxillae. Pedicellariae often valvate. Gonads confined to interradii. Three subfamilies in British seas but only one species of the subfamily Hipperasteriinae in shallow water.

Hippasteria phrygiana (Parelius) FIG. 15.3

Diameter of adults up to about 400 mm, usually smaller; reddish in colour. Large disc; five arms short, narrow, tapering (R/r *c.* 2). Upper surface closely covered with fine plates, the larger ones bearing a pedicellaria or stout spine in the centre; edges of large and small plates granular; oral interradial areas similar, the large spines bearing pedicellariae. Marginal plates similarly granular with one to three stout central spines. Along the ambulacra the furrow spines are in groups of two or three. The adambulacral ossicles have one outer spine and are flanked by a row of plates bearing obliquely arranged pedicellariae.

On mixed substrata, 20–800 m; rather rare, coasts of Scotland down to Northumberland in the east; also off Cornwall. Else-

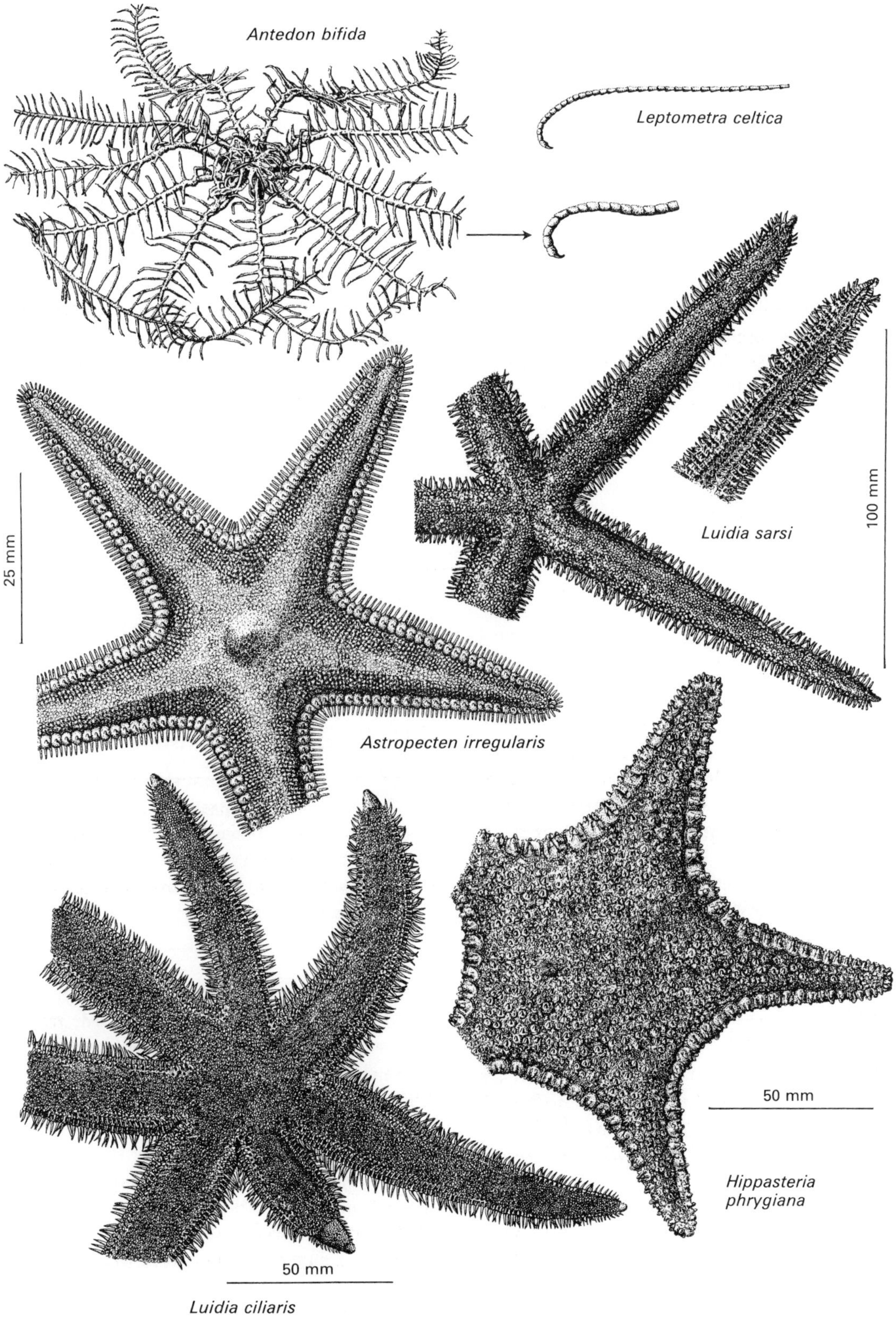
Antedon bifida
Leptometra celtica
Luidia sarsi
100 mm
25 mm
Astropecten irregularis
50 mm
Hippasteria
phrygiana
50 mm
Luidia ciliaris

Fig. 15.3

where predominantly northern in distribution: Greenland, Iceland, and Finmark.

4 PORANIIDAE

Phanerozonia with a distinctively near-naked upper surface; marginal plates somewhat reduced. No pedicellariae. There are some rare and deep water species in addition to the commoner one described here.

Porania pulvillus (O. F. Müller) FIG. 15.4

Adults up to a diameter of about 110 mm, colour red or yellowish white. Large disc; five short, wide, tapering arms (R/r 1.5–2). Upper surface smooth and free of any spines or pedicellariae; only scattered groups of papulae; oral interradial areas crossed by radiating furrows on otherwise smooth surface. Upper marginal plates smooth and narrow; lower marginals with three to five spines forming a fringe.

Muddy substrata, sublittoral, 10–300 (–1000) m. All British coasts except southern North Sea. Elsewhere distributed from Trondheim to Biscay.

5 ASTERINIDAE

Spinulosa with very short arms giving an almost pentagonal shape to the flattened body; edge sharp, formed of small indistinct marginal plates.

---

1. Very flat and thin, with radial thickenings corresponding to the ambulacra, 100–120 mm diameter **_Anseropoda placenta_**

   Slightly swollen, no radial crests, up to 50 mm **2**

2. Up to 50 mm, two spines on ventral mouth plates, no dark pattern on upper surface **_Asterina gibbosa_**

   Up to 15 mm, ventral mouth plate spines rare and only in larger specimens, dark star-shaped pattern on upper surface **_Asterina phylactica_**

Asterina gibbosa (Pennant) FIG. 15.4
[EMSON AND CRUMP 1979: 77]

Maximum diameter 50 mm, usually smaller; dark olive-green ranging through greeny brown to pale brown, sometimes mottled; small ones sometimes have dark central pattern but never on green background; infralittoral fringe specimens paler. Subpentagonal (R/r 1.2–2.0); lower side flat, upper surface swollen. Spines, arranged in groups of four to eight, arise from plates all over upper surface; on the oral interradii larger spines occur in groups of two or three. Adambulacral ossicles have two outer spines and three furrow spines. Large gonopores on oral side of sexually mature females (18–22 mm). Mouth plates bear two spines.

Intertidal rock-pools and sublittoral to about 125 m. Common on all British coasts, except North Sea. Distributed southwards to Azores and Mediterranean.

Asterina phylactica Emson and Crump FIG. 15.4
[EMSON AND CRUMP 1979: 77]

Maximum diameter 15 mm; background colour of upper surface grey-green; central area marked with dark star-shape, resulting from the strong red colour of spine groups in radial areas; underside pale. Very similar to *A. gibbosa* but smaller and distinctively coloured. Mouth plates rarely possess spines, whereas they bear two in *A. gibbosa*. Females are distinguishable from young *A. gibbosa* by presence of gonopores (sexually mature at diameter of only 6 mm).

Intertidal rock-pools and sublittoral to at least 18 m. Common on south and west coasts of British Isles, also Brittany and Mediterranean.

Anseropoda placenta (Pennant) FIG. 15.4

Maximum diameter about 200 mm; red above, yellow on underside. Subpentagonal (R/r *c.* 1.5), very thin and flat. No pedicellariae but covered on upper and lower surfaces by fine radiating rows of spine groups. Five or six furrow spines united by a basal web arise from each adambulacral plate.

On sand bottoms, sublittorally, 10–200 (–600) m. Common on all British coasts except southern North Sea. Distributed from Shetland to Mediterranean.

6 SOLASTERIDAE

Spinulosa with usually more than five arms and the disc rather large. Upper surface usually strengthened by a reticulum of ossicles. Marginal plates represented by paxillae.

---

1. The large marginal paxillae in a single row; three to five furrow spines on each adambulacral plate **_Crossaster papposus_**

   Marginal paxillae form a double row; two to three small furrow spines in each group **_Solaster endeca_**

Crossaster papposus (Linnaeus) FIG. 15.4

Up to 340 mm diameter; disc colour red, arms off-white with broad transverse red band. Disc large; 10–12 arms (rarely 8–13) tapering (R/r *c.* 2). Upper surface strengthened by a reticulum of ossicles with papulae distributed in the spaces. The brush-shaped paxillae are large but well-spaced. The marginal paxillae are large and arranged in a single row. Three to five furrow spines; adambulacral plates also possess series of five to nine larger spines. No pedicellariae.

Infralittoral fringe to 50 (–1200) m. Common on all British coasts. Distributed from the Arctic to the English Channel, also on both east and Pacific coasts of North America.

Solaster endeca (Linnaeus) FIG. 15.4

Up to 400 mm diameter; orange sometimes tinged violet. Disc quite large; nine or ten arms (rarely 7–13), long and tapering (R/r 2.5–3). Marginal paxillae small and forming a double row in which the upper ones are smaller. The upper surface is crowded with small paxillae bearing short spines. Adambulacral

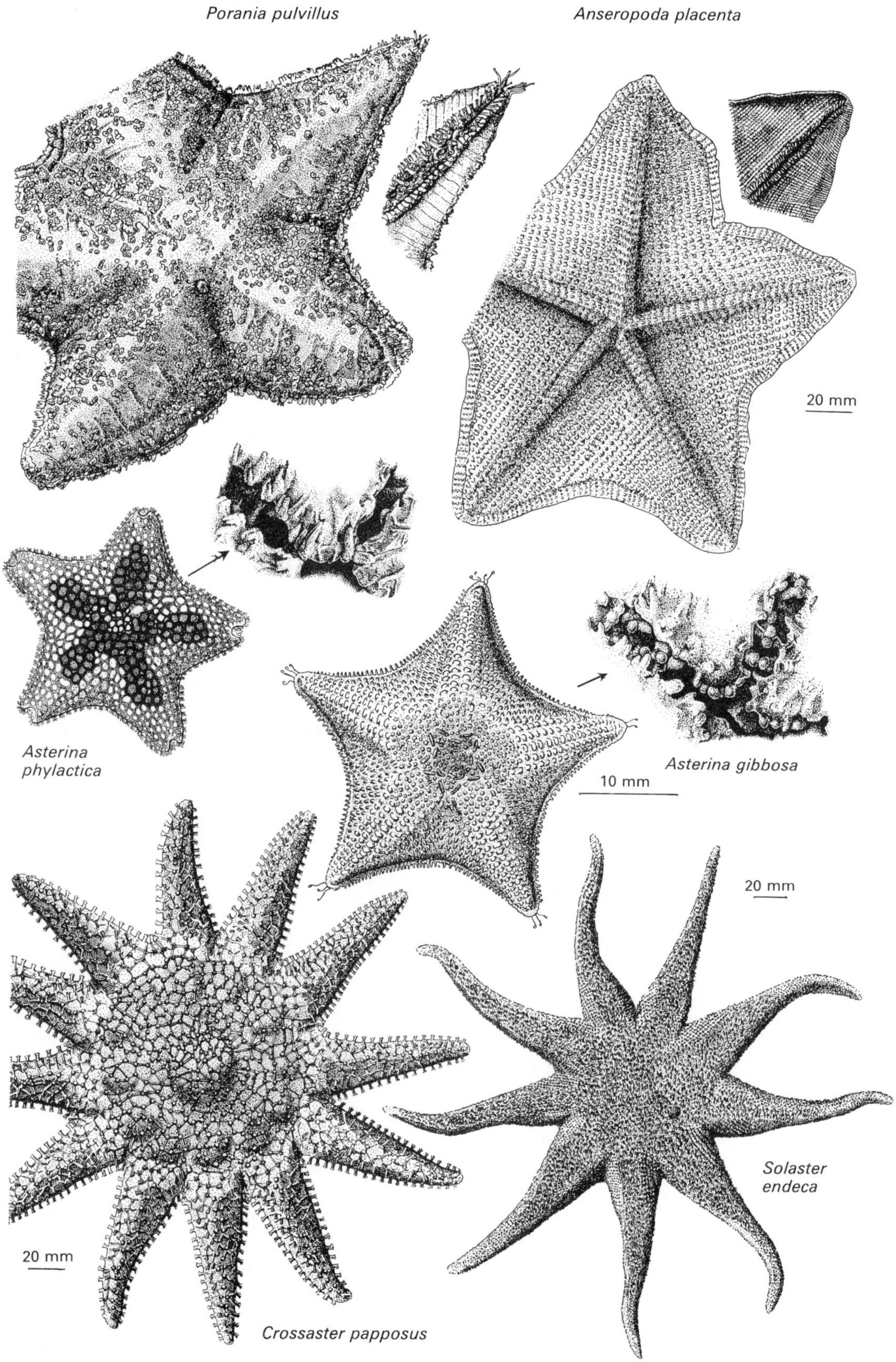
Porania pulvillus
Anseropoda placenta
20 mm
Asterina phylactica
Asterina gibbosa
10 mm
20 mm
Solaster endeca
20 mm
Crossaster papposus

Fig. 15.4

plates bear two to three furrow spines and rows of six to eight short outer spines.

Infralittoral fringe to about 450 m. Not uncommon on western and northern coasts, and in the Irish Sea but not in southern North Sea or the Channel. Distributed northwards to Spitzbergen and Greenland, also on the east and west coasts of North America.

7 ECHINASTERIDAE

Starfish of the order Spinulosa having a very small disc and five long cylindrical arms in which there is no distinct margin between upper and lower surface. Reticulate skeleton in body wall. No paxillae or pedicellariae.

1. Small groups of thin spines; some papulae on lower surface as well as upper surface **2**
   Singly arranged, fairly robust spines; no papulae on lower surface ***Echinaster sepositus***
2. Spines end in a cluster of long spikes not covered by skin ***Henricia sanguinolenta***
   Spines end in a cluster of rather short points covered by thick skin ***Henricia oculata***

Henricia sanguinolenta (O. F. Müller) FIG. 15.5

Maximum diameter 200 mm; red with whitish tips to arms. Very small disc; cylindrical tapering arms (R/r 4.5). Scattered groups of small spines on upper surface. Spines on the adambulacral ossicles form a single graded series, decreasing in size outwards. The larger spines tipped with a cluster of sharp spikes not covered by skin.

Infralittoral fringe and shallow water, down to 1000 (2450) m. Common on north, north-west and north-east costs; records from Biscay may result from confusion with *H. oculata* (see below). Distribution circum-arctic: Greenland, Iceland, Scandinavia, Baltic, north-west Atlantic, and north Pacific.

Henricia oculata (Pennant)

Similar size to *H. sanguinolenta*; red, orange, or yellow. Small disc, narrow arms. Distinguished from *H. sanguinolenta* by spines which are blunt, with numerous irregular points and covered with thick skin.

On coarse sand and gravel, infralittoral fringe to about 100 m. Common on south and west coasts including Irish Sea.

Echinaster sepositus (Retzius) FIG. 15.5

Maximum diameter 140 mm; scarlet. Usually five arms, occasionally six or seven.

Infralittoral fringe, and sublittoral to 200 m; no further north than Channel Isles; elsewhere southwards. Abundant in Mediterranean.

8 STICHASTERIDAE

Starfish of the order Forcipulata (with pedicellariae). Skeleton of upper surface in the form of longitudinal rows of overlapping plates. Adambulacral plates all similar. Four rows of tube feet. Crossed and straight pedicellariae (but not on adambulacrals). Species of several genera from deep water; only one within the depths dealt with here.

Stichastrella rosea (O. F. Müller) FIG. 15.5

Maximum diameter about 300 mm; colour orange or yellow-orange. Disc very small, five tapering arms, round in section (R/r *c.* 7). Plates of upper surface in rather irregular transverse and longitudinal rows; surface of plates nodular. Madreporite prominent, surrounded by nodules. Adambulacral plates with rows of three spines. Crossed pedicellariae on upper surface, straight ones flanking ambulacral groove.

On sandy substrata, sublittorally 4–400 (?–1300) m. All British coasts except south, uncommon; elsewhere from northern Norway to Biscay.

9. ASTERIIDAE

Starfish of the order Forcipulata having the body wall strengthened by reticulate pattern of ossicles and armed with longitudinal rows of single spines. Usually both crossed and straight pedicellariae present.

1. Many adambulacral spines bear a few sessile pedicellariae near their tips **2**
   Adambulacral spines do not bear pedicellariae ***Marthasterias glacialis***
2. Papulae usually occur in groups of three ***Asterias rubens***
   Most papulae occur singly although some small groups may also be found ***Leptasterias muelleri***

Asterias rubens Linnaeus FIG. 15.5

Diameter up to about 520 mm, but usually smaller (200–300 mm); commonly orange but ranges from pale brown to delicate violet, deep water specimens pale. Disc small, five arms (rarely four to eight, also often regenerating arms and other abnormalities) broad at base, tapering. R/r *c.* 4.5 (3.5–7). Body wall very flexible with numerous groups of papulae in soft areas. Major spines of upper surface often in one or more longitudinal rows, sometimes surrounded by bundles of straight and crossed pedicellariae. Ventrolateral spines occur just outside adambulacrals, in oblique rows of three or four (two in young ones). Straight pedicellariae occur on lower surface in ambulacral grooves and attached to furrow spines.

Occurs on a variety of mostly hard substrata; midlittoral, infralittoral fringe and sublittoral to 400 (–650) m. Abundant on all British coasts. Distributed from Iceland and the White Sea southwards to Senegal; not in the Mediterranean.

A caprellid amphipod, *Pariambus typicus* (Krøyer) is often found attached to spines, also a semi-parasitic copepod, *Scottomyzon gibbosum* (Scott).

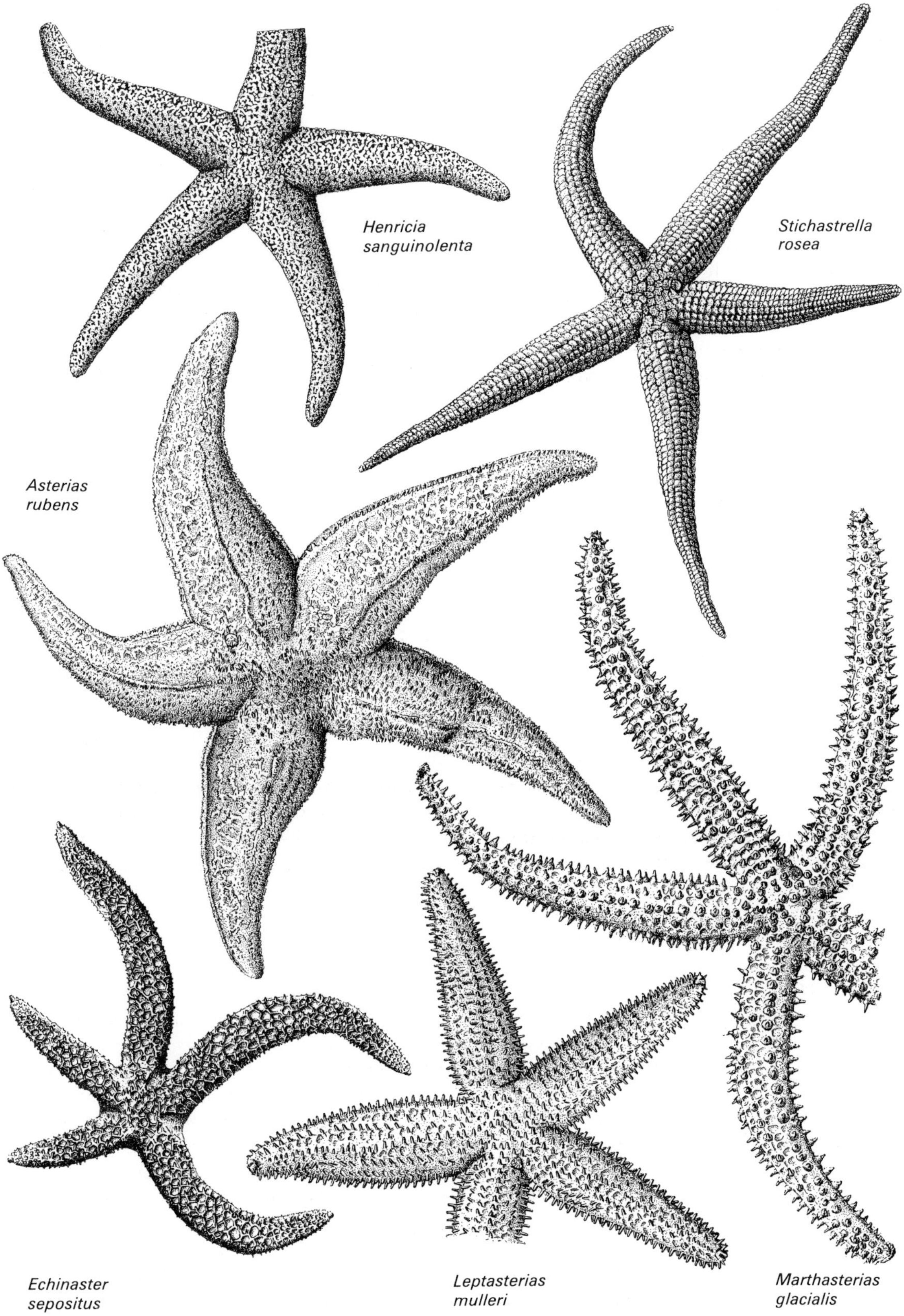
Henricia
sanguinolenta
Stichastrella
rosea
Asterias
rubens
Echinaster
sepositus
Leptasterias
mulleri
Marthasterias
glacialis

Fig. 15.5

Leptasterias muelleri (M. Sars) FIG. 15.5

Maximum diameter about 200 mm; pink or pale violet, grading to off-white at ends of arms. Disc very small, five tapering arms (R/r 4–7). Body wall tougher than *Asterias* with less space between ossicles. Papulae mostly singly placed in soft areas. Spines of upper surface often surrounded by bunches of crossed pedicellariae at the base. Ventral lateral spines single and furnished with straight pedicellariae near their tips.

Infralittoral fringe and sublittoral to about 800 m. Iceland, west coast of Shetland, North Sea; not on southern coasts. Distribution northwards to Spitzbergen, Greenland, and North America.

Marthasterias glacialis (Linnaeus) FIG. 15.5

Maximum diameter 700 mm (usually much smaller); delicately pale-coloured grey, green, tinged purple, or yellowish ranging to reddish. Disc very small, five evenly tapering long arms, increasing in relative length with age (R/r 5–9). Well-marked longitudinal rows of spines: mid-dorsal, upper marginal, lower marginal. Thick cushion of crossed pedicellariae at base of each major spine. Single series of adambulacral spines not bearing pedicellariae. Straight pedicellariae scattered on upper surface and sessile on adambulacral ossicles (but not on their spines).

Infralittoral fringe down to about 180 m. All British coasts except east coast of England and eastern half of Channel. Elsewhere Finmark and Iceland to Cape Verde Isles, Azores, and Mediterranean.

C. Class **Ophiuroidea**

The ophiuroids are the largest group of living echinoderms, comprising some 1800 species. In British seas they usually occur sublittorally or in the lower intertidal as epifauna on rock or soft substrata and also as an infaunal component of soft substrata, in truly marine areas.

Morphologically the adult consists of a dorso-ventrally flattened *disk* to which are attached five thin, articulate *arms*. These arms break off easily, hence the common name brittle star.

The dorsal (upper) surface of the disk has no apertures (as found in other echinoderms) but is covered with a series of scales (Fig. 15.1). The central scale is the *primary plate* surrounded by concentric rings of *secondary plates*. On the dorsal surface of the disk, adjacent to the base of each arm is a pair of *radial shields*. The dorsal plates of the disk, often with the exception of the radial shields, may be covered with scales, granules, or small spines. At the edge of this disk, near the base of each arm, may be found a row of papillae, the *arm comb*. The ventral (lower) surface of the disk (Fig. 15.1) contains a central *mouth* leading to the stomach. Food is ingested through the mouth and undigested remains are egested by the same route. The mouth is divided by five *jaws* corresponding to the *interradii*. At the apex of the jaw are the *teeth*. These may comprise a single vertical row of teeth (as in *Ophiura*), or a double row of *teeth papillae* (as in *Ophiothrix*). In some families, e.g., Amphiuridae, single or paired *infradental papillae* cover the outermost part of the tooth. Along the jaw edge, leading to the jaw angle, are usually found the *mouth papillae*. Moving from the jaw apex to the interradius, the first pair of plates are the *adoral shields*. These shields partially surround the *oral* (*mouth*) *shield* which is often the largest plate on the ventral disk. One of these oral shields houses the *madreporite*. The remaining ventral interradial area is continuous with the dorsal sculpture. From the jaw angle at the radius the first plate to occur is the first *ventral arm plate*. With few exceptions, to either side of this plate lies the *second foot* (*tentacle*) *pore*. This tentacle pore may or may not open into the angle of the jaw and is lined with a variable number of *tentacle scales*. The first tube foot is hidden within the mouth. Lying along either side of the arms of the central side of the disk are the radially orientated *genital slits*, occasionally lined with *genital papillae*. The arms consist of a series of one *dorsal*, one *ventral* and two *lateral plates*, a set of four plates thus forming an arm segment. The distal edge of the lateral plates may be fringed with a series of spines varying in number, shape, and size. On each side of the ventral plate lies the *tube foot pore* which may or may not be covered by a *tentacle scale*.

KEY BRITISH FAMILIES

1.	Arms simple, unbranched, moving horizontally. Articulation of arms by pits and processes. Distinct scales covering disk and arms	2
	Arms simple or branched, moving vertically. Articulation of arm segments by means of hourglass-shaped surface. No distinct scale covering on disk or arms	6
2.	Arm spines lying flat against arms; disk scales distinct	**1. Ophiolepidae**
	Arm spines erect	3
3.	Tooth papillae present on apex of jaw	4
	No tooth papillae, but a single vertical row of teeth	5

4. Mouth papillae present around sides of jaw	**2. Ophiocomidae**
No mouth papillae around jaw	**3. Ophiotrichidae**
5. Single infradental papilla on apex of jaw	**4. Ophiactidae**
Paired infradental papillae on apex of jaw	**5. Amphiuridae**
6. Arms branched. Hooks on outer part of arms	**6. Gorgonocephalidae**
Disk large. Slender arms	**7. Asteronychidae**

Order **Ophiurae**

1 OPHIOLEPIDAE

Scales on both sides of disk. Mouth papillae form a continuous series around jaw. Single vertical row of teeth. Arm spines appressed against arms. Single or double combs at base of arm. Other familiar members of this family, viz. *Ophiura ljungmani* (Lyman), *Ophiura carnea* M. Sars, *Ophiopleura aurantiaca* (Verrill), and *Ophiomusium lymani* Thomson are only found in deep water off the west coast of Ireland. All are described by Mortensen (1927).

1. Single arm comb forms a continuous series across base of arm	***Ophiocten sericeum***
Outer arm comb does not form a continuous series across base of arm	**2**
2. Ventral plates in proximal part of arm separated by pores. Numerous papillae (*c.* 30) in arm combs	***Ophiura ophiura***
No pores between proximal ventral plates. Seven to fifteen papillae in outer comb	**3**
3. One or two small tentacle scales; ventral plates widely separated	**4**
Three or more tentacle scales	***Ophiura albida***
4. Only three mouth papillae on jaw. Outer one plate-like. Inner arm comb meets in centre of arm	***Ophiura affinis***
Three or four small mouth papillae in jaw	***Ophiura robusta***

Ophiura ophiura (Linnaeus) FIG. 15.6

Disk up to 35 mm diameter, usually about 25–30 mm. Dorsal side of disk red/brown in life, ventral side of disk pale. Primary plate distinct. Radial shields large, about one-half disk radius; just touching above innermost dorsal arm plate. Two layers of arm combs above each arm. Outer comb consists of about 30 fine papillae. Proximal ventral arm plates separated by a pair of pores in mid-line of arm. Mouth shields occupy two-thirds of distance from disk edge to adoral plate. Four to six mouth papillae.

Common all round Britain. Mainly sublittoral on a variety of soft substrata. Distributed in the north-east Atlantic from northern Norway to Madeira, and in the Mediterranean.

Ophiura albida Forbes FIG. 15.6

Disk up to 12 mm diameter. Dorsal surface red/brown in life. Primary plates not distinct. Radial shields small. Innermost dorsal arm plates heart-shaped. Outer arm comb of ten to twelve short, stout spines. Inner comb indistinct. Dorsal and ventral arm plates with convex outer edge. Three arm spines equidistant from each other. Spines equal to one-half lateral plate length. Mouth shield small, approximately one-third disk radius. Three to five simple mouth papillae.

Common all round Britain. Mainly sublittoral on a variety of soft substrata. Distributed in the north-east Atlantic from northern Norway to the Azores, and in the Mediterranean.

Ophiura affinis Lütken FIG. 15.6

Maximum disk diameter 8 mm. Primary plates conspicuous. Radial shields small, wholly separate, similar in size to central plate. Second tentacle pore with only two tentacle shields. Outermost mouth papillae plate-like. Occasionally rudimentary arm combs may meet across base of arm.

Occurs on muddy sand, fine shell, and gravel. All British coasts, but not in the southern North Sea. Distributed from northern Norway to the Bay of Biscay, and also recorded from the Mediterranean.

Ophiura robusta (Ayres) FIG. 15.6

Maximum disk diameter 10 mm. Scales of aboral surface uniform in size. Radial shields very small and wholly separated. Arm combs consist of stunted papillae and side of arm base. Dorsal arm plates with strongly convex outer edge. Ventral plates have shallow concavity on outer edge, except first ventral plate which has three shallow concavities on outer edge. Three arm spines, the lower two short, the upper long and stout. Short broad mouth shields and three or four mouth papillae.

Found on a variety of substrata in the northern half of the British Isles. A predominantly Arctic/Boreal Arctic species, occurring in both the east and west North Atlantic.

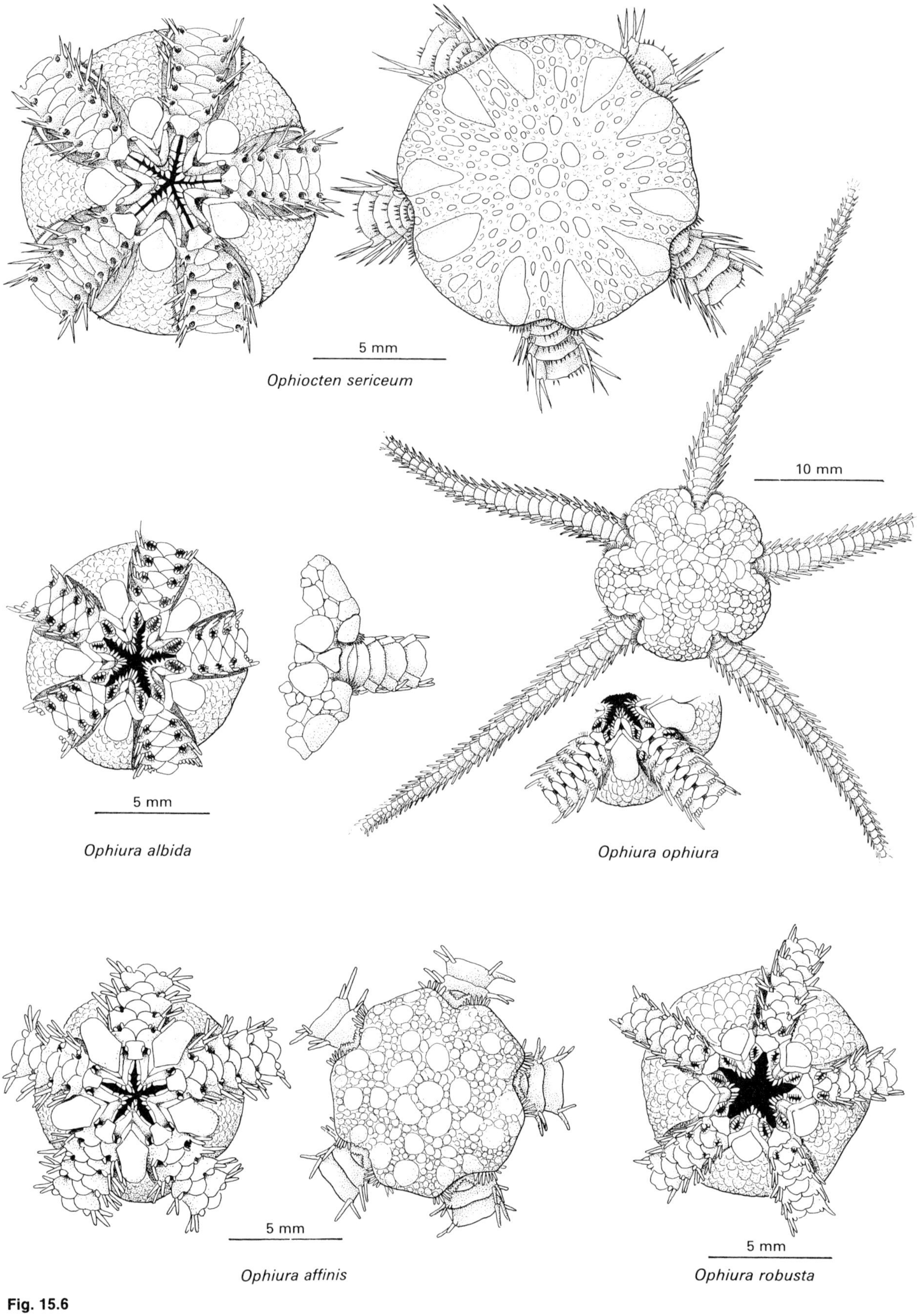
5 mm
Ophiocten sericeum
10 mm
5 mm
Ophiura albida
Ophiura ophiura
5 mm
5 mm
Ophiura affinis
Ophiura robusta

Fig. 15.6

Ophiocten sericeum (Forbes) FIG. 15.6

Maximum disk diameter 18 mm. Dorsal disk surface covered in scales which expose primary plates, and widely separated radial shields. Proximal dorsal arm plates with fine papillae along outer edge. Disk edge above the arm base also carries a row of short spines. Ventral plates very short and broad. Mouth shields almost broad as long; four or five mouth papillae.

This is a cold water species occurring in the Arctic and in deeper waters to the west of the British Isles.

2 OPHIOCOMIDAE

Disk with granules completely covering scales and radial shields. Erect arm spines. Well-developed dorsal and ventral plates. Mouth papillae present. Tooth papillae present over a series of strong teeth. Second pair of tube feet inside mouth edge.

1. Disk granulated. Tentacle scales simple **_Ophiocomina nigra_**

   Disk not granulated. Inner tentacle scale elongated crossing that of opposite tentacle scale **2**

2. Six to eight (commonly seven) arm spines. Radial shield distinct, small and widely separated **_Ophiopsila aranea_**

   Twelve arm spines. Radial shields not visible externally, wholly covered by scales **_Ophiopsila annulosa_**

Ophiocomina nigra (Abildgaard) FIG. 15.7

Disk up to 25 mm diameter. Colour in life black to grey. Fine scales on dorsal disk surface completely covered by granules, but only part of ventral interradius covered with granules. Mouth shields oval; 10–15 tooth papillae over single vertical row of teeth; four long, thin mouth papillae. Two tentacle scales, outer one larger than inner. Five to seven smooth arm spines.

A rheophilic species, occurring mainly on stones or rock. Common on all south and west coasts and occasionally in the North Sea. Ranges from western Norway to the Azores, and in the Mediterranean.

Ophiopsila aranea Forbes FIG. 15.7

Elongated tentacle scale equal to one arm joint. Spines short and flat with first arm spine equivalent to one arm joint. Each fifth segment with a surrounding stripe.

On hard substrata offshore. South-western coasts only, ranging south to the Azores and the Mediterranean.

Ophiopsila annulosa (M. Sars) FIG. 15.7

Elongated tentacle scales twice arm joint length. Spines long and flat with first ventral spine longer than arm joint.

Rare. Reported from west coasts of Ireland, the Mediterranean, and north-west Africa.

3 OPHIOTRICHIDAE

Large, conspicuous radial shields. Spines and thorny towers cover remainder of dorsal surface of disk. Dorsal and ventral arm plates well-developed. Well-developed tooth papillae. No mouth papillae. Second pair of tube feet inside mouth edge. Thorny erect spines.

Ophiothrix fragilis (Abildgaard) FIG. 15.7

Disk up to 20 mm diameter; showing great colour variation. Large, naked, conspicuous radial shields, equal to two-thirds of disk radius. Remainder of dorsal disk covered with spines or spinate towers, which also occur on ventral disk interradius. Keel on naked dorsal arm plates. One small tentacle scale. Seven, often serrate, arm spines. Gonads expand ventral side of disk between arms.

Very common on all British coasts, in lower littoral, and sublittorally on suitable hard bottoms including some sand/shell sediments. Rheophilic. Widely distributed in the eastern Atlantic, from northern Norway to the Cape of Good Hope.

The closely related *Ophiothrix luetkeni* Thomson is found off the west coasts of Ireland at depths of more than 150 m.

4 OPHIACTIDAE

Scales visible through spines and granules of dorsal disk surface. Mouth papillae present. Single series of square teeth overlain by infradental papillae. Second pair of tube feet inside mouth edge. Short, erect spines.

1. Dorsal plates of each arm surrounded by small plates **_Ophiopholis aculeata_**

   Dorsal plates not surrounded by small plates **2**

2. One mouth papilla on each side of jaw **_Ophiactis balli_**

   Two mouth papillae on each side of jaw **_Ophiactis abyssicola_**

Ophiopholis aculeata (Linnaeus) FIG. 15.7

Disk up to 20 mm diameter (usually <15 mm). Colour in life variable, especially red and purple. Radial shields concealed. Primary plates obvious, surrounded by granules and blunt spines, continuing on to ventral side. Dorsal arm plates oval, surrounded by ring of small plates. Ventral arm plates rectangular. Six to seven arm spines. Three mouth papillae; infradental papillae small; wedge-shaped, truncated, vertical row of teeth.

On hard substrata, offshore to 1880 m. Common round most of Britain although rarer in the south. A circumpolar cold water species, reaching its southern limit in the north-east Atlantic in the British Isles.

Ophiactis balli (Thompson) FIG. 15.8

Disk small, up to 5 mm diameter; colour variable, but reds especially occur. Some colour banding is usually present on the arms. Radial shields small, separate; the remainder of the dorsal disk surface covered by small scales. Spines occur along disk edge on ventral interradii. Mouth shields small. Single mouth papilla on each side of jaw; single series of teeth overlain by single infradental papilla at apex. Dorsal arm plates rounded

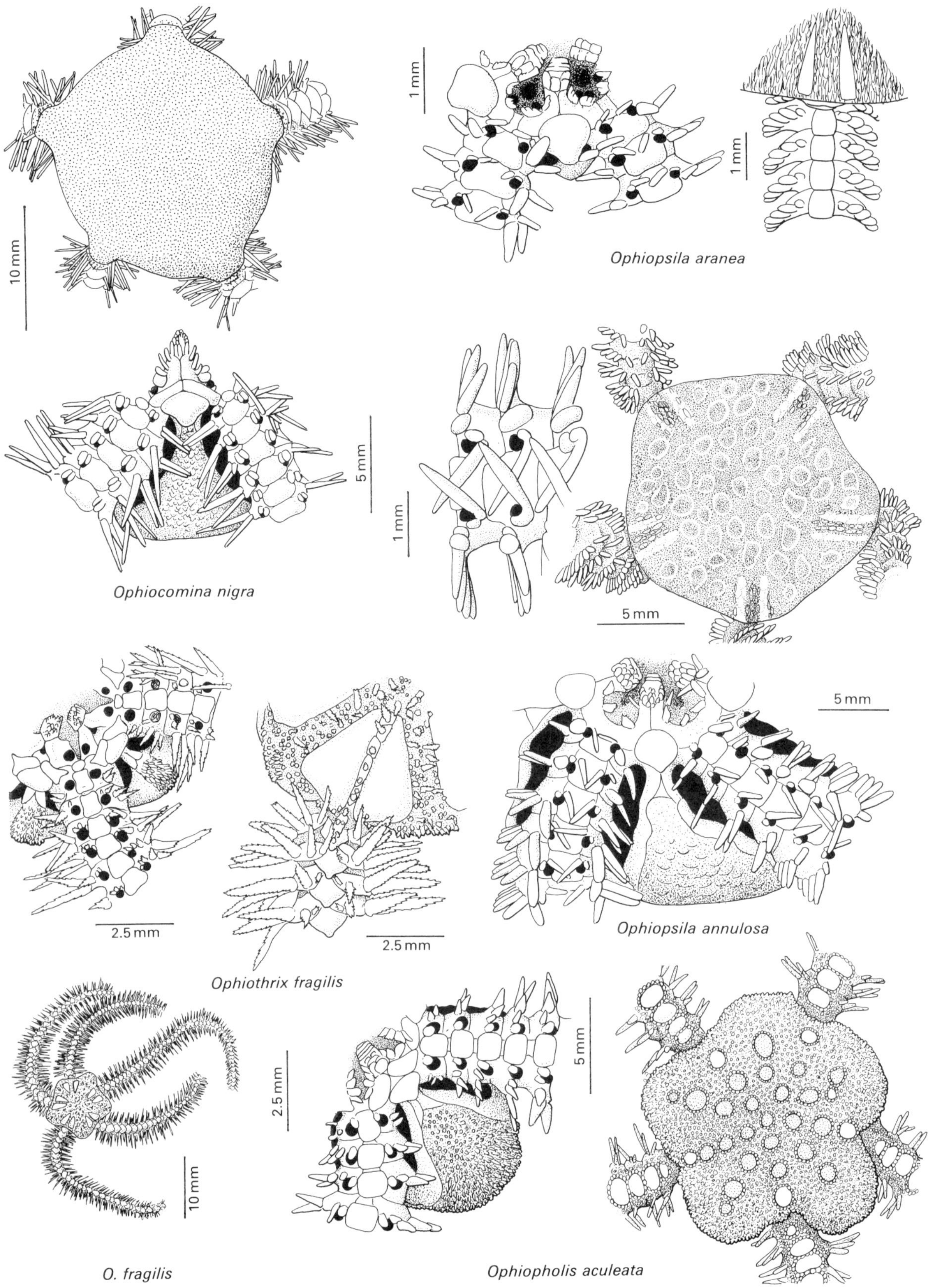
10 mm
1 mm
1 mm
Ophiopsila aranea
5 mm
1 mm
Ophiocomina nigra
5 mm
5 mm
2.5 mm
2.5 mm
Ophiopsila annulosa
Ophiothrix fragilis
2.5 mm
5 mm
10 mm
O. fragilis
Ophiopholis aculeata

Fig. 15.7

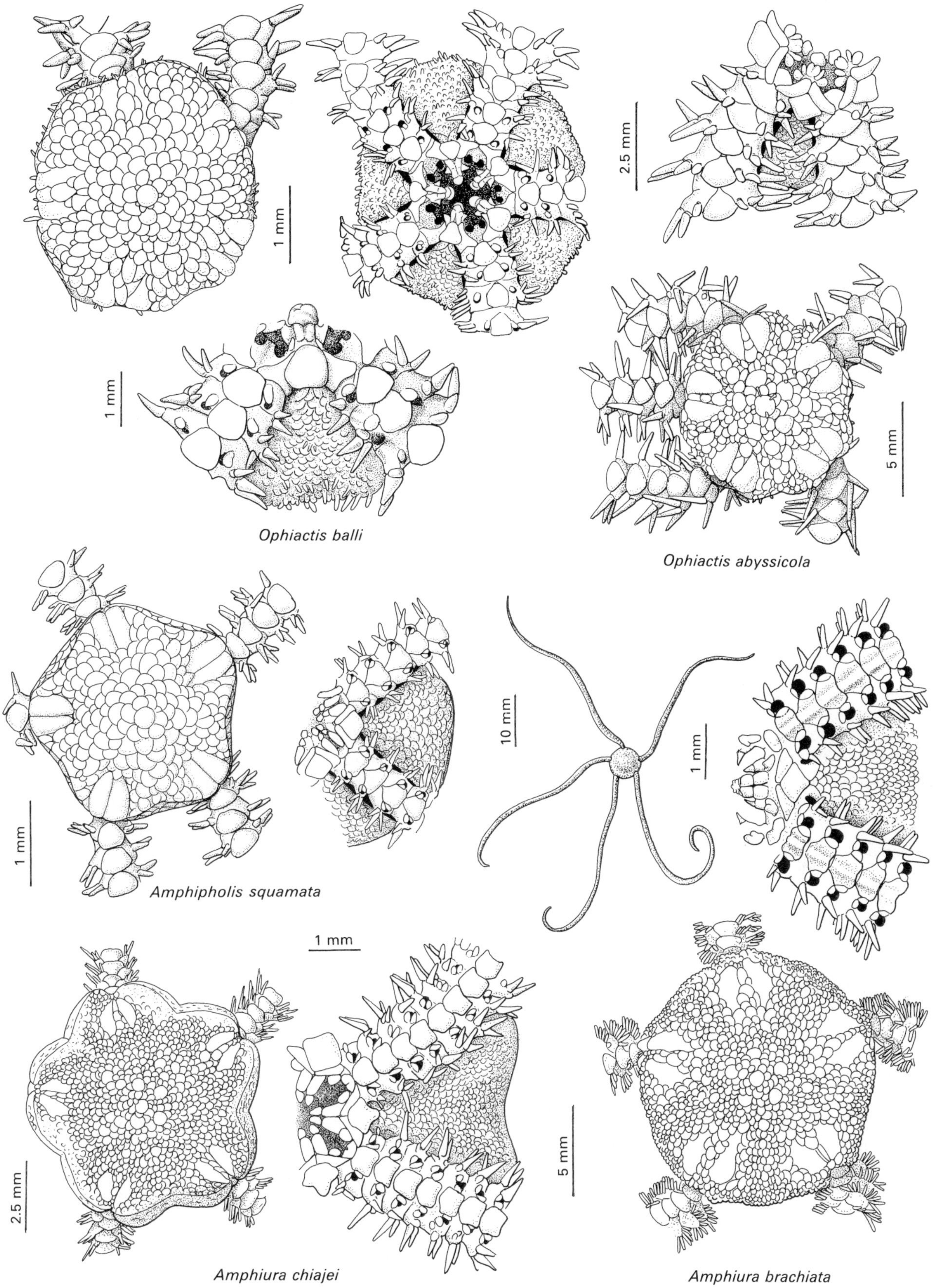
1 mm
2.5 mm
1 mm
Ophiactis balli
5 mm
Ophiactis abyssicola
10 mm
1 mm
1 mm
Amphipholis squamata
1 mm
2.5 mm
5 mm
Amphiura chiajei
Amphiura brachiata

Fig. 15.8

triangular, joined. Ventral arm plates rounded, with obtuse inner angle. One large tentacle scale. Four to five short, conical arm spines.

Found sublittorally off south and west coasts, and in the North Sea as far south as the Durham coast. Especially in small crevices in stones and shell, and also between the membranous layers of *Chaetopterus* tubes. Distributed in the north-east Atlantic from Norway to the Canary Isles.

Ophiactis abyssicola (M. Sars) FIG. 15.8

Disk up to 9 mm diameter, red in colour. Dorsal surface of disk with coarse scales, with occasional spines. Large radial shields. Two broad mouth papillae on each side of jaw.

This species has only been recorded from the west coast of Ireland in depths greater than 125 m. Ranges from Norway and Greenland to South Africa.

5 AMPHIURIDAE

One pair of infradental papillae on apex of jaw; a single series of square teeth, no tooth papillae. Mouth papillae on each side of jaw, well-separated from or contiguous with infradental papillae. Two pairs of tube feet within mouth. Arms are long, fine, and very flexible.

1. One outer mouth papilla on each side of jaw, not contiguous with infradental papillae — 2
   Two outer mouth papillae on each side of jaw, contiguous with infradental papillae — ***Amphipholis squamata***
2. Scales on ventral side, and margin of dorsal side, of disk each ending in a small tubercle. Radial shields with a transverse furrow — ***Acrocnida brachiata***
   Scales smooth. No transverse furrows on radial shield — 3
3. Two tentacle scales. Underside of disk wholly covered with scales — ***Amphiura chiajei***
   No tentacle scales. Underside of disk partly or wholly naked — 4
4. Dorsal side of disk wholly scale-covered. Outer mouth papillae conical — ***Amphiura filiformis***
   Dorsal side of disk partly or wholly naked. Outer mouth papillae small, scale-like — ***Amphiura securigera***

Amphipholis squamata (Delle Chiaje) FIG. 15.8

Disk very small, 5 mm diameter, greyish in colour. Dorsal disk surface covered by small scales. Radial shields equal to one-third disk radius and joined for almost entire length. Dorsal arm plates rounded, triangular, broader than long in proximal part of arm. Ventral plates triangular, truncated at proximal apex. Three to four short, conical arm spines. Mouth shields small rhomboids. Outer mouth papillae elongate and broad.

This species occurs commonly all round Britain in both the mid- and lower littoral, and sublittorally, mainly under stones and shell, and occasionally on sandy bottoms. Cosmopolitan in temperate and warm temperate seas.

Amphiura brachiata (Montagu) FIG. 15.8

Disk diameter up to 12 mm, with very long, thin flexible arms. Brown-grey in colour. Dorsal surface covered by fine scales with small primary plates and separate, small, radial shields. Mouth shields diamond-shaped. Mouth papilla broad and scale-like. Two infradental papillae cover teeth. Dorsal arm plates rhombic; ventral arm plates rectangular, with marked median longitudinal keel in proximal part of arm, and a small lateral keel on each side. Two tentacle scales in proximal part of arm.

Lives littorally and sublittorally in fine sand, buried in the sediment with only the distal parts of the arms protruding. Common where substratum is suitable; recorded from western coasts of Britain and Ireland, and from the western edge of the Dogger Bank.

Amphiura chiajei Forbes FIG. 15.8

Disk diameter up to 11 mm, with long, fine arms. Disk covered with fine scales on both dorsal and ventral sides. Primary plates distinct. Radial shields separate. Outer mouth papillae broad and flat. Mouth shields rhombic. Four to six arm spines. Two tentacle scales.

A. chiajei is a deposit feeder living especially in muddy substrata. Recorded off all British coasts. Distributed from western Norway to the Azores, and in the Mediterranean.

Amphiura filiformis (O. F. Müller) FIG. 15.9

Disk up to 10 mm diameter, red-grey in colour, with long, fine arms. Disk covered with fine scales, not extending to the ventral interradii. Radial shields usually separate, but may join at base. Mouth shields rounded pentagons. Mouth papillae conical, spine-like. Dorsal arm plates oval; ventral arm plates square, with convex outer edge and a small median keel. Five to seven arm spines, the second from bottom blunt. No tentacle scales.

Common sublittorally off all British coasts, in suitable muddy sand substrata. This species is a suspension feeder. Ranges from western Norway to the Mediterranean.

Amphiura securigera (Düben and Koren) FIG. 15.9

Maximum disk diameter 5 mm. Disk wholly naked. Mouth shields triangular. Radial shields long, narrow, parallel, and contiguous in their whole length. Second arm spine from below is axe-shaped.

Recorded sporadically off northern and western coasts; ranges northwards to Faroes and the Lofoten Isles.

Another small amphiurid brittlestar *Paramphiura punctata* (Forbes) has been recorded very rarely in British waters and is described by Mortensen (1927: 219).

Order Euryalae

6 GORGONOCEPHALIDAE

Gorgonocephalus caputmedusa (Linnaeus) FIG. 15.9

Disk diameter up to 90 mm. Disk covered with numerous, short,

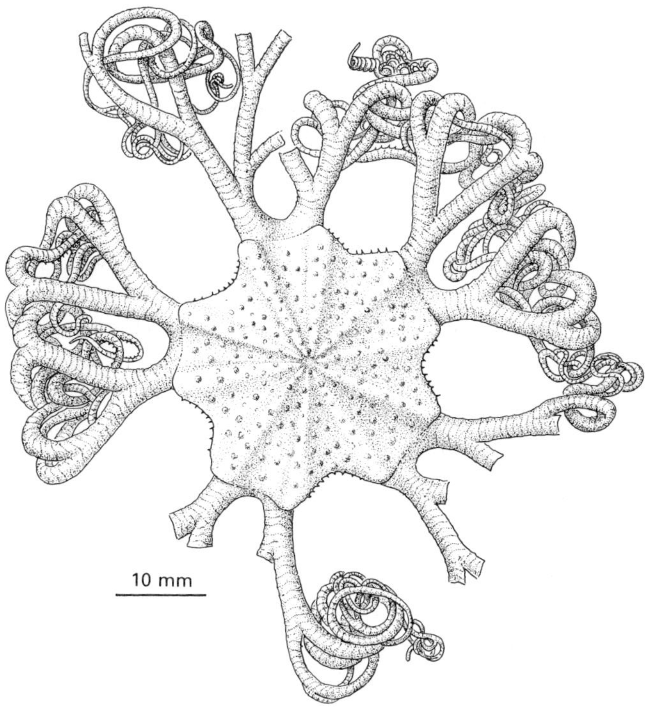

Gorgonocephalus caputmedusa

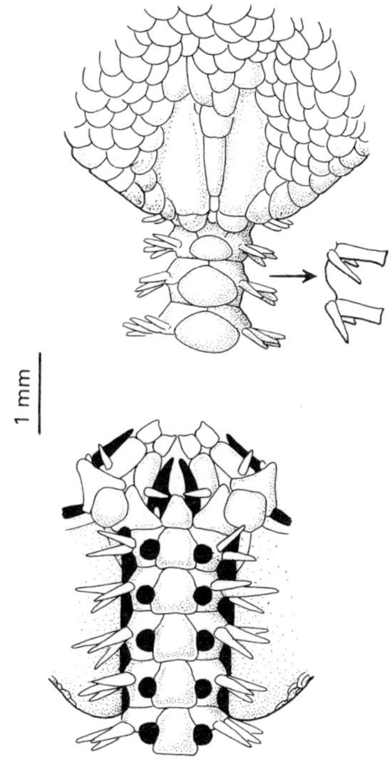

Amphiura filiformis

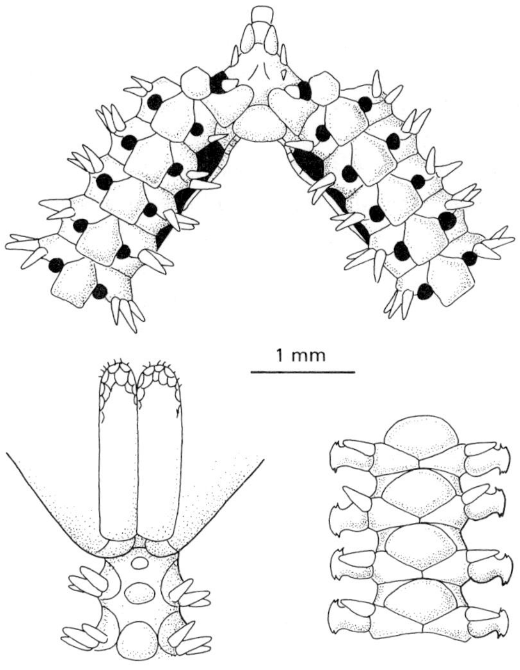

Amphiura securigera

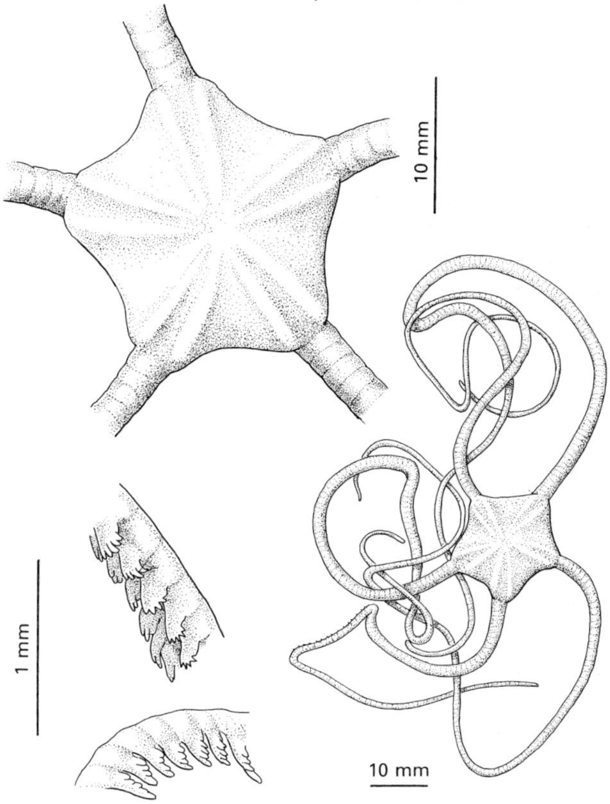

Asteronyx loveni

Fig. 15.9

stumpy conical spines. Radial shields prominent, forming a ridge. Distinctly branching arms covered with fine grains, arm spines consist of minute hooklets. Disk appears to overlay first arm joint.

Usually found in water deeper than 100 m; reported from scattered localities off northern and western coasts.

7 ASTERONYCHIDAE

Asteronyx loveni Müller and Troschel FIG. 15.9
Maximum disk diameter 35 mm. Dorsal surface with a few scattered, small, plates (especially in dried specimens) and long, narrow, radial shields which appear ridged. Disk covered with thick skin. No dorsal arm plates. Ventral plates present but covered in thick skin. Ventral arm spines club-shaped. Other arm spines small and hook-shaped.

Usually found only below 100 m depth, often commensal on pennatulids. Reported from scattered localities off northern coasts.

D. Class **Echinoidea**

Echinoderms with a more or less globular or disc-shaped rigid test consisting of 20 meridional rows of plates running from the apical region above, to the oral region on the underside. Of these rows, five pairs are *ambulacral* plates and five *interambulacral*. The ambulacral plates are each pierced by several pairs of pores, each pore-pair corresponding to one tube foot. Otherwise the test is densely furnished with long or very long mobile spines which articulate with fixed spine bosses on the test. Small, spherical, flexibly-stalked bodies called *sphaeridia*, thought to act as statocysts, may be present. More important are the three-jawed *pedicellariae* (Fig. 15.2) of up to five distinctive types, each with its particular distribution, specialized structure, and function. These are: firstly, the very small *triphyllous* forms; the larger *ophiocephalous*, with long flexible (snakelike) stalks; the long-jawed *tridactyle* pedicellariae, and lastly the taxonomically important *globiferous* pedicellariae. The latter are stiff-stalked with swollen glandular (toxin secreting) jaws whose blade-ossicles are distinctive. A fifth type sometimes occurs. The apical region consists of five radial *ocular* plates and five interradial *genital* plates; one of the latter being enlarged as the *madreporite*. Within the test in the region of the mouth there is in some groups a complex dental apparatus known as 'Aristotle's Lantern', which operates five teeth or jaws which project from the mouth. The buccal region surrounding the mouth is then flexible to facilitate the grazing action of the teeth. There may be ossicles embedded in the flexible buccal membrane but they are never rigidly connected. There are five pairs of buccal tube feet arising from the membrane and five pairs of branched gills. The latter arise at the edges of the buccal membrane in the angles between ambulacral and interambulacral plates, these angles being referred to as '*gill-cuts*'.

There are five existing orders of Echinoidea. Two are radially symmetrical, the Cidaroidea and Diadematoidea, of which only the second is represented in the British sea area; the secondarily bilateral Clypeastroidea and Spatangoidea are also found, but the Cassiduloidea do not occur here.

Order **Diadematoidea**

Test circular; spherical or flattened. Apical area with central anus. Peristome not covered by regular rows of ambulacral and interambulacral plates. Gills on peristome. Well developed dental apparatus ('lantern') present. Ambulacral plates usually compound, each section with one tube foot and one primary spine. Sphaeridia present. Four types of pedicellariae. Three suborders, of which only the Camarodonta occur in British waters (families Echinidae, Toxopneustidae, and Strongylocentrotidae).

Order **Clypeastroidea**

Test often more or less elongated and flattened. Mouth central. Anus also on underside posterior to mouth. No gills on peristome. Well-developed dental apparatus ('lantern'). Spines and tube feet small. Only three types of inconspicuous pedicellariae. Mostly tropical forms; only the family Fibulariidae represented in the region.

Order **Spatangoidea**

Test more or less elongated but not greatly flattened. Mouth anterior. Anus posterior. No gills on peristome. No dental apparatus. Ambulacra often petaloid on upper surface. Distinctive posterior interambulacrum modified on underside as sternum with its first plate forming *labrum* or lip to the mouth. Several different types of specialized spines for burrowing, sand moving, funnel building. Small, ciliated, club-shaped spines occur in distinctive rows known as *fascioles* and function to set up water currents. Tube feet also diversified, including long penicillate forms. Rostrate (beaked) pedicellariae in addition to usual types. Sphaeridia present.

Two suborders: Amphisternata and Meridosternata; only the former is represented in British waters, by the family Spatangidae.

KEY TO FAMILIES

1. Anus situated within apical system, test radially symmetrical, gills sometimes present on peristome **2**
   Anus outside apical system, test secondarily bilateral, no gills on peristome **7**

2. Peristome region includes both ambulacral and interambulacral plates; no gills on peristome; no sphaeridia; no compound ambulacral plates. Order Cidaroidea (not found in Britain)
   Peristome includes only reduced ambulacral plates but no interambulacral; ambulacral plates usually compound; gills on peristome; sphaeridia present Order Diadematoidea **3**

3. Lantern teeth lack inner keel. Central plate of each compound ambulacral triplet is the largest. Spine bosses perforated. Suborder Aulodonta. Test flexible **1. Echinothuridae**
   Teeth with an inner keel. Spine bosses not perforated **4**

4. Epiphyses of dental lantern do not form arcs above the teeth. Central plate of triplet largest. Suborder Stirodonta. Short primary spines; four or five periproct valves **2. Arbaciidae**
   Dental lantern includes arcs above the teeth; lowermost plate of triplet largest (Suborder Camaradonta) **5.**

5. Conspicuous slits on peristome mark position of gills, globiferous pedicellariae have dumbbell-shaped ossicles in skin but lack side teeth to jaws **4. Toxopneustidae**
   Slits marking gills inconspicuous; globiferous pedicellariae lack the dumbbell-shaped ossicles but sometimes have lateral teeth (see Fig. 15.2) **6**

6. Globiferous pedicellariae have head attached to stalk without neck; lateral teeth present below terminal tooth of each jaw **3. Echinidae**
   Globiferous pedicellariae have a flexible neck and their jaws lack lateral teeth **5. Strongylocentrotidae**

7. Well-developed teeth and 'lantern' apparatus; mouth central; test usually flattened; some fine tube feet on interambulacra. (Order Clypeastroidea) **6. Fibulariidae**
   No teeth or 'lantern'; mouth often slightly forward of centre, tube feet only on ambulacra **8**

8. Mouth central or slightly forward; interambulacral regions all similar; no fascioles: Order Cassiduloidea (not found in Britain).
   Mouth anterior; unpaired posterior interambulacrum forming a lip to the mouth, followed by two large plates forming a 'plastron'; fascioles present (Order Spatangoidea). Plastron plates side by side: (Suborder Amphisternata). Subanal fasciole present **7. Spatangidae**

1 ECHINOTHURIDAE

Test flattened, soft and flexible; peristome covered with ambulacral plates; gills vestigial. Strongly coloured deep sea forms.

Araeosoma fenestratum (Wyville Thomson) FIG. 15.11

Test up to 280 mm (horizontally). Generally brownish-violet. Test very flexible, leathery, and lacking the fully calcified plates of other echinoids; often collapses on collection.

Depth range 145–900 m. Off the western coasts of British Isles. A related, smaller, bright red species, (*A. hystrix* (Wyville Thomson)) occurs in deeper waters.

2 ARBACIIDAE

Test compact, rigid. Spines short, smooth and solid. Ambulacral plates usually of three primary plates. Three pore-pairs.

Arbacia lixula (Linnaeus) FIG. 15.10

Test up to 50 mm diameter; pink with ten red ambulacral lines; spines up to 30 mm, solid and sharp-pointed, black. Test low and characterized by very large peristomial orifice. Three pore-pairs per ambulacral plate.

Often associated with *Paracentrotus* in the Mediterranean but of no culinary virtue. Lowest shore level to 50 m depth. Besides Mediterranean found on Atlantic coasts of Portugal and North Africa.

3 ECHINIDAE

Ambulacral plates with three or more pore-pairs. Test without depressions. Jaws of the globiferous pedicellariae have one or two pairs of teeth below the single terminal tooth.

1. Three pore-pairs to each ambulacral plate **2**
   Five or six pore-pairs to each ambulacral plate ***Paracentrotus lividus***

2. Peristome membrane set with thick plates; globiferous pedicellariae with grooved jaw blades and a row of strong teeth along each side ***Psammechinus miliaris***
   Peristome membrane with a few thin plates; globiferous pedicellariae have tubular jaw-blades with only one or two lateral teeth **3**

3. Ambulacra have primary tubercles (and spines) on only every second plate at most **4**
   All the ambulacral plates bear a primary tubercle. Globiferous pedicellariae with two or three lateral teeth to jaw-blades ***Echinus elegans***

4. Test globular reddish, tubercles white; secondary spines numerous, similar in length to primaries; buccal ossicles bear some small spines ***Echinus esculentus***
   Test sub-conical; red colour not uniform, often broken by vertical white bands; secondary spines smaller than primaries, no spines on buccal ossicles ***Echinus acutus***

Paracentrotus lividus (Lamarck) FIG. 15.10

Test up to 70 mm diameter; spines dark violet to dark olive, rarely brownish; test greenish. Test rather low with comparatively small peristome orifice; primary tubercles very regular. Primary spines fairly long if not worn down. Five or six pore-pairs per ambulacral plate. Globiferous pedicellariae have grooved jaw blades but these do not have a row of sharp spines each side as seen in the next species.

Grazes on algae; breeds in summer. Gonads have commercial value as seafood delicacy in Mediterranean. Intertidal and shallow sublittoral (to 30 m). In rock pools and other holes bored in softer rocks, often in large groups; also amongst *Zostera* and coralline strands. Common south and west Ireland. Rare Devon and Cornwall. Elsewhere southwards to Morocco, Mediterranean, Canaries, and Azores.

Psammechinus miliaris (Gmelin) FIG. 15.10

Test up to 50 mm diameter; greenish with violet tips to spines. Test robust, slightly flattened; spines mostly of similar shortish size and closely packed. Ambulacral plates have three pore-pairs and one primary tubercle. Globiferous pedicellariae abundant, their jaw-blades grooved and bearing a row of sharp spines along each side. The flexible buccal membrane closely set with ossicles.

Omnivorous, minor pest of oyster beds. Breeds early summer. Said to form hybrids with *Echinus esculentus*. Frequently harbours polychaete *Flabelligera affinis* M. Sars amongst spines. Under stones and rocks, and amongst *Zostera*; intertidal, occasionally down to about 100 m. Common on all coasts of British Isles. Distributed elsewhere from Iceland, Scandinavia, Baltic, North Sea, southwards to Morocco and the Azores. Not present in Mediterranean, where a separate species (*P. microtuberculatus* (Blainville)) occurs and extends to Portugal.

Echinus esculentus Linnaeus FIG. 15.10

Test up to 160 mm at about 8 years old, maximum 176 mm; approximate age of small specimens: 55 mm = 2 yrs, 80 mm = 3 yrs, 100 m = 4 yrs. Colour of test red or sometimes pale purple verging greenish; spines reddish with violet tips. Test large and globular. Primary and secondary spines and consequently their tubercles (bosses), of similar size but distinguishable. Ambulacra have a primary tubercle on only every second or third plate. Globiferous pedicellariae large, their jaw-blades bearing only one lateral tooth each side. Buccal plates bear some small spines in addition to pedicellariae.

Omnivorous, feeding on *Laminaria*, Bryozoa, barnacles, etc. Gonads regarded as a delicacy in Portugal and elsewhere. Exploitation of the dried tests as souvenirs and lamp shades has lead to concern for their conservation. On rocky substrata from infralittoral fringe (low tide), especially at 10–40 m and down to 1200 m. Common on all coasts of British Isles. Elsewhere: Iceland, Finmark, to Portugal. Not Mediterranean.

Echinus acutus Lamarck FIG. 15.10

Adult test up to 150 mm diameter; white with brownish stripe down each row of plates; alternating dark and light growth bands to larger plates. Spines usually reddish with white tips. Test less

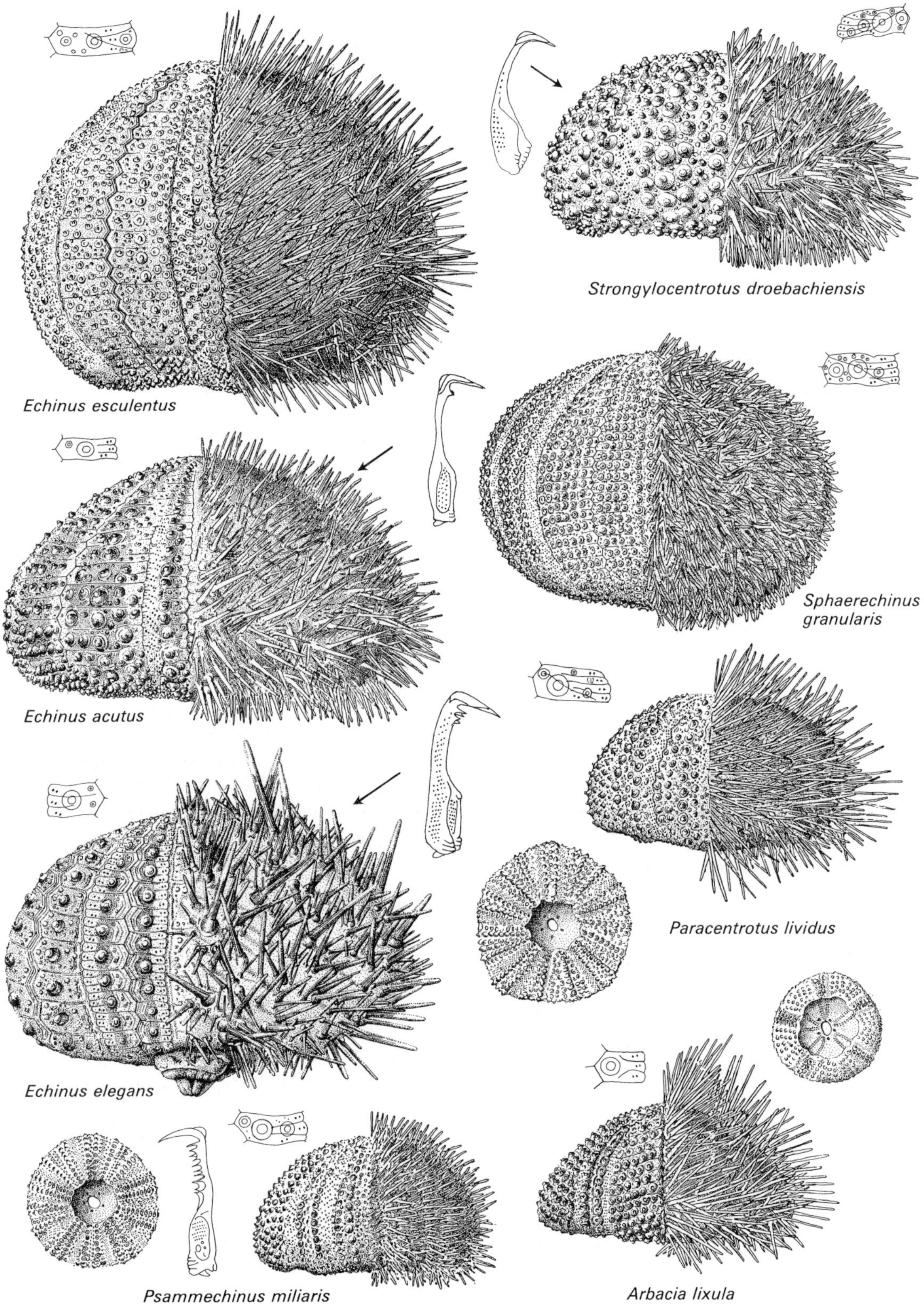
Echinus esculentus
Strongylocentrotus droebachiensis
Echinus acutus
Sphaerechinus granularis
Echinus elegans
Paracentrotus lividus
Psammechinus miliaris
Arbacia lixula

Fig. 15.10

globular than *E. esculentus* and slightly conical. Rather sparsely covered by spines on upper half; secondary spines shorter and more slender than primaries. Only alternate ambulacral plates bear primary spines. Globiferous pedicellariae with one pair of lateral teeth. Pedicellariae, but no spines, present on buccal ossicles.

On mixed bottoms, 200–1000 m (20–1200 m). Off the south-west and north coasts of British Isles, common. Distributed elsewhere from Bear Island to North Africa and the Mediterranean.

Echinus elegans Düben and Koren FIG. 15.10

Adult test up to about 80 mm; white with reddish tinge on top (occasionally violet); spines pink with white tip. Test slightly depressed. All the ambulacral plates bear a primary tubercle and these are arranged in very distinctly marked rows. The primary spines are larger than those of other species and longer than the secondaries. Globiferous pedicellariae with two or three pairs of spines on jaw-blades.

Deep water species (50–2000 m). Uncommon; off north and west coasts of British Isles, east coast of Scotland; elsewhere southern Norway, southern Iceland to Biscay and the Azores.

4 TOXOPNEUSTIDAE

Full radial symmetry. Three or more pore-pairs per ambulacral plate. Gill-cuts well marked. Globifeous pedicellariae jaw-blades lack lateral teeth. Principally an Indo-Pacific family; in Europe, only one species.

Sphaerechinus granularis (Lamarck) FIG. 15.10

Test up to about 130 mm diameter; dark purple; spines violet with white tips. Low test (height about two-thirds of diameter), covered by closely packed, short thick spines. All ambulacral plates bear a primary spine. Globiferous pedicellariae with rather squat jaws, their blades lacking lateral teeth; the spicules of these pedicellariae are C-shaped. Buccal plates lack spines.

Infralittoral fringe and shallow water, 0–50 m amongst *Zostera*. The Channel Is. represent its northern limit; ranges southwards to Cape Verde, Azores, and Mediterranean where it is very common.

5 STRONGYLOCENTROTIDAE

Full radial symmetry. More than three pore-pairs per ambulacral plate. Gill-cuts small and inconspicuous. Globiferous pedicellariae lack lateral teeth to jaw-blades which are distinctively tubular. Only one genus known. One species in the British Isles.

Strongylocentrotus droebachiensis (O. F. Müller) FIG. 15.10

Test up to about 80 mm in diameter; greenish brown; spines variably reddish, greenish, or even violet, tips usually white. Low symmetrical test; usually five or six pore-pairs to each ambulacral plate. Usually two of the ocular plates abut on periproct. Primary and secondary spines of similar length. Globiferous pedicellariae large with numerous spicules arranged in two rows in each jaw. Buccal plates lack spines.

Mid- and infralittoral fringe, occurring down to 1200 m. Not common around British Isles, confined to east coasts; not found on west coasts. A northern, circumpolar species occurring off Greenland, Spitzbergen, and south to New Jersey; in the North Pacific southwards to Vancouver and Korea.

6 FIBULARIIDAE

Very small clypeastroids in which the ambulacra are only slightly petaloid. The anus is situated on the lower surface a short distance behind the mouth. One British species.

Echinocyamus pusillus (O. F. Müller) FIG. 15.11

Test not more than about 15 mm, grey or greenish; very small, flattened, slightly elongated, bilaterally symmetrical. Six to nine pore-pairs per ambulacal plate. The four genital pores are larger than the ocular pores.

In coarse sand or fine gravel on sheltered shores, and sublittorally down to 1250 m; common off all British coasts. Elsewhere Finmark, Baltic, Iceland, and south to West Africa, Azores, and the Mediterranean Sea.

7 SPATANGIDAE

The order Spatangida is divided into two suborders, the Amphisternata and Meridosternata. The latter are exclusively deep-water forms often of extraordinary shape; the suborder Amphisternata is divided into five families of which only one, the Spatangidae, occurs in the British sea area. This family is characterized by a ring of ciliated spines (the 'sub-anal fasciole') situated below the anus and enclosing a group of 'sanitary-drain-building' tube feet. There are six British species.

1. Only one ring of ciliated spines present (the sub-anal fasciole) **2**
   Two rings of ciliated spines **3**
2. Sub-anal ring about three times wider than height; test low ***Spatangus purpureus***
   Sub-anal ring width less than twice height; test high ***Spatangus raschi***
3. Ring of ciliated spines enclosing anterior ambulacrum (inner fasciole) **4**
   Ring of ciliated spines enclosing whole petaliferous area (peripetalous fasciole) on upper surface ***Brissopsis lyrifera***
4. Anterior ambulacrum recessed as far as apical system; tube foot pore-pairs of anterior ambulacrum in close double row ***Echinocardium cordatum***
   Anterior ambulacrum flush with interambulacrum **5**
5. Labral plate not extending beyond middle of adjacent ambulacral plate; no large tubercles on upper inter-ambulacral areas ***Echinocardium pennatifidum***
   Labral plate pronounced, reaching second adjacent ambulacral plate; large tubercles on all inter-ambulacra of upper surface ***Echinocardium flavescens***

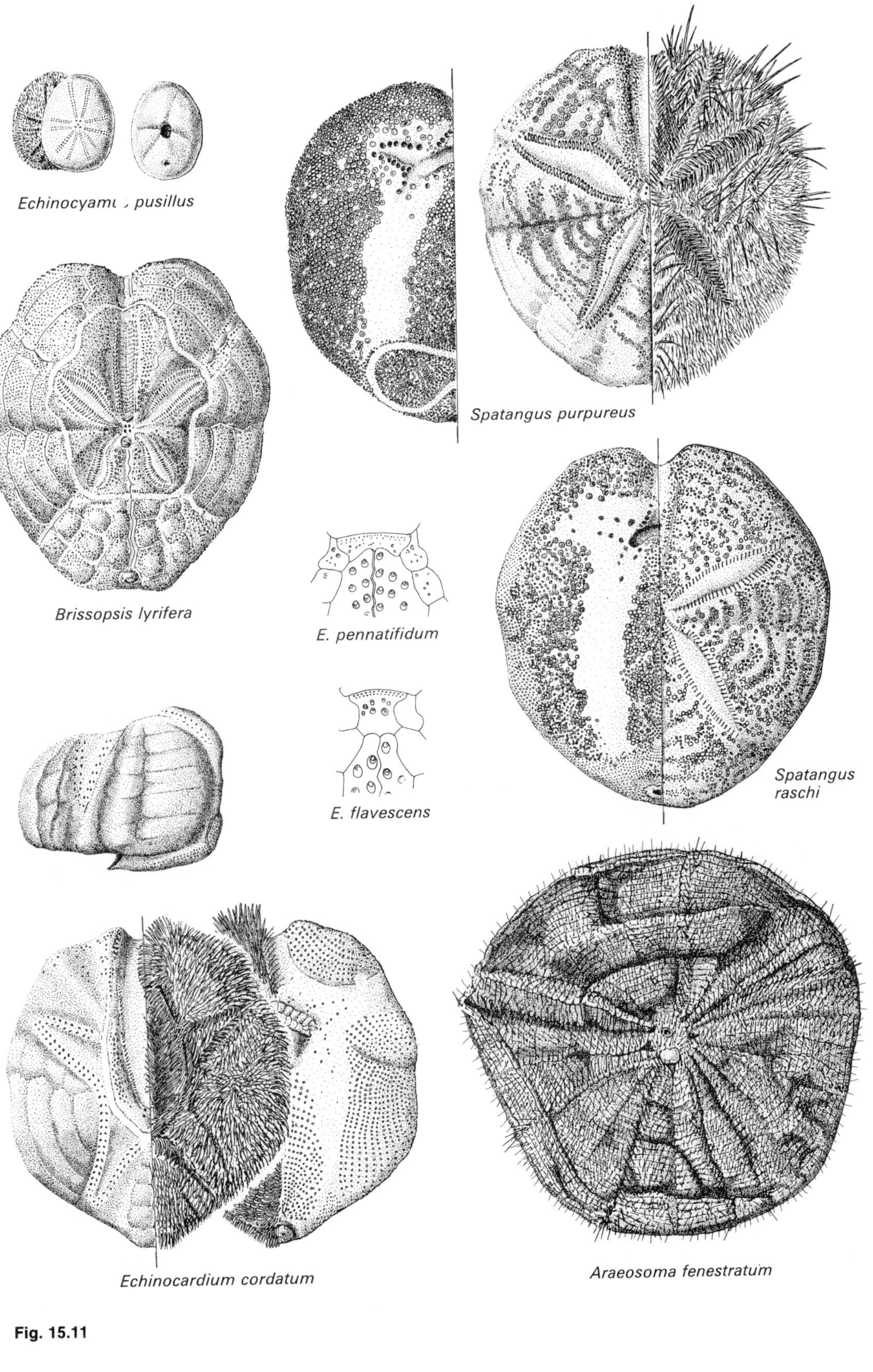
Echinocyamus pusillus
Spatangus purpureus
Brissopsis lyrifera
E. pennatifidum
E. flavescens
Spatangus raschi
Echinocardium cordatum
Araeosoma fenestratum

Fig. 15.11

Spatangus purpureus O. F. Müller FIG. 15.11

Test length up to about 120 mm, deep violet; some of long upper spines white; test broad and low, heart-shaped. Labrum rounded. The sub-anal fasciole about three times as broad as high. The four petaloid ambulacral areas on upper surface not enclosed by a fasciole. Large spine tubercles present on the interambulacral, but not on ambulacral areas.

Occurs shallowly buried in coarse sand or gravel (rarely mud), in sheltered areas in the infralittoral fringe, down to about 900 m. Locally common around coasts of British Isles; distributed elsewhere from North Cape, Norway, to North Africa, the Azores, and the Mediterranean.

The bivalve mollusc *Montacuta substriata* (Montagu) is often found attached to spines.

Spatangus raschi Lovén FIG. 15.11

Size and colour similar to *S. purpureus* but upward sloping at front. Mouth somewhat sunk but labrum more pronounced than in *S. purpureus*. The sub-anal fasciole (ring of ciliated spines) less than twice as wide as high. Some large spine tubercles present in the ambulacra beyond the petal-shaped areas.

Occurs in muddier conditions than *S. purpureus*. In the British Isles only off the west coast of Ireland, often abundant at 150–1000 m. Distributed elsewhere from Norway and Shetland, to the Azores. Not in the Mediterranean.

Echinocardium cordatum (Pennant) FIG. 15.11

Adults usually about 60 mm in length, but up to 90 mm; fawn in colour, spines yellowish. Anterior ambulacrum slightly recessed up to and including part of the apical area, both being surrounded by the ring of ciliated spines, the 'inner fasciole'. Highest part of test posterior to apical system. Some large spines and their prominent tubercles scattered on the anterior interambulacra. The two series of tube feet in the anterior ambulacrum each forms a double row as seen by the pore-pairs on the test.

Occurs buried about 80 mm deep (to 150 mm) in sand, mainly intertidal (lower midlittoral and infralittoral fringe) but also sublittoral to 230 m. Common to abundant on all British Isles coasts. Elsewhere, almost cosmopolitan, Norway to South Africa, Mediterranean, Australia, and Japan.

The bivalve mollusc *Tellimya ferruginosa* (Montagu) is often associated with the burrow near the spines of the sub-anal fasciole. The amphipod crustacean *Urothöe marina* (Bate) is another common commensal.

Echinocardium flavescens (O. F. Müller) FIG. 15.11

Up to about 70 mm, colour yellowish or sometimes pink. Almost circular; anterior ambulacrum not recessed. Highest point of test at front of inner fasciole. Some larger spine tubercles scattered in all the interambulacra. The tube foot pore-pairs of the anterior ambulacrum form two simple rows. Labrum reaching adjacent second ambulacral plates. Valves of triphyllous pedicellariae nearly smooth.

Less common than *E. cordatum*; sublittoral down to 300 m, rarely found intertidally. All British coasts; distributed elsewhere from Finmark and Iceland south to the Azores and the Mediterranean.

Echinocardium pennatifidum Norman FIG. 15.11

Up to 70 mm, yellowish in colour. Almost circular, rather flattened; anterior ambulacrum not recessed. Highest point about half-way back. No larger spine tubercles in any of the interambulacra on the upper surface. Anterior ambulacrum has the pore-pairs in two single files. Labrum very short, not extending beyond the first adjacent ambulacral plate. Valves of triphyllous pedicellariae serrated along whole edge.

Intertidal (Herm, Channel Islands) to 250 m. Locally common, all coasts of British Isles except east. Distributed from Norway to the Mediterranean.

Brissopsis lyrifera (Forbes) FIG. 15.11

Adult test 40–70 mm in length; colour reddish brown. Test rather fragile, slightly longer than wide, notched in position of anterior ambulacrum; all ambulacra slightly recessed above; highest point of test behind the apical system. On the upper surface the ambulacra form distinctive petal-shaped areas. The petaliferous fasciole of ciliated spines encloses all five ambulacral petals. The posterior petals diverge.

Lives buried in mud, sublittorally 5–500 m. All British coasts except south, sometimes abundant. Elsewhere from Norway and Iceland to South Africa and the Mediterranean, also east coast of North America; not Greenland. Another species, *B. atlantica* Mortensen, with parallel posterior 'petals', occurs in deeper water.

E. Class **Holothurioidea**

The holothurians, or sea-cucumbers, are elongate sausage-shaped animals. In cross-section they may be round or slightly dorso-ventrally flattened, the ventral surface being recognized by three longitudinal rows of tube feet, the *trivium*, and the dorsal surface by two longitudinal rows, the *bivium* (Fig. 15.2). The *mouth* is at the anterior end, occasionally situated either slightly dorsally or ventrally. Surrounding the mouth are the *tentacles* which may be bush-, shield-, or feather-shaped, or simply digitate. The tentacle shape is constant at order level and therefore of taxonomic importance. The mouth leads to the oesophagus which is surrounded by a *calcareous ring*, usually consisting of five radial and five interradial pieces but which may show considerable structural variation. Behind the calcareous ring is the *water vascular ring* from which radial water vessels issue, running forwards inside, then posteriorly outside, the calcareous ring. These water vessels give off *tentacle canals* from which may arise *tentacle ampullae* that, if present, lie free in the body cavity. Further details of the water vascular system may be found in Hyman (1955). The gut is a long, looped, simple tube ending in the *cloaca* (rectum). From the cloaca issue the *respiratory trees* into which water is pumped by the cloaca. These structures are well-developed in the Aspidochirotea but totally lacking in the synaptids. The skin of holothurians is usually thick, except in the Apoda where it is thin and transparent. There is no rigid skeleton, as found in the other classes of echinodems, but the skin of holothurians contains numerous delicately structured microscopic *calcareous bodies* (Fig. 15.2). These bodies include *tables, disks, wheels, anchors*, and some amorphous shapes. These are primary taxonomic characters. Under the epidermis is a layer of circular muscle with five *radial muscles* which run the full length of the body. *Retractor muscles* issue from the radial muscles and attach to the calcareous ring. They retract the anterior part of the body (the *introvert*) including the mouth and tentacles. *Tube feet* may occur in the trivium or bivium or all over the body, but never in a furrow as in other echinoderms. In some subfamilies the tube feet are entirely lacking, or occur as a series of papillae round the anus. The gonads are found only in the mid-dorsal interradius.

Holothurians generally require preliminary preparation before they can be identified. It is essential that specimens be relaxed by narcotization in order for the morphology of tentacles and tube feet to be observed. Large specimens should be left overnight, in cool conditions, in a bowl of seawater to which a scattering of menthol crystals has been added. Small specimens usually relax in 7% magnesium chloride, but up to 8 hrs may be necessary before narcotization is complete. Alternatively, propylene phenoxetol, pipetted slowly into a bowl of seawater to form a globule 2 cm in diameter, has been used successfully to achieve rapid narcotization. Dissection to view internal anatomy of relaxed specimens requires a single, longitudinal, dorso-lateral cut. The calcareous deposits present in the body wall of most holothurians are especially important in identification. To observe these, a small portion (ca. 1 cm^2) of body wall should be removed and digested in domestic bleach, in a solid watchglass. Once the tissue has dissolved the bleach should be pipetted out, and the residue washed with several changes of distilled water. The deposits may be viewed directly, using a high power binocular microsocpe with dark ground illumination, or transferred to glass cavity slides and permanently mounted. Fine details are then revealed by high power compound microscopes.

KEY TO FAMILIES

1.	Tube feet well-developed; body form various, but usually stout, not worm-like	2
	Tube feet absent; body form slender and worm-like (Order Apoda)	**6. Synaptidae**
2.	Tentacles shield-shaped. No retractor muscles (Order Aspidochirota)	3
	Tentacles bush-shaped. Retractor muscles present (Order Dendrochirota)	5
3.	Tentacle ampullae absent	**1. Synallactidae**
	Tentacle ampullae present	4
4.	Gonads on both sides of dorsal mesentery	**2. Stichopididae**
	Gonads only on left side of dorsal mesentery	**3. Holothuriidae**
5.	Body cylindrical or uniform without well-defined ventral sole	**4. Cucumariidae**
	Body more or less flattened with, obvious ventral sole	**5. Psolidae**

Order Aspidochirota

1 SYNALLACTIDAE

Only one species occurs in shallow water around the British Isles.

Mesothuria intestinalis (Ascanius and Rathke) FIG. 15.12

Body cylindrical, up to 300 mm long, with 20 tentacles around a subventral mouth. Tube feet best developed in the ventrolateral ambulacra with only a few on the dorsal surface. The superficially lying calcareous deposits are quadri-radiate tables, the spire composed of four rods ending in a crown of thorns. Disk of table contains an inner circle of primary holes surrounded by smaller holes.

Found off the west coasts of Britain on muddy bottoms often covered in shell fragments. Distributed from Barents Sea southwards to the Azores and the Mediterranean.

2 STICHOPODIDAE

Only one species known to occur in relatively shallow water round the British Isles.

Stichopus tremulus (Gunnerus) FIG. 15.12

Body up to 500 mm long, cylindrical with a flattened ventral surface. The mouth subventral; 20 tentacles surrounded by a fringe of anterior papillae united to form a web. Tube feet generally small. Numerous calcareous deposits consisting of tables with serrated disks, the rods in the spire containing two sets of cross-beams. Deeper deposits may be star- or rod-shaped.

Found mainly off northern and western coasts of Britain. Distributed from northern Norway to Biscay and the Canary Isles.

3 HOLOTHURIIDAE

Only one species occurs in British coastal waters.

Holothuria forskali Delle Chiaje FIG. 15.12

A large species, up to 200 mm long, with a flattened ventral side bearing numerous tube feet in three or four series. The dorsal surface is black with the ventral surface either brown or yellow. Usually 20 tentacles around a terminal mouth, and well-developed tentacle ampullae. Dorsal surface covered with conical papillae. Calcareous deposits are scarce, consisting of a small disk with four holes in the skin, branched rods in the tube feet, and curved rods in the tentacles.

Generally found in shallow water on the west and south-west coasts of Britain. Present on all Atlantic coasts of Europe, and in the Mediterranean.

Order **Dendrochirota**

4 CUCUMARIIDAE

1. Ten tentacles **2**

   Fifteen to thirty tentacles **11**

2. Tube feet in distinct series along length of body **3**

   Tube feet scattered over whole body **8**

3. Body elongate, attenuating, curving upwards anteriorly and posteriorly ***Trachythyone elongata***

   Body not elongate or attenuating or curving upwards **4**

4. Very large, up to 500 mm; thick skin with a few spicules ***Cucumaria frondosa***

   Small forms, rarely longer than 150 mm; with numerous calcareous deposits **5**

5. Calcareous deposits smooth **6**

   Calcareous deposits mostly tuberculate **7**

6. Deposits large fenestrated plates, no star-shaped plates in outer skin ***Paracucumaria hyndmani***

   Deposits small, button-like, with four main holes or holes at either end. Small star-shaped bodies in outer layer of skin. ***Pawsonia saxicola***

7. Deposits in outer layer of skin cup-shaped ***Aslia lefevrei***

   Very small forms, less than 40 mm long. Tube feet scarce, if present in zigzag rows. Deposits in outer layer of skin as irregular stars ***Ocnus lactea***

8. Calcareous deposits in skin over the whole of the body **9**

   Calcareous deposits absent, or in posterior part of body only **10**

9. Deposits with a spire of two rods ***Thyone fusus***

   Deposits with no spire but smooth holes. Body with posterior prolongation ***Thyone raphanus***

10. Deposits present in posterior body only. Skin thin, transparent, rosy ***Thyone inermis***

    No deposits at all except in terminal disk of tube feet. The ten tube feet surrounding anal opening contain simple rods. Skin thick ***Thyone roscovita***

11. Tube feet confined to radii ***Neopentadactyla mixta***

    Tube feet spread all over body **12**

12. Body wall thin, transparent. Few large tube feet. Tentacles of inner circle close together ***Thyonidium pellucidum***

    Body wall thick. Numerous fine tube feet. Tentacles of each pair of inner circle distinctly separate ***Thyonidium commune***

Cucumaria frondosa (Gunnerus) FIG. 15.12

Body cylindrical, dark brown, up to 500 mm long. Tentacles thick, bushy. Tube feet in numerous series. Skin very thick, with few deposits; fenestrated plates present in young specimens, but older animals almost completely lacking them.

Lower shore and shallow sublittoral, to perhaps 200 m. Often common among kelps. West Scotland, from the Clyde Sea to Shetland. Distributed northwards to Arctic waters.

Trachythyone elongata (Düben and Koren) FIG. 15.12

Body elongate, tapered at both ends, up to 150 mm; dark brown in colour. Tentacles short. Tube feet in five distinct rows; not completely retractile. Three types of calcareous deposits: large, smooth, irregular plates, with numerous holes, in deepest layers of skin; small, rounded plates with few holes in intermediate layers of skin, small, cup-shaped deposits close to skin surface.

Shallow sublittoral, to 65 m, on muddy bottoms where it is commonly partly buried in the sediment. On east coasts, as far south as Durham, on west from Shetland to the western end of the Channel. Distributed from western Norway to the Mediterranean.

Paracucumaria hyndmani (Thompson) FIG. 15.12

Body short and stout, barrel-shaped, tapered posteriorly, up to 50 mm long, greyish-white to pale red. Tube feet in conspicuous double rows, almost completely retractile. Skin thick and

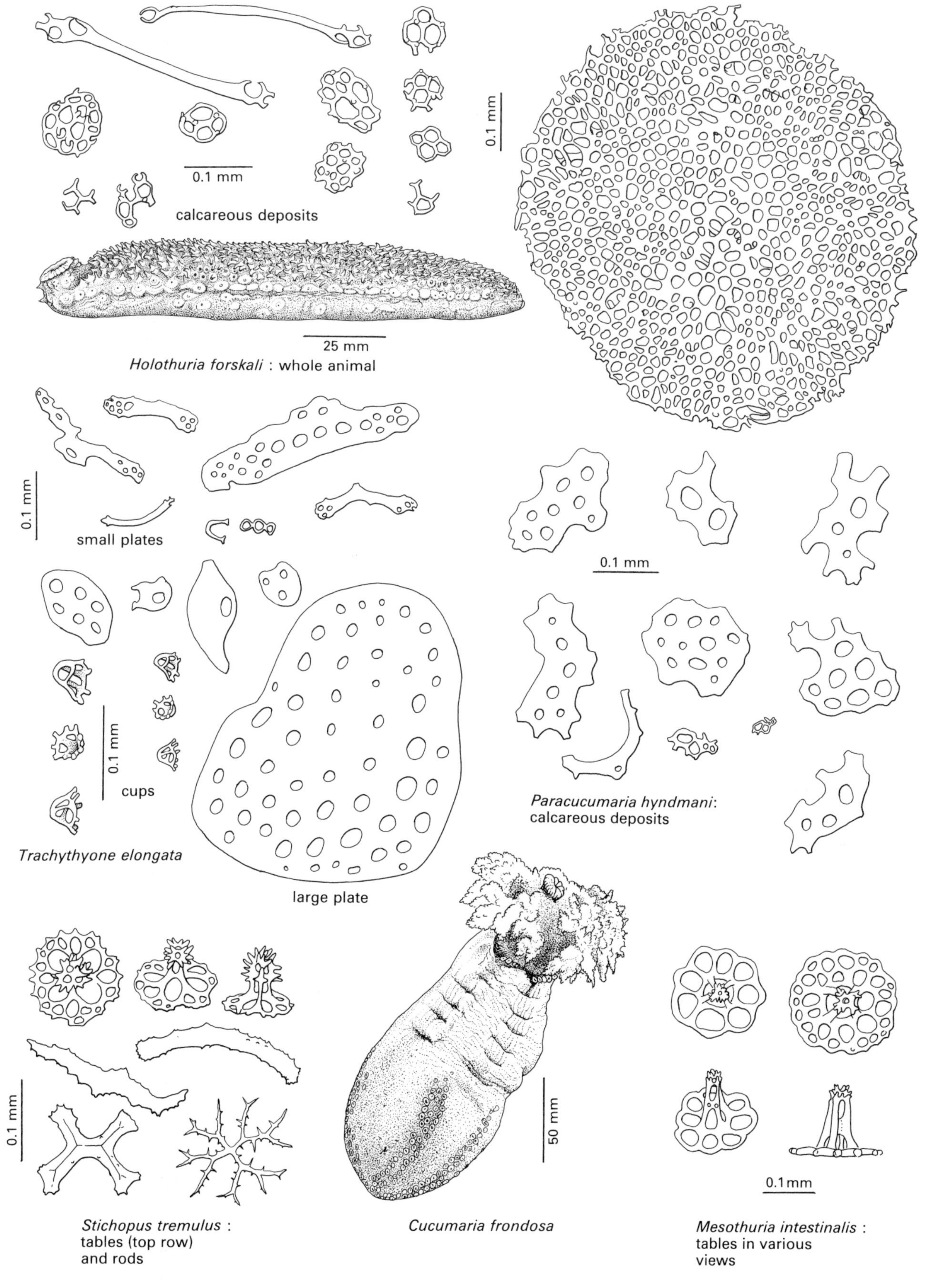
0.1 mm
calcareous deposits
0.1 mm
25 mm
Holothuria forskali : whole animal
0.1 mm
small plates
0.1 mm
0.1 mm
cups
Trachythyone elongata
large plate
Paracucumaria hyndmani:
calcareous deposits
0.1 mm
50 mm
0.1mm
Stichopus tremulus :
tables (top row)
and rods
Cucumaria frondosa
Mesothuria intestinalis :
tables in various
views

Fig. 15.12

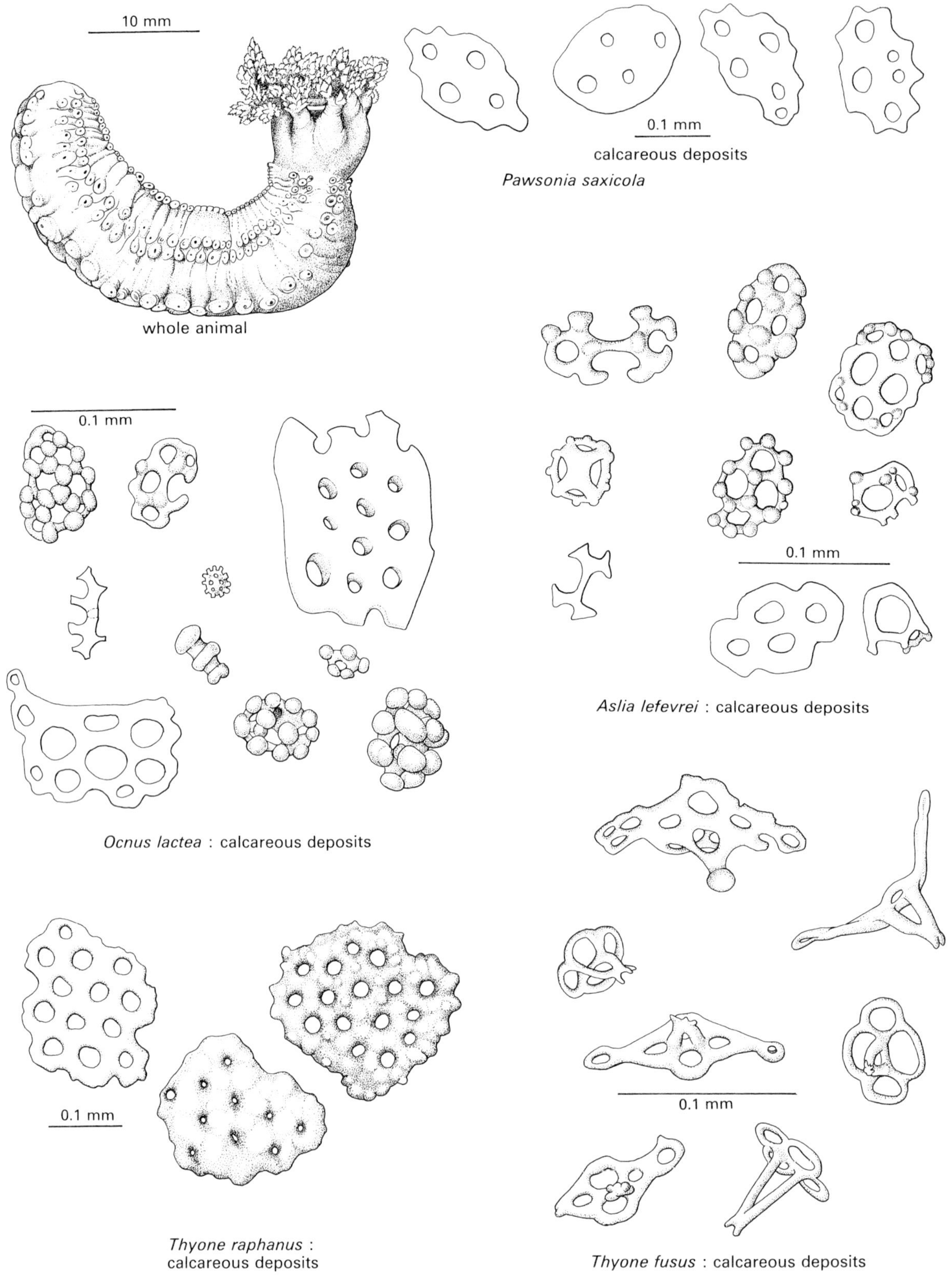
10 mm
whole animal
0.1 mm
calcareous deposits
Pawsonia saxicola
0.1 mm
Aslia lefevrei : calcareous deposits
0.1 mm
Ocnus lactea : calcareous deposits
0.1 mm
Thyone raphanus :
calcareous deposits
0.1 mm
Thyone fusus : calcareous deposits

Fig. 15.13

smooth, with large fenestrated deposits, closely packed and tending to overlap when the animal is completely contracted.

On hard grounds, offshore, to below 1000 m, usually covered with shell-fragments and gravel. On west coasts from Shetland to the west Channel, on the east as far south as St Andrews. Distributed from west Norway to the Mediterranean.

Pawsonia saxicola (Brady and Robertson) FIG. 15.13

Body cylindrical, white, with dark tentacles, up to 150 mm long. Tube feet in double longitudinal rows ventrally, reduced in size dorsally and arranged in zigzag series. Skin thin and smooth, with small, button-like deposits, each with four principal holes centrally and one or more small ones at each end; with small, star-shaped spicules scattered in outer layers of skin.

Intertidal and shallow sublittoral, in crevices, kelp holdfasts, and under stones. Common on south and west coasts of Britain and Ireland. Distributed southwards to the Azores.

Aslia lefevrei (Barrois) FIG. 15.13

Body cylindrical, off-white to brownish, up to 150 mm. Tube feet in distinct double rows, and sometimes also occurring in interradii. Skin tough, leathery, and wrinkled, with continuous layer of cup-shaped spicules close to surface, and larger deposits, each with four holes and twelve prominent nodules, in deeper skin layers. Tube feet spicules each with two or more parallel rows of holes.

Intertidal and shallow sublittoral; not common, reported from scattered localities on south-west and western coasts of Britain and Ireland. Reported elsewhere from Atlantic coasts of France.

Ocnus lactea (Forbes and Goodsir) FIG. 15.13

Body cylindrical, white to brown, up to 40 mm long. Tube feet thinly distributed, in zigzag rows, not completely retractile. Skin thick, smooth, with abundant deposits: typically with four holes and numerous nodules, or large irregular and platelike; small cup-shaped spicules in outer layers of skin.

Lower shore and sublittoral, to 100 m; on hard bottoms, among stones, shell, and calcareous algae. On west coasts of Britain and Ireland, and in the east as far south as Durham. Distributed from west Norway to Brittany.

Thyone fusus (O. F. Müller) FIG. 15.13

Body ovoid, tapered at both ends, up to 200 mm long, white to pinkish. Tube feet in longitudinal series, more or less distinct. Skin very delicate, ruptures readily on handling; with few, widely dispersed deposits, typically small, smooth tables, with four holes and a spire of two rods, fused distally and sometimes bearing apical thorns.

On shell gravels, characteristically covering itself with shell fragments. West coasts of Britain and Ireland, on the east as far south as Northumberland. Distributed from west Norway to Madeira and the Mediterranean.

Thyone raphanus Düben and Koren FIG. 15.13

Body ovoid, tapered posteriorly to form a slender 'tail'; typically curved in a U-shape; yellowish brown, up to 60 mm long. Tube feet only thinly distributed on dorsal side, and absent completely from posterior part of 'tail'. Skin thick, with numerous, closely spaced, often overlapping, deposits: large, smooth, fenestrated plates, often tuberculate.

Shallow sublittoral, to 1000 m, on sandy or muddy bottoms, typically buried in the sediment with only the posterior 'tail' projecting. All British coasts. Distributed from western Norway to the Mediterranean.

Thyone inermis Heller Not illustrated

Body tapered at both ends, pinkish coloured, up to 80 mm long. Skin thin, transparent, and lacking calcareous deposits, except for a few small disks, with spines of two rods, at posterior end of animal. Tube feet lacking spicules, but with well-developed terminal disks.

Offshore. Reported from Shetland, and perhaps distributed generally off western coasts of Britain and Ireland. Distributed southwards to the Azores and the Mediterranean.

Thyone roscovita Hérouard Not illustrated

Body up to 80 mm, grey to pinkish. Skin thick, opaque; entirely lacking deposits, except for terminal disks of tube feet, and some simple rods in circum-anal tube feet.

Intertidal and shallow sublittoral. Rare; known only from the French Channel coast, southwards to the southern Bay of Biscay.

Neopentadactyla mixta (Östergren) FIG. 15.14

Body elongate, cylindrical, up to 200 mm long, attenuated at the ends; yellowish-violet; twenty tentacles, ten small alternating with ten larger. Tube feet form six regular series. Calcareous deposits are very numerous, composed of tables with a large smooth-edged disk and a spire composed of four columns; in the tentacles the calcareous deposits are slender rods with perforated ends.

On muddy and sandy bottoms, offshore to 200 m. West coasts of Britain and Ireland. Distributed from Norway to Brittany.

Thyonidium pellucidum (Fleming) FIG. 15.14

Body fusiform, up to 120 mm long, reddish coloured. With few relatively large tube feet. The body wall is thin and rather transparent. The twenty tentacles consist of five outer pairs of larger, and an inner circle of five small tentacles, the latter being united basally. Calcareous deposits are tables with a large fenestrated serrated disk and a low spire. These are common in young animals but tend to be found only in the skin of the introvert and posterior part of the body in older specimens.

On sandy/muddy bottoms usually covered with shell fragments. South-west and west coasts and on the east coast as far south as Northumberland. Ranging northwards to Arctic waters.

Thyonidium commune (Forbes) FIG. 15.14

Body fusiform, up to 200 mm long, dull white with darker tentacles. Fine, numerous tube feet arranged in series. Skin thick. No calcareous deposits except for small tables with tall

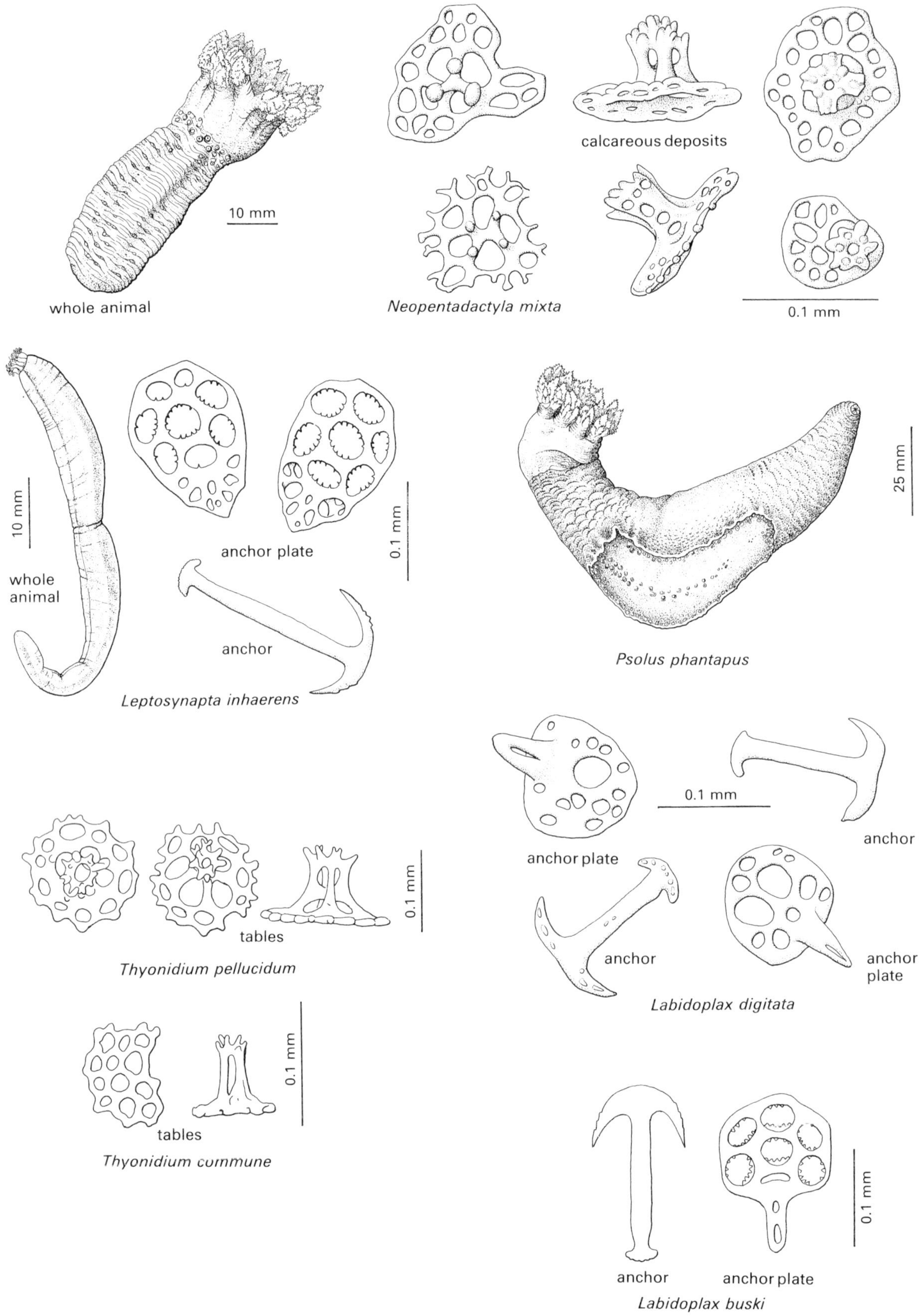
10 mm
whole animal
calcareous deposits
Neopentadactyla mixta
0.1 mm
10 mm
whole animal
anchor plate
0.1 mm
anchor
Leptosynapta inhaerens
25 mm
Psolus phantapus
0.1 mm
anchor plate
anchor
anchor
anchor plate
Labidoplax digitata
tables
0.1 mm
Thyonidium pellucidum
tables
0.1 mm
Thyonidium commune
anchor
anchor plate
0.1 mm
Labidoplax buski

Fig. 15.14

spines in the introvert. Tentacles similar in structure to *T. pellucidum*, but the inner ring not joined basally.

Ecology and distribution similar to *T. pellucidum*.

5 PSOLIDAE

Only one known British species in this family.

Psolus phantapus (Strussenfelt) FIG. 15.14

Body arched upwards at either end, posterior with a long conical tail-like prolongation; up to 150 mm long, yellowish brown to black, with orange tentacles. Ten tentacles around anterior mouth. Well-developed ventral rectangular sole containing numerous tube feet but very few calcareous deposits, consisting of either small cups or larger spherical/oval bodies of complicated structure. Apart from the sole, tube feet are found only on the soft-skinned introvert.

Lower shore to 400 m. North west coasts of Britain and Ireland, on the east coast as far south as Yorkshire. Distributed northwards to the Arctic.

Order Apoda

6 SYNAPTIDAE

1. Calcareous deposits as anchor plates with a distinct narrow handle **2**
   Anchor plate not having a narrow handle ***Leptosynapta inhaerens***
2. Eleven tentacles, each with a long terminal and one pair of smaller lateral digits ***Labidoplax buski***
   Twelve tentacles; no terminal digit, but two pairs of lateral digits at end of each tentacle. ***Labidoplax digitata***

Leptosynapta inhaerens (O. F. Müller) FIG. 15.14

Body slender, soft, pink; up to 300 mm long. Fairly transparent. 12 tentacles each with five to seven pairs of digits, the large outermost digit being unpaired. The calcareous ring comprises 12 pieces. Calcareous deposits consist of an anchor and its plate, the latter without a narrow handle. The anchor plate also has six (occasionally seven) serrate holes around a central hole.

From lower shore to 50 m. Lives buried in sand or mud. South and west coasts from Plymouth to the Shetlands; on the east coast as far south as Northumberland. Distributed from northern Norway to Brittany.

Labidoplax buski (McIntosh) FIG. 15.14

Small, up to 30 mm long, colourless. Eleven tentacles, each with one pair of lateral digits and a long terminal digit. The calcareous ring consists of 11 pieces, with two interradii in the mid-dorsal interradius. The calcareous deposits are anchors, and the anchor plate has a pronounced handle and six serrate holes and one or two slit-like holes. The anchor is narrower than the anchor plate.

Lives in muddy bottoms off west Scotland. Also reported from southern Norway.

Labidoplax digitata (Montagu) FIG. 15.14

Body up to 300 mm long, red or brown, darkest dorsally. Twelve tentacles, each with two pairs of lateral digits and a rudimentary terminal digit. The calcareous ring consists of 12 pieces with perforate radials. Anchor plates racquet-shaped with a handle and four large smooth central holes. Anchors up to 0.35 mm long with serrate arms.

Lives in muddy sandy bottoms from the lower shore down to 70 m. All western coasts, ranging southwards to the Mediterranean.

16 HEMICHORDATA AND UROCHORDATA

PHYLUM HEMICHORDATA

(Adelochordata, Stomochordes)

HEMICHORDATES comprise worm-like Enteropneusta (= Balanoglossida) and tentaculate Pterobranchia. Enteropneusts have gill slits very like those of *Amphioxus* (Ch. 17) so are linked with chordates, particularly because they also have a dorsal, sometimes tubular, nerve cord in the *collar* (a short region just behind the mouth). They lack a notochord, however, so are placed in a phylum separate from the Chordata. The 'tornaria' larvae of one family are like some echinoderm larvae. Pterobranchia are closely related to enteropneusts (see e.g. Horst 1927–39), but the only pterobranchs known from Britain belong to a minute colonial genus, atypical of hemichordates.

In Enteropneusta ('acorn-worms') the preoral *proboscis* may resemble an acorn, whilst the *collar* looks like the the acorn's cup. The long *trunk* which follows consists, first of a branchial region extending for 5–20 per cent of the total body length, marked externally by dorsolateral *gill pores*. Each pore is the opening of a gill pouch, which is associated with a *gill slit* in the pharynx wall. Each gill slit is U-shaped, because of downgrowth of a *tongue bar* from its dorsal margin, but this cannot usually be seen unless the pharynx is opened so as to view the gill slits from the inside (Fig. 16.1F,J). In Ptychoderidae and some Spengeliidae the tongue bars are joined to neighbouring gill bars by many short *synapticulae*. The second part of the trunk is often termed the oesophageal or genital region, though the gonads generally occur throughout the branchial region as well. In *Saccoglossus* and in Spengeliidae there is on each side a series of *oesophageal pores*, like reduced gill pores. Behind the oesophagus the gut has *hepatic pouches*, which form external protuberances in Ptychoderidae. Finally there is a long *intestinal region*, which is laden with muddy sand and usually tears under its own weight if the worm is not supported by water.

All British enteropneusts burrow in sand or mud. When dug up they are remarkably inactive, but very soft-bodied, so entire specimens are rarely obtained. Some species smell characteristically of iodoform.

LITERATURE

Brambell, F. W. R. and Cole, H. A. (1939). *Saccoglossus cambrensis* sp.n., an enteropneust occurring in Wales. *Proceedings of the Zoological Society, London* B, **109**, 211–36.

Brambell, F W. R. and Goodhart, C. B. (1941). *Saccoglossus horsti* sp.n., an enteropneust occurring in the Solent. *Journal of the Marine Biological Association of the U.K.*, **25**, 283–301.

Burdon-Jones, C. (1953). Records of British Ptychoderidae. *Nature, London*, **172**, 342–3.

Burdon-Jones, C. (1956). Observations on the enteropneust *Protoglossus koehleri* (Caullery and Mesnil). *Proceedings of the Zoological Society, London* B, **127**, 35–58.

Burdon-Jones, C. and McIntyre, A. D. (1960). *Stereobalanus*, a genus new to the old world. *Nature, London*, **186**, 491–2.

Burdon-Jones, C. and Patil, A. M. (1960). A revision of the genus *Saccoglossus* (Enteropneusta) in British waters. *Proceedings of the Zoological Society, London*, **134**, 635–45.

Dilly, P. N. (1985). A note on the status of *Rhabdopleura* from Bermuda. *Journal of the Marine Biological Association of the U.K.*, **65**, 987–91.

Horst, C. J. van der (1927–39). Hemichordata. *Bronn's Klassen und Ordnung das Tierreichs*, **4**, Abt. 4, Buch 2, Teil 2.

Lankester, E. R. (1884). A contribution to the knowledge of *Rhabdopleura*. *Quarterly Journal of Microscopical Science*, **24**, 622–47.

Meek, A. (1922). *Glossobalanus marginatus*, a new species of Enteropneust from the North Sea. *Quarterly Journal of Microscopical Science*, **66**, 579–94.

Stebbing, A. R. D. (1970*a*). The status and ecology of *Rhabdopleura compacta* Hincks (Hemichordata) from Plymouth. *Journal of the Marine Biological Association of the U.K.*, **50**, 209–21.

Stebbing, A. R. D. (1970*b*). Aspects of the reproduction and life cycle of *Rhabdopleura compacta* (Hemichordata). *Marine Biology*, **5**, 205–12.

KEY TO FAMILIES

1. Microscopic colonies of zooids with pinnate tentacles and interconnecting stalks, in branching annulated tubes attached to shells or old coral (PTEROBRANCHIA)
   1. Rhabdopleuridae

   Soft non-colonial worm burrowing in sand or mud, with preoral proboscis, short postoral collar and long trunk bearing (anteriorly) two rows of gill pores (ENTEROPNEUSTA) **2**

2. Proboscis not much longer than collar. Trunk bearing genital folds or wings anteriorly and protuberant hepatic pouches in a middle region **3. Ptychoderidae**
   Proboscis may be longer than collar. Trunk without prominent protuberances **3**

3. Circular muscle fibres absent from body wall of trunk **2. Harrimaniidae**
   Circular muscle fibres present in trunk, on inner side of longitudinal musculature **Spengeliidae**

The Spengeliidae are not dealt with here because they are not recorded from Europe, apart from *Glandiceps talaboti* dredged off Marseille.

Class **Pterobranchia**

1 RHABDOPLEURIDAE

Minute colonial zooids, each with a creeping cephalic shield (representing the proboscis of Enteropneusta), an inconspicuous collar bearing a pair of pinnate ciliated tentacles, a saclike body with U-shaped gut and antero-dorsal anus, and a long stalk connecting with other zooids. The branching tubes (diameter about 0.2 mm) mostly adhere to substrata and bear obliquely transverse ridges, which are interrupted in a criss-cross pattern along the free side. Erect ends of branches, from which undisturbed zooids protrude their tentacles, bear ridges which form complete rings, with no obliquity.

1. Tubes in close contact with one another, forming compact encrustations only a few mm across ***Rhabdopleura compacta***
   Tubes directed away from each other, to form straggling system of main branches and side branches, often several cm long ***Rhabdopleura normani***

Rhabdopleura compacta Hincks FIG. 16.1A
[STEBBING 1970*a*, 209]

Zooids lemon-yellow, spotted with reddish-brown and light green. Long regarded as a junior homonym of the following species, but primary zooids do not form 'embryonal rings' (Stebbing 1970*b*).

On concave (lower) surfaces of shells (valves) of *Glycymeris;* 20–100 m; near Belfast, Plymouth, and Roscoff.

Rhabdopleura normani Allman
[DILLY 1985, 989; LANKESTER 1884, 622]

Dilly regards straggling form of colony (see key) as most practicable distinction from foregoing species, because the embryonal ring of stolon (which marks colony's point of origin) is very difficult to find. Lankester described zooids from Norway as spotted with orange and black.

On stones and dead branches of coral *Lophelia;* 160–900 m; North Atlantic from Barents Sea to Azores. On coralligenous concretions at 20–50 m in Mediterranean (Banyuls).

Class **Enteropneusta**

2 HARRIMANIIDAE

Up to 300 mm long. Gill bars without synapticulae. Hepatic caeca inconspicuous, not distending body wall. Trunk without circular muscles. Yolky eggs without planktotrophic larvae.

1. Overall length 20–50 times breadth, which is less than 5 mm. Branchial region gonads all lie ventral to gill pores, which open to surface near dorsal midline (Fig. 16.1G,H) **2**
   Length about 12 times breadth, which is about 7 mm. Gonads not only ventral but also in a row dorsal to gill pores, which are thus enclosed between genital folds (Fig. 16.1E,L) **5**

2. Proboscis not more than twice length of collar. Tongue bars of gill slits short (Fig. 16.1F) ***Protoglossus koehleri***
   Proboscis at least three times length of collar. Tongue bars long (Fig. 16.1J) **3**

3. Up to 30 mm long, with nine to 22 pairs of gill pores. Collar yellow ***Saccoglossus pygmaeus***
   Much more than 30 mm long when mature, with more than 40 pairs of gill pores. Collar red, brown, or orange **4**

4. Length up to 300 mm, with 100–140 pairs of gill pores. proboscis white, yellowish, or pink, up to 20 mm long. When burrowing only one peristaltic bulge passes down proboscis at a time. Trunk lacks red spots ***Saccoglossus horsti***
   Length up to 200 mm, with 50–95 pairs of gill pores. proboscis pink, up to 30 mm long. When burrowing each bulge appears at tip before preceding one has reached posterior end of proboscis. Pale brownish trunk bears occasional small red spots ***Saccoglossus ruber***

5. Gill pores dorsolateral (Fig. 16.1E) ***Harrimania kupfferi***
   Gill pores lateral (Fig. 16.1L) ***Stereobalanus canadensis***

Harrimania kupfferi (von Willemoes-Suhm) FIG. 16.1E
[HORST 1927–39, 672]

Length 80–90 mm overall. Proboscis 9 × 6 mm. Collar 3.5 × 7 mm. Branchial region about 15 mm long. Branchial skeleton is shown dense black in transverse section illustrated, particularly within tongue bars. Proboscis and collar pale during life. Gonads reddish. Nerve tracts brown. Gut of semi-transparent hepatic region brownish anteriorly, then greenish.

In mud; 20–30 m; Kattegat; Greenland.

Protoglossus koehleri (Caullery and Mesnil) FIG. 16.1F
[BURDON-JONES 1956, 35; HORST 1927–39, 658]

Length 40–75 mm. Proboscis 3 × 1.25 mm. Collar 1.5 × 1.5 mm. Gill pores up to 40 pairs. Tongue bars (figured) fairly short, without skeletal thickening. General colour yellowish white. Branchial region greenish dorsally. Gut brownish in hepatic region. White testes or rose ovaries. Eggs large, 0.4 mm diameter.

In fine sand near low water; Holyhead; Cap de la Hague.

Saccoglossus horsti Brambell and Goodhart FIG. 16.1J,K
[BRAMBELL AND GOODHART 1941, 283]

Very like *S. ruber*, but proboscis relatively shorter (see key). Proboscis 13–18 × 4–5 mm, a transverse section through it (not figured) revealing about nine rings of longitudinal muscle fibres. Length of tongue bars (see by cutting pharynx ventrally and viewing from within, as in Fig. 16.1J) is reduced suddenly in posterior gill silts, so that only the last two have bar length shorter than its breadth (Fig. 16.1K). Egg diameter 0.25 mm.

In mud or muddy sand near low water. Lymington-on-Solent, Whitstable, Poole Harbour.

Saccoglossus pygmaeus Hinrichs and Jacobi
[HORST 1927–39, 669]

Up to 30 × 1 mm overall. Proboscis 3–7 mm long, milky white. Trunk yellow anteriorly, otherwise bearing a few red spots. Tongue bars short. Eggs large, soon hatching to produce young 0.6 mm long.

In '*Amphioxus* gravel', 4–5 miles north-east of Helgoland.

Saccoglossus ruber (Tattersall) FIG. 16.1G,H
[BRAMBELL AND COLE 1939, 211; BURDON-JONES AND PATIL 1960, 635]

Proboscis pink, up to 30 × 3 mm. Collar red or orange, 4 × 3–3.5 mm. Trunk red behind collar and then brown, paler posteriorly, with a few small red spots. Transverse section through proboscis (not figured) shows four to six rings of longitudinal muscle fibres around a less regular central mass. Lengths of tongue bars decrease gradually near hind end of pharynx (i.e. length less than breadth in last eight bars).

In sand at LWST on sheltered beaches in Ireland, Scotland, and Wales. Sublittoral in Firth of Clyde.

Stereobalanus canadensis (Spengel) FIG. 16.1L,M
[BURDON-JONES AND MCINTYRE 1960, 491; HORST 1929–39, 673]

About 80 mm overall. Proboscis 15 × 7 mm. Collar 3 × 7 mm. General colour golden-yellow. Equally large dorsolateral and ventrolateral genital ridges envelop the wide lateral gill pores.

In black mud at 20 m in Scottish sea lochs, but not south of Mull.

3 PTYCHODERIDAE

Length may exceed 0.5 m, but only 2–3% is proboscis. Gill bars joined by synapticulae. Oesophageal pores absent. Numerous body wall protuberances associated with hepatic caeca. A dorsolateral septum, with which the gonads are associated, divides the trunk coelom on each side, within genital folds. A layer of circular muscle usually lies outside the longitudinal muscles of trunk. Eggs small, developing into tornaria larvae.

1. Dorsolateral genital folds rounded, diverging anteriorly and not roofing over any gill pores (Fig. 16.1D) ***Glossobalanus sarniensis***

   Genital folds thin, forming 'wings' which approach one another in the dorsal mid-line, so as to meet over the anterior gill pores **2**

2. Dorsal margins of genital wings remain close together throughout much of branchial region. Gonopores (g in Fig. 16.1C) open along inner faces of wings, near margins ***Glossobalanus marginatus***

   Dorsal margins of wings meet anteriorly just behind collar, but then diverge widely. Gonopores open near gill pores, at base of wings ***Balanoglossus clavigerus***

Balanoglossus clavigerus Delle Chiaje FIG. 16.1B
[HORST 1927–39, 704]

Length 0.5 m or more, breadth 4–9 mm. Proboscis 7 × 4 mm. Collar 5 × 5 mm. Gill region up to 25 mm long. Genital wings fused with posterior edge of collar. Anterior hepatic caeca pale, closely and regularly arranged. Main mass of caeca greenish, densely packed and irregular.

In sand, LWST and sublittoral; Brittany, Mediterranean. Only British records are of the larva 'Tornaria krohni'.

Glossobalanus marginatus Meek FIG. 16.1C
[HORST 1927–39, 703; MEEK 1922, 579]

Proboscis 7 × 5 mm. Collar 5 × 5 mm. Gill region about 15 mm long, figured here in transverse section. Thin fringes of genital wings, distal to gonopores, inspired name.

North Sea near Farne Islands; 100 m; Skagerrak, particularly in deep muddy areas. Larva probably 'Tornaria mielcki', which is common in North Sea.

Glossobalanus sarniensis Koehler FIG. 16.1D
[BURDON-JONES 1953, 342; HORST 1927–39, 695]

May exceed 500 × 10 mm. Proboscis egg-shaped, about 10 mm

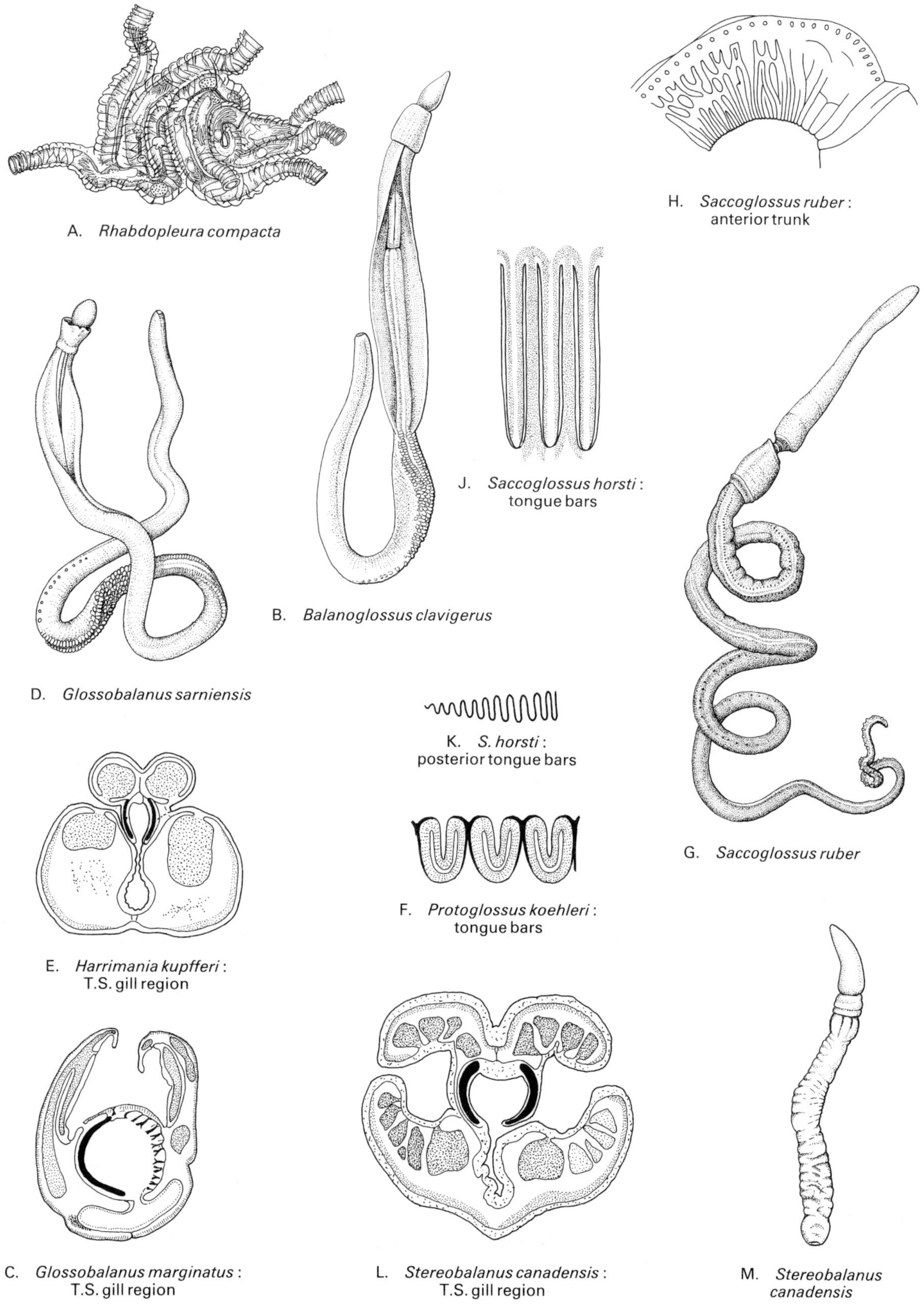
A. *Rhabdopleura compacta*
H. *Saccoglossus ruber*: anterior trunk
J. *Saccoglossus horsti*: tongue bars
B. *Balanoglossus clavigerus*
D. *Glossobalanus sarniensis*
K. *S. horsti*: posterior tongue bars
G. *Saccoglossus ruber*
F. *Protoglossus koehleri*: tongue bars
E. *Harrimania kupfferi*: T.S. gill region
C. *Glossobalanus marginatus*: T.S. gill region
L. *Stereobalanus canadensis*: T.S. gill region
M. *Stereobalanus canadensis*

Fig. 16.1

long. Genital wings separated from hepatic region by a transitional region which lacks prominent protuberances. (In the Mediterranean *G. minutus* this region is lacking, the wings extending 1–2 mm into the hepatic region.)

Occasionally in sand at LWST at Channel Isles and Salcombe; Tresco (Isles of Scilly), Strangford Lough. Skagerrak, but less common than *G. marginatus* and in shallower water. Larva probably 'Tornaria bournei', which is common in Celtic Sea.

PHYLUM CHORDATA

Subphylum **Tunicata** (=**Urochordata**)

Class **Ascidiacea**

The tunicates constitute the largest subphylum of 'protochordates'. The body is contained within a *test* or tunic secreted by the underlying *mantle*. This test may be thin, thick, clear, opaque, clean, or covered with sand or living organisms. Tunicates are well-represented on most rocky seashores by the ascidians, which are sessile and either colonial or unitary. Some are indefinite in shape, and so hidden within their tunics that they superficially resemble sponges or some of the more fleshy bryozoans or coelenterates (e.g. *Alcyonidium* or *Alcyonium*). When cut open, however, they display a more complicated internal organization (Fig. 16.2), particularly a filter-feeding pharynx or *branchial sac* resembling that of *Amphioxus* (Ch. 17) in having a peripharyngeal ciliated band near the mouth, from which a ventral *endostyle* and a *dorsal lamina* extend posteriorly. The dorsal lamina primitively has the form of a series of slender lobes (*languets*) but in some ascidians becomes a continuous membrane. The pharynx walls bear numerous perforations, the *stigmata*, which differ from gill slits in being generally elongated antero-posteriorly. They are arranged in transverse rows, however, each of which may be regarded as a subdivided gill slit; transverse blood-vessels pass between adjacent rows. In some genera the stigmata are arranged in spiral groups.

The feeding current enters the test via an *oral* (or branchial) opening or *siphon* and passes into the pharynx through a ring of *oral tentacles*. Flow is generated by cilia, which line the stigmata and beat towards the surrounding *atrium* or *mantle cavity*. From there the water passes to the exterior via a dorsal atriopore, or *excurrent siphon*, which is usually more constricted by sphincter muscles than is the mouth. Positive pressure therefore builds up inside the atrium, itself kept narrow by fine strands connecting the pharynx wall to the peripheral mantle. The atrium consequently expands whilst the strands drag out the wall of the pharynx, which becomes distended with water entering through the gaping mouth. On disturbance, contraction of mantle muscles forces out jets of water through the oral and atrial sphincters, hence the term 'sea squirts' popularly applied to larger ascidians. Solitary ascidians may be 1 cm long or more, and the largest British forms reach about 15 cm.

Colonial ascidians are made up of numerous *zooids* formed by *budding*. Each zooid is like a small unitary ascidian with few stigmata, but rarely exceeding a few mm in length. The colonies may be larger (sometimes, though rarely, as much as 10–20 cm). The zooids may arise from stolons or a basal mat, or project from a common test, or be wholly immersed. If the zooids are embedded within the test they generally are arranged in *systems*; i.e. the atria do not open individually to the exterior but internally with others into a *common cloaca*. The atrial openings may wholly lack siphons or there may be a *languet* (tongue) projecting from its adoral margin. The common cloaca communicates with the exterior through an irregular, circular or elevated aperture.

Ascidians are hermaphrodite but the location of gonads within the body varies. In order Enterogona (*enteron*, gut; *gone*, seed) they are essentially associated with the gut loop, though displaced into a *post-abdomen* in the long, slender zooids of polyclinids. In order Pleurogona (*pleura*, side) there is no abdomen and the gut loop is attached to the mantle wall on one or other side of the pharynx; the gonads are also associated with the mantle wall, usually on both sides of the body. The largest ascidians (including many unitary forms) shed their eggs into the sea, whilst the smallest (nearly all colonial species and many small unitary ones) incubate their embryos. These will be found in the atria of most colonial species but in the basal test of didemnids. Nearly all hatch as *tadpole larvae* with chordate tails. In viviparous species not only the primary zooid (*oozooid*) may be well-developed in the larva but also the first bud (*blastozooid*).

Ascidians contain no skeletal parts and their identification depends upon features of their internal morphology, e.g. position of the gut and gonads, and structure of the pharynx. Identification must therefore usually be preceded by dissection. Since contraction caused by death or fixatives distorts the body shape, narcotization is usually essential if diagnostic characters are to be observed. Dissection without fixation is usually impossible so that narcotized specimens have also to be fixed.

Collected specimens should immediately be placed in water, in plastic bags of appropriate size, to which is added a few crystals of menthol. Narcotization will usually occur within a couple of hours (and can be verified in larger specimens by gently touching the open siphons with a needle: if they contract, narcotization is incomplete). Fix the specimens by adding one-tenth the sea-water volume of commercial formalin (40% formaldehyde). Fixed specimens can then be stored in a less noxious preservative such as 1% propylene phenoxytol in distilled water, or transferred back to sea-water for immediate examination. The colour of specimens is likely to fade in preservative and should be recorded before fixing.

Note (draw if necessary) the external characteristics, particularly the arrangement of systems found in many colonial species. Unitary specimens should be dissected under water in a wax-based dish, observing through an arm-mounted dissecting microscope if necessary. Place the animal on its right side (i.e. its atrial siphon to your right); pin as and when necessary. Remove the test. This may be accomplished either by making a circumferential cut above the base (Fig. 16.2), so allowing the test to be turned inside out over the siphons, or by cutting with scissors in the mid-line, starting at the oral siphon, proceeding down the line of the endostyle, around the base, and up to the atrial siphon; thence back to the oral siphon. Carefully separate the left half of the test from the mantle and remove it. Cut and open the mantle by a dorsal cut from the base to the atrial siphon, carefully freeing the mantle from the branchial sac by cutting the connecting strands (Fig. 16.2).

Compound ascidians should be handled in a petri dish or watch glass on the stage of a stereo-microscope. If there are separate zooids, open them with fine scissors, cutting down from the oral siphon. In others (e.g. many polyclinids) zooids are embedded in, but easily dissected out of, the common test. Handling small zooids requires the use of fine needles, needle knives, or sharp, small bladed scalpels, and watch-maker's forceps; spring-bow scissors are also a help. The minute, close-packed zooids of didemnids and similar colonies are best recovered from a few vertical slices rather than by individual dissection. Didemnid colonies may contain densely aggregated granules: some should be saved for high-power microscopy but the colony may be cleared without the generation of gas bubbles using EDTA (ethylene diamine tetra-acetic acid). This is prepared as a solution of 250 g in 1750 ml distilled water buffered to pH 7 by addition of sodium hydroxide pellets.

There are two orders of Ascidiacea: Enterogona and Pleurogona. The former has two suborders.

Order 1 **Enterogona**

The larva has left and right atrial openings which meet and fuse medially during metamorphosis to form the atrial siphon. Gonads are associated with the intestinal loop, or are post-abdominal.

Suborder **Aplousobranchia**

Branchial sac without internal longitudinal vessels (*haploos*, simple), except in families Cionidae and Diazonidae. Body divided into thorax, abdomen and, sometimes, post-abdomen. Gut abdominal; gonads abdominal or post-abdominal. Colonial except Cionidae, placed here (Hawkins *et al.* 1983) rather than in the Phlebobranchia (Berrill 1950).

Suborder **Phlebobranchia**

Branchial sac with internal longitudinal vessels (*phleps*, vein). Body not divided into thorax and abdomen. Gut loop and gonads beside the branchial sac, left or right. Unitary except for Perophoridae. *Perophora* has stolonate colonies and rather small, non-immersed zooids; rows of papillae project from the transverse vessels into the lumen of the pharynx. In other genera the papillae bifurcate and fuse with their neighbours fore and aft, to form the inner longitudinal vessels. These vessels are evenly distributed around the walls of the pharynx, which are not longitudinally folded.

Order 2 **Pleurogona**

Suborder **Stolidobranchia**

The atrial opening forms as a single, median invagination. Gut loop alongside the pharynx; gonads also lateral, developing in the mantle wall, usually on both sides. Branchial sac (except botryllines) with longitudinal folds (*stolidotos*, hanging in folds) and with inner longitudinal vessels usually distributed unevenly, mostly along the lateral infoldings. Unitary or colonial.

LITERATURE

Ascidiologists are well served by Berrill's (1950) Ray Society monograph (although it is unfortunate that the illustrations lack scales). A few additions and deletions are necessary. More recent and concise (sometimes too brief to be easy for the beginner) is Millar's (1970) synopsis. The same author's *Marine invertebrates of Scandinavia* volume includes maps showing point records in eastern Britain and the overall distribution, but unfortunately only for species occurring in the North Sea. During the 'seventies a major series of papers on the Didemnidae of Brittany was published by Francoise Lafargue (1968–77). It is impossible yet to assess the impact of her studies on the taxonomy of British species, but ecologists should be alert for the discovery of unfamiliar species, especially in the south-west. A useful manual for scuba divers, illustrated with underwater photographs, has been prepared by Picton (1985).

Berrill, N. J. (1950). *The Tunicata, with an account of the British species.* Ray Society, London.

Carlisle, D. B. (1954*a*). Notes on the Didemnidae (Ascidiacea) III. A comparison of *Didemnum maculosum, D. candidum, D. helgolandicum*, and *Trididemnum alleni*. *Journal of the Marine Biological Association of the U.K.*, **33**, 313–24.

Carlisle, D.B. (1954*b*). *Styela mammiculata* n. sp., a new species of ascidian from the Plymouth area. *Journal of the Marine Biologial Association of the U.K.*, **33**, 329–34.

Crothers, J. H. (ed.) (1966). Dale Fort Marine Fauna, 2nd edn. *Field Studies*, **2** (Supplement).

Hawkins, C. J., Kott, P., Parry, D. L., and Swinehart, J. H. (1983). Vanadium content and oxidation state related to ascidian phylogeny. *Comparative Biochemistry and Physiology*, **76B**, 555–8.

Kott, P. (1952). Observations on compound ascidians of the Plymouth area, with descriptions of two new species. *Journal of the Marine Biological Association of the U.K.*, **31**, 65–83.

Lafargue, F. (1968). Les peuplements sessiles de l'Archipel de Glénan, II. Les Didemnidae. Systématique, écologie. *Vie et Milieu* A. **19**, 353–446.

Lafargue, F. (1970). Peuplements sessiles de l'Archipel de Glénan, I—Inventaire Ascidies. *Vie et Milieu* B, **21**, 729–42.

Lafargue, F. (1975*a*). Revision taxonomique des Didemnidae des côtes de France (Ascidies composées): Description des espèces de Ban-

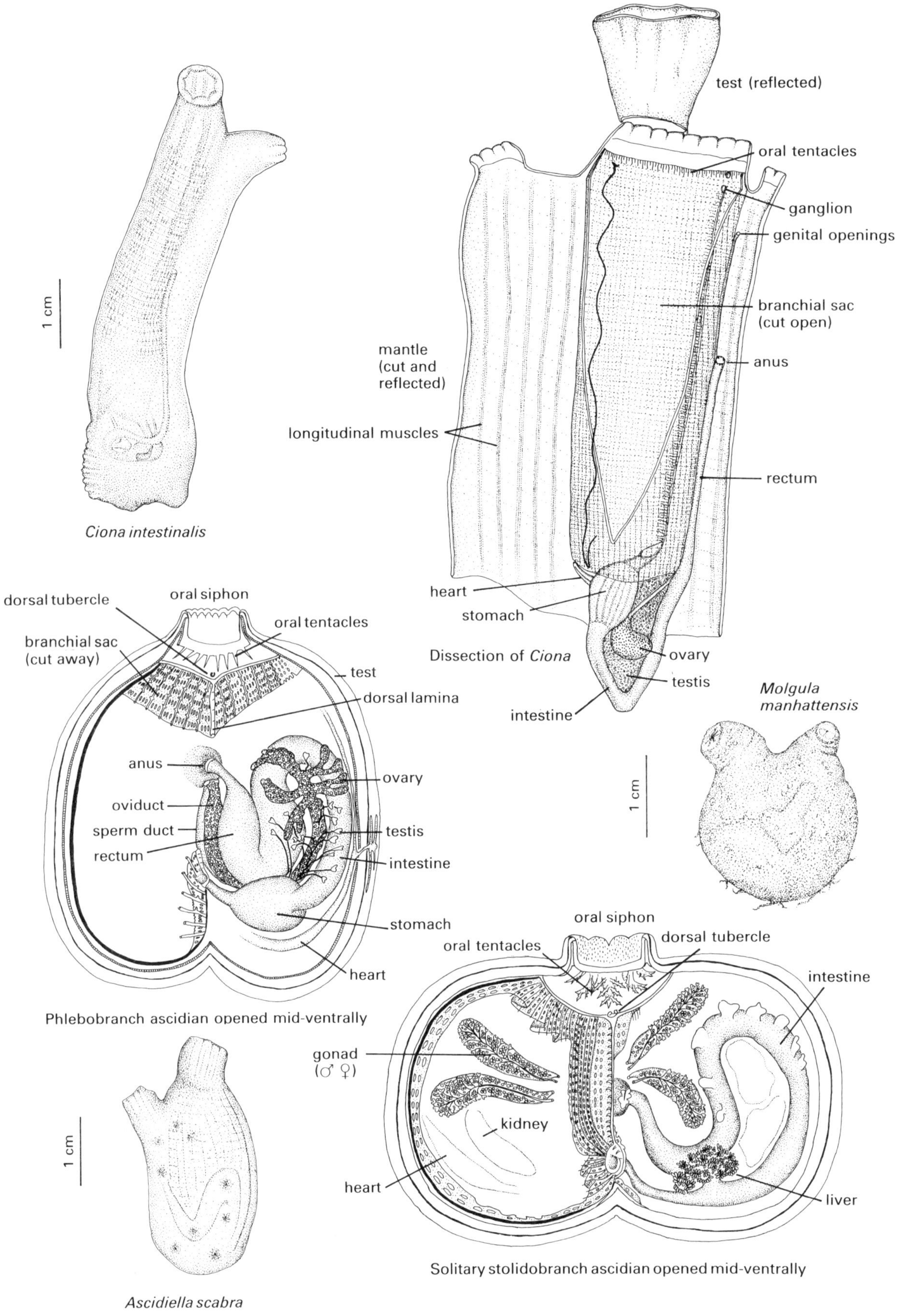
1 cm
Ciona intestinalis
test (reflected)
oral tentacles
ganglion
genital openings
branchial sac (cut open)
anus
mantle (cut and reflected)
longitudinal muscles
rectum
heart
stomach
Dissection of Ciona
ovary
testis
intestine
Molgula manhattensis
1 cm
dorsal tubercle
oral siphon
oral tentacles
branchial sac (cut away)
test
dorsal lamina
anus
ovary
oviduct
sperm duct
testis
rectum
intestine
stomach
heart
Phlebobranch ascidian opened mid-ventrally
oral siphon
oral tentacles
dorsal tubercle
intestine
gonad (♂ ♀)
kidney
heart
liver
1 cm
Solitary stolidobranch ascidian opened mid-ventrally
Ascidiella scabra

Fig. 16.2

yuls-sur-Mer. Genre *Didemnum*. Première partie: larves a trois ventouses. *Annales de l'Institute Océanographique, Monaco*, **51**, 173–94.
Lafargue F. (1975*b*). Revision taxonomique des Didemnides des côtes de France (Ascidies composées): Description des espèces de Banyuls-sur-Mer. Genre *Polysyncraton*. *Vie et Milieu* A. **25**, 133–64.
Lafargue, F. (1976). Revision taxonomique des Didemnidae des côtes de France (Ascidies composées): description des espéces de Banyuls-sur-Mer. Genre *Didemnum*. Deuxiéme partie: larves a deux ventouses. *Annales de l'Institute Océanographique, Monaco*, **52**, 259–81.
Lafargue, F. (1977). Revision taxonomique des Didemnidae des côtes de France (Ascidies composées): Description des espéces de Banyuls-sur-Mer. Synthése des resultats principaux. *Annales de l'Institute Océanographique, Monaco*, **53**, 135–53.
Lafargue, F. and Wahl, M. (1987). The didemnid ascidian fauna of France. *Annales de l'Institute Océanographique, Monaco*, **63**, 1–46.
Lane, D. J. W. (1976). The marine fauna of Lundy. Ascidiacea (sea squirts). *Report of the Lundy Field Society*, **27**, 48–52.
Médioni, A. (1970*a*). Ascidies du benthos rocheux de Banyuls-sur-Mer. Didemnidae (Ascidies Composées). *Vie et Milieu*, A, **21**, 25–48.
Médioni, A. (1970*b*). Ascidies du benthos rocheux de Banyuls-sur-Mer. Polyclinidae (Ascidies Composées). *Vie et Milieu*, A, **21**, 287–307.
Millar, R. H. (1960). Ascidiacea. *Fauna of the Clyde Sea Area*, Millport.
Millar, R. H. (1966). Tunicata Ascidiacea. *Marine Invertebrates of Scandinavia*, **1**, 1–123.
Millar, R. H. (1970). British ascidians. *Synopses of the British Fauna*, N.S. **1**, 1–92.
Nelson-Smith, A. and Case, R. M. (1984). Aspects of the shore and sublittoral ecology of the Daucleddau estuary (Milford Haven). *Zoological Journal of the Linnean Society*, **80**, 177–90.
Picton, B. E. (1985). *Ascidians of the British Isles. A colour guide*. Marine Conservation Society.
Rowe, F. W. E. (1972). The marine fauna and flora of the Isles of Scilly: Enteropneusta, Ascidiacea, Thaliacea, Larvacea and Cephalochordata. *Journal of Natural History*, **6**, 207–13.

KEY TO FAMILIES

1.	Gut loop posterior to pharynx, in an abdomen which is more or less distinct from thorax. Mostly colonial, but zooid patterns not of colours contrasting with surrounding test (APLOUSOBRANCHIA)	2
	Gut loop alongside pharynx; body not divided into thorax and abdomen. Some colonial forms have zooid patterns of contrasting colours	9
2.	Unitary	**1. Cionidae**
	Colonial	3
3.	Zooids united only basally, with distal parts separate from each other	4
	Zooids embedded within common tunic of colony	6
4.	Zooids several cm long, half enclosed in common test with distal halves protruding	**2. Diazonidae**
	Zooids much smaller and arising separately from a stolon or thin basal layer	5
5.	Aperture rims smooth. Usually seven or more rows of stigmata	**3. Clavelinidae**
	Aperture rims 6-lobed. Only three or four rows of stigmata	**4. Polycitoridae**
6.	Gonads behind the gut loop in a posterior process, the post-abdomen. Colonies fleshy, often elongated	**6. Polyclinidae**
	Gonads alongside gut loop. No post-abdomen. Colonies fleshy or encrusting	7
7.	Colonies thin, encrusting, gelatinous or hard and sometimes white, the tunic being full of calcareous, often stellate, granules. Zooids of species with such granules having saucer-shaped 'lateral organs', one on each side of the pharynx. Embryos brooded within general cavity of the colony, outside the parent zooids	**7. Didemnidae**

Colonies fleshy, forming small rounded masses, without granules in tunic or lateral organs in zooids. Embryos brooded in atrial cavity or in a special pouch developed from it **8**

---

8. Zooids with both apertures 6-lobed and atriopores opening separately. Incubation in atrium. Larvae with three attachment papillae in a vertical row **4. Polycitoridae**

Zooids forming systems surrounding a common excurrent aperture (cloaca). Oral aperture with six shallow lobes and a single languet formed by its anterior lip. Incubation in a brood sac developing as an outgrowth from the atrium. Larvae with one ventral and two dorsal papillae **5. Holozoidae**

---

9. Unitary, or if colonial, then colonies not with embedded zooids. Gonads within gut loop on one side of the pharynx. Pharynx wall usually with inner longtiudinal vessels, which are distributed evenly around the pharynx, but without longitudinal folds (PHLEBOBRANCHIA) **10**

Unitary or colonial, sometimes forming coloured patterns of zooids embedded within encrusting colonies. Gonads in mantle wall, usually on both sides. Pharynx wall usually with longitudinal infoldings, along which the inner longitudinal vessels are concentrated (STOLIDOBRANCHIA) **12**

---

10. Colonial, with zooids budded from a thin stolon **8. Perophoridae**

Unitary (sometimes clustered or aggregated) **11**

---

11. Gut loop on right of pharynx. Stigmata crescentic and arranged in whorls **9. Corellidae**

Gut loop on left of pharynx. Stigmata straight, not forming whorls **10. Ascidiidae**

---

12. Unitary or colonial. Buccal tentacles (round posterior margin of buccal cavity) always simple. Pharynx with no more than four folds. Stigmata straight. Stomach distinct, rounded, with longitudinal ridges and a single small 'liver' diverticulum **11. Styelidae**

Unitary. Buccal tentacles branching in larger species. Pharynx with five to seven folds on each side. Stigmata straight or spiralling. Stomach narrow and small, not marked off from the rest of the gut but with many diverticula forming a complex liver **13**

---

13. Tunic usually thick and leathery, may be sculptured, spiny, or sandy. Oral aperture 4-lobed. Stigmata straight and parallel to one another. No conspicuous renal sac **12. Pyuridae**

Tunic thin and translucent, though usually covered with sand, gravel or mud. Oral aperture 6-lobed. Stigmata curved and arranged in spirals. A renal sac containing a yellowish or purplish concretion can usually be seen through the tunic on the right side, if adhering sand particles are scraped away **13. Molgulidae**

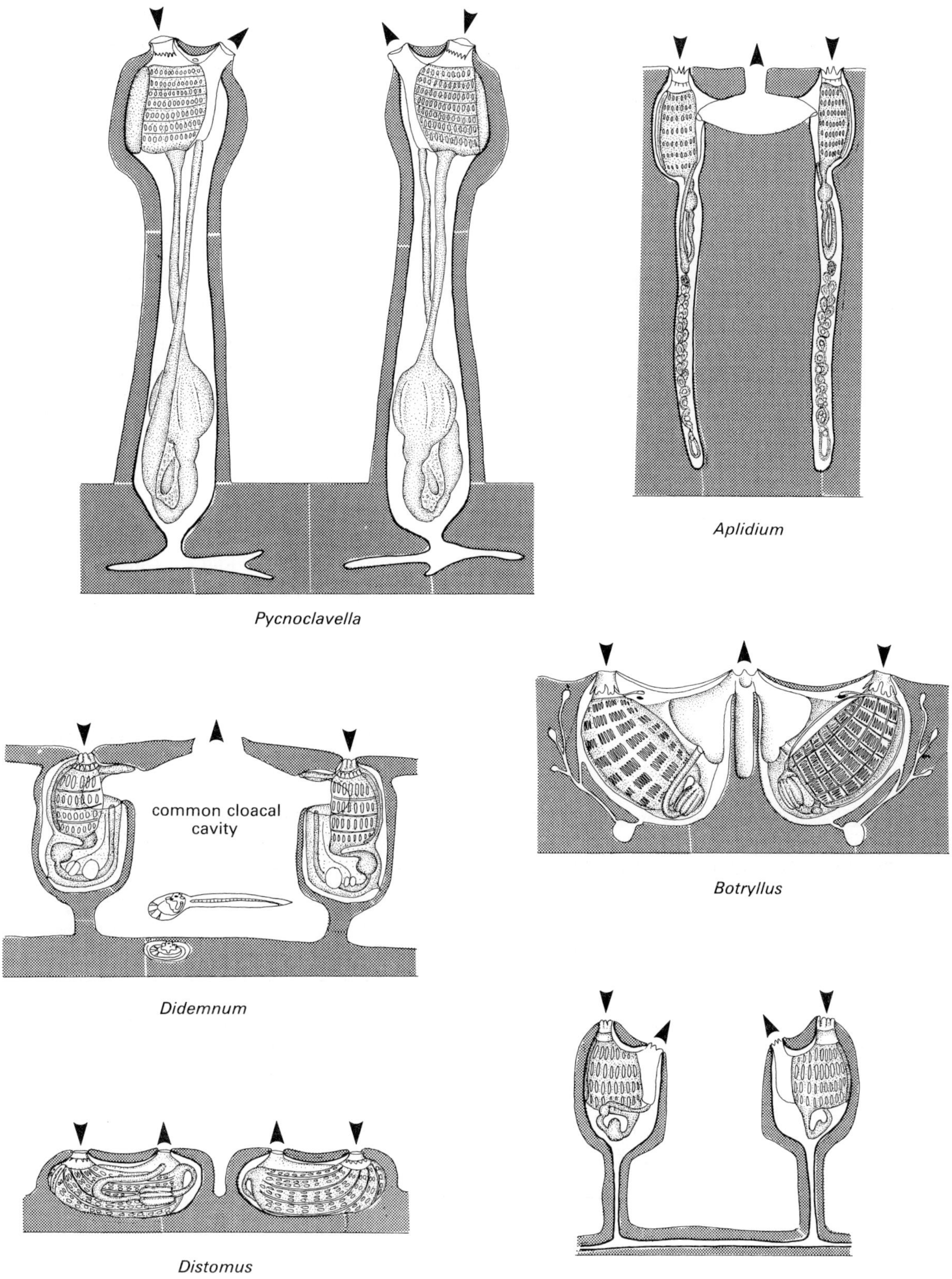

TYPES OF COLONIAL ORGANIZATION

Fig. 16.3

Order **Enterogona**

Suborder **Aplousobranchia**

1 CIONIDAE

Unitary. Pharynx with inner longitudinal vessels and many rows of stigmata. One British representative.

Ciona intestinalis (Linnaeus) FIG. 16.2

[BERRILL 1950, 132; MILLAR 1970, 39]

Test soft and translucent. Body reaching 100–150 mm, nearly cylindrical, pale yellow-green, gelatinous and contractile, with five longitudinal muscle bands clearly visible in the mantle in each side. Apertures on siphons, which may have bright yellow margins and small red pigment spots between the lobes; eight marginal lobes on the oral and six on the atrial siphon. About 70 oral tentacles. Oviparous. Eggs surrounded by radiating, elongate follicle cells.

On rocks and piles, LWST and shallow water; common in harbours; all coasts (map: Millar 1966, 39).

2 DIAZONIDAE

Colonial. Pharynx with inner longitudinal vessels and many rows of stigmata. One British representative.

Diazona violacea Savigny FIG. 16.4

[BERRILL 1950, 136; MILLAR 1970, 401]

Colonies 100–300 mm across. Zooids soft, translucent, pale green, like a bunch of small *Ciona* with their proximal ends embedded in a common basal mass. Oviparous, unlike other colonial ascidians.

Sublittoral and mostly in deep water, but common at 60 m in some Scottish sea lochs, e.g. Little Loch Broom; Outer Hebrides and western Ireland to Adriatic.

3 CLAVELINIDAE

Zooids arising freely from a basal stolon or encrustation. Thorax transparent but the ciliated tracts marked by lines of bright opaque pigment. Pharynx without inner longitudinal vessels. Apertures on short siphons, with unlobed margins. Larvae brooded in atrium; with triangular arrangement of three adhesive papillae.

1. Zooids up to 20 mm long, arising from stolons or thin encrustations. Up to 17 rows of stigmata **Clavelina lepadiformis**

   Zooids up to 6 mm long, arising from an encrustation about 5 mm thick. Up to nine rows of stigmata **Pycnoclavella aurilucens**

Clavelina lepadiformis (Müller) FIG. 16.4

[BERRILL 1950, 70; MILLAR 1970, 121]

White or yellow lines, very conspicuous through the transparent test during life, mark the endostyle, peripharyngeal band, and dorsal lamina. About 12 oral tentacles. Stomach brown with cream-white markings. Eggs and larvae reddish in the atrial cavity. Tadpoles with two rows of stigmata.

On rocks etc., from LW pools to about 50 m; all coasts but local; Norway to Adriatic (map: Millar 1966, 12).

Pycnoclavella aurilucens Garstang FIG. 16.4

[BERRILL 1950, 74; MILLAR 1970, 13]

White, yellow, or orange pigments mark the pharyngeal feeding tracts. Oral tentacles few. Basal test greenish-brown, sometimes sand encrusted. Tadpoles with four rows of stigmata.

On current-swept rocks at 10–30 m from Plymouth, south-west England, Lundy, Skomer, and the west of Ireland to Galway Bay (Picton 1985). May invest stems of *Eunicella* and *Nemertesia*; Brittany and perhaps south to Mediterranean.

4 POLYCITORIDAE

Zooids embedded basally in sand-covered encrustations, but more or less separate distally and not grouped around a common cloaca. Oral and atrial apertures free, each with six marginal lobes. Pharynx with three or four rows of stigmata and no inner longitudinal vessels. Tadpoles with three attachment papillae in a vertical row, borne on an antero-ventral stalk and accompanied by ampullae.

1. Zooids mostly free from base, each with three rows of stigmata **Archidistoma aggregatum**

   Zooids mostly embedded, each with four rows of stigmata **Polycitor searli**

Archidistoma aggregatum Garstang FIG. 16.4

[BERRILL 1950, 67; KOTT 1952, 65; MILLAR 1970, 15]

Ramifying base covering area 10 to many mm across. Greenish or brown zooids, sand encrusted, projecting 5–10 mm from it in clumps of two to four, back to back, with tests partially fused but distal siphons free, though very short. Tadpole with eight ampullae associated with its three adhesive organs, all arising from a narrow mid-ventral stalk; developing three rows of stigmata.

On stones and shells, 0–30 m, near Plymouth.

Polycitor searli Kott FIG. 16.4

[KOTT 1952, 66; MILLAR 1970, 16]

Colonies up to 6 mm across, rounded, squat or subcylindrical, sometimes with a constriction where a transparent distal part of the common test protrudes from the sandy basal part. Zooids white, red or orange, four to twelve per colony, each 4–5 mm long, wholly embedded except for the distal apertures, and forming a single clump with atrial apertures centrally orientated but separate. Tadpoles with two ampullae alternating with three adhesive organs; developing four rows of stigmata.

In rock crevasses at 8–10 m, in Plymouth Sound.

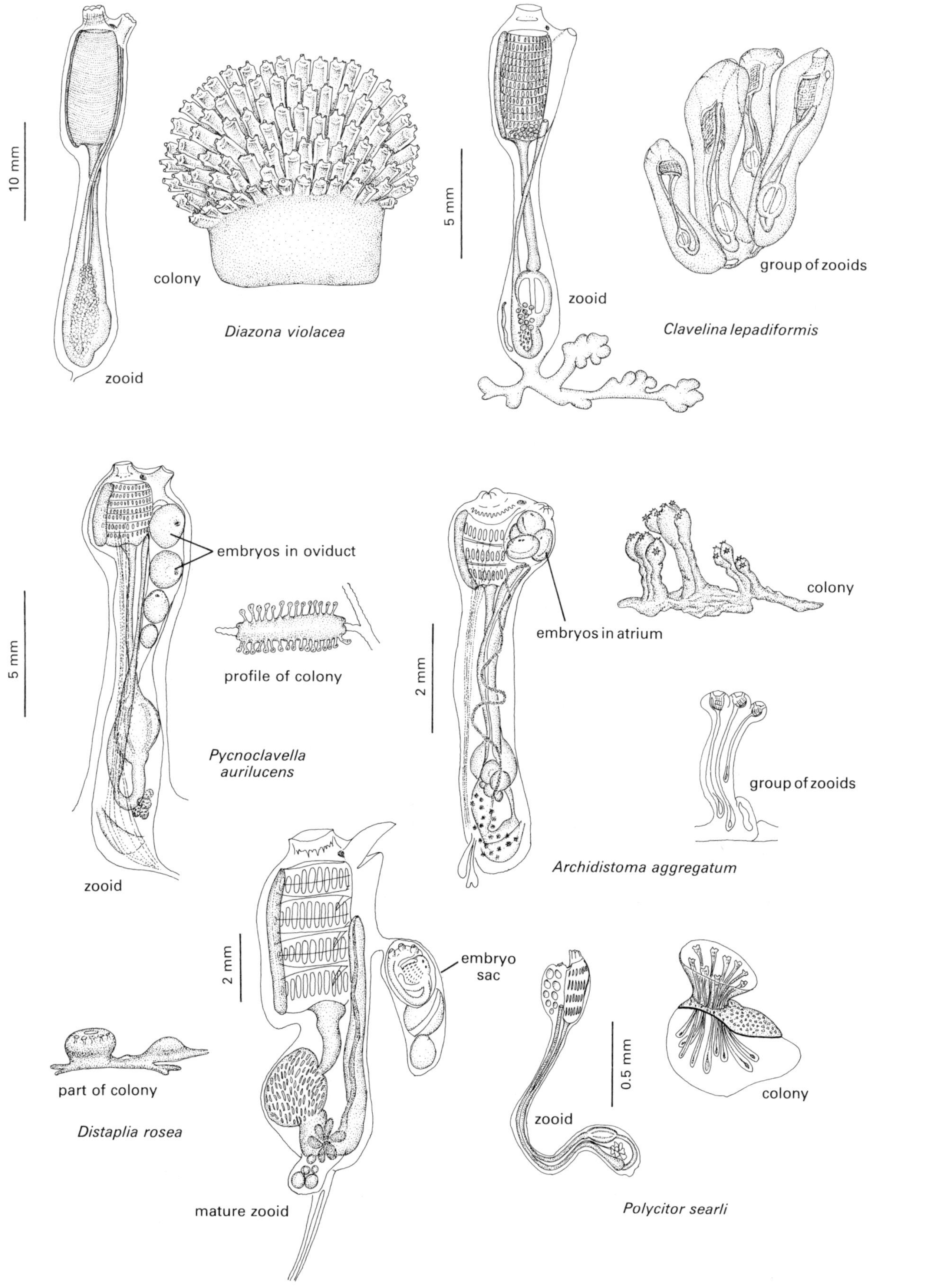
10 mm
zooid
colony
Diazona violacea
5 mm
zooid
group of zooids
Clavelina lepadiformis
embryos in oviduct
profile of colony
5 mm
zooid
Pycnoclavella aurilucens
embryos in atrium
colony
2 mm
group of zooids
Archidistoma aggregatum
2 mm
embryo sac
part of colony
Distaplia rosea
mature zooid
0.5 mm
zooid
colony
Polycitor searli

Fig. 16.4

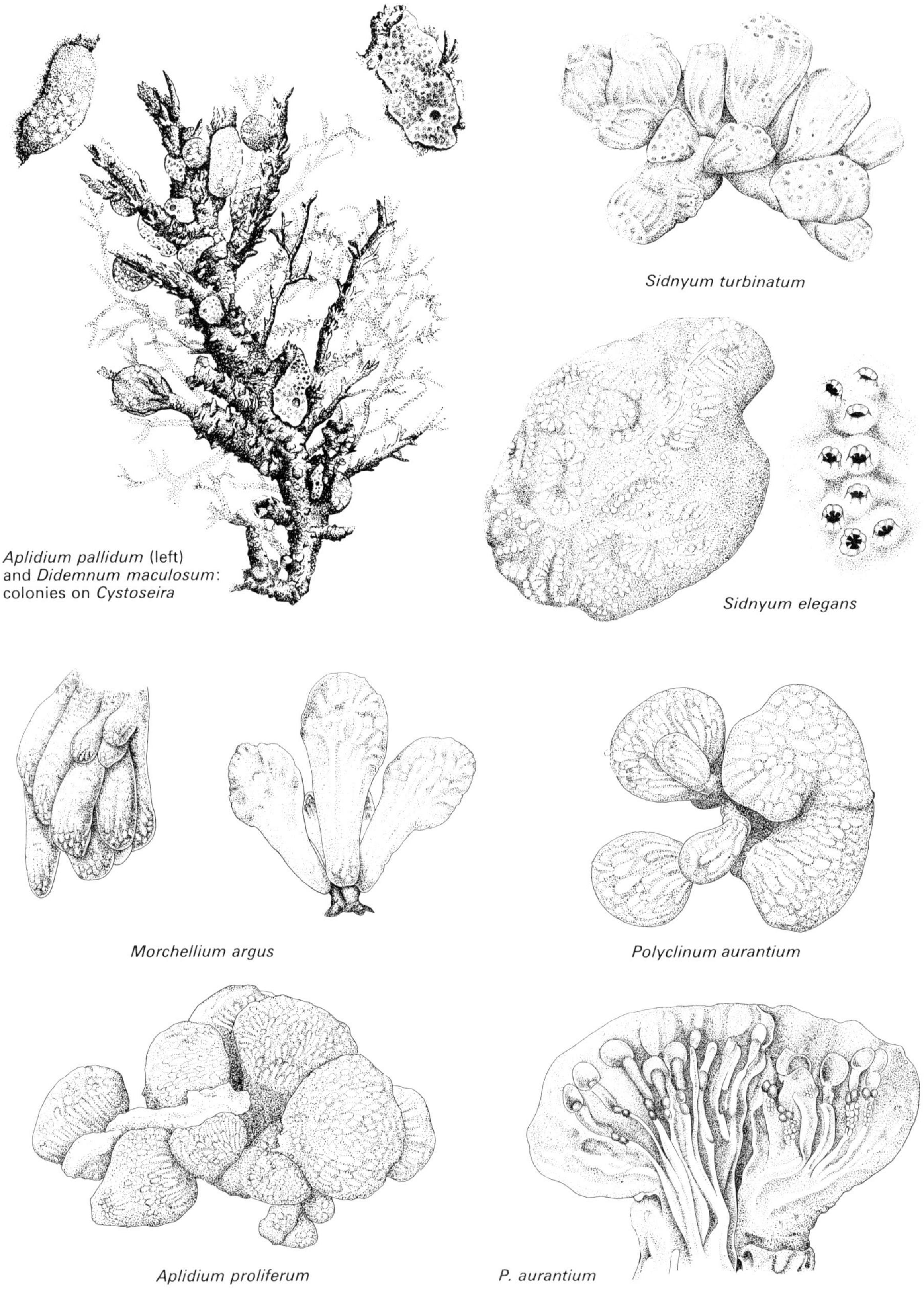
Aplidium pallidum (left)
and Didemnum maculosum:
colonies on Cystoseira
Sidnyum turbinatum
Sidnyum elegans
Morchellium argus
Polyclinum aurantium
Aplidium proliferum
P. aurantium

Fig. 16.5

5 HOLOZOIDAE

Fleshy colonies with zooids embedded in groups, each group surrounding a common atriopore. Gonads alongside gut loop. Tadpoles with three attachment papillae forming a triangle. One British representative (Millar 1970).

Distaplia rosea Della Valle FIG. 16.4

[BERRILL 1950, 78; MILLAR 1970, 14]

Colonies in form of rounded, dome-shaped or lobate masses 10–30 mm across, arising from basal plates or strands. Zooids with long atrial languet, four rows of stigmata, and a stomach wall usually having many short longitudinal folds. Tadpoles developing four rows of stigmata.

On stones and shells, 0–100 m, from Scotland to the Mediterranean.

6 POLYCLINIDAE

Colonies more or less mound-shaped or capitate, with the zooids wholly immersed and forming systems. Zooids elongate, divisible into thorax, abdomen, and post-abdomen; branchial sac with numerous rows of stigmata; heart and gonads displaced into the post-abdomen. Development viviparous. The British species are classified into five genera. The key to species uses characters of zooids, which therefore should be dissected out of the colonies with cuts through the test parallel to them. Prior narcotization with menthol is important.

---

1. Oral siphon with six lobes **2**
   Oral siphon with eight lobes **4**

2. Zooids 'wasp-waisted' between abdomen and post-abdomen; stomach smooth ***Polyclinum aurantium***
   Zooids not constricted between abdomen and post-abdomen; stomach surface corrugated or tuberculate **3**

3. Stomach with rounded papillae or markings ***Synoicum pulmonaria***
   Stomach with continuous or interrupted, linear corrugations (*Aplidium*) **6**

4. Stomach with tubercles ***Morchellium argus***
   Stomach with continuous or interrupted, linear corrugations (*Sidnyum*) **5**

5. Colony lobes few and broad; >10 rows of stigmata; stomach with >17 longitudinal folds ***Sidnyum elegans***
   Colony lobes narrow (<12 zooids); <10 rows of stigmata; stomach with 10–15 folds ***Sidnyum turbinatum***

6. Atrial opening without languet; colonies cushion-shaped ***Aplidium pallidum***
   A prominent languet present over the atrial opening; colonies lumpy or capitate **7**

7. Stomach with six folds ***Aplidium punctum***
   Stomach with more than six folds **8**

8. Stomach with >20 folds ***Aplidium proliferum***
   Stomach with <20 folds **9**

9. Southern; upper face of heads sand encrusted; atrial languet simple ***Aplidium densum***
   Northern; upper surface of heads clean; atrial languets often with lateral lobes ***Aplidium glabrum***

Aplidium glabrum (Verrill) FIG. 16.6

[BERRILL 1950, 109; MILLAR 1970, 31]

Colony composed of upright lobes, the sides sand coated, the tops flat and uncoated; 10–20 mm high. Oral siphon 6-lobed; branchial sac with 10–12 rows of 16–18 stigmata; languet present, with long median and small lateral lobes. Stomach with (10) 12–15 (20) folds.

An Arctic-Boreal species, in Britain confined to Scottish waters; 0–400 m; typically associated with other ascidians; rare (map: Millar 1966, 26).

Aplidium densum (Giard) FIG. 16.6

[BERRILL 1950: 108—no illustration; MEDIONI 1970*b*: 197, as *Amaroucium*]

Colony composed of sessile cushions, each with a single common cloacal opening; sides and upper surface heavily encrusted with sand; 6–10 mm across. Oral siphon 6-lobed; branchial sac with 11–18 rows [commonly 12 (Berrill) or 15 (in Mediterranean: Médioni)] of 18–20 stigmata per side. Atrial opening opposite rows 3–9, with a languet. Stomach with eight to ten vertical folds.

Rock faces, etc., *Laminaria* zone down to 50 m; western coasts including Irish Sea up to Northern Ireland; southwards to Mediterranean.

Aplidium pallidum (Verrill) FIGS. 16.5, 16.6

[BERRILL 1950, 97; MILLAR 1970, 27]

Colonies small, 15–20 (40) mm across, thin, flattish, attached by a broad base; yellowish-buff. The surface mostly free of sand. Systems with six to eight zooids, the small colonies comprising a single system. Zooids small, <4 mm long, with short post-abdomen. Oral siphon 6-lobed: atrial siphon without languet. Five to eight rows of 12–14 stigmata; stomach with 10–12 undivided folds. Larvae lack ampullae and, until late in development, vesicles; in the atrium, June–September.

Lower shore to 200 m, especially on algae and *Zostera*; common on south-western coasts.

Aplidium proliferum (Milne Edwards) FIGS. 16.5, 16.6

[BERRILL 1950, 120; KOTT 1952, 74; MILLAR 1970, 29]

Colony capitate with yellowish zooids in a transparent test or cushion-shaped with reddish zooids. The flatter colonies (to 20 mm high by 60 mm across) with three or four distinct well-separated systems; capitate colonies to 50 mm high, with closer packed, longer zooids and less distinct systems. Oral siphon 6-lobed; branchial sac with 9–14 rows of 12–16 stigmata; atrial

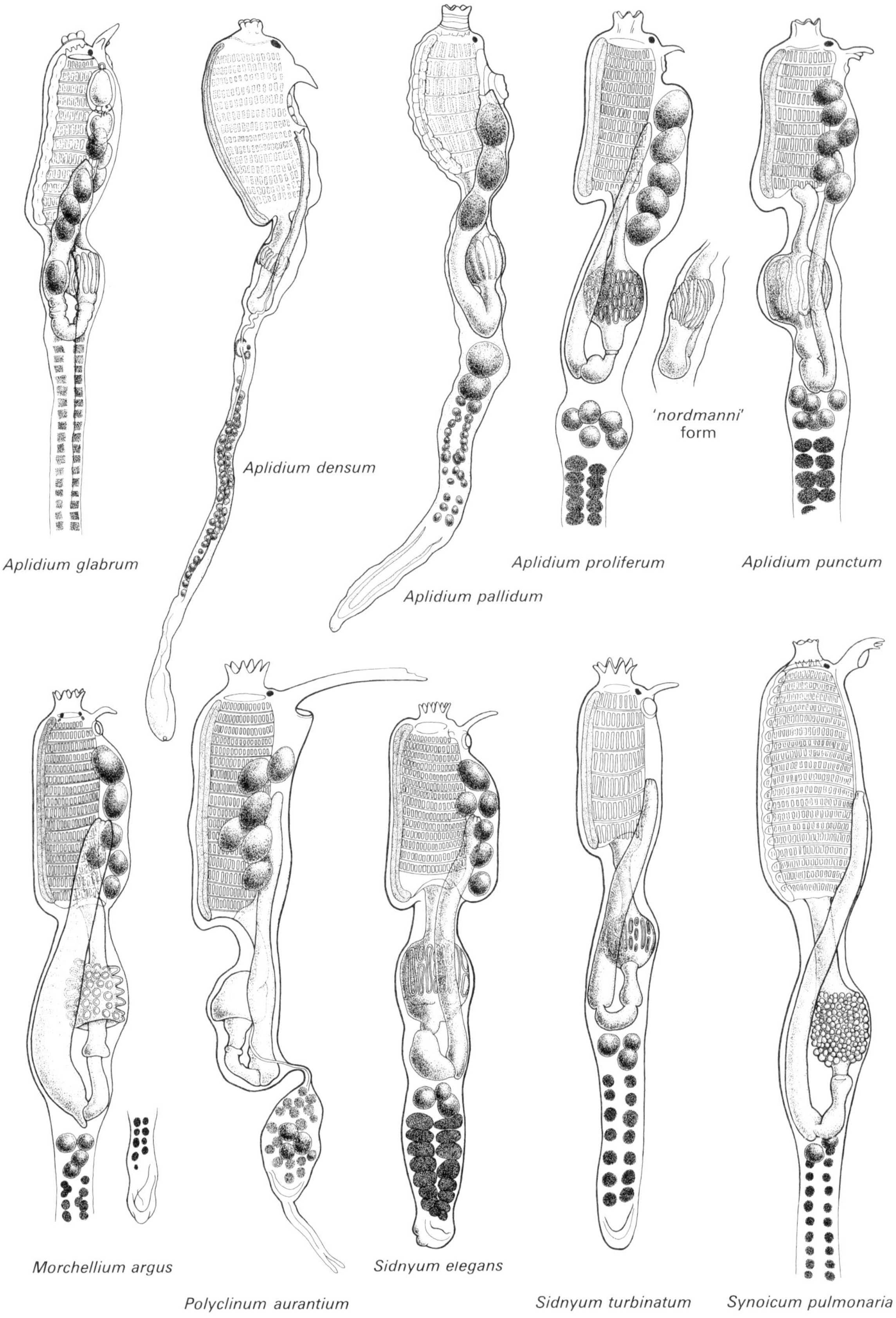
Aplidium glabrum
Aplidium densum
Aplidium pallidum
Aplidium proliferum
'nordmanni' form
Aplidium punctum
Morchellium argus
Polyclinum aurantium
Sidnyum elegans
Sidnyum turbinatum
Synoicum pulmonaria

Fig. 16.6

languet present and simple; stomach with about 30 longitudinal folds which may be simple or transversely divided even within the same colony. Larvae large (trunk up to 1 mm), with numerous small vesicles and three adhesive papillae; in the atrium (August).

It is unresolved whether the flat-topped colony form which rather resembles *Sidnyum elegans*, should be treated as a separate species, *A. nordmanni* (Milne Edwards) (e.g. Berrill 1950, 105) or as a form of *A. proliferum* (e.g. Kott 1952). Pending clarification, it would seem wise for observers to record which form they have observed.

On the lower shore and in shallow water; on and under rocks etc., also on shells and sponges. On all except North Sea coasts, commonest in the south-west; Shetland to the Mediterranean.

Aplidium punctum (Giard) FIG. 16.6
[BERRILL 1950, 100; MILLAR 1970, 30]
Distinctive clavate or capitate colonies, usually 6–12 mm long but occasionally up to 40 mm; arising from a creeping, sand-encrusted base. Test clear; zooids whitish, not obviously in systems; up to 40 or more in each lobe. Oral siphon 6-lobed; branchial sac with (9)10–11(12) rows of stigmata, a red spot over the anterior end of the endostyle. Atrial opening with a trifid languet; stomach with six thick folds; post-abdomen very long. Larvae in the atrial cavity, late summer.

Attached to algae, or in rock-crevices; lower shore and shallow water; western English Channel.

Morchellium argus (Milne Edwards) FIG. 16.5, 16.6
[BERRILL 1950, 95; KOTT 1952, 71; MILLAR 1970, 22]
Colony of reddish capitate lobes up to about 40 mm in length (longer when hanging below rock faces); the stalks sand-encrusted, the heads translucent, flocculent creamy with numerous fine red spots. Zooids long, filamentous; oral siphon 8-lobed; branchial sac with (10)12–14(20) rows of about thirty stigmata and four bright carmine spots anteriorly; atrial languet present, simple. Stomach with numerous, tubercle-like swellings. Ovary behind gut loop, testis occupying the length of the post-abdomen.

The zooids and capitate colony lobes regress in autumn. New colonies develop in early spring with a single central head and basal stolons: the post-abdomina strobilate into the stolons, and rounded systems protrude by midsummer. From July, colony spread stops but the capitula elongate as the zooids continue to lengthen. Larvae are present in the atria from late summer. The life cycle at Plymouth and structure of the larvae have been described by Kott (1952). Larvae have a double row of vesicles on either side of the trunk.

Lower shore and shallow water, under boulders and overhangs, sometimes impressively abundant. West and south-west Britain to France (Brittany).

Polyclinum aurantium Milne Edwards FIGS. 16.5, 16.6
[BERRILL 1950, 87; MILLAR 1970, 18]
Colony composed of round, pear-shaped or flat-topped heads up to 30 mm across; yellow-grey or yellow-brown, the test sand-encrusted. Each head with zooids in one or few systems, arranged around circular common cloacal openings. Zooids distinctive: post-abdomen with slender attachment to abdomen. Oral siphon 6-lobed; branchial sac with 10–16 rows of stigmata, the transverse vessels minutely papillate; stomach wall smooth, the intestinal loop below the stomach twisted. Larvae in May–October; trunk about 0.5 mm long with three adhesive papillae and four pairs of lateral ampullae.

On stones, rock, and algae; lower shore to about 100 m; mainly southern and western coasts; distributed from Norway to the Mediterranean (map: Millar 1966, 17).

Sidnyum elegans (Giard) FIGS. 16.5, 16.6
[BERRILL 1950, 110; KOTT 1952, 70; MILLAR 1970, 26]
Colonies cushion-shaped, narrowed below but rarely capitate; 5–10(20) mm high, up to 60 mm across; pinkish white in colour. Zooids in irregular but conspicuous systems, arranged along common cloacal canals, with somewhat indeterminate common openings; systems oval in small colonies. Zooids relatively short; oral siphon 8-lobed; branchial sac rose-coloured, siphons white; 10–15 rows of 10–14 stigmata; atrial languet with short lateral lobes. Stomach with 18–22 longitudinal folds, which are rarely broken. Larvae in the atrium August–November (Plymouth), trunk reaching 0.8 mm in length; three suckers. The larva has been further described by Kott (1952).

Low on shore and in shallow water; mainly English Channel.

Sidnyum turbinatum Fleming FIGS. 16.5, 16.6
[BERRILL 1950, 112; KOTT 1952, 68; MILLAR 1970, 24]
Colony composed of several lobes rising from encrusting stolons. Intertidally the lobes, containing two to four systems each of six to ten zooids, reach a maximum length of 20 mm and width of 10–15 mm in late summer. Subtidal specimens reach 40 mm in height, are broader and contain more systems. Test semi-transparent, yellow, grey, and white. Common cloacal openings with raised undulating margins. Oral siphon 8-lobed; three or sometimes four red pigment spots (the latter as in *Morchellium argus*); atrial opening with languet. Branchial sac with seven to nine rows of nine to ten stigmata. Stomach with 10–15 folds, continuous or variously broken into short lengths. Larvae in atrial cavity; trunk up to 0.6 mm long, with three adhesive papillae and a row of vesicles on each side meeting dorsally.

On stones, rocks and algae; lower shore to 200 m; all British coasts, rarely North Sea; Norway to Mediterranean (map: Millar 1966, 28).

For nomenclature, see Kott (1952).

Synoicum pulmonaria (Ellis and Solander) FIG. 16.6
[BERRILL 1950, 91; MILLAR 1970, 20]
Colonies at first pear-shaped, later rounded, to about 100 mm in diameter; semi-transparent, yellowish grey, with or without sand. Many systems, with seven to eight zooids in a circle around each common cloacal opening. Oral siphon 6-lobed; atrial siphon short, with broad languet terminating in three processes. Branchial sac with eight to 20 rows of 30 or more stigmata; stomach wall with rounded markings or low papillae; intestine twisted. Larvae in atrial cavity; trunk 0.6–1.0 mm long; three adhesive papillae, with median and paired lateral ampullae.

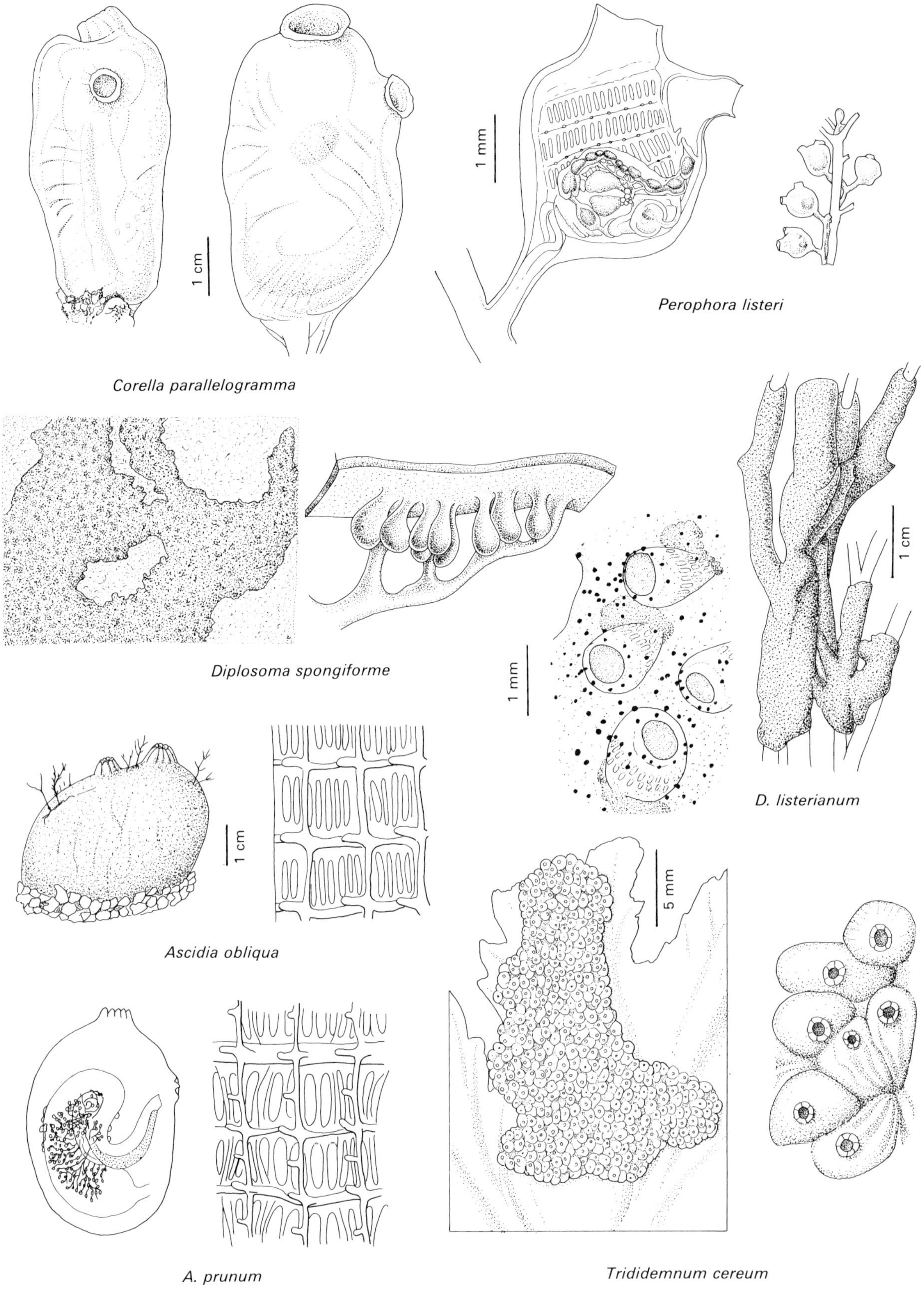
1 cm
Corella parallelogramma
1 mm
Perophora listeri
Diplosoma spongiforme
1 mm
1 cm
D. listerianum
1 cm
Ascidia obliqua
5 mm
A. prunum
Trididemnum cereum

Fig. 16.7

Sublittoral; 20–600 m on hard sandy bottoms; mainly North Sea but reaching English Channel, deep cold water off western Scotland and Northern Ireland; Arctic-Boreal (map: Millar 1966, 19).

7 DIDEMNIDAE

Colonies thin and incrusting, the base coextensive with the surface; gelatinous or packed with fine calcareous, stellate *granules*. Zooids very small, grouped around common cloacal openings constricted below the pharynx and abdomen; no post-abdomen. Often what is generally, but probably incorrectly, termed a retractor muscle extending postero-obliquely from the thorax. Branchial aperture 6-lobed; atrial aperture variable. Three or four rows of stigmata. Saucer-shaped *lateral thoracic organs* (LTOs), one each side of the thorax, produce the calcareous granules in those species which have them. Gonads in the intestinal loop. Eggs large, developing one at a time. Testis single or subdivided; sperm duct straight or coiled spirally (anticlockwise from the apex) around the testis; when the latter, the rectum crosses to the left of the oesophagus. Tadpole larvae with two or, more usually, three adhesive papillae and with anterior ampullae; developing in the common test toward the base of the colony; sometimes with precocious budding. Colonial budding oesophageal: at its simplest, two buds grow from the waist of the zooid and, as the zooid itself divides, the anterior bud forms a new abdomen for the thorax and the posterior bud forms a new thorax for the abdomen. Neither part of the parent zooid regresses during this process.

Features of very transparent didemnids may be observed direct. Spicules can be removed by treatment with EDTA but usually it is necessary to cut thin vertical slices through a previously narcotized and fixed colony. Major accounts of this family are given by Lafargue (1968) and Lafargue and Wahl (1987), the latter including keys to reproductive and non-reproductive colonies, good colour photographs, and scanning electron micrographs of spicules.

---

1. Sperm duct spiralled around the testis **2**
   Sperm duct straight (atrium usually to left in diagrams) **11**

2. Three rows of stigmata **3**
   Four rows of stigmata **5**

3. No granules in test; no LTOs ***Didemnopsis translucidum***
   Test with granules; LTOs present (*Trididemnum*) **4**

4. Colony whitish-grey, granules dense. LTOs as high as the stigmata, placed centrally on each side between the second and third rows of stigmata ***Trididemnum cereum***
   Colony small, translucent; granules rare. LTOs less than the height of the stigmata, situated level with the base of the third row ***Trididemnum delesseriae***

5. Atrial opening with a tubular siphon ***Leptoclinides faeroensis***
   Atrial opening simple; no siphon **6**

6. Testis undivided (*Didemnum*) **7**
   Testis with two or more lobes (*Polysyncraton*) **10**

7. Tadpole larvae with two fixatory papillae **8**
   Larvae with three fixatory papillae **9**

8. Colony orange, extensive; granules dense, with <15 rays. LTOs vertical, in centre of fourth row of stigmata ***Didemnum fulgens***
   Colony variably coloured (white, beige, reddish, violet, black); granules dense, with >15 rays. LTOs vertical, between the third and fourth rows of stigmata ***Didemnum maculosum***

9. Colony translucent, thinly encrusted with sand; pinkish-brown or honey-yellow; granules few, with >20 rays, each ray a cluster of acicular crystals ('needles'). LTOs small, vertical, between the second and third rows ***Didemnum lahillei***
   Colony smooth or rugose; granules dense, with <15 non-acicular rays. LTOs horizontal, below fourth row of stigmata ***Didemnum coriaceum***

10. Colony fleshy, velvety red. Testis with five or six lobes. Atrial opening very large, exposing almost all the stigmata; with a variably developed languet ***Polysyncraton lacazei***
    Colony leathery, orange. Testis bilobed. Atrial opening not exposing the stigmata and without a languet ***Polysyncraton bilobatum***

11. Test containing granules; zooids with LTOs (*Lissoclinum*) **12**
    Test without granules; no LTOs (*Diplosoma*) **13**

12. Colony leathery, white or variously coloured. No atrial languet. Testis undivided, to the right of the sperm duct ***Lissoclinum perforatum***
    Colony soft, white spotted with brown. Atrial opening with a large, cleft languet. Testis bilobed ***Lissoclinum weigelei***

13. Colony thin, 2 mm or less, transparent. Gonads to left of gut; two testicular lobes; ovary with several eggs; larvae <0.6 mm, with a blastozooid. Retractor muscle present ***Diplosoma listerianum***
    Colony thin, 2 mm or less, transparent, zooids orange-yellow. Gonads to right of gut; testis undivided; ovary with several eggs; larvae <0.6 mm, without a blastozooid. Retractor muscle absent ***Diplosoma singulare***
    Colony thick, >2 mm, opaque. Gonads to left of gut; two testicular lobes; ovary with one egg; larvae large, trunk >0.6 mm, with a blastozooid. Retractor muscle present ***Diplosoma spongiforme***

Didemnopsis translucidum Lafargue FIG. 16.9
[LAFARGUE 1968, 364]
Colonies encrusting, becoming extensive, 4–5 mm thick; test translucent, without granules; with sinuous systems, each a double row of zooids back to back (as *Botrylloides*). Common cloacal apertures few, subcircular, the margin not raised. Zooids about 1.5 mm long. Eight oral tentacles; three rows of stigmata; atrial siphon short, centred midway between the second and third rows of stigmata. No LTOs. Retractor muscle present but small. Middle intestine horizontal. Gonads on the left. Testis single, with spiral sperm duct making six to twelve turns. Larvae with three fixatory papillae; no internal bud at the time of release.

On rocks among *Laminaria digitata*, just subtidal. So far known only from the Brittany Peninsula (France).

Didemnum coriaceum (Von Drasche) FIG. 16.8
[CARLISLE 1954*a*, 313; LAFARGUE 1968, 386; MILLAR 1970, 35; all as *D. helgolandicum*; LAFARGUE 1975*a*, 188]
Colonies thin, whitish to violet; rigid from densely packed granules (and indistinguishable from *D. maculosum*). Zooids about 1 mm, with four rows of stigmata and a large atrial opening. Granules with usually fewer than five rays, the summits rounded. LTOs wing-like, horizontal, level with the base of the fourth row of stigmata. Retractor muscle long. Larvae large, 0.5 mm in length, and with three fixatory papillae (cf. *D. maculosum*); present May–September (France); the first bud present at the time of release.

On rocks; in Britain recorded only from Plymouth (Carlisle 1954); elsewhere from Brittany to western Norway and the Faroes.

Didemnum fulgens (Milne Edwards) FIG. 16.8
[LAFARGUE 1968, 395; 1976, 277]
Colony thin (1–2 mm) but extensive, smooth; orange. Common cloacal openings numerous and visible from their darker colour. Granules very numerous; with <15 pointed (45°) rays. Zooids about 1 mm; thorax slightly larger than abdomen. Four rows of stigmata. LTOs in the centre of the fourth row of stigmata; vertical. No atrial languet. Retractor muscle long, inserted ventrally at the level of the 'waist'. Gonads to the left; testis undivided, sperm duct making about nine turns. Larvae small, about 0.37 mm long; with two fixatory papillae, four to eight pairs of ampullae.

On rock faces down to 40 m; small colonies under stones or on algae. Recent records are from Brittany but should be present in south-west Britain. Berrill (1950) included *D. fulgens* as orange *D. maculosum*.

Didemnum lahillei Hartmeyer FIG. 16.8
[LAFARGUE 1968, 382; 1975*a*, 186; KOTT 1952, 76, as *D. gelatinosum*]
Colony small, about 1 mm thick, translucent beige or dull to deep yellow; colony margins clear, devoid of granules. Zooids visible, in circular or elongate systems; common cloacal openings small and inconspicuous, fairly numerous and disposed toward the edge of the colony. Granules sparse and characteristically clustered around the zooids away from the surface; with >20 lamellar or straight spines made up from acicular crystals. Zooids 1.5 mm. Four rows of stigmata. Cloacal opening small, level with the first three rows of stigmata (but smaller when contracted); the stigmata not exposed. LTOs vertical, near the endostyle, between the third and fourth rows of stigmata (second and third rows in Kott's description). Retractor muscle long. Gonads to the left; testis undivided. Larvae to 0.70 mm; three fixatory papillae; four to six pairs of ampullae; no internal buds at time of release. Present March–August at Plymouth.

Common intertidally in the Plymouth area (Kott), also Brittany; on stones.

Didemnum maculosum (Milne Edwards) FIGS. 16.5, 16.8
[CARLISLE 1954, 313; LAFARGUE 1968, 389; MILLAR 1970, 34; all as *D. candidum;* LAFARGUE 1976, 260]
Colony encrusting, smooth or rugose, hard, packed with granules; white, grey, violet, or whitish marked with violet. Granules generally with >15 rays, summits pointed (45°). Zooids about 1 mm but contracting strongly on preservation; four rows of stigmata; atrial opening simple, large, and exposing some of the stigmata. Three of the six oral lobes are elongate in var. *dentata*. LTOs vertical, on the edge of the atrial opening, between the third and fourth rows of stigmata. Retractor muscle long, inserted ventrally at the level of the waist. Gonads to the left; testis undivided, sperm duct spiral, with seven to ten turns. Larvae small, 0.4 mm, with two fixatory papillae and four to six pairs of ampullae.

On rocks, stones, *Laminaria* holdfasts, *Cystoseira, Delesseria*, etc. from the lower shore and sublittoral; southern and western coasts, with an extensive further distribution. Now regarded as distinct from *D. candidum* Savigny. Commoner than *D. coriaceum* (3:1 at Plymouth).

Diplosoma listerianum (Milne Edwards) FIG. 16.8
[BERRILL 1950, 125; LAFARGUE 1968, 406; MILLAR 1970, 37]
Colonies thin (<2 mm), sheeting and moderately extensive, transparent, sometimes with brown or grey pigmented cells. Common cloacal apertures slightly raised in life, scarcely visible from above. Zooids about 2 mm, with four rows of stigmata; atrial opening simple, large, almost completely exposing the stigmata; no languet. No LTOs. Retractor muscle well-developed. Gonads to the left of the gut. Testis usually bilobed; sperm duct not proximally coiled, aligned with the rectum. Ovary containing several eggs. Larval trunk 0.4–0.6 mm, with three fixatory papillae; larval zooid and first budded zooid present at the time of release. Reproductive season long, March–September in the Clyde, continuing through October on the Brittany coast.

Intertidal and in shallow water, especially on algal fronds, *Zostera* leaves, etc. Apparently widely distributed but in Britain not distinguished from the two following species.

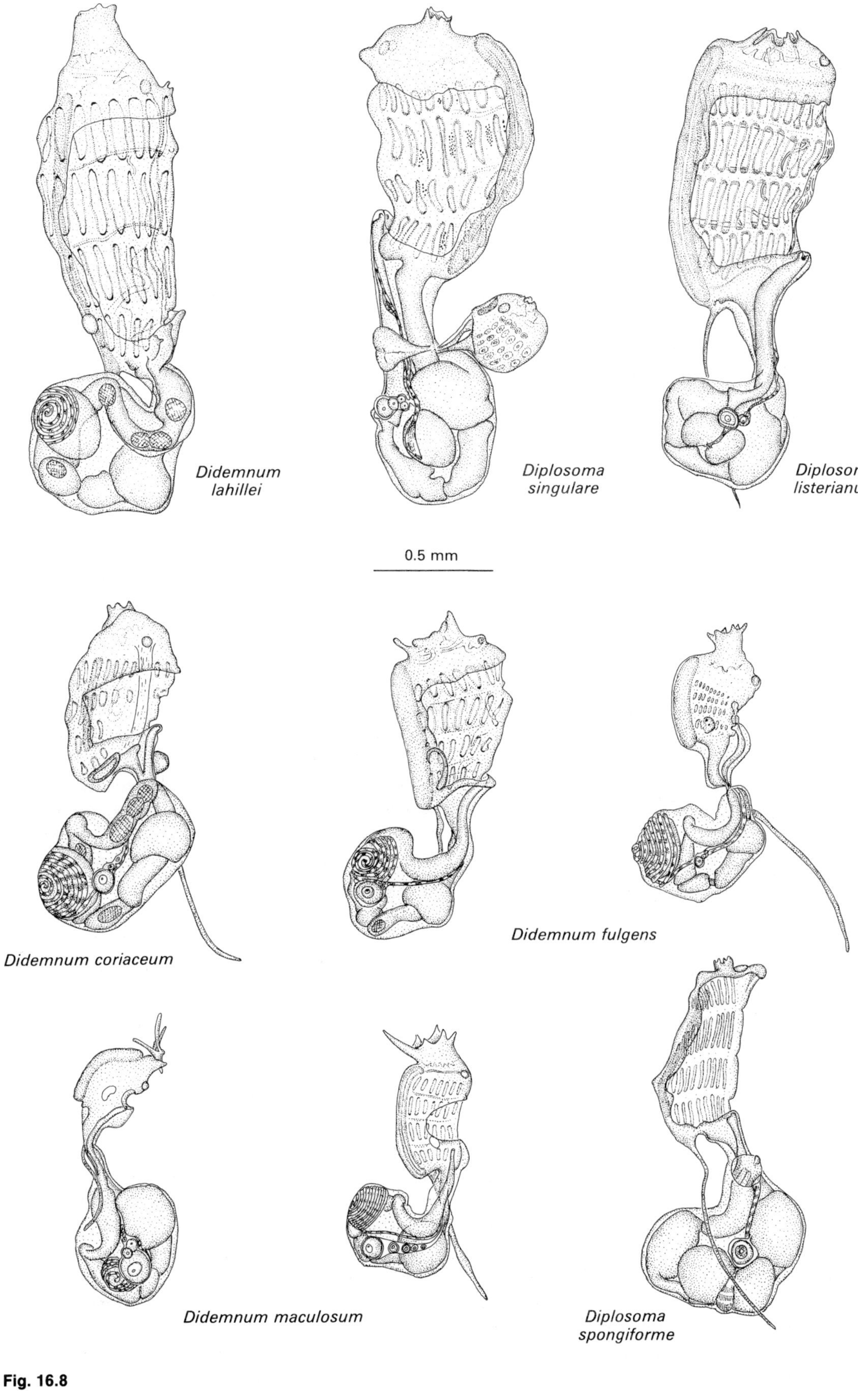
Didemnum
lahillei
Diplosoma
singulare
Diplosoma
listerianum
0.5 mm
Didemnum coriaceum
Didemnum fulgens
Didemnum maculosum
Diplosoma
spongiforme

Fig. 16.8

Diplosoma singulare Lafargue FIG. 16.8

[LAFARGUE 1968, 409]

Colonies thin (<2 mm), transparent, soft, rather small; the zooids, especially in the branchial sac, testis, and posterior of the abdomen, yellow-orange (the colour fading in preserved material). Zooids readily visible in life but the test opaque after preservation. Zooids large, about 2 mm; unordered; common cloacal openings scarcely visible in preserved material. Oral siphon with six lobes; at least twenty-four oral tentacles; four rows of stigmata. Atrial opening extensive. Neither LTOs nor retractor muscle. Zooid bent at a right angle at the 'waist' (not indicated in the illustration). Gonads to the right of the gut (cf. other species of *Diplosoma*). Testis undivided; sperm duct straight, aligned with the rectum. Ovary containing several eggs. Larval trunk 0.55 mm, with three fixatory papillae and four (rarely five) pairs of ampullae in two series; four rows of stigmata by the time of release but no bud. Mature zooids and larvae have been recorded in September (in Brittany).

Known only from a few collections off the Brittany peninsula; on red algae (*Delesseria, Phyllophora*).

Diplosoma spongiforme (Giard) FIG. 16.8

[KOTT 1952, 79, as *Lissoclinum cupuliferum*; LAFARGUE 1968, 413, as *D. cupuliferum*]

Colonies thick (commonly about 6 mm), soft but tough, pigmented; forming extensive, loosely attached crusts (up to 0.25 m^2) on rock faces, often overgrowing other encrusting species. Greyish, containing white, non-calcareous bodies; the zooids sometimes with black speckles. The surface and basal test joined by strands which branch toward the surface and enclose zooids in clusters of four to six around the generator zooid. Zooids about 1.5 mm. Oral siphon with six lobes and 12–16 tentacles; four rows of stigmata. Atrial opening large. No LTOs. Retractor muscle well-developed. Gonads to the left of the gut. Testis bilobed; common sperm duct straight, aligned with the rectum. Ovary containing only one oocyte at a time. Larvae large, the trunk 0.6–0.8 mm in length, with three to eight (commonly four) fixatory papillae and two to seven (commonly three) irregularly disposed pairs of ampullae; larval zooid and first bud present at the time of release. Reproduction April–October in Brittany.

Sublittoral, on rock faces (rarely algae) in the *Laminaria digitata* zone, down to 40 m; western coasts, common (Picton 1985). Brittany; also Mediterranean.

Leptoclinides faeroensis Bjerkan FIG. 16.9

[MILLAR 1966, 32; 1970, 36]

Colonies thick, grey or white. Granules mainly in the surface test, stellate, to 0.05 mm diameter. Zooids to 2 mm; with long backwards-directed atrial siphon; four rows of stigmata. Gonads on left of gut; testis undivided; sperm duct coiled, with four to six turns. Larval trunk about 1.2 mm long, with three fixatory papillae in vertical row, and four pairs of bifid ampullae.

An arctic-boreal species, in Britain known only from Looe (Cornwall).

Lissoclinum perforatum (Giard) FIG. 16.9

[MILLAR 1950, 38; LAFARGUE 1968, 422; as *L. argyllense*]

Colonies small, rarely exceeding 30 mm across, about 2 mm thick; white or creamy with dark spots marking the oral openings and larger slits opening from the extensive common cloaca; surface and basal test readily separable. Zooids not clearly in systems. Colony border distinct, without zooids. Granules densely packed: rays <15, blunt ended; about 0.03 mm diameter. Zooids about 2 mm; four rows of stigmata; a large atrial opening, no languet. LTOs wing-like, vertical, on the edge of the atrial opening between the second and third rows of stigmata. No retractor muscle. Gonads on right of gut; testis single, sperm duct straight, aligned with the rectum. Larval trunk about 0.6 mm long; three fixatory papillae in vertical row, four pairs of ampullae; oozooid with four rows of stigmata, no blastozooid. Larvae present May–September in Brittany, July–August in Scotland.

Southern and western coasts, common (Picton 1985); Brittany. On *Laminaria digitata* in the shallow sublittoral and on rocks, especially in strong tidal flow, to 30 m.

Lissoclinum weigelei Lafargue FIG. 16.9

[LAFARGUE 1968, 426]

Colonies encrusting, soft, about 2 mm thick, small, often about 30 mm in largest dimension; white with brown spots and circular to oval common cloacal openings; border distinct, without zooids. Granules densely packed; 0.02–0.03 (–0.04) mm diameter; rays >40, truncate, the ends toothed. Zooids 1–2 mm; four rows of stigmata; atrial opening extensive, with a large bifid languet. LTOs wing-like, vertical, on the edge of the atrial opening between second and third rows of stigmata. No retractor muscle. Gonads on the left; testis bilobed; common sperm duct straight, aligned with the rectum. Larval trunk about 0.7 mm long, with three fixatory papillae in line; up to nine ampullae on each side of the papillae; oozooid with four rows of stigmata, no blastozooid. Mature July–October (Brittany).

Under rocks and in crevices; 5–30 m; Brittany.

Polysyncraton bilobatum Lafargue FIG. 16.9

[LAFARGUE 1968, 401; MEDIONI 1970, 39]

Colonies thin (1–2 mm), often extensive (up to 50 cm^2), with a distinct border. Zooids irregularly dispersed, not in distinct systems. Golden-yellow or orange in colour. Granules numerous, densely and evenly distributed; 20–50 μm diameter; with about 15 rays which taper at about 60°, summits obtuse or dentate. Zooids <1 mm when contracted; four rows of stigmata; atrial opening fairly small, without languet. LTOs of variable size, wing-like, near the edge of the atrial opening, between the second and third rows of stigmata. Retractor muscle present. Gonads on the left; testis bilobed (two hemispheres); sperm duct proximally spiral, with seven to eight (nine) turns. Larval trunk to >0.7 mm, with three papillae and six to eight pairs of ampullae; oozooid with four rows of stigmata; two blastozooid buds (one concealed by the yolk mass) at the time of liberation. Mature zooids May–October (Brittany).

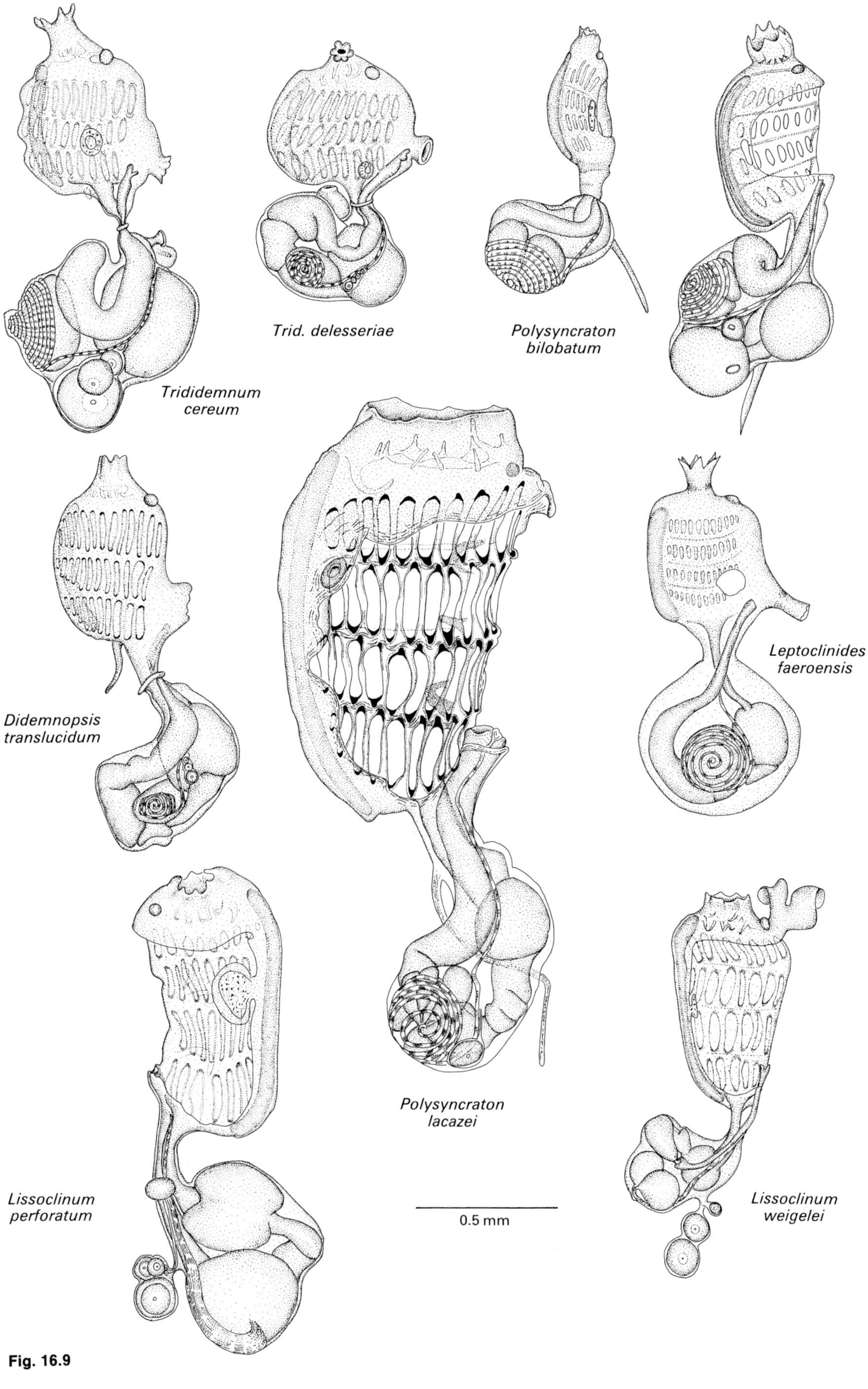
Trididemnum
cereum
Trid. delesseriae
Polysyncraton
bilobatum
Didemnopsis
translucidum
Polysyncraton
lacazei
Leptoclinides
faeroensis
Lissoclinum
perforatum
0.5 mm
Lissoclinum
weigelei

Fig. 16.9

On upper, lower and vertical surfaces of stones and rocks; sublittoral, to 40 m; Brittany.

This species is very similar to *D. fulgens* but is not found on algae. The surest distinction is by the size of the larvae and the number of fixatory papillae.

Polysyncraton lacazei (Giard) FIG. 16.9
[LAFARGUE 1968, 398; 1975*b*, 139]
Colonies variable in extent, thickness (1–5 mm), and colour (yellow-orange to wine-red). Granules (30–80 μm diameter) with about 20 rays per visible hemisphere, the summits dentate; smaller granules (15–20 μm), eccentric, the rays acicular. Distribution of granules irregular, most dense in the oral lobes which stand out as six white points, and around the common cloacal opening where they form five spots. Zooids not in recognizable systems; 2–2.5 mm; 24 oral tentacles; four rows of stigmata; atrial opening extensive, the stigmata exposed; languet variably developed. LTOs cup-shaped, situated near the endostyle towards the top of the second row of stigmata. Retractor muscle usually long. Gonads to the left; testis 5- or 6-lobed; sperm duct proximally spiral, making three to six turns. Larva large, the trunk 1.2 mm long; two papillae surrounded by numerous ampullae; oozooid and two blastozooids having four rows of stigmata at the time of release. Testes present May–September, ovaries and larvae June–September (Brittany).

Essentially a Mediterranean species; recorded rather rarely in Brittany from sublittoral rocks.

Trididemnum cereum (Giard) FIG. 16.9
[LAFARGUE 1968, 369; also BERRILL 1950, 116; MILLAR 1970, 33, as *T. tenerum*]
Colonies encrusting, up to 5 mm thick, of variable extent; semi-transparent, whitish, yellowish, or brownish black, with black pigment spots. Granules generally rather few, concentrated between the basal and surface test; when abundant in the surface test the colony resembles *Didemnum maculosum*. Granules very variable, the rays acicular, linear, or tapering. Zooids with six oral lobes (though Berrill's Fig. 35 shows eight), three rows of stigmata, and an atrial siphon level with the lowest row of stigmata. Often a dark spot anteriorly over the endostyle. LTOs conspicuous, mid-lateral, between the second and third rows of stigmata. Retractor muscle small. Gonads on the left; testis undivided; sperm duct proximally spiral, with seven to twelve turns. Larval trunk 0.6 mm long; three papillae in vertical row and four pairs of anterior ampullae; oozooid with three rows of stigmata, LTOs visible; no blastozooids. Mature zooids April–October (Brittany).

On rocks, algae (e.g. haptera of *Laminaria*), and other ascidians; lower shore down to about 200 m; south and west coasts of Britain; also Norway (Millar 1966) and Mediterranean. According to Lafargue (1968) the European *T. cereum* is distinct from the American *T. tenerum*, although they have often been synonymized.

Trididemnum delesseriae Lafargue FIG. 16.9
[LAFARGUE 1968, 374]
Colonies thin (about 1 mm), small (rarely exceeding 1 cm^2), often aggregated in a mosaic; soft, transparent, uncoloured, almost devoid of granules. Zooids clearly visible, about 1 mm; three rows of stigmata; atrial siphon midway between the second and third rows, LTOs small, level with the bottom row of stigmata. Retractor muscle small. Gonads to the left; testis undivided; sperm duct proximally spiral, with seven to twelve turns. Larval trunk 0.5 mm long; three papillae and four pairs of pedunculate ampullae; oozooid with three rows of stigmata, first blastozooid no more than an undifferentiated bud. Mature zooids May–November.

On red algae, particularly *Delesseria*, 20–30 m; Brittany.

Suborder **Phlebobranchia**

8 PEROPHORIDAE

Colonial, with zooids united by basal stolons. Branchial sac with or without longitudinal vessels. Gut to left of branchial sac; gonads in the intestinal loop.

Perophora listeri Forbes in Forbes and Hanley FIG. 16.7
[BERRILL 1950, 145; MILLAR 1970, 41]
Colonies of well-separated zooids arising from creeping stolons. Zooids shortly stalked, about 4 mm high, somewhat laterally compressed; transparent. Siphons short; four rows of stigmata; no inner longitudinal vessels. Ovary surrounded by radiating testis follicles. Larvae develop in atrial cavity; trunk about 0.3 mm long, with three papillae in line, and four rows of stigmata.

On stones, hydroids, algae, etc., lower shore to about 30 m; Irish Sea, south and west coasts; further distribution southwards to Mediterranean.

9 CORELLIDAE

Unitary. Gut to right of branchial sac. Gonads in or around the intestinal loop. Oviparous.

Corella parallelogramma (Müller) FIG. 16.7
[BERRILL 1950, 149; MILLAR 1970, 42]
Body subquadrate and laterally compressed, up to about 50 mm high, attached by a small part of the base. Test hard, thin, smooth, and transparent; mantle with yellow, white, and red markings, Branchial sac with up to 27 longitudinal vessels on each side but no papillae at the intersections with transverse vessels. Stigmata of variable length, curved, in subquadrate spirals.

On stones, shells, *Laminaria*, etc. in clear, shallow water down to about 200 m; all coasts; also Norway and south to the Mediterranean (map: Millar 1966).

10 ASCIDIIDAE

Unitary, often large. Oral siphon with six to eight lobes, atrial six. Gut to left of branchial sac. Stigmata straight. Gonads within the loop of or spreading over the intestine. Oviparous.

1. Internal longitudinal vessels papillate at intersections with transverse vessels **2**

   Internal longitudinal vessels smooth (*Ascidiella*) **3**

2. Surface of test mammillate (with smooth rounded swellings); posterior part of branchial sac extending beyond the stomach and recurved upon itself. Southern coasts only ***Phallusia mammillata***

   Surface of test smooth, with irregular swellings, or encrusted with shell; posterior part of branchial sac not recurved (*Ascidia*) **4**

3. Oral tentacles <40, fewer than internal longitudinal vessels (about 80). Dorsal tubercle C-shaped, with the horns incurved. Body up to 100 mm in length ***Ascidiella aspersa***

   Oral tentacles 30–110, always exceeding the number of internal longitudinal vessels. Dorsal tubercle U-shaped. Body not exceeding 50 mm ***Ascidiella scabra***

4. Atrial siphon less than one-third down body; rectum long ***Ascidia virginea***

   Atrial siphon more than one-third down body; rectum short **5**

5. Test thick, smooth, often pink ***Ascidia mentula***

   Test thin, greenish; smooth, roughish, or covered with particles ***Ascidia conchilega***

Ascidia conchilega Müller FIG. 16.10

[BERRILL 1950, 162; MILLAR 1970, 48]

Body elliptic-oblong, attached along the left side; oral siphon terminal, atrial siphon two-thirds down body; test thin, translucent, greenish, usually with small rough projections; naked or covered with shell fragments; up to 60 mm in length. In branchial sac the right face of the dorsal lamina with small irregular papillae. Gonads covering the sides of the intestinal loop; eggs small, pink, visible through the test.

On stones and shells; lower shore to deep water; all coasts, from the Faroe islands and Norway to the Mediterranean (map: Millar 1966).

Ascidia mentula Müller FIG. 16.10

[BERRILL 1950, 160; MILLAR 1970, 47]

Body elongate, usually 50–180 mm; test thick, cartilaginous, translucent, usually pink, raised into irregular swellings; attached mostly by the left side; oral siphon terminal, atrial siphon one-half to two-thirds down body. In branchial sac the right face of the dorsal lamina without papillae. Ovary compact, in the intestinal loop; testis diffuse, over the surface of the intestine.

On rock, stones, and shells; from the lower shore to about 200 m; all coasts, from Norway to the Mediterranean (map: Millar 1966).

Ascidia virginea Müller FIG. 16.10

[BERRILL 1950, 165; MILLAR 1970, 49]

Body subrectangular, attached by part or all of one side; usually 30–80 mm. Test smooth, cartilaginous, translucent, usually pink. Oral siphon terminal, atrial siphon a short distance from it. Rectum long, extending beyond the intestinal loop. Ovary on the left inner margin of the loop, testis spread over the right side.

On stones and shells; 30 to about 400 m; all except North Sea coasts; Norway to the Mediterranean (map: Millar 1966).

Ascidiella aspersa (Müller) FIG. 16.10

[BERRILL 1950, 154; MILLAR 1970, 44]

Body ovoid, usually attached by the base; 50–130 mm. Test cartilaginous, greyish, surface rough (finely papillate). Oral siphon terminal, atrial siphon about one-third down body. Dorsal tubercle with elliptical opening, both horns incurved. Oral tentacles 30–40, well-separated, fewer than the inner longitudinal vessels. No papillae at intersections of longitudinal and transverse vessels. Maturing at length 40–50 mm. Ovary on left side of the intestinal loop; testis on both sides. Eggs with large follicle cells, floating if placed in still water.

The very similar *A. scabra* may be attached more by the side and does not exceed about 50 mm in length; its atrial siphon is closer to the oral; and the oral tentacles are more numerous than the longitudinal vessels. The follicle cells are small and the eggs sink in still water. (The difference in the follicle cells can be seen in any mature specimen, living or preserved: Berrill 1950.)

On stones, algae, and submerged structures, often aggregated in masses; from the lower shore to about 50 m; all coasts; Norway to the Mediterranean (map: Millar 1966).

Ascidiella scabra (Müller) FIG. 16.10

BERRILL 1950, 58; MILLAR 1970, 45]

Body ovoid, usually attached by the left side; 20–50 mm. Test cartilaginous, smooth or wrinkled, with small red papillae round the siphons; semi-transparent, with red pigment on the upper mantle often visible. Oral siphon terminal, atrial siphon close to it (about one-quarter down body). Dorsal tubercle U-shaped, with the horns variously coiled. Oral tentacles 30–110, always more numerous and closer spaced than the inner longitudinal vessels; no papillae at the intersections of the longitudinal and transverse vessels. Eggs with numerous, small follicle cells; sinking in still water.

On algae, stones, and shells; lower shore to about 300 m; all coasts; Norway to the Mediterranean (map: Millar 1966).

Phallusia mammillata (Cuvier) FIG. 16.10

[BERRILL 1950, 171; MILLAR 1970, 52]

Body ovoid, broader at the base, up to 140 mm in length. Test thick, cartilaginous, smooth, raised into rounded mammillae; whitish to brown. Oral siphon terminal, atrial siphon one-third to one-half down body. Posterior part of branchial sac recurved upon itself. Renal vesicles large and abundant in the region of the gut loop, obscuring the ovary and partly covering the testis. Breeding takes place throughout the year.

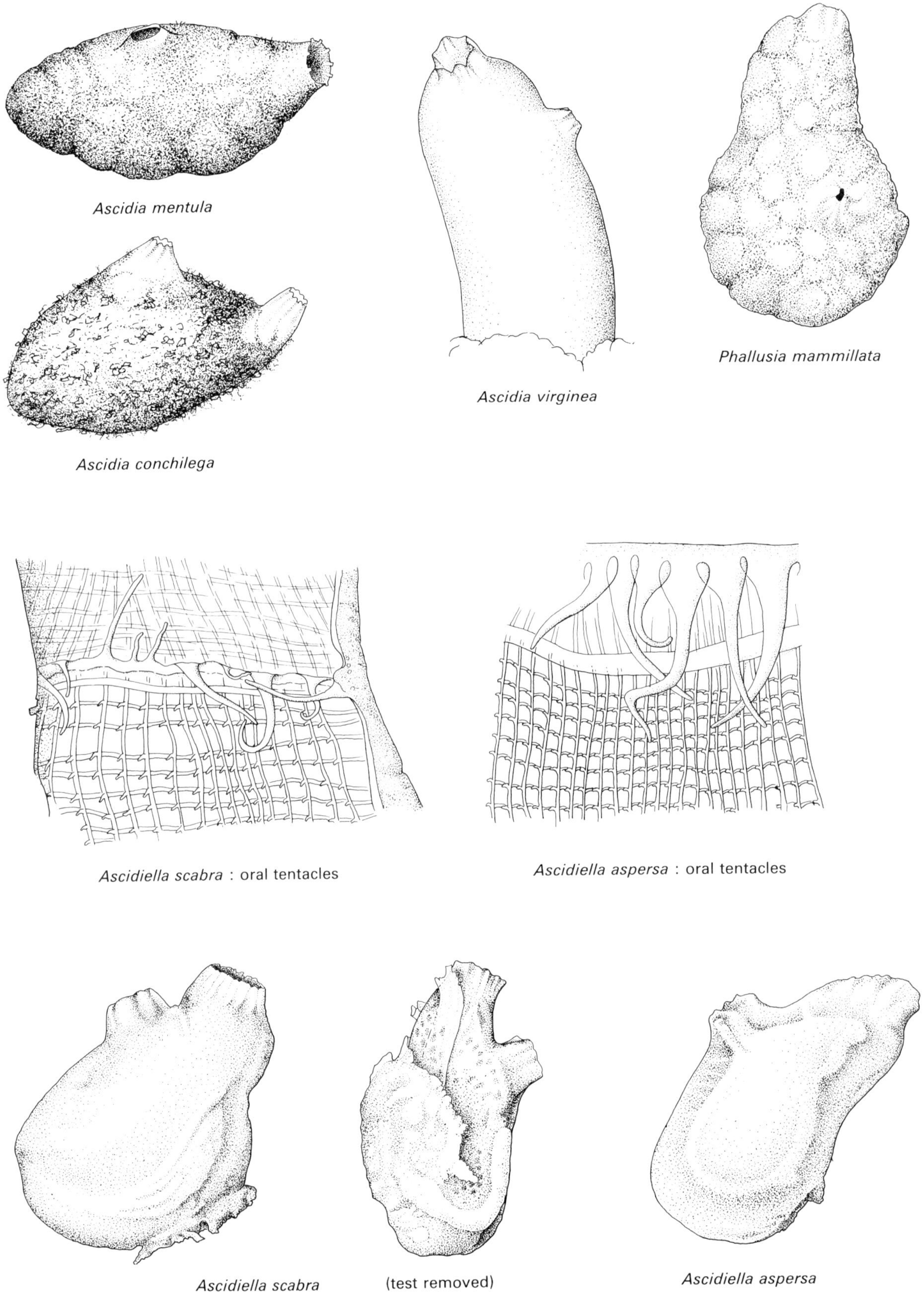

Ascidia mentula
Ascidia virginea
Phallusia mammillata
Ascidia conchilega
Ascidiella scabra : oral tentacles
Ascidiella aspersa : oral tentacles
Ascidiella scabra
(test removed)
Ascidiella aspersa

Fig. 16.10

Attached to stones but often in silty or muddy situations; lower shore and shallow water down to 180 m; south coast of Britain and south-west Ireland (possibly further north); southwards to the Mediterranean.

Order **Pleurogona**

Suborder **Stolidobranchia**

11 STYELIDAE

Unitary or colonial. Siphons smooth edged or 4-lobed; oral tentacles simple. Branchial sac with never more than four folds each side; stigmata straight; dorsal lamina smooth-edged. Stomach with ridges and pyloric caecum, but no 'liver'. Gonads in the body wall, alongside the branchial sac; variously developed, single, paired, several, or many; those in the form of multiple small bodies, embedded in the body wall or projecting from it, being termed *polycarps*. Tadpoles with ocellus reduced or absent.

The family is conveniently divided into three subfamilies:

A. STYELINAE. Unitary but may occur in clusters. Both siphons 4-lobed. Branchial sac usually with four folds per side. Gonads on one or both sides, usually multiple.

B. POLYZOINAE. Forming loose colonies or clonal aggregations by budding from the mantle wall; without systems. Branchial sac with or without folds.

C. BOTRYLLINAE. Colonial, with systems. No branchial folds.

KEY TO SUBFAMILIES

1. Unitary or colonial; zooids not in systems — 2
   Colonial; zooids in stellate or linear systems — **11C Botryllinae**

2. Colonial but each zooid with its own atrial opening to the exterior; zooids embedded in a flat lamina of test or on stolons — **11B Polyzoinae**
   Unitary but may occur in clusters resulting from aggregated settlement — **11A Styelinae**

[Confusion is possible between *Distomus* (Polyzoinae) and densely aggregated *Dendrodoa* (Styelinae). If doubtful, *Distomus* has the ovary on the right side and several horizontal testicular follicles on the left; *Dendrodoa* has an elongate ovotestis on the right side only.]

11A *STYELINAE*

1. Body club- or bottle-shaped — 2
   Body various but not club-shaped — 3

2. Bottle-shaped ascidian living in sand or mud, the siphons at the narrow end; branchial sac without folds; northern — ***Pelonaia corrugata***
   Club-shaped ascidian attached to firm substrata by its narrow end, siphons at the broad end; branchial sac with four folds; southern — ***Styela clava***

3. Gonad comprising central ovary and separate but adjacent testicular follicles (*Styela*) — 4
   Ovary and male follicles combined in a single, not necessarily compact, organ — 5

4. One gonad on each side — ***Styela coriacea***
   Two gonads on each side — ***Styela partita***

5. Gonad on the right side only — ***Dendrodoa grossularia***
   Gonads both sides — 6

6. Branchial folds rudimentary; several parallel, elongate gonads on each side — ***Cnemidocarpa mollis***
   Four branchial folds each side; gonads consisting of many small 'polycarps' scattered on both sides (*Polycarpa*) — 7

7. Surface of test smooth or wrinkled, not fibrillar, not heavily coated with particles — 8
   Surface of test with fibrillar processes, heavily coated with mud, shell, or sand — 9

8. Surface of test rough, wrinkled — ***Polycarpa pomaria***
   Surface of test smooth, red — ***Polycarpa rustica***

   Polycarps scattered over each wall — ***Polycarpa fibrosa***
   Polycarps in a single row on each side of the endostyle — ***Polycarpa gracilis***

Cnemidocarpa mollis (Stimpson) FIG. 16.11
[BERRILL 1950, 186; MILLAR 1970, 58]
Body ovoid to spherical, up to 15 mm diameter. Test with fibrils,

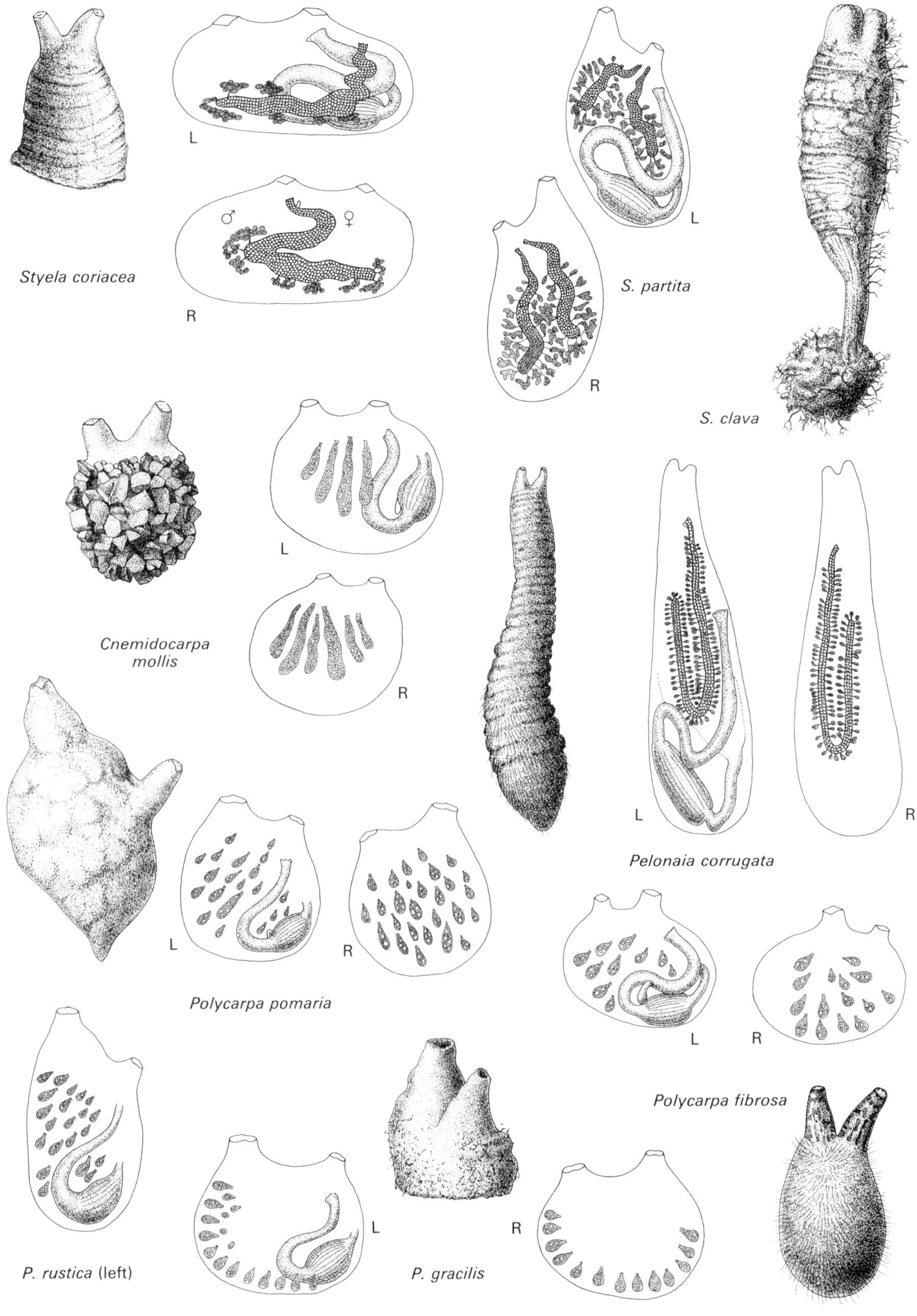
L
♂ ♀
R
Styela coriacea
L
R
S. partita
S. clava
L
R
Cnemidocarpa mollis
L
R
Pelonaia corrugata
L
R
Polycarpa pomaria
L
R
Polycarpa fibrosa
L
R
P. rustica (left)
P. gracilis

Fig. 16.11

coated with sand and shell particles. Siphons terminal, fairly close; branchial sac with four rudimentary folds each side. Gonads united as cylindrical ovotestes, three to six on the left, four to eight on the right. Oviparous.

Loosely attached in sand and mud; shallow water to at least 100 m; east and north Scotland, rare; also North America (map: Millar 1966).

Dendrodoa grossularia (van Beneden) FIG. 16.12
[BERRILL 1950, 198; MILLAR 1970, 64]
Body depressed, dome-shaped, on a spreading base, or upright; aggregated or solitary; reddish brown; up to about 20 mm in height, not more than 15 mm in diameter. Branchial folds rudimentary, their position marked by groups of inner longitudinal bars. Gut on the left, stomach horizontal, with longitudinal folds. One gonad only, on the right, cylindrical, parallel with the endostyle; ovarian follicles surrounded by testis; viviparous, the tadpoles developing in the atrium. Larval trunk subglobular, with three small, simple papillae surrounded by a ring of ampullae; with black otolith but no ocellus; tending to settle near the adults, leading to clustering.

On rock, stones, shells, etc. from the lower shore into deep water; all coasts, in aggregations; particularly in the south and west; locally common; North America, Arctic south to Isles of Scilly and Brittany peninsula (Glénan archipelago) (map: Millar 1966).

Pelonaia corrugata Forbes and Goodsir FIG. 16.11
[BERRILL 1950, 180; MILLAR 1950, 54]
Body large, up to 140 mm long; club-shaped, with the siphons terminal and close together at the narrow end. Test transversely wrinkled, with fibrils towards the lower end, particle encrusted. Branchial sac without folds. Gut on the left, stomach vertical. Gonads linear, the testis follicles marginal to the ovary, U- or J-shaped, the arm to the oviduct being the longer. Oviparous; development without tadpole.

Lives buried in sand or mud; northern waters only (Clyde to north-west England); North Sea to the Arctic (map: Millar 1966).

Polycarpa fibrosa (Stimpson) FIG. 16.11
[BERRILL 1950, 196; MILLAR 1970, 63]
Body subglobular or ovoid (vertically or horizontally), to about 35 mm high. Test tough, fibrillar, coated with sand or shell fragments, or with a thick layer of mud, except for the terminal siphons. Branchial sac with four folds on each side. Gut on the left, stomach horizontal. Ten to twenty drop-shaped polycarps on each side. Oviparous.

In sand or mud; LWST to deep water; all coasts; North America and Arctic southwards to Strait of Gibraltar (map: Millar 1966).

Polycarpa gracilis Heller FIG. 16.11
[BERRILL 1950, 191; MILLAR 1970, 61]
Body vertically or horizontally ovoid, to about 20 mm. Test thin, fibrillar, particle encrusted. Branchial sac with four folds each side. Gut on the left, stomach horizontal. Polycarps in a single (or partially double) series on each side of the endostyle.

On stones, etc., in shallow water; local but possibly overlooked; English Channel and southwards to the Mediterranean.

Polycarpa pomaria (Savigny) FIG. 16.11
[BERRILL 1950, 188; MILLAR 1970, 60]
Shape variable, ovoid, subglobular or conical, up to 70 mm long. Test leathery, wrinkled, and mammillated; brownish; not particle encrusted. Branchial sac with four low folds each side. Gut on the left, stomach horizontal. Polycarps drop-shaped, numerous, scattered each side of the body. Viviparous (at least sometimes).

Generally attached to stones and shells, frequently clustered; shallow water down to about 450 m; all coasts; locally common; Arctic to Mediterranean (map: Millar 1966).

Polycarpa rustica (Linnaeus) FIG. 16.11
[BERRILL 1950, 193; MILLAR 1970, 62]
Body tall, ovoid to cylindrical; to about 20 mm. Oral siphon terminal, atrial siphon slightly dorsal to it. Test smooth, red or pink (can superficially resemble *Dendrodoa grossularia*). Branchial sac with four folds each side. Gut on the left, stomach oblique. Polycarps drop-shaped, numerous, scattered on each side of the body. Probably viviparous.

On stones, etc., LWST to shallow water; Irish Sea and English Channel; also Brittany (its geographic range apparently very limited).

Styela clava Herdman FIG. 16.11
[MILLAR 1970, 57; CARLISLE 1954*b*, as *S. mammiculata*]
Body elongate, inverted club-shaped, tapering to a narrow stalk; up to 120 mm long, the stalk contribution about one-third. Siphons terminal, close together. Test leathery, with wrinkles and mammillae. Branchial sac with four folds each side. Gut on the left, stomach longitudinal. Gonads two to four on the left, five to eight on the right; the ovary long and narrow, surrounded by the testis follicles.

On stones, pilings, etc. from the lower shore and shallow water; south coasts from Portsmouth to Devon; Milford Haven (Nelson-Smith and Case 1984), Swansea docks. Introduced to Britain from the Pacific.

Styela coriacea (Alder and Hancock) FIG. 16.11
[BERRILL 1950, 182; MILLAR 1970, 55]
Body depressed, dome-shaped, up to 26 mm across (in strongly flowing water) to cylindrical, up to 30 mm high (in still water). Test finely wrinkled and tuberculate, with or without encrusting sand (can resemble *Dendrodoa grossularia*). Branchial sac with four folds each side. One gonad each side, the ovary elongate, sinuous; testes follicular, below the posterior two-thirds of the ovary. Oviparous (perhaps sometimes viviparous).

On stones, shells, *Polycarpa pomaria*, etc., from just below tidemarks to about 600 m; generally distributed in Britain though scarce on the east coast (but common elsewhere in the North Sea); North America, and Arctic south to Brittany.

Styela partita (Stimpson) FIG. 16.11

[BERRILL 1950, 184; MILLAR 1970, 56]

Body ovoid-conical, the base expanded; to 20 (–30) mm. Test rough and warty. Branchial sac with four folds each side. Gut on the left, stomach horizontal. Two gonads on each side, sinuous ovaries surrounded by testicular follicles. Oviparous.

On rocks, etc., from just below tidemarks to 30 m; south coast of England only, thence southwards to the Mediterranean and North Africa.

11B *POLYZOINAE*

1. Colony encrusting, the zooids closely packed; no branchial folds but the inner longitudinal bars in three groups on each side **Distomus variolosus**

   Colony composed of more or less separated zooids budded from stolons; branchial sac with folds **Stolonica socialis**

Distomus variolosus Gaertner FIG. 16.12

[BERRILL 1959, 203; MILLAR 1970, 66]

Colony a crust of zooids united basally, and to some extent laterally, by test; ovoid, depressed (or sometimes taller with the common base intermittent); about 10 mm long; reddish; siphons terminal, opening independently. Branchial folds rudimentary, their position indicated by groups of longitudinal bars. Ovaries six to ten, each with a single large egg and short oviduct, restricted to the right side; testes about 12, ovoid-elongate, only on the left. Viviparous; larval trunk about 0.7 mm long, with triangular arrangement of three papillae surrounded by a ring of about 16 small ampullae, and a complex sense-organ (ocellus + otolith). Budded zooids develop at the end of atrial stolons (outgrowths from the thoracic body wall), which subsequently atrophy.

On rock etc., but especially around the haptera and up the stipe of *Laminaria hyperborea*, often washed up following autumn storms; local, south Devon, Cornwall, Lundy (Lane 1976), Skomer (Crothers 1966), Wexford (Picton 1985), Galway Bay (JSR); Irish Sea and west of Scotland records require confirmation on account of confusion with *Dendrodoa grossularia* (Millar 1970). South to Portugal.

Stolonica socialis Hartmeyer FIG. 16.12

[BERRILL 1950, 209; MILLAR 1970, 68]

Zooids in clusters, upright with terminal siphons, to about 20 mm, united basally by stolons; yellowish or orange. Branchial sac with two or three folds each side. Gonads in three series, two ventral, paralleling the endostyle, the third intestinal. Posterior gonads in the left endostylar series comprise ovary, producing one egg, and testicular follicles; all the rest are small rosettes of testicular follicles. Viviparous; tadpoles large (trunk about 1 mm), with triangular arrangement of three small papillae, irregularly developed ampullae, and a compound sense-organ (as *Distomus*).

Subtidal rocks, 5–35 m, along the south coasts of Devon, Cornwall, and Ireland; Lundy (Lane 1976), Skomer (Crothers 1966), Donegal (Picton 1985): also Channel Islands and Brittany.

11C *BOTRYLLINAE*

1. Zooids appearing subcircular-ovate, in biserial linear systems **Botrylloides leachi**

   Zooids appearing slender obovate, grouped in star-shaped systems **Botryllus schlosseri**

Botrylloides leachi (Savigny) FIG. 16.12

[BERRILL 1950, 224; MILLAR 1970, 71]

Colony flat, encrusting, usually grey, orange, or red-brown. Zooids in two parallel rows with the common cloaca in between them, the openings elongate-oval; sometimes the systems obscured by crowding. Vascular ampullae fringing the colony. Atrial opening hooded, the upper lip at first tapering but its free end becoming broader than the zooid itself. Branchial sac with nine to 15 rows of stigmata and three internal longitudinal bars; no folds. Gut to the left; stomach with nine to 11 furrows; pyloric caecum small. Ovary with single ovum, posterior to the lateral, mulberry-like testis on each side. Larvae developing in a brood pouch, escaping by rupture into the common cloaca. The tadpole and budding as in *Botryllus*.

On algae, stones, ascidian tests, etc., on the lower shore and in shallow water; all coasts; distributed from northern Norway to the Mediterranean (map: Millar 1966).

Botryllus schlosseri (Pallas) FIG. 16.12

[BERRILL 1950, 216; MILLAR 1950, 70]

Colony flat, encrusting, or lobate; the test of almost any colour, though blue is commonest. Zooids contrasting in colour with the test, in stellate systems of three to twelve. Atrial siphons long, tubular, tapering, opening into the common cloaca. Vascular ampullae fringing the colony. Branchial sac with seven to 15 rows of stigmata and three internal longitudinal bars; no folds. Gut to the left, stomach with shallow grooves; a hooked pyloric caecum. Testes mulberry-like, on each side; ovaries lateral to the testes, with one to four ova. Eggs develop in cup-like brood sacs formed from the oviducts. Buds develop on each side, anterior to the gonad, becoming linked to the colonial vascular system; they become zooids in phase throughout the colony, the parent cohort of zooids then degenerating. Tadpole larvae about 0.5 mm, with a triad of papillae surrounded by a ring of ampullae. There is a photolith (complex sense organ).

On algae, ascidian tests and inert surfaces; lower shore and shallow water; all coasts; from northern Norway to the Mediterranean (map: Millar 1966).

12 PYURIDAE

Unitary, often large, sometimes stalked. Test leathery. Siphons both 4-lobed; oral tentacles branched. Branchial sac with five to eight folds on each side; dorsal lamina continuous or broken

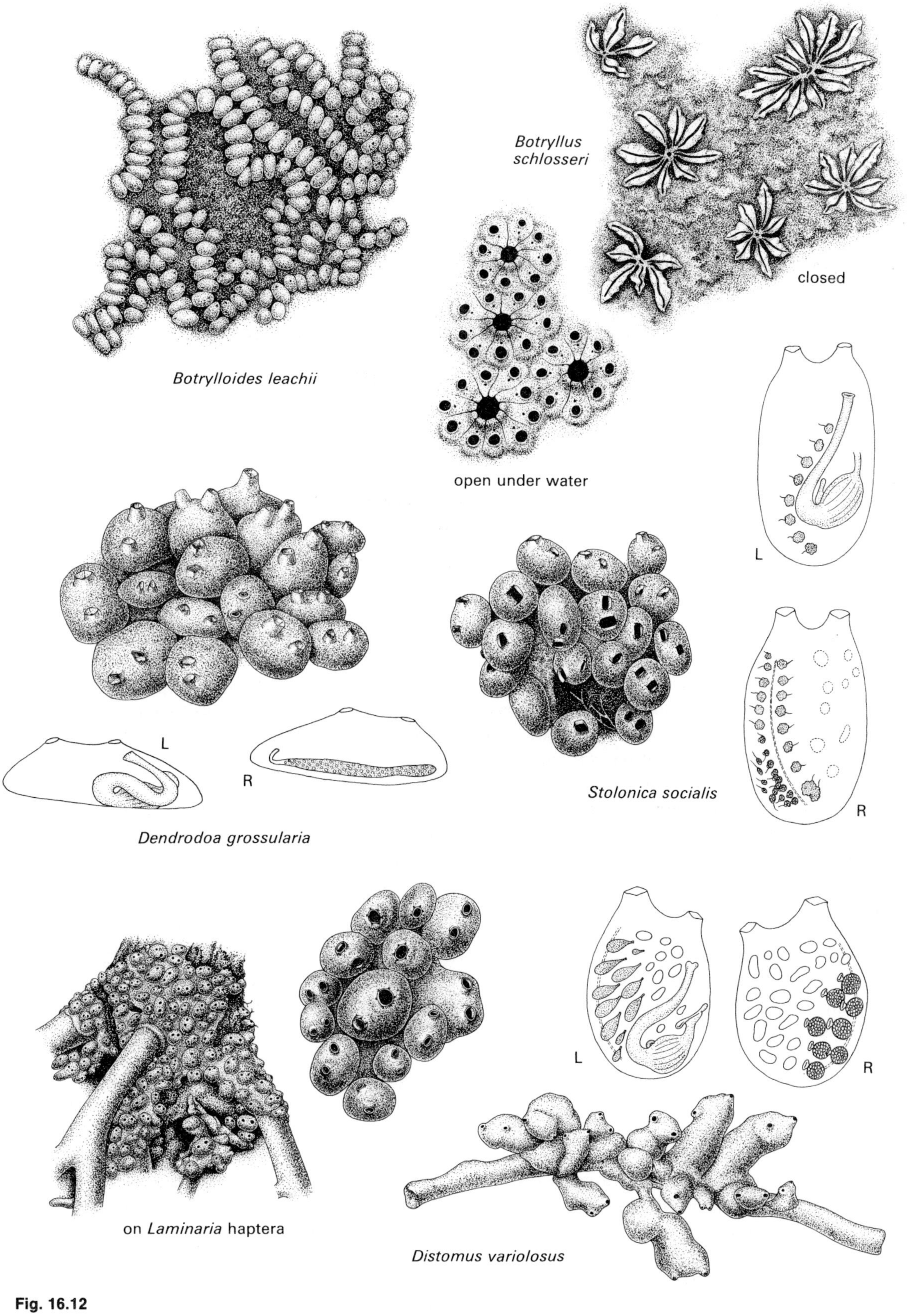
Botryllus schlosseri
closed
Botrylloides leachii
open under water
L
R
L
R
Dendrodoa grossularia
Stolonica socialis
L
R
on Laminaria haptera
Distomus variolosus

Fig. 16.12

into languets. Stigmata small and straight. Gut on left; stomach bearing hepatic tubules, the 'liver'. Gonads mixed, on both sides of the body; oviparous.

---

1. Stigmata transverse (crossed by the longitudinal vessels); test bearing radially branched spines **Boltenia echinata**

   Stigmata longitudinal; test not spiny **2**

2. Dorsal lamina continuous **Microcosmos claudicans**

   Dorsal lamina broken into languets (*Pyura*) **3**

3. Body flattened, siphons far apart; test tessellate (a mosaic of platelets); four branchial folds per side **Pyura tessellata**

   Body subspherical; test not tessellate; more than four folds per side **4**

4. Seven branchial folds per side; two gonads on the right side, one on the left **Pyura microcosmos**

   Six branchial folds; one gonad each side **Pyura squamulosa**

Boltenia echinata (Linnaeus) FIG. 16.13

[BERRILL 1950, 233; MILLAR 1970, 73]

Body spherical, fixed basally, up to 20 mm across; test bearing radially branched spines; siphons short, terminal. Branchial sac with six folds per side; stigmata transverse to the longitudinal bars. A distinctive and easily recognized species.

On hydroids, stones, shells, other ascidians, etc. from LWST to about 350 m; a northern species not found in southern Britain (map: Millar 1966).

Microcosmos claudicans (Savigny) FIG. 16.13

[BERRILL 1950, 235; MILLAR 1970, 74]

Body subspherical, fixed basally, up to 40 mm across; test thick, wrinkled, with fine hairs and adhering debris; siphons prominent. Branchial sac with seven to eight folds on the right, eight to nine on the left. Stigmata straight; dorsal lamina a continuous membrane. One gonad each side.

On rock and stones in shallow water (to 40 m); east and south coasts; southwards to Mediterranean (map: Millar 1966).

Pyura microcosmos (Savigny) FIG. 16.13

[BERRILL 1950, 237; MILLAR 1970, 77]

Body ovoid, attached by its broad base, up to 30 mm in length. Test tuberculate, each tubercle with a scale, reddish (mulberry-like); siphons well-separated. Branchial sac with seven folds each side; dorsal lamina represented by languets. One gonad on the left, within the intestinal loop; two parallel gonads on the right; oviparous.

On rock, stone, and shells; lower shore and shallow water; all except North Sea coasts; distributed southwards to the Mediterranean.

Pyura squamulosa (Alder) FIG. 16.13

[BERRILL 1950, 240; MILLAR 1970, 76]

Body subspherical, attached basally, up to about 25 mm across. Test rather smooth, covered with small scaly platelets (but not conspicuously tessellated); siphons terminal, moderately separated. Branchial sac with six folds each side; dorsal lamina represented by languets. One elongate gonad on each side, the left one within the intestinal loop; oviparous.

On rock and stones; lower shore and shallow water; all except North Sea coasts; distributed southwards to the Mediterranean (map: Millar 1966).

Pyura tessellata (Forbes) FIG. 16.13

[BERRILL 1950, 242; MILLAR 1970, 75]

Body rather flat, ovate in plan, the siphons far apart; to about 10 mm in length. Test strongly tessellated (i.e. a mosaic), divided into oblong or hexagonal, concentrically marked platelets. Branchial sac with not more than four folds per side; dorsal lamina represented by languets. One elongate gonad on each side, the left one within the intestinal loop; oviparous.

On stones, shells, other ascidians, etc. from the lower shore to about 300 m; all except North Sea coasts; distributed from Faroes and Norway to the Mediterranean (map: Millar 1966).

13 MOLGULIDAE

Unitary (the most highly evolved family of unitary ascidians), often globular in shape and dwelling in soft substrata. Test fairly thin, usually bearing filamentous processes (helping to anchor the animal in particulate substrata). Oral siphon 6-lobed; oral tentacles branched; atrial siphon 4-lobed (both Berrill (1950: 60) and Millar (1970: 10) inadvertently state that both the siphons are 6-lobed). Stigmata in spirals. Gut on the left; stomach bearing hepatic tubules, the 'liver'. A large, sausage-shaped renal organ attached to the body wall on the right side. Gonads mixed, on both sides of the body; oviparous or viviparous (the larvae sometimes tailless or *anuran*).

---

1. Branchial sac without folds **Eugyra arenosa**

   Branchial sac with folds (*Molgula*) **2**

2. Six folds per side **3**

   Seven folds per side **4**

3. Oviduct directed posteriorly **Molgula manhattensis**

   Oviduct directed anteriorly **Molgula complanata**

4. Oviduct directed anteriorly **Molgula complanata**

   Oviduct directed posteriorly **5**

5. Attached to firm substrata, the test naked or with few attached particles; oviduct long (about as long as the ovary) **Molgula citrina**

   Unattached, the test heavily covered with particles; oviduct much shorter than the ovary **6**

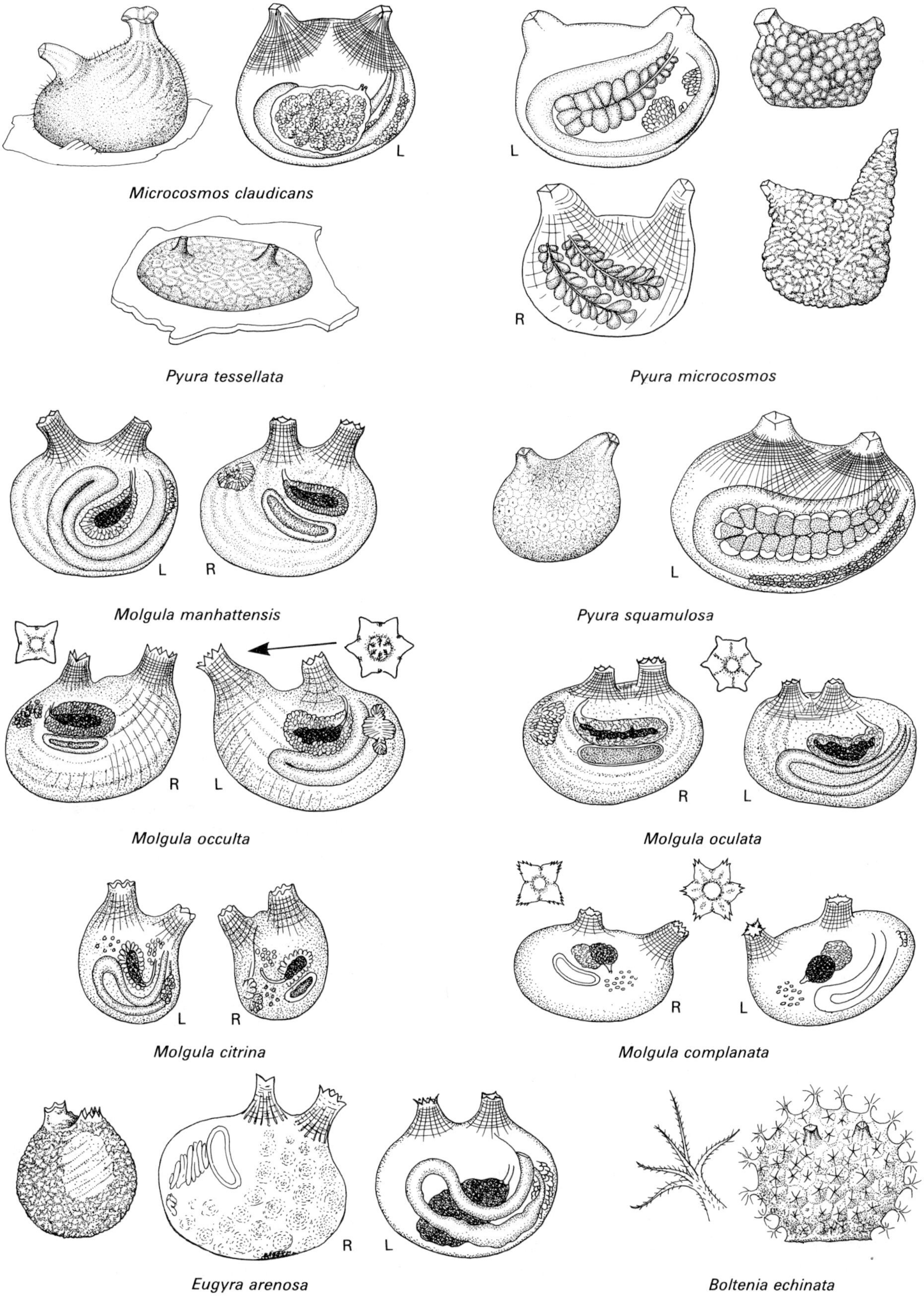
L
Microcosmos claudicans
L
R
Pyura tessellata
Pyura microcosmos
L
R
Molgula manhattensis
L
Pyura squamulosa
R
L
Molgula occulta
R
L
Molgula oculata
L
R
Molgula citrina
R
L
Molgula complanata
R
L
Eugyra arenosa
Boltenia echinata

Fig. 16.13

6. Body up to 80 mm; a naked area of test between the siphons; 10–11 longitudinal bars per branchial fold; renal sac with purple contents (in life); larva a tadpole **Molgula oculata**

   Body not above 30 mm; no naked area between the siphons; five to six longitudinal bars per branchial fold; renal sac with yellow-brown contents (in life); larva anuran **Molgula occulta**

Eugyra arenosa (Alder and Hancock) FIG. 16.13

[BERRILL 1950, 265; MILLAR 1970, 84]

Body globular or nearly so, non-attached; up to 20 mm diameter. Test with filaments to which particles adhere; in the absence of encrustation then translucent, the viscera visible. Siphons close together. Branchial sac without folds; the spiral stigmata involuted as 'infundibula'; dorsal lamina plain edged. Renal organ on right, gonad only on left; oviparous.

Unattached, in sand, shell-gravel, or mud; 20–100 m; off all coasts; distributed from Norway to the English Channel and south-west Ireland (map: Millar 1966).

Molgula citrina Alder and Hancock FIG. 16.13

[BERRILL 1950, 258; MILLAR 1970, 82]

Body ovoid, subglobular, or slightly depressed, to about 18 mm in length, basally attached. Test hard, often greenish, naked or with some adhering sand grains. Siphons short, well-separated. Branchial sac with seven folds each side. One gonad each side, that on the left lying in the upper curve of the intestine; the duct as long as the gonad. Renal organ ovate, below the right gonad. Viviparous; larva a tadpole.

A similar attached species known from Brittany is *M. bleizi* (Lacaze-Duthiers), which differs in having anuran development (see Berrill 1950: 257).

On rock, stones, pier piles, other ascidians, etc., occurring in clusters; lower shore and shallow (generally <50 m) water; all coasts; Arctic to Western Approaches, also North America (map: Millar 1966).

Molgula complanata Alder and Hancock FIG. 16.13

[BERRILL 1950, 262; MILLAR 1970, 82]

Body subspherical, conical, ovoid, or somewhat compressed, up to about 15 mm in height and 20 mm in greatest dimension, but usually smaller. Test with fibrils, to which are attached sand and shell particles. Siphons well-separated. Branchial sac with six to seven small folds each side. Gonads, with ovary anterior to testis, on both sides; the left one above the loop of the intestine, the right one above the curved renal organ; the short ducts leading anteriorly. Mature at a body length of 3 mm. Viviparous, larva a tadpole.

In crevices and depressions on stones and shells on a rocky or sandy bottom, probably widely overlooked on account of its small size; lower shore and shallow water; all coasts; Arctic to Brittany (map: Millar 1966).

M. bleizi, from Brittany, is smaller but has anuran development (Berrill 1950: 257).

Molgula manhattensis (De Kay) FIGS. 16.2, 16.13

[BERRILL 1950, 248; MILLAR 1970, 79]

Body rounded, 10–30 mm. Test grey or greenish, covered with fibrils to which particles may be attached. Siphons basally close together, diverging, short or long (up to one-half body height). Branchial sac with six folds each side. Gut sharply looped, disposed in a C. Gonads club-shaped, the left one almost encircled by the gut; the right one above the elongate, slightly curved renal organ; the long axis of the left gonad approaching vertical, that of the right nearer horizontal; ducts short, directed to the base of the atrial siphon. Oviparous.

On stones, pilings, algae, etc., often in aggregations; also lying loose in sand; lower shore and shallow water, including harbours and estuaries, to about 90 m; all coasts, including south-west Wales (Crothers 1966); distributed from northern Norway to Portugal, also North America south to the Gulf of Mexico (map: Millar 1966).

Molgula occulta Kupfer FIG. 16.13

[BERRILL 1950, 254; MILLAR 1970, 81]

Body ovoid to globular, up to 30 mm long; test (apart from the siphons) completely covered with sand or shell particles; siphons short. Branchial sac with seven folds each side; five to six longitudinal bars on each fold; stigmata spiral on the folds, almost straight between them. The gut almost horizontal; left gonad immediately above it, testis surrounding the ovary; renal vesicle slightly curved, immediately below the right gonad; duct rather short, leading straight upwards toward the atrial siphon. When living, the renal sac contains a yellowish concretion lying in brownish fluid. Oviparous; development anuran.

Unattached, on mud, sand, and shell gravel; from just below LWST to about 100 m; south and west coasts; distributed from Norway to the Mediterranean (map: Millar 1966).

Molgula oculata Forbes FIG. 16.13

[BERRILL 1959, 252; MILLAR 1970, 80]

Body ovoid to subglobular; to 80 mm in greatest dimension; test heavily coated with sand or shell particles except for the short siphons and the space between them. Branchial sac with seven folds, each with 10–11 longitudinal bars; stigmata spiral on the folds. Gut almost horizontal; left gonad above it, the ovary enveloped by testis; renal vesicle slightly curved, immediately below the right gonad; duct moderately short, leading straight upwards towards the atrial siphon. When living, the renal sac contains a purplish fluid. Oviparous, developing to a tadpole.

Unattached, on soft substrata; LWST to about 80 m; known from Shetland, north-east England, the western English Channel, and Milford Haven (Crothers 1966); also Biscay (more restricted than *M. occulta*).

17 CEPHALOCHORDATA AND EUCHORDATA

Subphylum **Cephalochordata**

THE subphylum Cephalochordata comprises the lancelets, which are placed in the single order Amphioxi. In lancelets the *notochord* persists throughout life and there is no bony or cartilaginous skeleton. The body is slender, tapering at both ends and superficially fish-like. The mouth opens ventrally between a pair of lateral expansions, which together form the *oral hood*, the free edge of which bears numerous *cirri*. The pharynx has the form of a perforated *branchial sac* and in the adult is surrounded by the *atrial chamber*; this occupies the anterior two-thirds of the body and opens to the exterior by a small posterior aperture, the *atriopore*. The ventral surface below the atrium is flat, with the edges produced as *metapleural folds* which begin at the mouth and terminate behind the atriopore. A median *ventral fin* extends from the atriopore to join the *caudal fin*, while the *dorsal fin* extends the entire length of the body. There are no eyes or nostrils. One family is present in British waters.

1 BRANCHIOSTOMIDAE

Mouth in the mid-line, surrounded by the cirri. Pharynx with gill slits opening into the atrium on both sides. Gonads developing segmentally on both sides of the atrium. Metapleural folds terminating symmetrically behind the atriopore. A single species.

Branchiostoma lanceolatum (Pallas). Lancelet, Amphioxus
FIG. 17.2

Rostral process, including the anterior end of the notochord, projecting a short distance beyond the oral hood. Caudal fin pointed, its dorsal and ventral lobes higher than the dorsal and ventral fins. Anus situated slightly behind origin of ventral lobe of caudal fin. About 60 myotomes, less than 40 of them anterior to the atriopore. To a maximum of 6 cm.

In shallow water, burrowing in sand and shell-gravel. Off all coasts; local.

Subphylum **Euchordata (Vertebrata)** **Pisces**

Fish are sufficiently familiar to need little general introduction. It should be realized, however, that 'Pisces' does not represent a natural group but, corresponding to the word 'fish', embraces the jawless cyclostomes, the cartilaginous sharks and their allies (*Chondrichthyes*) and the bony fishes (*Osteichthyes*), each of these groupings constituting a separate and distinctive vertebrate class. Many of the technical terms refer primarily to bony fishes, but the names of fins and parts of the body apply equally to cartilaginous fishes.

The gross external features of a fish's body are indicated in Fig. 17.1. The generalized shape is termed *fusiform*, but the body may be variously elongated or flattened laterally (*compressed*) or dorso-ventrally (*depressed*). Fins in the mid-line are the *dorsal* (D), the *anal* (A) and the *caudal* (C); while paired fins supported by the pectoral and pelvic girdles are referred to as the *pectorals* (P) and the *pelvics* or *ventrals* (V) respectively. In cartilaginous and primitive bony fishes the pelvic fins are *abdominal* and situated near the vent; but in advanced bony fishes, in which the pectorals are placed higher on the flanks, the pelvics have moved forward to below or in front of the pectorals, being then respectively *thoracic* or *jugular*. The fins are supported by rays of various types, horny in Chondrichthyes but bony in Osteichthyes.

The morphology of the caudal fin varies considerably between the classes. In Chondrichthyes it is basically unequal or *heterocercal*, with the vertebral column entering and supporting the upper fin lobe. In most Osteichthyes the tail is outwardly symmetrical or *homocercal*, but in actuality the end of the vertebral column is upturned and almost the entire fin has been derived from what was originally the lower lobe. Most of the rays articulate with the modified haemal arches (called *hypurals*) of the last few vertebrae.

In primitive bony fishes the finrays are *soft*, being flexible, articulated and—in general—branched; in advanced fishes the anterior rays of the dorsal and anal fins, and the outer ray of the pelvics, are unjointed and unbranched. Such *spiny rays* may be thin and flexible but are more often stiff and pointed. When summarizing the number of finrays for convenience in formula form, spiny rays are denoted by Roman and soft rays by Arabic numerals: thus, D V/10–11 indicates a dorsal fin composed of five spiny rays succeeded by ten or eleven soft rays. Where there is more than one dorsal fin, each is indicated by a preceding numeral, i.e. 1D, 2D.

Also important in identifying fishes is *squamation*, or the structure and arrangement of scales. In sharks the scales are *placoid* or tooth-like, but in modern bony fishes the scales are thin flakes of bone. *Cycloid* scales (Fig. 17.1), found in most soft-rayed bony fishes are disciform, with the exposed rear margin entire; *ctenoid* scales, characteristic of many spiny-rayed

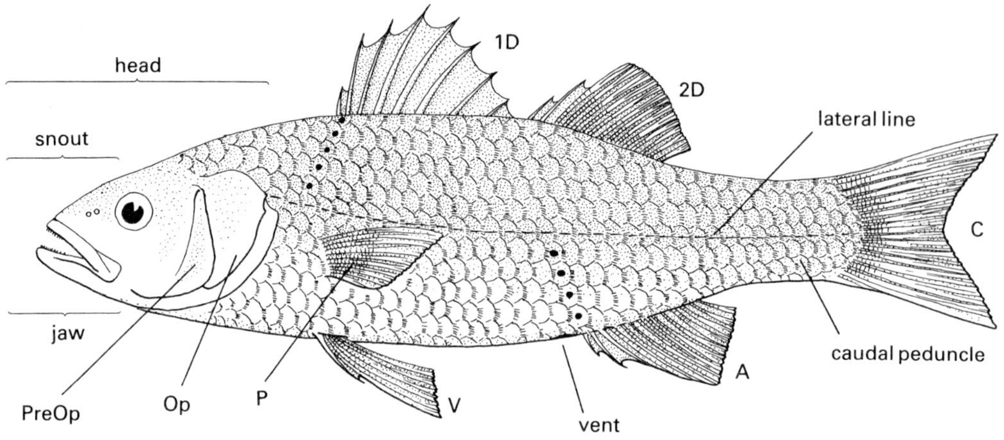

General teleost features *(Dicentrarchus)*

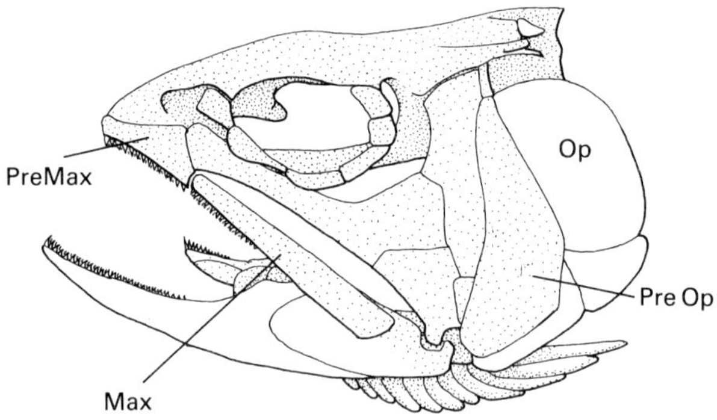

Skull of *Salmo*

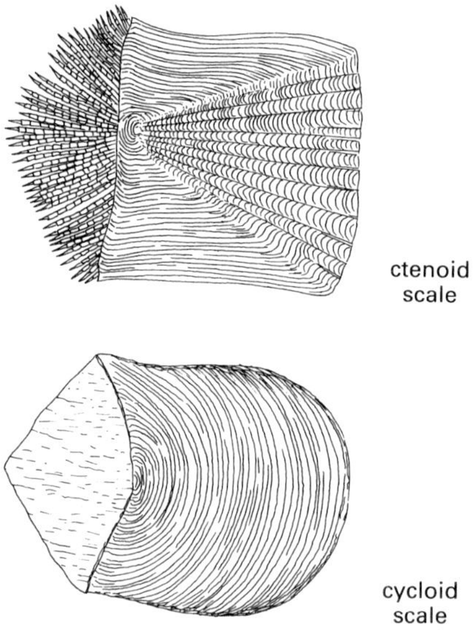

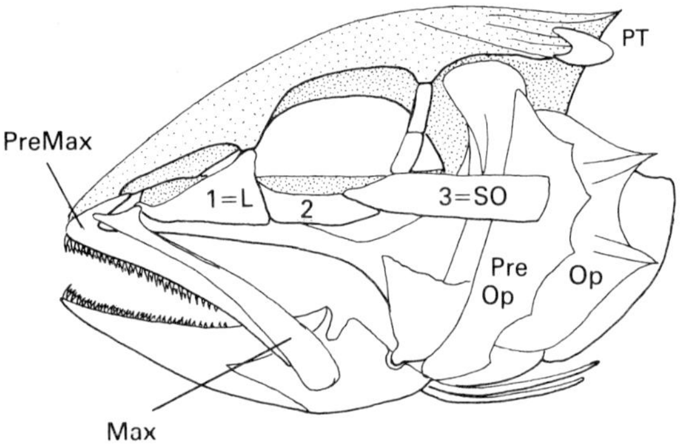

Skull of mail-cheek (scorpaeniform)

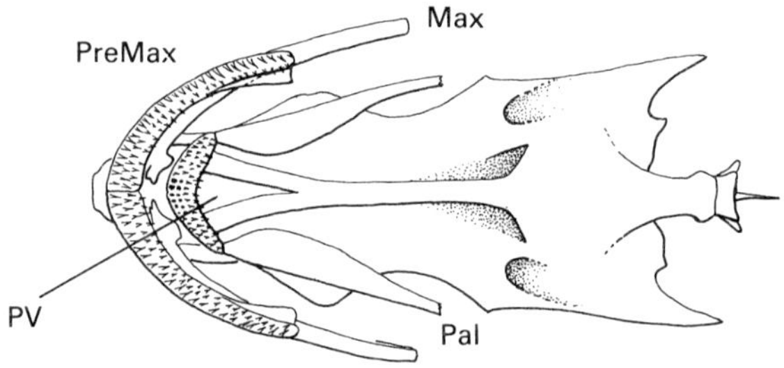

Skull of *Gadus*, palatal view

1 = L = lachrymal (1st suborbital)
2 = 2nd suborbital
3 = SO = 3rd suborbital
1–3 form the suborbital stay
PT = post-temporal
SO = suborbital

Op = operculum
PreOp = preoperculum
PreMax = premaxilla
Max = maxilla
Pal = palatine
PV = prevomer (also known as vomer)

A = anal fin
C = caudal fin
1D = 1st dorsal fin
2D = 2nd dorsal fin
P = pectoral fin(s)
V = ventral (=pelvic) fin(s)

Fig. 17.1

fishes, have a toothed rear margin. Scale counts are helpful, and used most frequently is the number of scales along the lateral line (e.g. lat. sc. 50–60) counting from the free edge of the gill cover to the origin of the caudal fin, ignoring any scales which lie actually on the base of the fin itself. If the number of scale rows is required, count downwards and backwards from the commencement of the (first) dorsal fin, and upwards and forwards from the commencement of the anal fin, reaching—but not including—the scale over the lateral line. Thus, in the descriptions, 5/11 indicates five longitudinal rows of scales above the lateral line and eleven below.

The external gill openings of cartilaginous fishes are separate, and show as a lateral series of slits; in bony fishes the gills lie together in a common cavity, overlain by the gill-cover, with a single exterior opening on each side. Less conspicuous are the olfactory openings. Cyclostomes possess a single nostril on top of the head. In sharks and rays there is on each side a single, ventral nostril closed by a valvular flap. The olfactory organs in bony fishes have (usually) two dorsal nostrils on each side, placed most often just in front of the eyes.

The skull of a bony fish is complex, and the form and disposition of bones has been modified considerably during the course of evolution. The few bones commonly referred to in the descriptions of species are labelled in Fig. 17.1. In addition to the cranium, notice the jaws and the flat bones constituting the gill cover, viz. *preoperculum*, *operculum*, *suboperculum*, and *interoperculum*. Below these, and supporting the jugular continuation of the opercular membrane, is a series of slender *branchiostegal rays* (br.) attached to the bones of the hyoid arch (particularly the ceratohyal). The number of rays may be helpful in taxonomy. In fairly primitive bony fishes, such as salmonids, skull features include the delimitation of the upper jaw by two tooth-bearing bones, *premaxilla* and *maxilla*, and the presence of a series of small *circumorbital bones* around the eye. In more highly evolved fishes, such as gadids, the maxilla lies behind the premaxilla; the maxilla provides the upper jaw pivot, but the premaxilla—which in some fish orders is protrusible—alone is toothed and defines the upper side of the gape. The maxilla, however, delimits the upper jaw in external view, and jaw measurements are made to its posterior end. The first circumorbital bone (*preorbital* or *lachrymal*) may extend forwards as far as the premaxilla, while in one order the third circumorbital or *suborbital* is elongated and articulates with the preoperculum. Teeth may be present not only on the jaws but in the roof of the mouth, on the *palatines* and/or *vomer*, and in the floor of the mouth, on the terminal bones of the hyoid arch.

The branchial skeleton in bony fishes consists of four gill-bearing arches. Typically each arch comprises four paired elements, of which the middle two bear gill-filaments, and an unpaired basal element lying in the mid-line. The number of *gill rakers*, the bony knobs or bristles opposing the gills in each arch, is frequently of systematic importance. The first gill arch is utilized and should be removed in its entirety prior to counting. Pharyngeal or gill-teeth may also be present and taxonomically helpful, as in wrasses. *Hyperpharyngeal teeth* are situated on the upper elements of the gill arches; *hypopharyngeal teeth* are borne on bones representing the lower elements of the fifth branchial arch, the components usually being fused into a single structure.

The pectoral girdle of bony fishes is intimately associated with the skull (Fig. 17.1), and is joined to it dorsally by the *post-temporal* bone. The largest constituent bone on each side is the *cleithrum*. In fishes with thoracic or jugular ventral fins the pelvic girdle is anchored to the cleithra.

The nomenclature of orders employed in this book, with the ending-formes, follows the system of the Russian ichthyologist L. S. Berg (1940); and their sequence and grouping into superorders, which may be unfamiliar to some readers, follows the proposals of Greenwood *et al.* (1966). Full definitions of these higher taxa frequently depend upon details of osteology which, as being inappropriate to this book, have been generally abbreviated or omitted. The classification is as follows.

Superclass **Agnatha**

Craniates without jaws.

Class I. **Cyclostomata (Marsipobranchii)**

Notochord persistent. Gills in the form of pouches supported by a complex cartilaginous skeleton. Elongated body; no paired fins.

Order 1. **Petromyzoniformes**

Adults with a terminal sucker surrounding the mouth. Scaleless body. Series of gill openings on each side. Lampreys.

Subphylum **Gnathostomata**

Craniates with jaws.

Class I. **Chondrichthyes (Elasmobranchii, Selachii)**

Skull and skeleton cartilaginous. Paired fins present. Gill filaments attached throughout their length to the septa; five or more branchial openings each side. Skin usually rough, with placoid scales ('dermal denticles'). No air bladder. Intestinal spiral valve present. Males with copulatory organs ('claspers').

Subclass **Euselachii**

Sharks and rays.

Order 1 **Lamniformes**

Two spineless dorsal fins. Anal fin present.

Order 2 **Squaliformes**

Two dorsal fins, each preceded by a spine. No anal fin.

Order 3 **Rajiformes**

Body depressed; gill slits opening ventrally. Skates and rays.

Order 4 **Torpediniformes**

Electric rays.

Class II **Osteichthyes**

Skeleton bony. Paired fins present. Gills filamentous; a single, operculate gill opening. Air bladder or lung usually present. Scales bony.

Subclass **Actinopterygii**

Ray-finned fishes.

Infraclass **Teleostei**

The most advanced and diverse bony fishes, with symmetrical (homocercal) tails and thin scales. Probably evolved from Holostei in at least three separate lines.

Superorder **Elopomorpha**

Primitive teleosts with toothed maxillae forming part of gape and no spiny finrays; generally of eel-like form and having a leptocephalus larva. Differ from Clupeomorpha in many characters of the skull.

Order 1 **Anguilliformes**

Eels. Larva a 'leptocephalus' (flat, leaf-like, transparent, and planktonic).

Superorder **Clupeomorpha**

Primitive teleosts with compressed silvery bodies and caducous scales. Toothed maxillae form part of gape. Pectoral fins low on the flank, pelvics abdominal; no spiny finrays. Diverticula of the swimbladder extend into the head and end in a characteristic arrangement of bullae around the ears. No lateral line pores on trunk.

Order 2 **Clupeiformes**

Characters of the superorder.

Superorder **Protacanthopterygii**

Primitive teleosts, probably ancestral to the remaining superorders. Maxilla included in gape. Pectoral fins low on the flank, pelvics abdominal; no spiny finrays. A rayless (adipose) fin frequently present.

Order 3 **Salmoniformes**

Adipose fin present.

Superorder **Paracanthopterygii**

Advanced marine teleosts, with or without spiny finrays, characterized (and distinguished from Acanthopterygii), e.g. by non-protrusible jaws, the presence of a particular jaw muscle (m. levator maxillae superioris) and the structure of the caudal skeleton. Pelvic fins thoracic or jugular.

Order 4 **Gobiesociformes**

Pelvic fins modified into a complex sucker. Scaleless. No spiny rays in the median fins. No swimbladder. Clingfishes.

Order 5 **Lophiiformes**

First dorsal finray modified as a lure. The pectoral fins modified to facilitate crawling. Scaleless. Angler fishes.

Order 6 **Gadiformes**

Nektonic fishes. Scales cycloid. Fins without spiny rays. Swimbladder well-developed. Cods and their relatives.

Superorder **Atherinomorpha**

A small superorder of osteologically distinct fishes with protrusible jaws of a characteristic type.

Order 7 **Atheriniformes**

Characters of the superorder. Garfish and sand smelts.

Superorder **Acanthopterygii**

The largest superorder, comprising advanced teleosts of varied form and habit, characterized by: protrusible jaws having a distinctive mechanism, the absence of the m. levator maxillae superioris, the exclusion of the maxilla from the gape, the structure of the caudal skeleton, ctenoid scales, presence of spiny finrays, and the thoracic or jugular position of the pelvic fins.

Order 8 **Zeiformes**

Spiny-finned teleosts with deep, compressed bodies and highly extensible jaws.

Order 9 **Gasterosteiformes**

Dorsal fin preceded by separate spines. Leading ray of pelvics spiny. Sticklebacks.

Order 10 **Syngnathiformes**

Tubular-mouthed fishes with pelvic fins abdominal (or absent). First dorsal fin with spiny rays, but no free spines preceding it. Pipefishes.

Order 11 **Scorpaeniformes**

Differ from Perciformes by the presence of an enlarged suborbital bone which extends posteriorly to articulate with the preoperculum (Fig. 17.1).

Order 12 **Perciformes**

A vast order of spiny-finned teleosts, central to the Acanthopterygii. Leading ray(s) of dorsal, anal, and pelvic fins usually spiny.

Order 13 Pleuronectiformes

Benthic, strongly compressed teleosts (flatfishes), with asymmetric head having both eyes placed on the same side of the body.

The species which are described in this book are those found typically in coastal waters. They include shore fish, those that can be trawled, and those caught by rod and line with the omission of the more oceanic sport fishes. On rocky shores fish can be found under boulders (rocklings, blennies, butterfish, and *Nerophis lumbriciformis*) and in pools; netting in the fringing algae may produce *Crenilabrus melops* and *Gobiusculus flavescens*, while *Lipophrys pholis* and *Gobius paganellus* are more likely to be hiding among stones on the pool bottom or in crevices. At extreme low water the holdfasts of *Saccorhiza polyschides* may conceal clingfishes and *Liparis montagui*. On sandy shores push-netting at low-tide collects young flatfish, *Syngnathus rostellatus*, and *Pomatoschistus minutus;* if *Zostera marina* is present *Spinachia spinachia*, pipefishes, wrasses, and many others will be found. Different species occur when the water is brackish.

Quinaldine has an anaesthetic effect on fish, and Gibson (1967) has described how it can be used to facilitate the study of rock-pool species; it is, however, expensive. Specimens may be preserved in formalin or 70% ethanol, or be deep frozen, but colour patterns should be recorded before storage in preservative. The measurements employed in fish identification should be largely self explanatory. Fishery biologists measure length as from the tip of the snout to the mid-point of the caudal fin (*fork length*), but the *standard length* utilized by systematists is the length excluding the caudal fin. *Head length* is measured from the snout tip to the rear edge of the operculum, and the snout is taken to end at the anterior edge of the orbit.

LITERATURE

Of the general biological accounts available, Norman and Greenwood's *History of fishes* (1974) and Bone and Marshall's *Biology of fishes* (1982) are exceptionally good. There is unfortunately no modern monograph for the specialist on British fishes, that of Day (1880–4) being inevitably out of date in many respects. The still earlier treatises of Yarrell (1836) and Couch (1862–5) will be of interest mainly to serious students of the group; Couch's volumes contain the most complete set of good colour illustrations. Wheeler's semi-popular *Fishes of the British Isles and north-west Europe* (1969) is comprehensive and easy to use, though some of the illustrations are indifferent; it has largely removed the necessity of consulting monographs and scientific papers for routine purposes, although a few of the most useful are listed below. The same author's field key (Wheeler, 1978) is highly recommended. The small book by Travis Jenkins (1936), for long the only one available, is incomplete and somewhat dated. For northern waters Andriyashev (1964) has been the standard work, but the whole north-eastern Atlantic area (including the Mediterranean) is now covered in a new 3-volume monograph (Whitehead *et al.*, 1984–6). Species that enter fresh or brackish water are also covered by Maitland (1972). Russell (1976) is comprehensive for eggs and larvae. Nomenclature in this chapter follows Hureau and Monod (1979).

Anderson, M. (1982). The identification of the British grey mullets. *Journal of Fish Biology*, **20**, 33–8.

Andriyashev, A. P. (1964). *Fishes of the northern seas of the U.S.S.R.*, Jerusalem.

Bone, Q. and Marshall, N. (1982). *Biology of fishes*, Glasgow.

Berg, L. S. (1940). Classification of fishes both Recent and fossil. *Travaux de l'Institut Zoologique de l'Academie des Sciences de l'U.R.S.S.*, **5**, 87–517.

Chang, H.-W. (1951). On *Callionymus reticulatus* C. & V. and its distribution in European seas. *Journal of the Marine Biological Association of the U.K.*, **30**, 297–312.

Clark, R. S. (1926). Rays and skates: a revision of the European species. *Scientific Investigations of the Fishery Board of Scotland*, **1926**, No. 1: 1–66.

Conseil Permanent International pour l'Exploration de la Mer (1938). *Faune ichthyologique de l'Atlantique nord*, Copenhague.

Couch, J. (1862–5). *A history of the fishes of the British Islands*. London, 4 vols.

Day, F. 1880–4. *The fishes of Great Britain and Ireland*. London and Edinburgh, 2 vols.

Gibson, R. N. (1967). The use of the anaesthetic quinaldine in fish ecology. *Journal of Animal Ecology*, **36**, 295–301.

Greenwood, P. H., Rosen, D. E., Weitzman, S. H., and Myers, G. S. (1966). Phyletic studies of teleostean fishes, with a provisional classification of living forms. *Bulletin of the American Museum of Natural History*, **131**, 339–456.

Hureau, J. C. and Monod, T. (eds.) (1979). *Check-list of the fishes of the north-eastern Atlantic and of the Mediterranean*, 2nd imp. with Supplement. Paris, 2 vols.

Jenkins, J. T. (1936). *The fishes of the British Isles*, 2nd edn., London.

Lythgoe, J. and Lythgoe, G. (1971). *Fishes of the sea*, London.

Maitland, P. S. (1972). Key to the freshwater fishes of the British Isles. *Freshwater Biological Association Scientific Publications*, **27**.

Miller, P. J. (1961). The external appearance and systematic position of *Gobius forsteri* (Teleostei-Percomorphi). *Proceedings of the Zoological Society of London*, **137**, 539–51.

Miller, P. J. (1969). Systematics and biology of the leopard-spotted goby, *Gobius ephippiatus* (Teleostei-Gobiidae), with description of a new genus and notes on the identity of *G. macrolepis* Kolombatovic. *Journal of the Marine Biological Association of the U.K.*, **49**, 831–55.

Miller, P. J. (1974). A new species of *Gobius* (Teleostei: Gobiidae) from the western English Channel, with a key to related species in the British and Irish fauna. *Journal of Zoology, London*, **174**, 467–574.

Miller, P. J. and El-Tawil, M. Y. (1974). A multidisciplinary approach to a new species of *Gobius* (Teleostei: Gobiidae) from southern Cornwall. *Journal of Zoology, London*, **174**, 539–74.

Norman, J. R. and Greenwood, P. H. (1974). *A history of fishes*, 3rd edn., London.

Poll, M. (1947). *Faune de Belgique: Poissons marins*, Bruxelles.

Russell, F. S. (1976). *The eggs and planktonic stages of British marine fishes*, London.

Sears Foundation (1948–82). *Fishes of the western North Atlantic*, Vols. 1–8. New Haven.

Svetovidov, A. N. (1963). *Fauna of the U.S.S.R. Fishes: Gadiformes*, Jerusalem.

Svetovidov, A. N. (1962). *Fauna of the U.S.S.R. Fishes: Clupeidae*, Jerusalem.

Trewavas, E. and Ingham, S. E. (1972). A key to the species of Mugilidae (Pisces) in the northeastern Atlantic and Mediterranean, with explanatory notes. *Journal of Zoology, London*, **167**, 15–29.

Webb, C. J. (1980). Systematics of the *Pomatoschistus minutus* complex (Teleostei: Gobioidei). *Philosophical Transactions of the Royal Society* Ser. B, **291**, 201–41.

Webb, C. J. and Miller, P. J. (1975). A redescription of *Pomatoschistus norvegicus* (Collet, 1903) (Teleostei: Gobioidei) based on syntype material. *Journal of Fish Biology*, **7**, 735–47.

Went, A. E. J. and Kennedy, M. (1969). *List of Irish fishes*, 2nd edn., Dublin.

Wheeler, A. (1960). *Gobius capito* as a British fish. *Annals and Magazine of Natural History*, (**13**) **3**, 177–81.

Wheeler, A. (1969). *The fishes of the British Isles and north-west Europe*, London.

Wheeler, A. (1970). *Gobius cruentatus*—a fish new to the northern European fauna. *Journal of Fish Biology*, **2**, 59–67.

Wheeler, A. (1978). *Key to the fishes of northern Europe*, London.

Whitehead, P. J. P., Bauchot, M.-L., Hureau, J.-C., Nielsen, J., and Tortonese, E. (1984–6). *Fishes of the North-eastern Atlantic and the Mediterranean*, Vol. 1 (1984), Vol. 2 (1986*a*), and Vol. 3 (1986*b*) Paris.

Yarrell, W. (1836). *A history of British fishes*, 2 vols.

KEY TO FAMILIES

1.	More than one gill opening each side	2
	A single gill opening each side (OSTEICHTHYES: TELEOSTEI)	8
2.	Seven gill openings each side. Paired fins absent (CYCLOSTOMATA)	**1. Petromyzonidae**
	Five gill openings each side. Paired fins present (CHONDRICHTHYES: EUSELACHII)	3
3.	Gill openings ventral. Body depressed	4
	Gill openings lateral. Body elongated or fusiform	6
4.	Mouth terminal	**5. Squatinidae**
	Mouth ventral	5
5.	Disk rounded anteriorly; caudal fin well-developed (electric rays)	**7. Torpedinidae**
	Disk pointed anteriorly; caudal fin absent or rudimentary (skates and rays)	**6. Rajidae**
6.	Anal fin absent; both dorsals preceded by spines	**4. Squalidae**
	Anal fin present; dorsals without spines	7
7.	Anterior dorsal above or behind pelvics	**2. Scyliorhinidae**
	Anterior dorsal above gap between pectorals and pelvics	**3. Carcharinidae**
8.	Body long and slender, covered by a rigid exoskeleton in the form of bony rings. Head drawn out into a long tubular snout (pipefish)	**22. Syngnathidae**
	Body not covered by a rigid exoskeleton of bony rings. Head without tubular snout	9
9.	Body eel-like, supple. Pelvics absent. Scales hidden or absent. No barbel. (Eels)	10
	The above characters not present in combination	11
10.	Dorsal fin commencing well behind pectorals. Lower jaw more prominent	**8. Anguillidae**
	Dorsal fin commencing above free end of pectorals. Upper jaw more prominent	**9. Congridae**

11.	Pelvic fins abdominal	12
	Pelvic fins thoracic, jugular, or absent	15
12.	Body elongated. Jaws produced as a long beak	**18. Belonidae**
	Body fusiform. Jaws not produced as a long beak	13
13.	Single dorsal fin present	**10. Clupeidae**
	Adipose fin present behind the rayed dorsal fin	14
14.	Tail deeply forked	**12. Argentinidae**
	Tail square or slightly forked	**11. Salmonidae**
15.	Fish lying on their sides, with asymmetrical head and both eyes on the same side (flatfish)	16
	Symmetrical fish, not lying on their sides	18
16.	Eyes placed on the left side	**42. Bothidae**
	Eyes placed on the right side	17
17.	Preopercular bone with a free edge. Mouth terminal.	**43. Pleuronectidae**
	Edge of preopercular bone covered or hidden by skin. Mouth subterminal	**44. Soleidae**
18.	Anterior dorsal finray situated on the head and transformed into a 'fishing line' (angler fish)	**14. Lophiidae**
	Anterior dorsal finray not situated on the head and not transformed into a fishing line	19
19.	Spiny finrays generally absent. Pelvics with 5–17 soft rays. One or two anal fins	20
	Some spiny rays present. Pelvic fins with a maximum of five soft rays (or sometimes transformed into a ventral sucker). Only one anal fin	22
20.	Anterior dorsal and anterior anal finrays spiny. Fish compressed	**20. Zeidae**
	Only soft finrays present	21
21.	Two dorsal fins. No barbel on lower lip	**16. Merlucciidae**
	Two or three dorsal fins. Usually (always when there are only two dorsal fins) a barbel on the lower lip	**15. Gadidae**
22.	First dorsal fin represented by series of separated spines	**21. Gasterosteidae**
	First dorsal fin continuous, not represented by separated spines	23
23.	Pelvic fins subjugular, widely spaced, separated by an adhesive disk. Dorsal fin lacking spiny rays	**13. Gobiesocidae**

	Adhesive disk (if present) formed from the pelvic fins. Anterior dorsal fin (or part of fin) composed of spiny rays	**24**
24.	Rear part of dorsal and anal fins broken up into a series of separate finlets	**41. Scombridae**
	These fins not broken up into finlets	**25**
25.	Pelvic fins not recognizable, fully transformed into an adhesive organ	**26**
	Pelvic fins not transformed into an adhesive organ	**27**
26.	Body dumpy, tuberculate. Visible dorsal fin short	**26. Cyclopteridae**
	Body elongate, not tuberculate. Dorsal fin long	**27. Liparidae**
27.	Head armoured with bony plates	**28**
	Head unarmoured	**30**
28.	Lower pectoral finrays separate; pelvics with five soft rays	**23. Triglidae**
	Lower pectoral finrays not separate; pelvics with fewer than five soft rays	**29**
29.	Body to a large extent naked	**24. Cottidae**
	Body covered by a series of bony plates	**25. Agonidae**
30.	Pelvic fins united, forming a disk	**40. Gobiidae**
	Pelvic fins independent, sometimes reduced or absent	**31**
31.	Fish red, with widely separated dorsal fins; pelvic fins attached below pectorals; two long barbels on the lower jaw	**30. Mullidae**
	Not with the above characters in combination	**32**
32.	Dorsal fins widely separated. Pelvic fins behind the point of insertion of the pectorals	**33**
	Dorsal fins feebly separated or contiguous at the base or fused. Pelvic fins thoracic or jugular	**34**
33.	Anterior dorsal with four spiny rays; anal with three spines	**31. Mugilidae**
	Anterior dorsal with five to nine spiny rays; anal with one spine	**19. Atherinidae**
34.	Lateral line covered by a series of bony scutes. Anal fin preceded by two separated spines	**29. Carangidae**
	Lateral line not covered by a series of bony scutes. Anal fin not preceded by two isolated spines	**35**
35.	First dorsal fin very short, second dorsal separate and much longer	**36**
	Dorsal fin(s) not of this form	**37**

36. Head somewhat laterally compressed. Anterior dorsal with five to seven spiny rays, posterior dorsal with 21–32 rays **33. Trachinidae**
Head strongly depressed. Anterior dorsal with four spiny rays, posterior dorsal with nine to ten rays **39. Callionymidae**

37. Pelvic fins thoracic **38**
Pelvic fins jugular or absent **39**

38. Lips normal. Palatal teeth present; pharyngeal teeth absent. Scales ctenoid **28. Moronidae**
Lips relatively thick. Teeth on the jaws but not on the palate; pharyngeal teeth present. Scales cycloid **32. Labridae**

39. Caudal fin forked; pelvics absent **38. Ammodytidae**
Caudal fin rounded or tapering, pelvics small but usually present **40**

40. Caudal fin completely fused with dorsal and anal; dorsal fin with a depressed region posteriorly **17. Zoarcidae**
Caudal fin distinct **41**

41. Dorsal fin more or less distinctly divided into two regions **34. Blenniidae**
Dorsal fin uniform, made up of spiny rays only **42**

42. Pelvic fins rudimentary; dorsal marked with a series of black spots **37. Pholidae**
Pelvic fins jugular; dorsal not spotted **43**

43. Head with two pairs of orbital tentacles **35. Stichaeidae**
Head without orbital tentacles **36. Lumpenidae**

Class Cyclostomata

Order Petromyzoniformes

1 PETROMYZONIDAE

Eel-like body; scaleless. Mouth jawless, rounded, suctorial. Single dorsal nasal opening. Eyes present. Seven branchial openings each side. Distinctive larva ('ammocoetes').

Lampetra fluviatilis (Linnaeus). FIG. 17.2
Lampern, Lamprey [WHITEHEAD et al. 1984: 65]
Body eel-like, without paired fins. Oral disk subcircular, provided with sharp, horny teeth (which are not aligned in 11–12 rows on each side of the mouth, as in the larger *Petromyzon marinus* L.); two well-separated supraoral teeth (closely approximated in *P. marinus*). Eye rather small, followed by seven rounded gill openings. Tripartite dorsal fin and single, posterior ventral fin. Dark blue above with silvery flanks or grey above and creamy below. To 50 cm.

Parasitic on other fish, but also taken in pelagic townets; coastal; anadromous, entering rivers during September–October prior to spring spawning; all coasts except the extreme north.

Class Chondrichthyes

Order Lamniformes

2 SCYLIORHINIDAE

Small, bottom-living sharks. Mouth ventral. Spiracle present; five gill slits. Dorsal fins spineless, the first placed above or behind the pelvics; anal fin present; caudal fin with shallow ventral lobe. Eye without nictitating membrane. Oviparous.

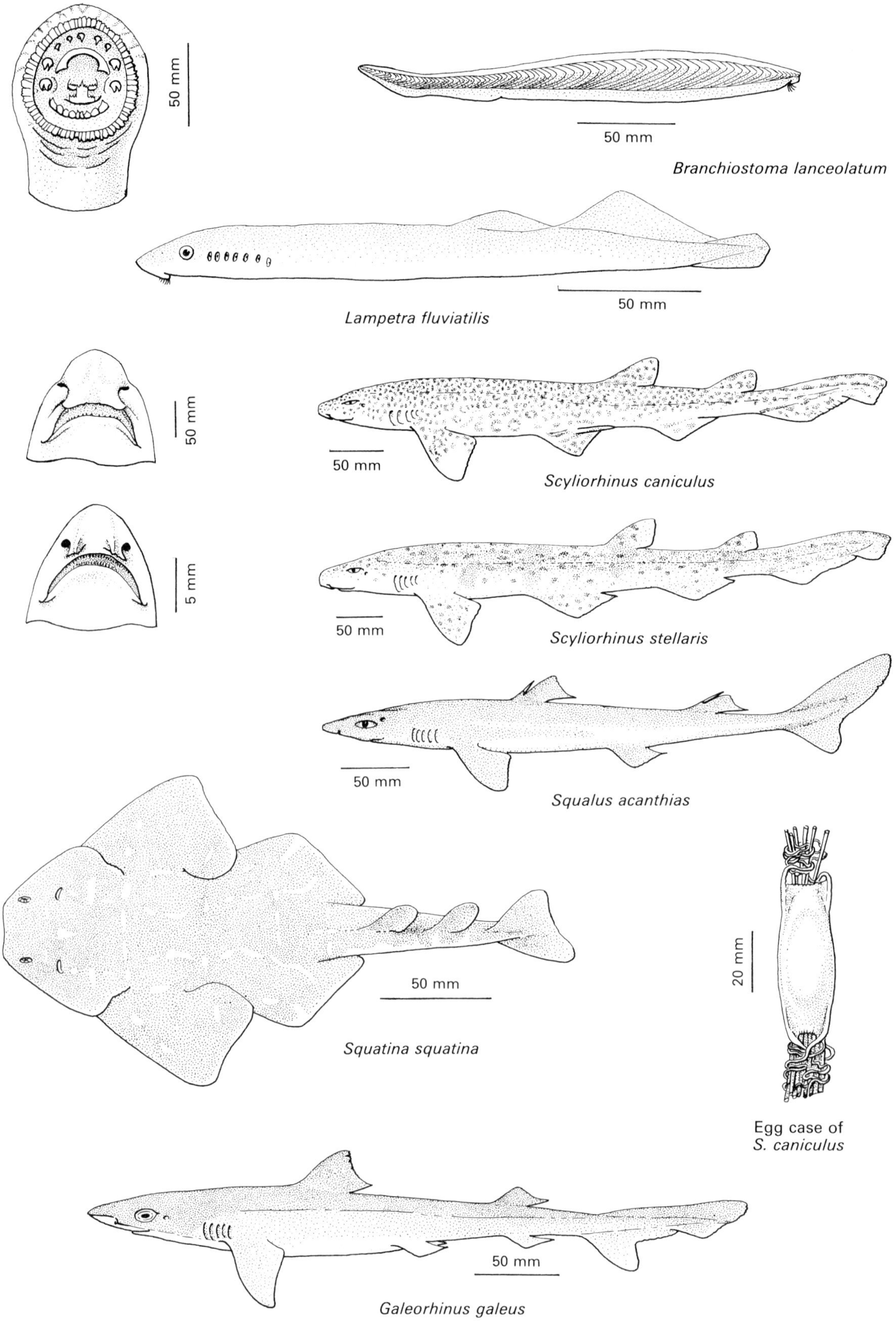

50 mm
50 mm
Branchiostoma lanceolatum
50 mm
Lampetra fluviatilis
50 mm
50 mm
Scyliorhinus caniculus
5 mm
50 mm
Scyliorhinus stellaris
50 mm
Squalus acanthias
50 mm
Squatina squatina
20 mm
Egg case of
S. caniculus
50 mm
Galeorhinus galeus

Fig. 17.2

1. Markings consisting of large brown spots on a light background. Nasal flaps separated by a wide gap **_Scyliorhinus stellaris_**

   Markings consisting of small brown spots on a light background. Nasal flaps continuous across the front of the mouth **_S. caniculus_**

Scyliorhinus caniculus (Linnaeus). Dogfish FIG. 17.2
[WHITEHEAD et al. 1984: 99]
Body long, slender, and rough. Origin of 1D behind pelvics; origin of 2D level with hind end of A. Nasal flaps hardly separated; almost joined in front of mouth. Colour sandy, with small brown spots; pale below. To 75 cm. Egg capsules not more than 6–7 cm long, pale brown, with a twisted tendril at each corner.

In shallow water, generally near the bottom. Off all coasts; common.

Scyliorhinus stellaris (Linnaeus). Nursehound FIG. 17.2
[WHITEHEAD et al. 1984: 100]
Body long, slender, and rough. Origin of 1D level with hind edge of pelvics; origin of 2D in front of hind end of A. Nasal flaps separated by a wide gap, not joined. Colour sandy, with large brown spots; pale below. To 150 cm. Egg capsules at least 10 cm long, brown, with a twisted tendril at each corner.

Inshore and in deeper water, over rough ground. Off all coasts, but most frequent in the south and west.

3 CARCHARINIDAE

Sharks. Snout elongated; mouth crescentic, ventral. Spiracle present or not. Teeth triangular, with serrated edges. First dorsal fin spineless, placed above the interspace between pectorals and pelvics; anal fin present. Eye with nictitating membrane.

Galeorhinus galeus (Linnaeus). Tope FIG. 17.2
[WHITEHEAD et al. 1984: 118]
Shark-like form, no lateral keel on the caudal peduncle; skin rather rough. Lower lobe of tail moderately developed. Spiracle present. Teeth unicuspid, serrated on the inner side, in three to four rows. Grey above, white below. To 150 cm. Ovoviviparous.

Bottom living in shallow water over sand, coming inshore during summer; off all coasts.

Order Squaliformes

4 SQUALIDAE

Spiny sharks. Snout elongated, mouth ventral, with a deep groove on either side. Spiracles present. Two dorsal fins, each preceded by a spine. No anal fin.

Squalus acanthias (Linnaeus). Spurdog FIG. 17.2
[WHITEHEAD et al. 1984: 146]
Body long and slender; origin of 1D not in front of rear edge of pectorals, 2D placed well behind pelvics; caudal with well-developed lower lobe. Teeth unicuspid. Grey above, with a row of white spots along the flank; pale below. To more than 1 m. Ovoviviparous.

Pelagic, forming shoals; may occur inshore. Off all coasts; common.

5 SQUATINIDAE

Body depressed; pectoral fins large and overlapping pelvics, not attached to the head. Mouth terminal, flanked by nostrils which have valvular flaps. Spiracles large. Two dorsal fins; no anal.

Squatina squatina (Linnaeus). Monkfish FIG. 17.2
[WHITEHEAD et al. 1984: 150]
Distinctive body shape, broad anteriorly; rough on both surfaces. Nostrils with a skinny flap and barbel-like prolongations, the barbels being only slightly branched (much branched in related species found south of the British Isles). Colour sandy above, with darker and lighter markings; white below. To about 2 m. Ovoviviparous.

Bottom living on sandy and gravelly sea-beds; inshore or in fairly shallow water. Off all coasts, but especially in the south and west; commonest during summer.

Order Rajiformes

6 RAJIDAE

Skates and rays. Body depressed, comprising a rhombic or circular disk and a slender tail. Pectoral fins ('wings') greatly expanded, joined to the snout. Two dorsal fins; caudal fin rudimentary or absent. Oviparous; the egg case with a curved, pointed horn arising from each corner. The long-snouted species are customarily termed skates and the short-snouted species rays. Specific identification may be difficult. Infraspecific variation is frequently extensive, with morphometry, spinulation, and colour changing during growth. In general the spinules on the wings develop with age, while the median spines may become less obvious. Males and females may differ (i.e. apart from the presence or absence of claspers). All species here described belong to genus *Raja*. Detailed descriptions were given by Clark (1926).

1. Snout long; under surface greyish (skates) **_Raja batis_**

   Snout short; under surface white (rays) **2**

2. Wings rounded laterally **3**

   Wings more or less pointed laterally, their front and rear edges meeting at an angle of roughly 90° **5**

3. A prominent black and yellow ocellus (round marking) approximately in the centre of the upper-side of each wing **_R. naevus_**

   No ocelli **4**

4. Upper surface bearing large, broad-based, striated spines; brownish in colour with dark and white spots but no stripes. Northern **_R. radiata_**

Spines on the upper surface small, not broad-based, unstriated; brownish in colour, marked with dark brown stripes and white spots. South-western **R. undulata**

5. Upper surface marked with stripes, with or without spots **6**

   Upper surface spotted or of uniform colour, not striped **7**

6. Stripes lighter than the background colour **R. microocellata**

   Stripes darker than the background colour **R. clavata**

7. Spots dark (often sparsely developed in juveniles), and not extending to the margins of the wings; skin smooth and slimy except for the spines or spinules on the tail, around the snout and on the wing tips **R. montagui**

   Spots dark or light, sometimes extending to the margins of the wings; skin rough **8**

8. Skin very rough and (in adults) bearing large spines with swollen bases; upper surface usually grey with dark and/or light spots **R. clavata**

   Skin with texture of fine sandpaper, spines small and without swollen bases; upper surface light brown covered overall with small dark spots **R. brachyura**

Raja batis Linnaeus. Skate FIG. 17.3
[WHITEHEAD et al. 1984: 175]
Snout long; leading edge of the wing concave; tips obtusely pointed. Median tail spines present in adults, including one or two between the dorsal fins. Wing spinulation developing with age. Dark or brownish-grey above, with lighter spots; grey below. Ventral mucus pores black. To 2 m.

The commonest of the deep water skates. Off all coasts except in the south-east.

Raja brachyura Lafont. Blond Ray FIG. 17.3
[WHITEHEAD et al. 1984: 184]
Short-snouted; leading edge of the wings fairly straight; tips roughly right-angled, obtusely pointed. Mid-line spines confined to the tail in adults, but on disk in young; none between the dorsal fins. Upper surface with the texture of fine sandpaper, spinulose in adults. Jaws with 60–90 rows of teeth. Fawn-coloured above, with many small dark spots extending all over the disk and tail; no ocellus; white below. To over 1 m.

Prefers water less than 100 m deep; young fish coming inshore. Common off south and west coasts.

Raja clavata Linnaeus. Thornback Ray FIG. 17.3
[WHITEHEAD et al. 1984: 185]
Short-snouted; leading edge of the wings fairly straight; the tips roughly right-angled, pointed. Mid-line spines from shoulder to tail or (adult males) restricted to the tail; adults with one or two spines between the dorsal fins. Large, broad-based spines on the back, which is rough and coarsely spinulose in all stages. Jaws with 36–44 rows of teeth. Dorsal ground colour grey, often with darker stripes and dark and/or light spots; a mottled ocellus may be present; white below. To 1 m.

In shallow water, often close inshore. Off all coasts; common.

Raja microocellata Montagu. Painted Ray FIG. 17.3
[WHITEHEAD et al. 1984: 186]
Short-snouted; leading edge of the wings fairly straight; tips nearly right-angled, obtusely pointed. Spines on the mid-line and tail sharply hooked, bent parallel to the body; none between the dorsal fins. Spinules present on the front of the disk, the remaining area smooth. Jaws with 44–52 rows of teeth. Light brown above with large white spots and white stripes parallel with the edges of the wings; no ocellus; white below. To 80 cm.

On sandy grounds in shallow water, often close inshore. Confined to south-west Britain and to the south and west of Ireland; locally common.

Raja montagui Fowler. Spotted Ray FIG. 17.3
[WHITEHEAD et al. 1984: 188]
Short-snouted; leading edge of the wings fairly straight; tips nearly right-angled; obtusely pointed. Mid-line spines closely packed, from the shoulders to the tail, with one or two between the dorsal fins. Upper surface smooth, spinules restricted to the front of the wings (especially in the young). Jaws with 38–60 rows of teeth. Light brown above, with small black spots sometimes grouped as an ocellus but not extending to the lateral margins of the disk; white below. To 75 cm.

In shallow to moderately deep water, especially on sandy grounds. South and west coasts; common.

Raja naevus Müller and Henle. Cuckoo Ray FIG. 17.4
[WHITEHEAD et al. 1984: 181]
Short-snouted; leading edge of the wings fairly straight, outer angles broadly rounded. Mid-line spines only in juveniles; adults with two rows of spines each side of the tail, extending on to the disk; dorsal fins closely set, no spines between them. Disk spinulose above, spiny anteriorly, with a bare patch on each wing. Jaws with 54–60 rows of teeth. Fawn above, with a conspicuous black and yellow ocellus on each wing; white below. To 70 cm.

In shallow water and from moderate depths; all coasts; moderately common.

Raja radiata Donovan. Starry Ray FIG. 17.4
[WHITEHEAD et al. 1984: 174]
Short-snouted; leading edge of disk distinctly sinuous, outer angles broadly rounded. Mid-line spines 12–19, large, curved, with broad, deeply ridged bases; dorsal fins closely set, no spines between them. Disk coarsely spinulose above. Jaws with 38–40 rows of teeth. Light brown above, with dark and pale spots, but no stripes; no ocellus; white below. To 75 cm.

From moderate to deep water. Off coasts of Scotland and north-east England only; there common.

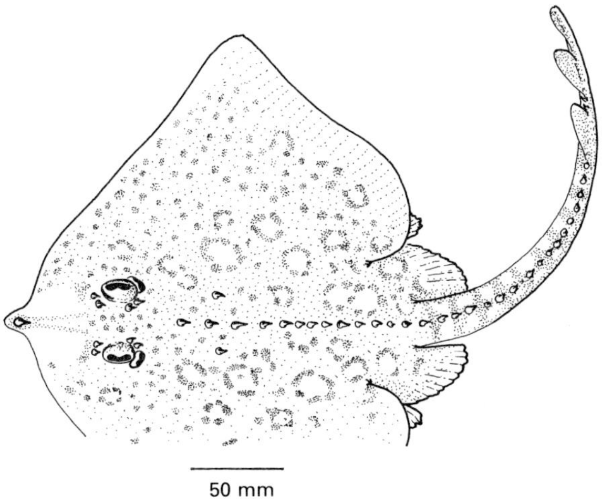

Raja clavata : dorsal

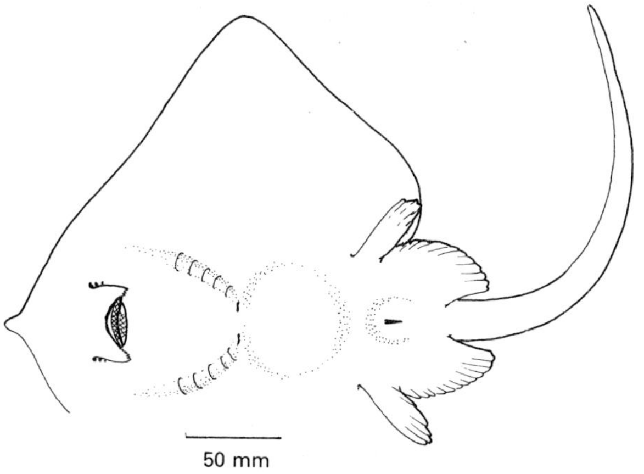

Raja clavata : ventral

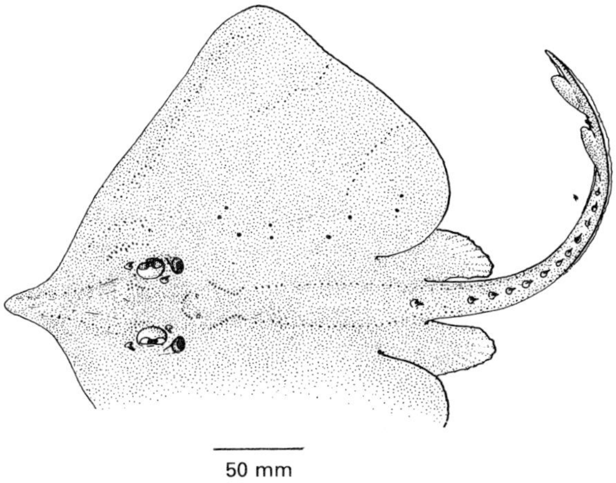

Raja batis

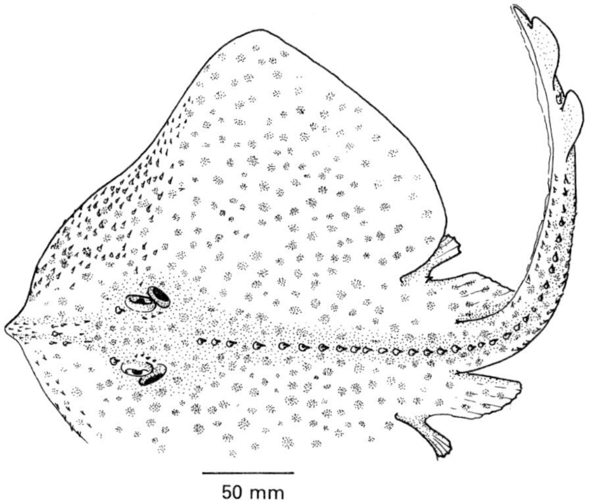

Raja brachyura

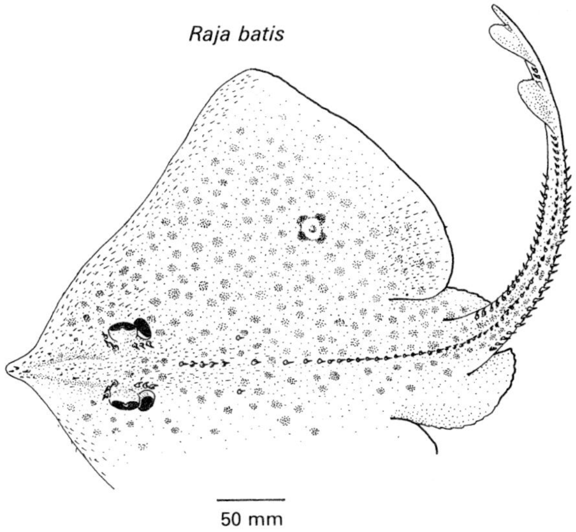

Raja montagui

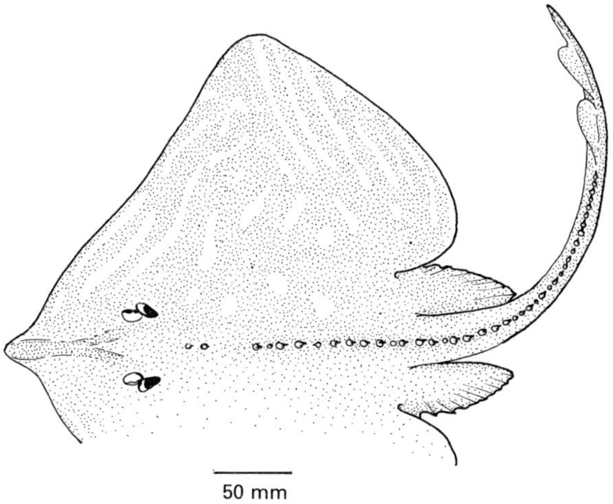

Raja microocellata

Fig. 17.3

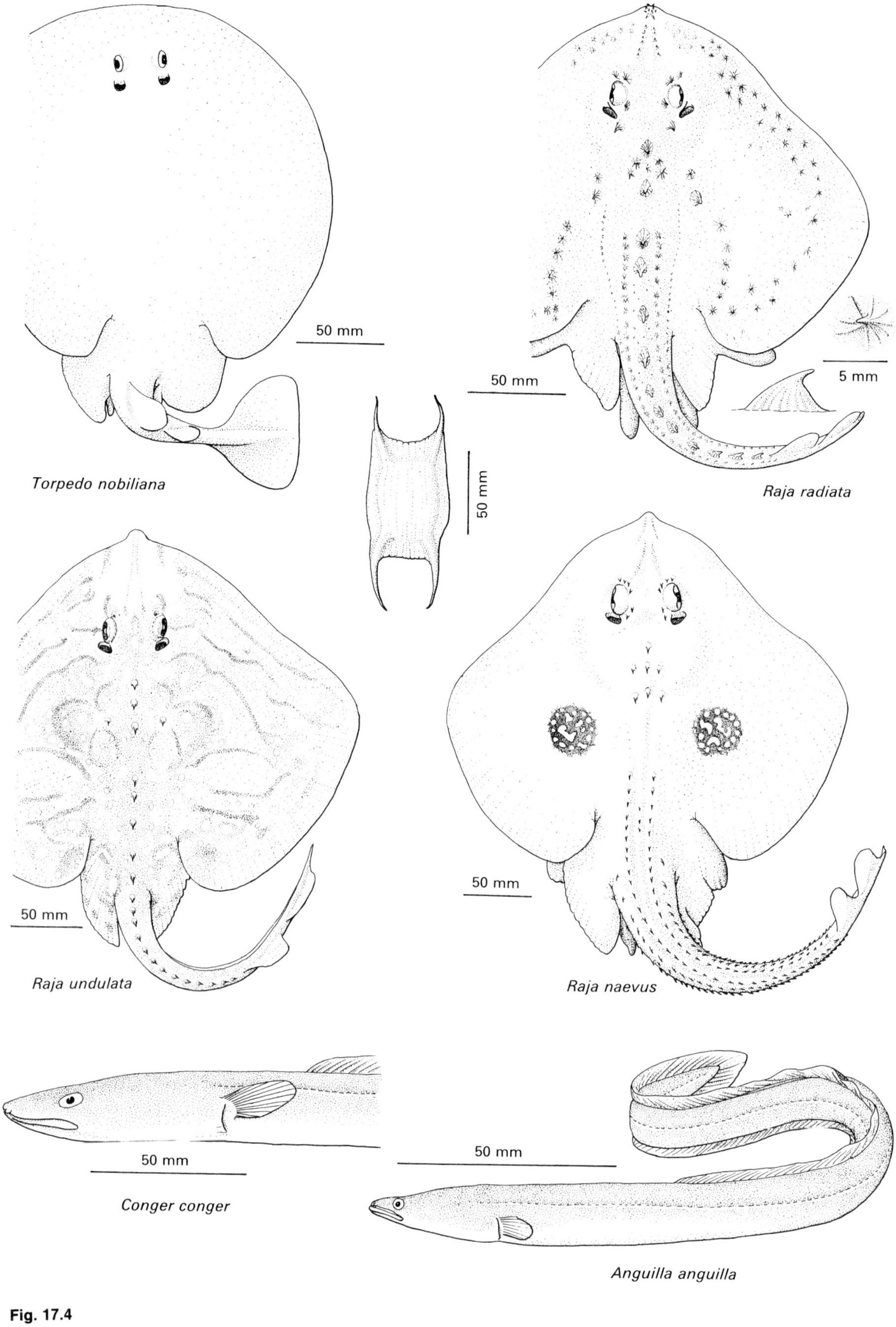
50 mm
Torpedo nobiliana
50 mm
5 mm
Raja radiata
50 mm
50 mm
Raja undulata
50 mm
Raja naevus
50 mm
Conger conger
50 mm
Anguilla anguilla

Fig. 17.4

Raja undulata Lacapède. Undulate Ray FIG. 17.4
[WHITEHEAD et al. 1984: 196]
Short-snouted; leading edge of wings distinctly sinuous, outer angles broadly rounded. Scattered mid-line spines and one or two lateral rows on the tail; usually two spines between the well-separated dorsal fins. Dorsal surface of disk spinulose anteriorly. Jaws with 45–50 rows of teeth. Light brown above with white spots and dark brown stripes; no ocelli; white below. To 1 m.

From shallow to moderately deep water; not inshore; to south-west of the British Isles only.

Order **Torpediniformes**

7 TORPEDINIDAE

Electric rays. Disk broad, anteriorly rounded, flabby, smooth skinned. Two dorsal fins; caudal fin well-developed. Electric organ (used primarily for stunning prey) filling most of each wing.

Torpedo nobiliana Bonaparte. Electric Ray FIG. 17.4
[WHITEHEAD et al. 1984: 160]
Disk rounded, smooth. First dorsal fin larger than second. Spiracles smooth. Dark brown or grey above; white below. To 180 cm. Large, freshly caught rays can discharge powerful electric shocks.

In shallow to moderately deep water; off southern and western coasts; frequent.

A second species, *T. marmorata* Risso, with subequal dorsal fins and spiracles fringed with papillae, reaches southern England in late summer.

Class **Osteichthyes**

Subclass **Actinopterygii**

Infraclass **Teleostei**

Superorder **Elopomorpha**

Order **Anguilliformes**

8 ANGUILLIDAE

Snake-like body; mouth terminal; dorsal and anal fins continuous; caudal fin rudimentary. Pectoral fins present; pelvics absent. Scales minute. Lateral line well-developed. Leptocephalus larva.

Anguilla anguilla (Linnaeus). Eel FIG. 17.4
[WHITEHEAD et al. 1986*a*: 535]
Lower jaw longer than upper. Pectoral fins small and rounded; dorsal originating far behind pectorals. D 245–275; A 205–235; P 17–18; vert. 139–143. Dark or black above, yellowish below (silvery when approaching sexual maturity). To 60 cm.

Catadromous. Leptocephali arrive in British waters in autumn and metamorphose into transparent elvers known as glass eels. Pigmented elvers (7 cm) ascend rivers in winter or early spring. Adults live in fresh water or in estuaries, amongst weed, in mud or in pools. Maturing 'silver' eels may be caught at river mouths at the start of their oceanic migration. All coasts; common.

9 CONGRIDAE

Snake-like body; mouth terminal; dorsal and anal fins continuous; caudal fin rudimentary. Pectoral fins present; pelvics absent. Scaleless. Lateral line well-developed. Leptocephalus larva.

Conger conger (Linnaeus). Conger FIG. 17.4
[WHITEHEAD et al. 1986*a*: 569]
Upper jaw longer than the lower. Pectoral fins obtusely pointed; dorsal originating at the level of the pectoral fin tip. Br. ix–x; D 270–300; A 205–230; P 16–19; vert. 149–156. Brown above, paler or white below. Larger than eel; to 2 m or more.

Found intertidally on rocky shores, hiding in crevices, and offshore; principally on the south and west coasts. Breeds in the eastern Atlantic, reaching British waters as leptocephali.

Superorder **Clupeomorpha**

Order **Clupeiformes**

10 CLUPEIDAE

Pelagic fishes frequently forming shoals. One dorsal fin; pectorals low on the flank, pelvics abdominal. Bodies completely covered with caducous cycloid scales, except on the head. No lateral line. No spiny finrays. Caudal fin with 19 rays.

1. Upper jaw projecting beyond lower; gape extending well behind rear of eye ***Engraulis encrasicolus***
   Lower jaw about as long as upper; gape not extending behind the eye **2**
2. Upper jaw notched in the mid-line, the lower jaw fitting inside it ***Alosa fallax***
   Upper jaw without notch **3**
3. Operculum with radiating ridges ***Sardina pilchardus***
   Operculum without radiating ridges **4**
4. Origin of D behind that of pelvics; 21–23 scales between the throat and the pelvic origin; pelvic fins with seven to eight rays ***Sprattus sprattus***
   Origin of D behind or slightly in front of that of pelvics; 27–30 scales between the throat and the pelvic origin; pelvic fins with eight to ten rays ***Clupea harengus***

Alosa fallax (Lacepède). Twaite Shad FIG. 17.5
[WHITEHEAD et al. 1984: 271]
Upper jaw notched, the lower jaw fitting inside it. Belly keeled and strongly serrated. 40–60 rakers on the first gill arch (cf. 80–

130 in the similar but less common Allis Shad, *A. alosa* (L.)). Br. viii; D 18–21; A 19–25; P 15–16; V 9; lat. sc. 60–70; vert. 55–59. Steel blue above, silvery white below; a large dark blotch on the shoulder, usually succeeded by a further five to six; golden yellow on the side of the head. To 50 cm.

Mainly on southern and western coasts; anadromous, entering rivers for spawning in May–June; quite common.

Clupea harengus Linnaeus. Herring FIG. 17.5
[WHITEHEAD et al. 1984: 273]
Lower jaw prominent, longer than the upper. Belly weakly keeled, not strongly serrated. Origin of D in front of base of pelvics, 27–30 scales on the belly between the throat and the pelvic fins; 13–15 between the pelvics and the vent. 18–23 pyloric caeca. Br. viii; D 17–20; A 16–18; P 17; V 8–10; lat. sc. 53–60; vert. 51–58. Dark blue above, silver-white below. To 40 cm.

Off all coasts; formerly in vast offshore shoals, now less common.

Engraulis encrasicolus (Linnaeus). Anchovy FIG. 17.5
[WHITEHEAD et al. 1984: 282]
Upper jaw much longer than the lower; the gape extending well past the eye. Belly rounded. D 15–18; A 16–20; P 16–17; V 7; lat. sc. 44–50; vert. 44–49. Greenish above, silver below. To 20 cm.

Pelagic, forming vast shoals close inshore; off southern coasts, particularly in the English Channel.

Sardina pilchardus (Walbaum). FIG. 17.5
Pilchard, Sardine [WHITEHEAD et al. 1984: 276]
Jaws of equal length; the gape not extending past the eye. Belly rounded; operculum marked with radiating ridges. Origin of D in front of base of pelvics; two posterior rays of A longer than the others. Br. vi–viii; D 17–18; A 17–18; P 16–17; V 6–8; lat. sc. 28–30; vert. 50–53. Greenish olive above, silver-white below. To 25 cm.

Pelagic and shoaling; common off south-western coasts, rarer northwards.

Sprattus sprattus (Linnaeus). Sprat FIG. 17.5
[WHITEHEAD et al. 1984: 280]
Lower jaw longer than the upper. Belly sharply keeled and strongly serrated. Origin of D behind base of pelvics. Anal with more rays than in herring, pelvics with fewer. 21–23 scales on the belly between the throat and the pelvic fins, 11–12 between the pelvics and the vent. Seven pyloric caeca. Br. vi–vii; D 16–19; A 18–20; P 16–17; V 7–8; lat. sc. 47–50; vert. 46–49. Blue or blue-green above, silver-white below. To 15 cm.

Pelagic, forming large shoals close inshore; all coasts; common.

Superorder **Protacanthopterygii**

Order **Salmoniformes**

11 SALMONIDAE

Two dorsal fins, the second *adipose* (rayless); pectorals low on the flank, pelvics abdominal. Body, but not the head, covered with cycloid scales. Jaws toothed. Vertebrae at base of caudal fin upturned; tail square or slightly forked. Caudal fin with 19 rays.

1. 10–13 rows of scales between rear edge of the adipose fin and the lateral line (count downwards and forwards) ***Salmo salar***
   13–16 rows of scales between the rear edge of the adipose fin and the lateral line ***S. trutta***

Salmo salar Linnaeus. Salmon FIG. 17.5
[WHITEHEAD et al. 1984: 381]
Lower jaw as long as the upper, or much longer and hooked. Maxilla not extending greatly behind rear of the eye. 17–24 rakers on the first gill arch. No teeth on the head of the vomer. 10–13 rows of scales from the rear edge of the adipose fin to the lateral line. Br. xi–xii; D 12–14; A 9–11; P 14; V 9; lat. sc. 120–125; vert. 57–60. Colour in the sea: blue-green above, flanks silvery, belly white; with small black spots on the back and flanks. To 150 cm.

Anadromous. Returns from the ocean to coastal waters during the spring and summer, entering estuaries and ascending rivers prior to spawning during the winter. All coasts, particularly the north and west.

Salmo trutta Linnaeus. Trout FIG. 17.5
[WHITEHEAD et al. 1984: 382]
Lower jaw as long as the upper, or hooked in old males. Maxilla extending behind rear of the eye. 13–18 rakers on the first gill arch. Three to four teeth on the head of the vomer. 13–17 rows of scales from the rear edge of the adipose fin to the lateral line. Br. x–xii; D 12–14; A 10–12; P 13–14; V 9; lat. sc. 115–130; vert. 56–61. Colour in the sea: brownish grey above, flanks silvery, belly white; with reddish and small black spots on the back and flanks. To 70 cm.

Sea trout are anadromous. Inshore waters and estuaries, ascending rivers for spawning; all coasts.

12 ARGENTINIDAE

Two dorsal fins, the second *adipose* (rayless); pectorals low on flank, pelvics abdominal. Body, but not the head, covered with cycloid scales. Teeth on vomer, but not on the jaws. Vertebrae at base of caudal fin not upturned; tail forked. 19 caudal finrays.

Argentina sphyraena Linnaeus. Argentine FIG. 17.5
[WHITEHEAD et al. 1984: 388]
Slender, silvery fish, with small terminal mouth, and strongly emarginate tail. D in front of pelvics. Br. vi; D 9–12; A 11–15;

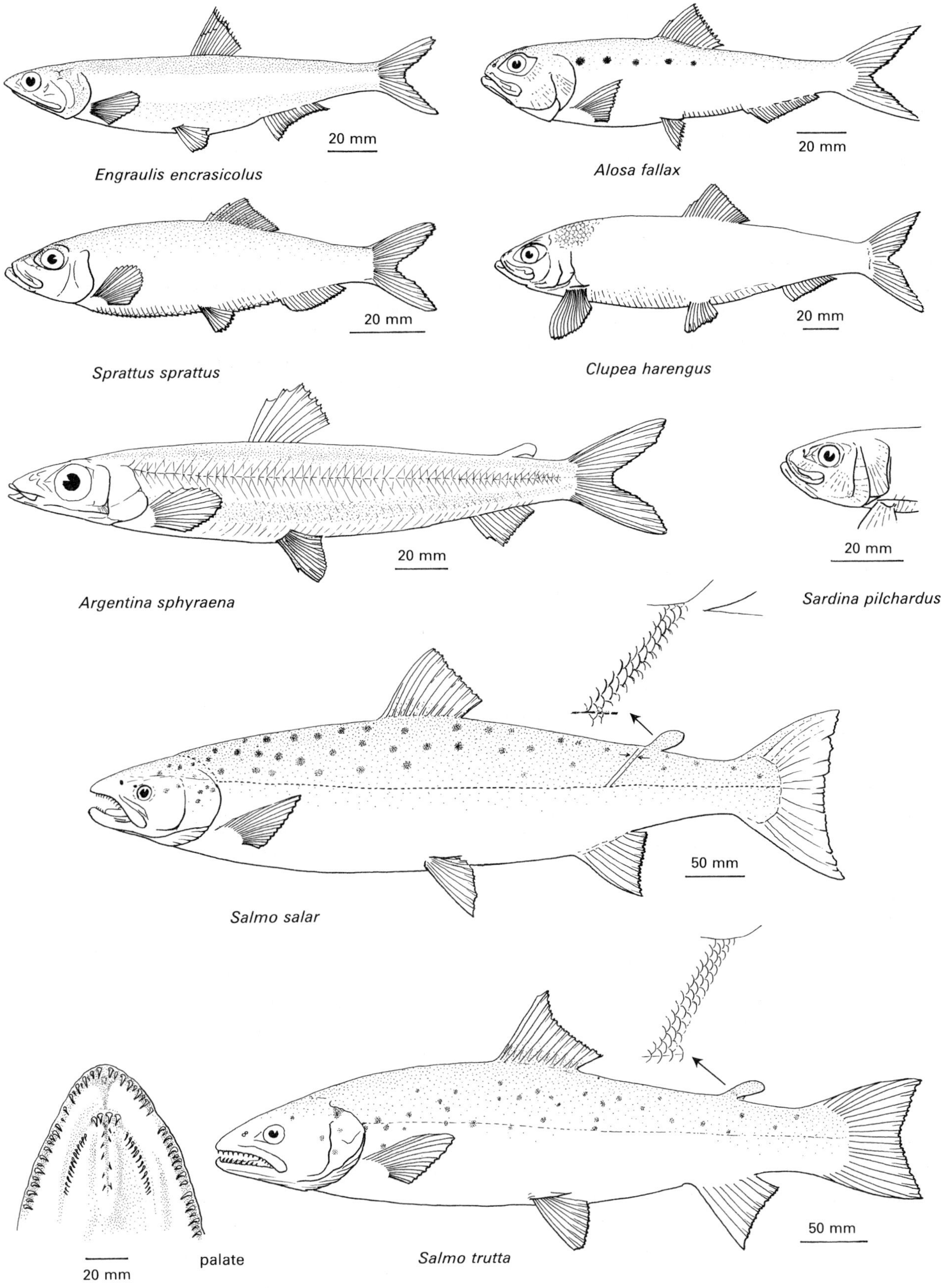
20 mm
Engraulis encrasicolus
20 mm
Alosa fallax
20 mm
Sprattus sprattus
20 mm
Clupea harengus
20 mm
Argentina sphyraena
20 mm
Sardina pilchardus
50 mm
Salmo salar
20 mm
palate
Salmo trutta
50 mm

Fig. 17.5

P 13–14; V 10–11; lat. sc. 50–54; vert. 49–52. Olive above, flanks silver, white below.

Mainly offshore, over a sandy or muddy sea-bed; west and north.

The Smelt *Osmerus eperlanus* (Linnaeus) [Whitehead et al. 1984: 401] is a somewhat similar fish, found locally in estuaries and inshore waters, mainly on the east coast of England. It differs from the Argentine in having a large mouth (extending past the eye), with teeth on the maxilla, and in having the dorsal fin set above the pelvics. Br. vii–viii; D 10–12; A 8–9; P 11–12; V 8–9; lat. sc. 60–65; vert. 55–62. Olive above, flanks silvery, white below.

Superorder **Paracanthopterygii**

Order **Gobiesociformes**

13 GOBIESOCIDAE

Clingfishes or suckers. Small, somewhat depressed fishes, having a complex ventral sucker (modified pelvic fins) below the pectorals; V I/4. Scaleless.

---

1. Dorsal fin with 15–20 rays, anal with nine to 12 rays **(*Lepadogaster*) 2**

   Dorsal and anal fins with four to seven rays **3**

2. Dorsal, caudal, and anal fins continuous ***L. lepadogaster***

   Dorsal and anal fins not continuous with the caudal fin ***L. candollei***

3. A small dark (not red) spot at the origin of the dorsal and anal fins (♀) or a large dark spot on each (♂); first anal finray below or in front of the second dorsal finray ***Apletodon dentatus***

   Dorsal and anal fins lacking a dark spot (sometimes with a red one); first anal finray below or behind the third dorsal finray ***Diplecogaster bimaculata***

Apletodon dentatus (Facciola) FIG. 17.6
[Syn. *A. microcephala* (Brook)].
Small-headed Clingfish [WHITEHEAD et al. 1986*b*: 1352]
A small clingfish with D and A short and similarly sized, both clearly separated from the caudal; first anal finray roughly below the first dorsal finray. Premaxillary teeth of two kinds, small and rounded in front, with one–three larger teeth at the sides. D 5–6; A 5–7; P 20–24; vert. 29–32. Brownish, with a small (female) or large (male) dark spot on the dorsal and anal fins. To 5 cm.

Lower shore, often in the hollow bases of *Saccorhiza*. On Atlantic coasts, but distribution uncertain.

Diplecogaster bimaculata (Bonnaterre). FIG. 17.6
Two-spotted Clingfish [WHITEHEAD et al. 1986*b*: 1354]
A small clingfish with D and A short and unequal, both clearly separated from the caudal; first anal finray behind the second dorsal finray. Teeth of one kind only. D 5–7; A 4–6; P 20–26; vert. 30–31. Reddish, often spotted; males with a large pectoral spot, purple outlined with yellow, partly hidden by the fins. To 5 cm.

Offshore, on shelly or gravelly grounds; western coasts; locally quite common.

Lepadogaster candollei (Risso). FIG. 17.6
Connemara Clingfish [WHITEHEAD et al. 1986*b*: 1356]
The largest British clingfish. Head depressed, of tapering profile. D and A moderately long, separated from the caudal. Front nostril bearing a very small tentacle. D 13–16; A 9–11; P 22–25. Colour variable, males reddish and spotted above, females greenish above; pale below. To 10 cm.

At or below low water mark on rocky shores. South and west Ireland, Devon, Cornwall, and Pembrokeshire; local.

Lepadogaster lepadogaster (Bonnaterre). FIG. 17.6
Cornish Clingfish [WHITEHEAD et al. 1986*b*: 1357]
Head depressed, of tapering profile. D longer than A, both continuous with the caudal. Front nostril bearing a conspicuous tentacle. Br. vi; D 16–20; A 9–11; P 20–25; vert. 32–33. Reddish, with a pair of conspicuous spots, blue outlined with red, on the back.

Between tidemarks on rocky, weed-covered shores; south and west; often common. The western European populations are referred to subspecies *purpurea*.

Order **Lophiiformes**

14 LOPHIIDAE

Body depressed; head large, jaws wide; the lower jaw projecting anteriorly. Jaws and vomers toothed. Gill slit opening behind pectoral fin. First dorsal fin spinous, preceded by three separate rays, the first of them modified as a lure. Base of pectoral fin elongated, arm-like ('pseudo-brachium'); pelvics stumpy, jugular.

Lophius piscatorius Linnaeus. Angler, Monk FIG. 17.6
[WHITEHEAD et al. 1986*b*: 1363]
Broad depressed body with huge mouth. Loose, scaleless skin, the whole fish flabby; with a lateral fringe of small flaps. Anteriormost dorsal finray long and slender, with a terminal flap. Br. vi; 1D II+I+III; 2D 10–13; A 9–11; C 8; P 23–28; V I/5; vert. 28–31. Brownish grey (changeable) above, white below. To nearly 2 m.

Occasionally found on the shore, but extends to the continental slope; all coasts; quite common. A second species, *L. budegassa* Spinola [Whitehead et al. 1986*b*; 1362] with 2D 9–10, has recently been recognized from the British Isles.

Order **Gadiformes**

15 GADIDAE

Body elongate, with small cycloid scales. Fins without spiny rays. Pectoral fins on the flank, pelvics thoracic or jugular. Lower

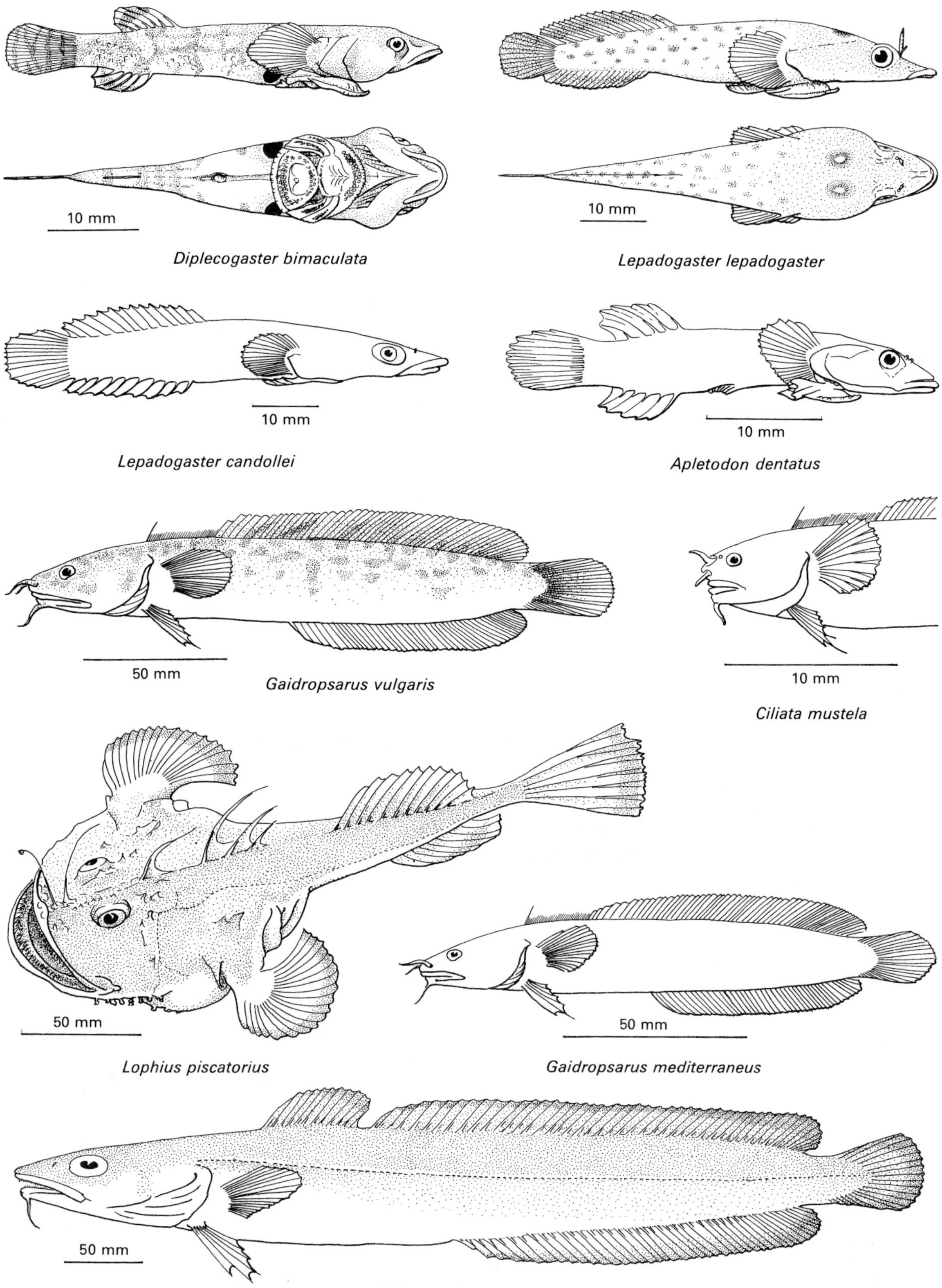
10 mm
10 mm
Diplecogaster bimaculata
Lepadogaster lepadogaster
10 mm
10 mm
Lepadogaster candollei
Apletodon dentatus
50 mm
Gaidropsarus vulgaris
10 mm
Ciliata mustela
50 mm
50 mm
Lophius piscatorius
Gaidropsarus mediterraneus
50 mm
Molva molva

Fig. 17.6

jaw usually with a barbel. Swimbladder present (reduced in rocklings). Branchiostegals seven.

---

1. Two dorsal fins (1D sometimes reduced); one anal fin **2**
   Three dorsal fins (1D, 2D, 3D); two anal fins (1A, 2A) **5**
2. 1D with normally developed rays; a single barbel present (on the lower jaw) ***Molva molva***
   1D with one long ray followed by a series of short, hair-like rays; 3–5 barbels on the jaws. (Rocklings) **3**
3. Five barbels present ***Ciliata mustela***
   Three barbels present **(*Gaidropsarus*) 4**
4. Uniform brownish colour; on the shore and in shallow water ***G. mediterraneus***
   Pinkish with dark brown markings; sublittoral ***G. vulgaris***
5. A barbel present on the lower jaw **6**
   Barbel absent or minute **10**
6. Upper jaw projecting **7**
   Lower jaw projecting **14**
7. 1A long, its origin below the middle of 1D ***Trisopterus luscus***
   1A shorter, its origin not or hardly in front of 2D **8**
8. Barbel short, less than half the diameter of the eye. A black patch below 1D; lateral line black ***Melanogrammus aeglefinus***
   Barbel longer, three-quarters or more of the eye diameter. No black patch below 1D; lateral line not black **9**
9. Pectoral fins reaching the origins of 2D and 1A; lateral line darker than the ground colour ***Trisopterus minutus***
   Pectoral fins not reaching the origins of 2D and 1A; lateral line lighter than the ground colour ***Gadus morhua***
10. Origin of 1A in front of 1D; dorsal fins separated by wide spaces ***Micromesistius poutassou***
    Origin of 1A not or hardly in front of 2D; dorsal fins adjacent or separated by narrow spaces **11**
11. Upper jaw projecting **12**
    Lower jaw projecting **(*Pollachius*) 13**
12. A black patch present above the pectoral fins; lateral line black ***Melanogrammus aeglefinus***
    A dark spot present at the base of the pectoral fins; lateral line brown ***Merlangius merlangus***
13. Origin of 1A below the middle of 1D; pectorals passing the origin of 1A; lateral line curved above origin of 1A ***P. pollachius***
    Origin of 1A not or hardly in advance of 2D; pectorals not reaching the origin of 1A; lateral line straight ***P. virens***
14. Pectoral fins with a dark spot at their base, and passing the origins of 2D and 1A ***Trisopterus esmarkii***
    Pectoral fins without a dark spot at their base, and not reaching the origins of 2D and 1A ***Pollachius virens***

Ciliata mustela (Linnaeus). FIG. 17.6
Five-bearded Rockling [WHITEHEAD et al. 1986*a*: 698]
Body elongate; 1D modified, with a single long ray anteriorly followed by a series of short hair-like rays. Five barbels: one on the lower jaw, two on the upper jaw and one on each anterior nostril. Maxilla reaching just behind level of the rear of the eye. Head about one-fifth of body. 1D 1+hairs; 2D 50–55; A 40–41; P 15–17; V 7; vert. 47. Dark brown above, white below. To 30 cm.

Littoral; under boulders and among weed on rocky shores; all coasts; common.

A second species *C. septentrionalis* (Collett) has been found off western and northern coasts. Its head is large, about one-quarter of body.

Gadus morhua Linnaeus. Cod FIG. 17.7
[WHITEHEAD et al. 1986*a*: 686]
Body fusiform, with three dorsal and two anal fins, 1A originating under the front of 2D. Pectoral fins not reaching 2D. Lower jaw shorter than upper; barbel long. 21–28 rakers on the first gill arch. 1D 14–15; 2D 18–22; 3D 17–20; 1A 19–23; 2A 17–19; P 16–21; V 6; vert. 49–53. Mottled golden-brown above, lateral line white, belly white. To 150 cm.

Inshore and in shallow water; all coasts; common.

Gaidropsarus mediterraneus (Linnaeus). FIG. 17.6
Shore Rockling [WHITEHEAD et al. 1986*a*: 700]
Body elongate; 1D modified, with a single long ray anteriorly followed by a series of short hair-like rays. Three barbels: one on the lower jaw and one on each anterior nostril. Pectoral fin with fewer rays than in *G. vulgaris*. 1D 1+hairs; 2D 53–63; A 43–53; P 15–17; V 5–7; vert. 47–49. Uniform dark brown colour above, paler below. To 25 cm.

Littoral; under boulders and among weed on rocky shores; western and southern coasts; common.

Gaidropsarus vulgaris (Cloquet). FIG. 17.6
Three-bearded Rockling [WHITEHEAD et al. 1986*a*: 700]
Body elongate; morphologically like *G. mediterraneus* but larger and with more rays in the pectoral fins. 1D 1+hairs; 2D 56–64; A 46–52; P 20–22; V 8; vert. 46–49. Ground colour pink with dark brown markings, paler below. To 50 cm.

Sublittoral; inshore, usually on hard grounds; off all coasts.

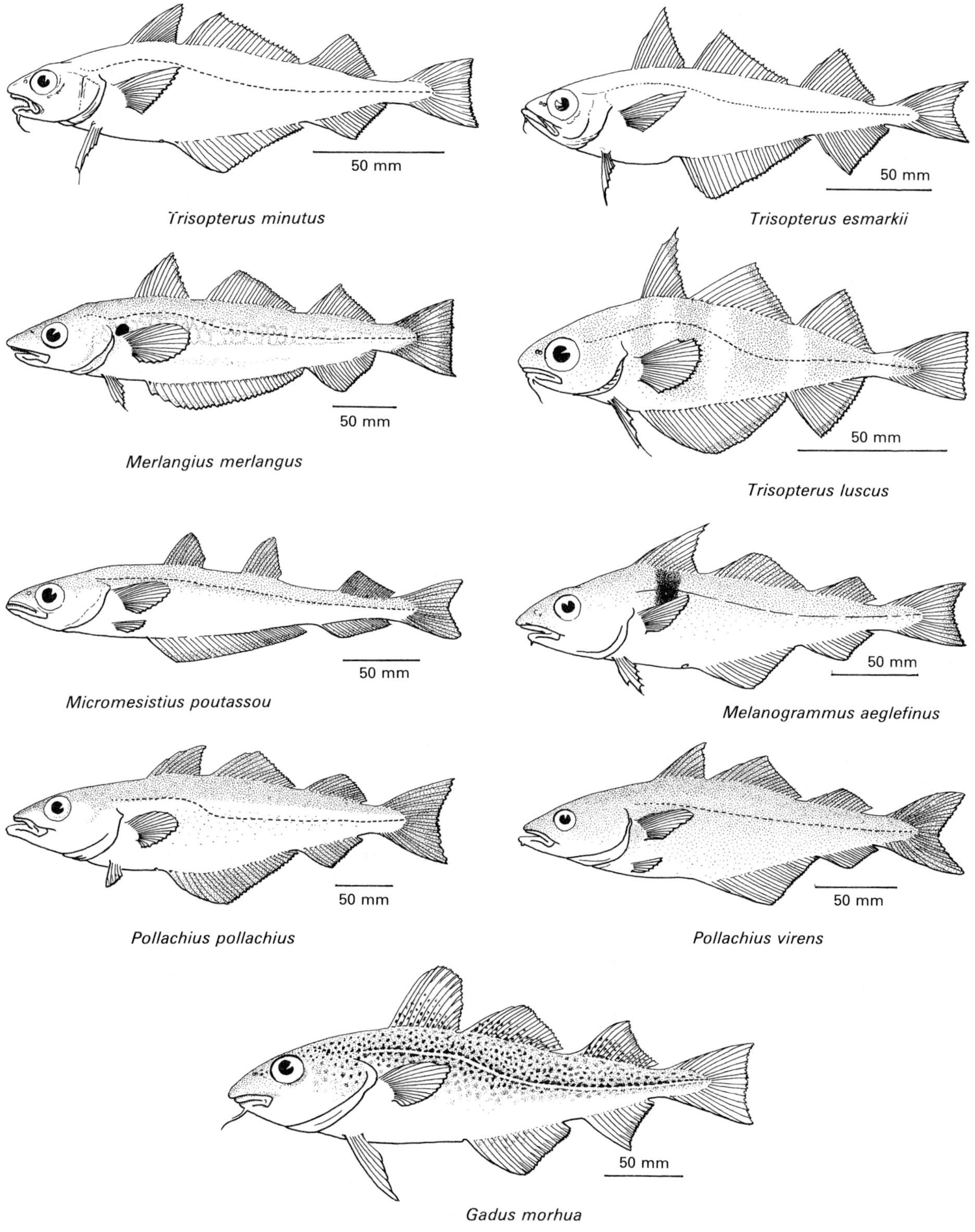

Fig. 17.7

Melanogrammus aeglefinus (Linnaeus). Haddock FIG. 17.7 [WHITEHEAD et al. 1986*a*: 687]
Body fusiform, with three dorsal and two anal fins, 1A originating under the first rays of 2D. Lower jaw shorter than upper; barbel small. 24–27 rakers on the first gill arch. 1D 14–16; 2D 19–24; 3D 19–22; 1A 21–25; 2A 20–23; P 19–20; V 6; vert. 52–57. Dark brown above, with a black oval patch above the pectoral fin; lateral line black; belly white. To 80 cm.

Inshore and offshore; off all coasts except in the English Channel.

Merlangius merlangus (Linnaeus). Whiting FIG. 17.7
[WHITEHEAD et al. 1986*a*: 688]
Body fusiform, with three dorsal and two anal fins, 1A originating under the central rays of 1D. Lower jaw slightly shorter than the upper; barbel rudimentary or absent. 21–22 rakers on the first gill arch. 1D 12–15; 2D 18–25; 3D 19–22; 1A 30–35; 2A 21–23; P 19–20; V 6; vert. 53–57. Brown or blue-green above, with a dark spot at the pectoral finbase; lateral line dark; silvery white below. To 70 cm.

In shallow water, sometimes inshore; off all coasts; abundant.

Micromesistius poutassou (Risso). Blue Whiting FIG. 17.7
[WHITEHEAD et al. 1986*a*: 689]
Body slender, with three well-separated dorsal fins and two anal fins, the first of them very long. Lower jaw longer than the upper; no barbel. 27–33 rakers on the first gill arch. 1D 12–14; 2D 12–14; 3D 23–26; 1A 33–39; 2A 24–27; P 20; V 6; vert. 56–60. Blue above, silvery below. To 40 cm.

Pelagic, forming huge shoals. Immatures sometimes inshore, but mainly oceanic; north and west of the British Isles.

Molva molva (Linnaeus). Ling FIG. 17.6
Body elongate, slender, with 1D short and 2D long; A long. Upper jaw longer than lower; with a long barbel. 1D 13–16; 2D 60–70; A 57–66; P 19–21; V 6–7; vert. 63–65. Dark brown above, paler below; a dark area on the posterior part of each dorsal fin. To 150 cm.

From moderate to deep water; immatures also inshore. A northern species largely absent from southern Britain. The account of *Molva* in Whitehead et al. 1986*a*: 702–3 has the illustrations of the two species transposed.

Pollachius pollachius (Linnaeus). Pollack FIG. 17.7
[WHITEHEAD et al. 1986*a*: 690]
Body fusiform, with three dorsal and two anal fins, 1A originating under the central rays of 1D. Lower jaw longer than the upper; no barbel. 26–27 rakers on the first gill arch. Lateral line dark, with a sharp curve below the first dorsal interspace. 1D 12–14; 2D 17–21; 3D 16–20; 1A 24–34; 2A 17–21; P 17–19; V 6; vert. 52–55. Dark olive-brown above, lighter and mottled on the flanks; belly white. To 80 cm.

Mainly inshore, over hard ground; sometimes in small shoals. Off all coasts, particularly the south and west.

Pollachius virens (Linnaeus). FIG. 17.7
Saithe, Coalfish, Coley [WHITEHEAD et al. 1986*a*: 691]
Body fusiform, with three dorsal and two anal fins. 1A originating below the rear end of 1D. Lower jaw slightly longer than the upper; barbel rudimentary. 35–40 rakers on the first gill arch. Lateral line light, almost straight (without a sharp curve below the first dorsal interspace). 1D 13–15; 2D 19–24; 3D 19–24; 1A 25–30; 2A 17–24; P 19–20; V 6; vert. 53–56. Dark greenish or blackish above, flanks paler, belly white. To 1 m or more.

Inshore and offshore, forming small shoals. Common along all coasts except in the south.

Trisopterus esmarkii (Nilsson). Norway Pout FIG. 17.7
[WHITEHEAD et al. 1986*a*: 693]
Body fusiform, with three dorsal and two anal fins, 1A originating below the rear half of 1D. Pectoral and pelvic fins long, extending well past the vent. Lower jaw longer than the upper; barbel well-developed. 33–43 rakers on the first gill arch. Lateral line gently curved anteriorly. 1D 14–15; 2D 23–26; 3D 23–27; 1A 26–31; 2A 24–30; P 19; V 6; vert. 51–55. Dark olive-brown above, belly silvery; a small dark spot at the pectoral finbase. To 25 cm.

Oceanic; abundant in northern waters, absent off southern coasts.

Trisopterus luscus (Linnaeus). Bib, Pout FIG. 17.7
[WHITEHEAD et al. 1986*a*: 693]
Body rather deep, with three dorsal and two anal fins, 1A originating below the middle of 1D. Pectoral and pelvic fins extending well past the vent. Lower jaw shorter than the upper, with a long barbel. 18–23 rakers on the first gill arch. Lateral line gently curved anteriorly. 1D 11–14; 2D 20–24; 3D 18–20; 1A 30–34; 2A 19–22; P 17; V 6; vert. 47–49. Bronze above, with several broad vertical bands of rather darker colour down the flanks (visible only in living or freshly caught fish); a black spot at the pectoral finbase. To 40 cm.

Inshore and coastal waters all round the British Isles; very common.

Trisopterus minutus (Linnaeus). Poor-cod FIG. 17.7
[WHITEHEAD et al. 1986*a*: 694]
Small, fusiform, with three dorsal and two anal fins, 1A originating below the first dorsal interspace. Pectoral and pelvic fins short, the latter scarcely reaching the vent. Lower jaw shorter than the upper, with a long barbel. About 28 rakers. Lateral line gently curved anteriorly. 1D 13; 2D 23–26; 3D 22–24; 1A 28–29; 2A 23–25; P 13–16; V 6; vert. 47–51. Brownish above, paler below; a small dark spot at the pectoral finbase. To 23 cm.

In shallow water, mainly offshore; common except off southeast coasts.

16 MERLUCCIIDAE

Similar to Gadidae in most respects. No barbel on lower lip. Two dorsal fins; single anal fin. Br. vii.

Merluccius merluccius (Linnaeus). Hake FIG. 17.8
[WHITEHEAD] et al. 1986*a*: 677]
Body elongate. Mouth large; two rows of teeth on lower jaw, one row on upper. Lateral line straight. 1D 9–10; 2D 36–40; A 36–40; P 12–14; V 7; lat. sc. 135–150; vert. 50–52. Dark blue-grey above, paler below; buccal cavity black. To 1 m.

Mid-water or demersal, offshore and in deep water during winter, but inshore during summer. Mainly in the south-west and west; absent from the southern North Sea.

17 ZOARCIDAE

Elongate to eel-shaped fishes. Scales cycloid or absent. Dorsal and anal fins long, merging with caudal. Pelvic fins jugular or absent. Jaws toothed. No swimbladder.

Zoarces viviparus (Linnaeus). FIG. 17.8
Eel Pout, Viviparous Blenny [WHITEHEAD et al. 1986*b*: 1149]

Body long. D long, with a depressed area near the tail where short weak spines replace soft rays; A long, both it and D continuous with caudal. Pectoral fins well-developed, pelvics jugular. Br. vi; D 72–85/V–XI/16–24; A 80–95; C 10–13; P 16–21; V 3; vert. 108–118. Brownish above, paler below; a row of dark spots along the back extending on to the dorsal fin, and irregular, somewhat coalesced blotches on the flank. To 35 cm.

Intertidal (though moving offshore during summer) to 10 m, and also in brackish waters; mainly along the northern east coast.

Superorder **Atherinomorpha**

Order **Atheriniformes**

18 BELONIDAE

Body elongate, sagittiform, covered with small cycloid scales. Jaws greatly produced, forming a beak. Pelvic fins abdominal.

Belone belone (Linnaeus). Garfish FIG. 17.8
[WHITEHEAD et al. 1986*a*: 605]

Body greatly elongate; lower jaw somewhat longer than the upper (markedly so in juveniles); jaws toothed. D and A posteriorly placed, neither with finlets. Lateral line low on flank. D II/15–19; A II/18–21; P I/11–12; V 6; vert. 80–84. Back dark blue-green, flanks silvery. To 75 cm.

Pelagic and oceanic, coming to coastal waters (and into estuaries) during summer; off all coasts; quite common. Spawns inshore and the distinctive large eggs (3.0–3.5 mm in diameter) provided with 60–80 long filaments may be found attached to coastal algae during early summer.

19 ATHERINIDAE

Slender, fusiform fishes, with large cycloid scales on the body. Jaws oblique. Two dorsal fins, the first spinous. Pectorals high on flank, anterior to pelvics.

Atherina presbyter Cuvier. Sand Smelt FIG. 17.8
[WHITEHEAD et al. 1986*b*: 1209]

Small, slender fish with two dorsal fins and an emarginate tail. 27–30 rakers on the first gill arch. 1D VII-VIII; 2D II/10–13; A II/13–16; C 17; P 13–15; V I/5; lat. sc. 57–62; vert. 50–52. Green above; a broad intense silvery band on the flank, white below. To 15 cm.

Inshore, including estuaries, in shoals; south and west coasts. Eggs less than 2.0 mm in diameter, with long filaments; attached to coastal algae during summer.

The very local *A. boyeri* Risso is similar but has 21–26 gill rakers, 50–58 lateral line scales and 43–47 vertebrae. Bamber, K. N. and Henderson, P. A. (1985), *Biol. J. Linn. Soc.*, **25**, 61–76, have suggested that all British specimens of *Atherina* are referable to a single, variable species, which should be known as *A. boyeri*.

Superorder **Acanthopterygii**

Order **Zeiformes**

20 ZEIDAE

Deep, compressed body; highly extensive jaws; large eyes; spiny rays in dorsal and anal fins.

Zeus faber Linnaeus. John Dory FIG. 17.8
[WHITEHEAD et al. 1986*a*: 772]

Very deep, strongly compressed body. Jaws large, protrusible. Dorsal fin spines long; interspinal membrane projecting, filamentous. Anal fin spines separated from the rays. Pelvic fins long. Br. vii; D IX–X/22–25; A III–IV/20–23; P 13: V I/6; vert. 32–33. Light brown, paler below; a large black spot, narrowly bordered with yellow, on the flank. To 66 cm.

Inshore and in shallow water; south and west coasts; frequent.

Order **Gasterosteiformes**

21 GASTEROSTEIDAE

Spiny finned fishes with elongate bodies. Dorsal fins preceded by a series of isolated spines, the rayed portion in shape and position symmetrical with the anal. Pectoral fins high on the flank, pelvics represented mainly by a single spine behind the pectorals. Scales absent, but flanks bearing bony scutes. Br. iii; V I/1. Males construct nests and guard the eggs.

---

1. Body long and thin; 15 isolated spines in front of the second dorsal fin; marine ***Spinachia spinachia***
   Body fusiform; three spines in front of the second dorsal fin; generally in brackish water ***Gasterosteus aculeatus***

Gasterosteus aculeatus Linnaeus. FIG. 17.8
Three-spined stickleback [WHITEHEAD et al. 1986*a*: 640]

Body hardly elongate, but with slender caudal peduncle. 2 D preceded by 3 spines, 2 long and 1 short. D III/8–14; A I/6–11; P 10–11; vert. 29–33. Back green or dark, flanks silver, belly white (underside of head and belly red in males during spring). To 10 cm.

In fresh and brackish water, occasionally on the shore; throughout and all round the British Isles.

The Ten-spined Stickleback *Pungitius pungitius* (Linnaeus), with 8–10 isolated spines preceding 2D, is essentially a freshwater fish, although occasionally found in brackish water.

Spinachia spinachia (Linnaeus). FIG. 17.8
Fifteen-spined Stickleback [WHITEHEAD et al. 1986*a*: 643]

Body elongate, the caudal peduncle long and slender. Dorsal fin preceded by 14–16 isolated spines. D XIV—XVI/6–7; A I/6–7; P 9–10; vert. 40–41. Brown or olive above, sometimes with dark bars on the flanks; paler below. To 15 cm.

Marine, but extending into brackish water; found inshore amongst seaweed and *Zostera*; all coasts, but uncommon in the south-east.

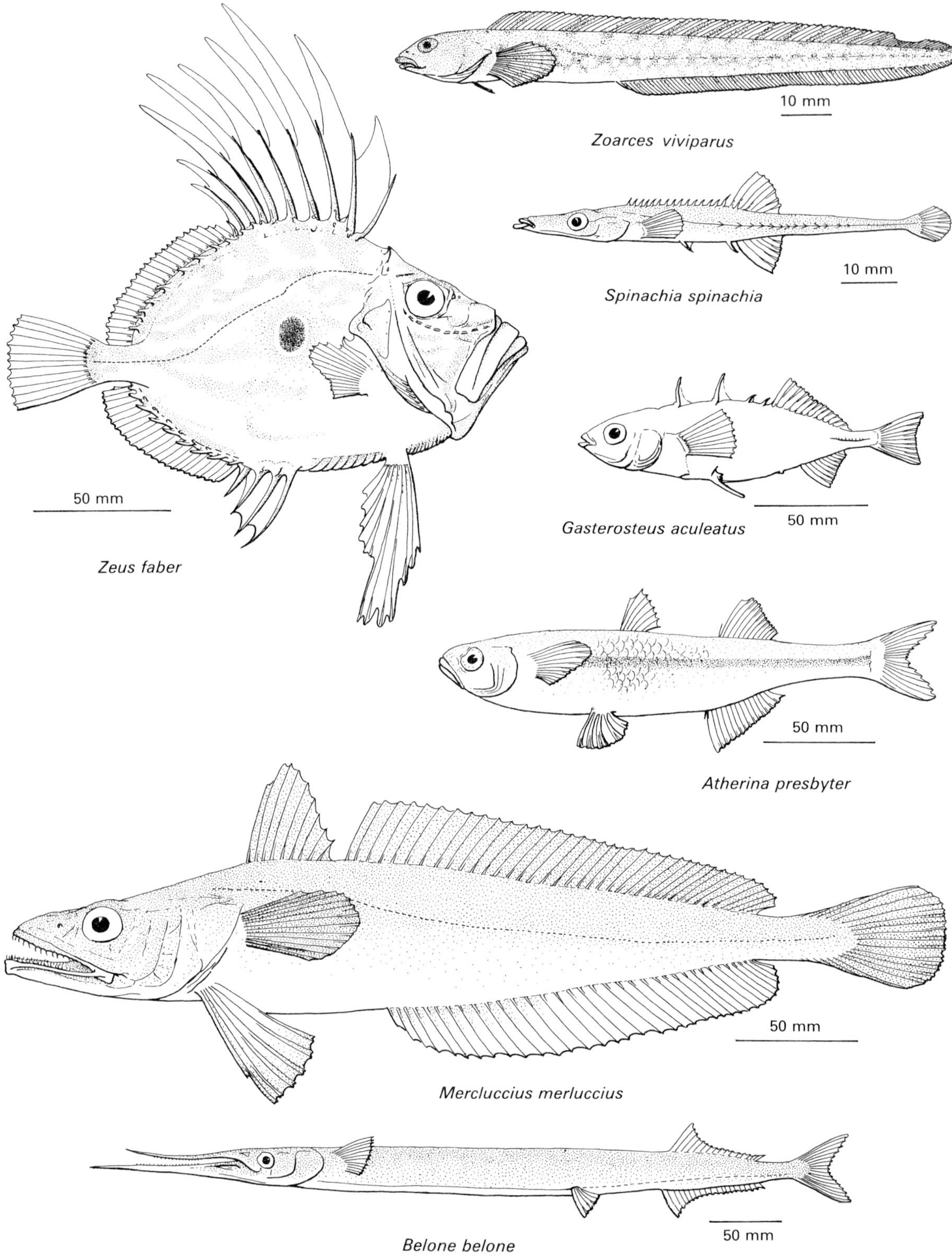
10 mm
Zoarces viviparus
10 mm
Spinachia spinachia
50 mm
Zeus faber
Gasterosteus aculeatus
50 mm
50 mm
Atherina presbyter
50 mm
Mercluccius merluccius
Belone belone
50 mm

Fig. 17.8

Order Syngnathiformes

22 SYNGNATHIDAE

Elongate fishes, long and thin (pipefishes) or with compressed, somewhat elevated trunk and prehensile tail (sea horses). Mouth small, at the end of a slender snout. Body covered by bony rings ('cristae'). Pelvic fins absent. Swim mainly by undulations of the dorsal fin. Eggs incubated by the males, adhering to the abdomen or within a caudal brood pouch. Live amongst weed.

---

1. Caudal fin absent or rudimentary; pectorals absent (except in young stages) **2**

   Caudal fin well-developed; pectorals present in all stages **(*Syngnathus*) 4**

2. Anus situated below the posterior half of the dorsal fin; a rudimentary caudal fin may be present ***Entelurus aequoreus***

   Anus situated below the anterior half of the dorsal fin; caudal fin absent **(*Nerophis*) 3**

3. Dorsal fin with 24–26 (–28) rays. (A common species under rocks on the shore) ***N. lumbriciformis***

   Dorsal fin with 34–40 rays ***N. ophidion***

4. Snout laterally compressed, nearly as deep as the postorbital portion of the head ***S. typhle***

   Snout rounded, less deep than the postorbital portion of the head **5**

5. 13–16 pre-anal body rings, 37–42 postanal; snout (tip of snout to anterior margin of orbit) about the same length as the rest of the head (anterior margin of orbit to posterior edge of operculum). (A common fish of sandy bays, not exceeding 17 cm in length) ***S. rostellatus***

   17–21 pre-anal body rings, 43–46 postanal; snout one and one third to two times as long as the rest of the head. (Usually a sublittoral fish growing to 45 cm in length) ***S. acus***

Entelurus aequoreus (Linnaeus). Snake Pipefish FIG. 17.9 (WHITEHEAD et al. 1986*a*: 629]

Body smooth. Snout long, greater in length than the rest of the head. 28–31 pre-anal rings, 60–69 postanal. D situated almost wholly in front of the vent. Anal and pectoral fins absent, caudal rudimentary. D 37—47; C 4–9. Eggs carried on the abdomen, not enclosed in a pouch. Yellowish brown, vertically banded. To 60 cm.

Oceanic; sometimes in coastal waters; off all coasts.

Nerophis lumbriciformis (Jenyns). FIG. 17.9
Worm Pipefish [WHITEHEAD et al. 1986*a*: 632]

Body smooth. Snout short, upturned. Vent in front of length centre point. 17–18 pre-anal rings, 46–54 postanal. D placed over the vent; anal, caudal, and pectoral fins absent. D 24–26. Eggs carried in a shallow abdominal groove. Very dark brown in colour, with white markings on the head and throat. To 15 cm.

Intertidal; found under boulders or amongst weed on rocky shores; all coasts; common where the habitat is suitable.

Nerophis ophidion Linnaeus. FIG. 17.9
Straight-nosed Pipefish [WHITEHEAD et al. 1986*a*: 633]

Body smooth. Snout about equal in length to the rest of the head (and longer than in *N. lumbriciformis*), straight above. Vent at about the length centre point. 28–32 pre-anal rings, 68–82 postanal. D placed over vent; anal, caudal, and pectoral fins absent. D 33–34. Eggs carried in a shallow abdominal groove. Green, darker dorsally, sometimes with pale blue markings on the head and underside. To 30 cm.

Coastal; amongst seaweed or *Zostera*; widespread but local.

Syngnathus acus Linnaeus. Greater Pipefish FIG. 17.9 [WHITEHEAD et al. 1986*a*: 635]

Body rings prominent. Snout long, greater in length than the rest of the head; round in section. 17–22 pre-anal rings; 39–43 postanal. D 36–45; A 3; C 10; P 11–13. Eggs carried in a postanal brood pouch. Brown above, paler below, with dark markings or bands. To 46 cm.

Costal; over hard or soft bottoms, amongst seaweed or *Zostera*; all coasts; fairly common.

Syngnathus rostellatus Nilsson. Lesser Pipefish FIG. 17.9 [WHITEHEAD et al. 1986*a*: 636]

Body rings prominent. Snout about the same length as the rest of the head; round in section. 13–17 pre-anal rings, 37–42 postanal. D 36–45; A 2–3; C 10; P 10–13. Eggs carried in a postanal brood pouch. Brown above, paler below. To 17 cm.

Inshore; a common fish of sandy bays in southern Britain and Ireland. Distribution in Scotland unknown.

Syngnathus typhle Linnaeus FIG. 17.9
[Syn. *Siphonostoma typhle*]. Deep-nosed Pipefish [WHITEHEAD et al. 1986*a*: 638]

Body rings distinct. Snout long, greater in length than the rest of the head, deep and compressed. 16–18 pre-anal rings, 33–39 postanal. D 28–41; A 2–4; C 9–11; P 12–16. Eggs carried in a postanal brood pouch. Brown or green, the head region marked with fine brown spots. To 30 cm.

Coastal; amongst seaweed or *Zostera*; widespread but local.

Order Scorpaeniformes

23 TRIGLIDAE

Body more or less fusiform, covered with scales. Head covered with rough bony plates; the preorbitals (lachrymals) extending to the tip of the snout. Spines may be present on the shoulder bone (post-temporal), on the preoperculum and operculum, and on the pectoral girdle (scapula or coracoid). Small teeth present on the jaws. Two dorsal fins, the first spinous. Pectoral fins large, the three lowest rays separated as slender tactile processes. Pelvic fins below pectorals. Br. vii.

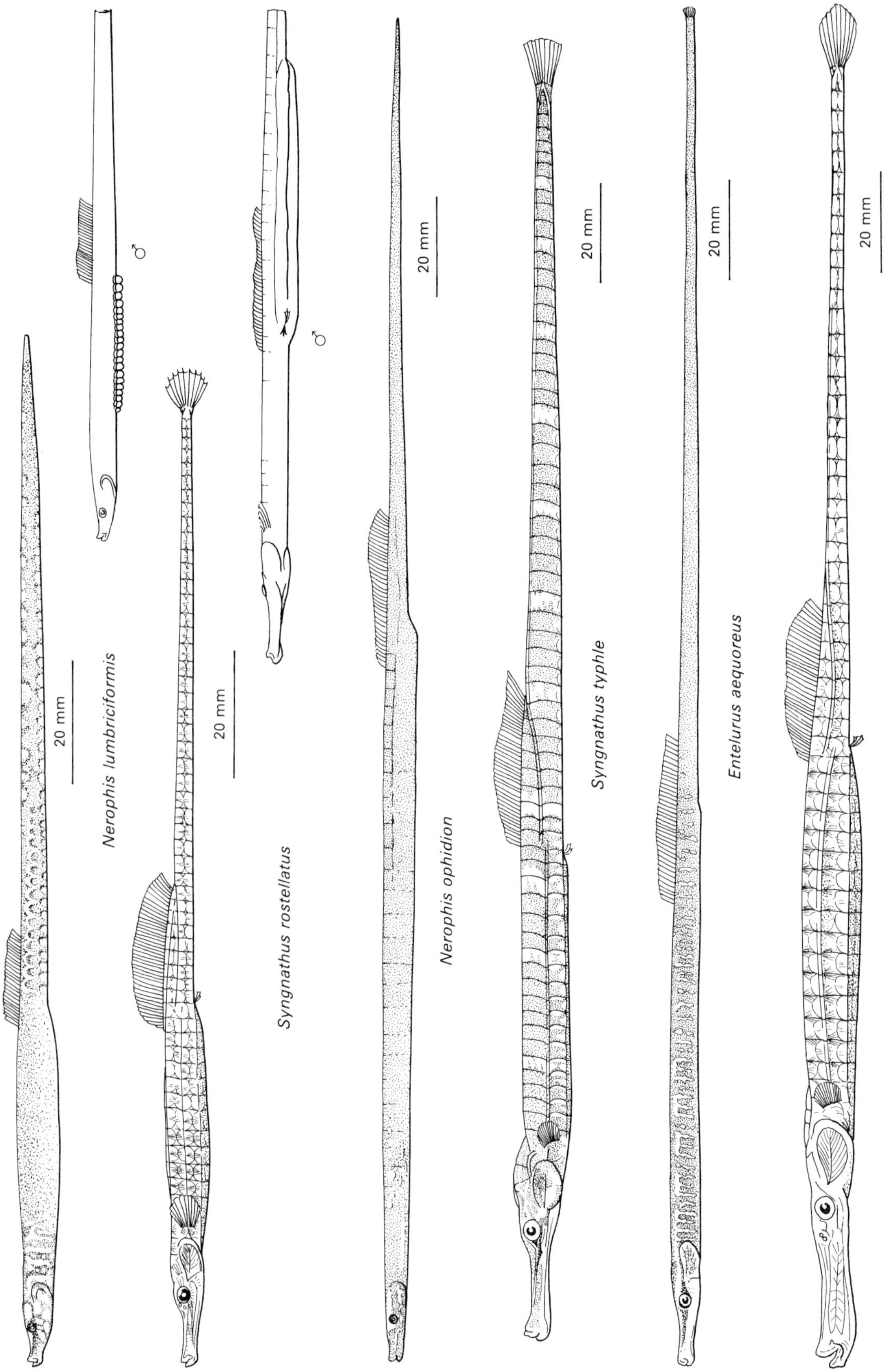

Nerophis lumbriciformis
♂
20 mm
Syngnathus rostellatus
20 mm
♂
Nerophis ophidion
20 mm
Syngnathus typhle
20 mm
Entelurus aequoreus
20 mm
Syngnathus acus
20 mm

Fig. 17.9

1. Pectoral fins not reaching the vent; anal fin with 18–20 rays; lateral line spiny **Eutrigla gurnardus**

   Pectoral fins reaching or extending beyond the vent; anal fin with 14–17 rays; lateral line usually not spiny **2**

2. Lateral line covered by soft, narrow, transversely elongated, smooth-edged scutes **Aspitrigla cuculus**

   Lateral line not covered by transversely elongated scutes **3**

3. Snout flattened and anteriorly produced; lateral line scales the same size as the remainder, not spiny; pectoral fins coloured intense blue on the hinder surface **Trigla lucerna**

   Snout not flattened, not anteriorly produced, lateral line scales somewhat larger than the remainder, weakly spinous posteriorly; body transversely ridged **Trigloporus lastoviza**

Aspitrigla cuculus (Linnaeus). Red Gurnard FIG. 17.10
[WHITEHEAD et al. 1986*b*: 1232]
Head one-fifth to one-quarter of the total length; snout length equalling one and a quarter to one and a half eye diameters, its upper profile slightly concave, descending rather abruptly, preorbital with acute angle and short pointed teeth anteriorly. Post-temporal roughened along its upper edge, ending in a short spine; angle of preoperculum shortly spinous, with two small projections below the spine; operculum with a well-developed spine; and a spine on the coracoid. Pectoral fin about as long as the head, reaching the anal fin. Lateral line scales unarmed but forming deep, linear plates, looking like short folds of skin. D IX–X/17–18; A 16–17; C 13; P 10+iii; V I/5; lat. sc. 70–73; vert. 37; pyloric caeca 8. Rose-coloured, paler below. To 30 cm.

In shallow water; off all coasts, but uncommon in the North Sea.

Eutrigla gurnardus (Linnaeus). Grey Gurnard FIG. 17.10
[WHITEHEAD et al. 1986*b*: 1233]
Head one-quarter to one-third of the total length; snout length equalling about two eye diameters, its upper profile nearly straight; preorbital with acutely pointed angle, toothed anteriorly. Post-temporal spine shorter than opercular spine; angle of preoperculum with 2–3 teeth; operculum with a large spine and a short prominence above it; a strong spine on the coracoid. Pectoral fin much shorter than the head, not reaching the anal fin. Lateral line scales shortly spinous. D VII–IX/18–19; A 17–19; C 15; P 10+iii; V I/5; lat. sc. 72–77; vert. 38; pyloric caeca 7–9. Grey, brownish or (rarely) dull red, whitish below; back and flank often white spotted; freqently a large black blotch on 1D. To 45 cm.

Inshore and shallow waters, off all coasts, common.

Trigla lucerna Linnaeus. FIG. 17.10
Tub, Sapphirine Gurnard [WHTEHEAD et al. 1986*b*: 1235]
Head about one-quarter of the total length; snout length equalling about two eye diameters, its upper profile slightly concave; preorbital with acutely pointed angle, toothed anteriorly. Post-temporal roughened, ending in a short, sharp spine; angle of preoperculum acute; operculum with a spine in juveniles; a ridge and a spine on the coracoid. Pectoral fin as long or longer than the head, overlapping the anal fin. Lateral line smooth. D IX–X/15–17; A 14–16; C 12; P 10+iii; V I/5; lat. sc. 70; vert. 34; pyloric caeca 10. Brownish red above, white below; pectoral fins reddish externally with bluish bands and/or margins, bright or dark blue internally with light spots or darker bands. To 60 cm.

Inshore and into moderate depths; common around the southern half of the British Isles, rare off northern coasts.

Trigloporus lastoviza (Brünnich) FIG. 17.10
[Syn. *T. lineata*]. Streaked Gurnard
[WHITEHEAD et al. 1986*b*: 1237]
Head one-quarter to one-fifth of the total length; snout length equalling one to one and a half eye diameters, its upper profile straight, descending abruptly, in the adult almost vertically; preorbital with obtuse angle, rough anteriorly. Post-temporal roughened along its upper edge, ending in a strong spine; angle of preoperculum blunt; operculum shortly spined; a strong spine on the scapula. Pectoral fin as long or longer than the head, reaching the anal fin. Lateral line scales weakly spined, with vertical ridges spreading from them dorsally and ventrally. D IX–XI/16–17; A 14–16; C 12; P 10–11+iii; V I/5; lat. sc. 66–71; vert. 35; pyloric caeca 10. Red above, white below; pectorals dark, with deep blue spots almost forming transverse bands. To 36 cm.

In shallowish waters, mainly off the south and south-west (less often west) coasts.

24 COTTIDAE

Body fusiform, the head broad and depressed; scaleless, but with covering bony or scale-like plates; spines on the head. Two dorsal fins, the first spinous; anal with spiny rays. No swimbladder. Males often with long, pointed urinogenital papilla.

1. Branchiostegal membranes from each side joining ventrally, forming a continuous free posterior margin; lateral line without bony plates **Myoxocephalus scorpius**

   Branchiostegal membranes from each side not joining ventrally, but broadly separated; lateral line marked by a row of small spinous scutes **Taurulus bubalis**

Myoxocephalus scorpius (Linnaeus). FIG. 17.11
Bull Rout, Father Lasher [WHITEHEAD et al. 1986*b*: 1254]
Head large, about one-third of total length, wide, depressed, covered with soft skin. Preoperculum with two posterior spines, the upper longer but not exceeding the diameter of the eye;

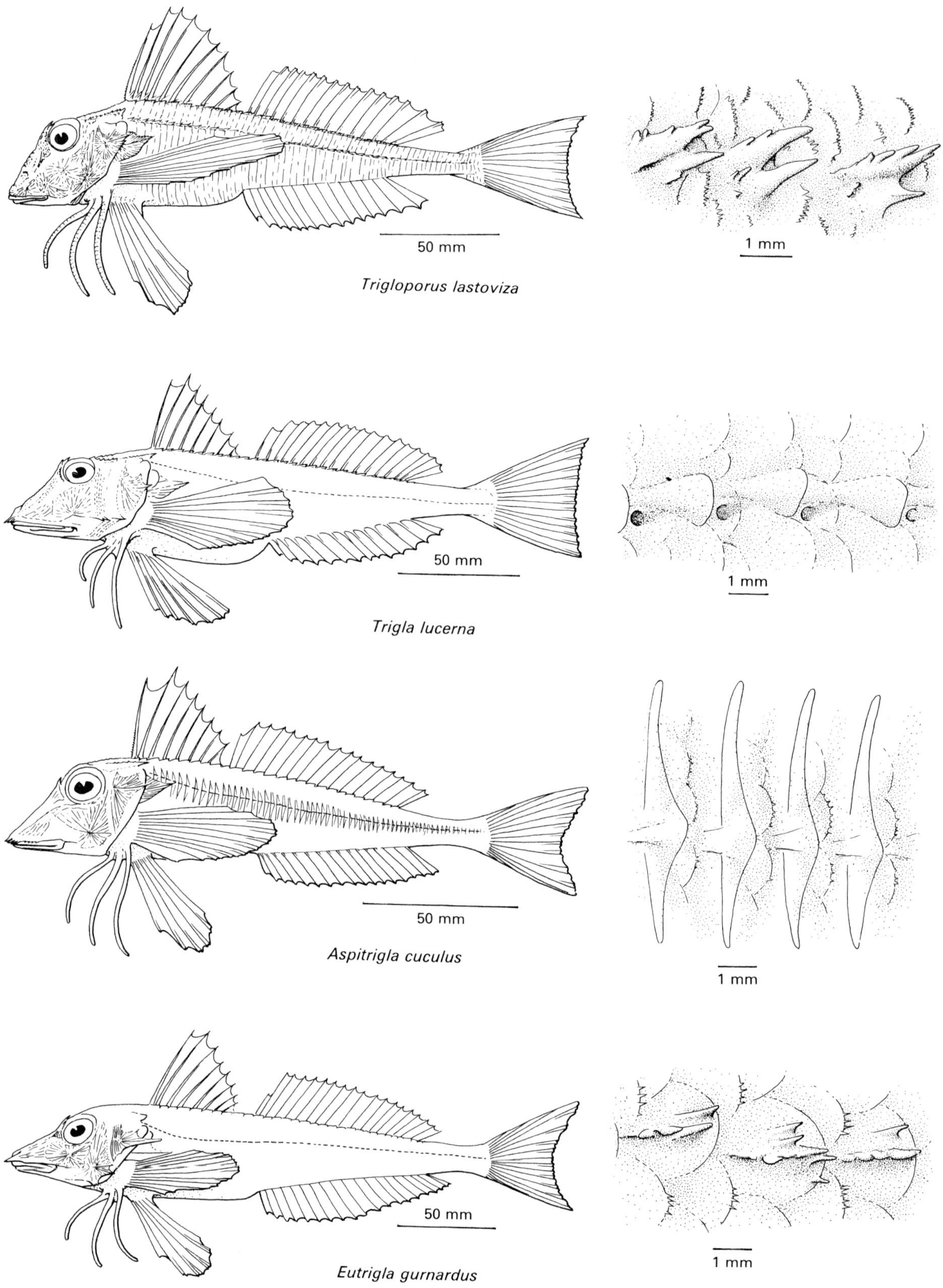
50 mm
1 mm
Trigloporus lastoviza
50 mm
1 mm
Trigla lucerna
50 mm
1 mm
Aspitrigla cuculus
50 mm
1 mm
Eutrigla gurnardus

Fig. 17.10

one small anterior spine. One spine each on post-temporal, operculum, suboperculum, and scapula. Skin smooth; lateral line unarmoured, but usually with small spiny plates (best developed in males) above and below it. Branchiostegal membranes joined ventrally across the 'isthmus', so that a single flap crosses the entire ventral surface. Male without urinogenital papilla. Br. vi; 1D VII–IX; 2D 12–17; A 10–14; P 14–18; V I/3; vert. 34–35. Light brown with dark bands and blotches; belly pale. To 30 cm (males smaller).

Inshore and in shallow water; off all coasts; not (at least in southern Britain) found intertidally.

Taurulus bubalis (Euphrasen). Sea Scorpion FIG. 17.11
[WHITEHEAD et al. 1986*b*: 1256]

Head large, about one-third of total length, wide, depressed, covered with soft skin. Preoperculum with four spines, the uppermost the longest, its length exceeding the diameter of the eye and almost reaching to the end of the opercular spine; the lowest directed downwards and forwards. Spines also at the back of the head and on the post-temporal; one on the operculum and two on the suboperculum, the lower directed downwards and forwards, and one on the scapula. Skin smooth; lateral line armed with spiny plates. Branchiostegal membranes from each side not joined ventrally across the 'isthmus'. Male with urinogenital papilla. Br. vi; 1D VII–IX; 2D 10–13; A 8–10; P 14–16; V I/3; lat. sc. 32–35; vert. 29–30. Light brown with dark bands and blotches; pale below. To about 15 cm.

Intertidal on rocky shores and inshore on hard grounds; all coasts; common.

The very similar *Micrenophrys lilljeborgi* Collett [Whitehead et al. 1986*b*: 1252], with V I/2, spinelets above the lateral line, and the longest preopercular spine not exceeding the eye diameter, has occasionally been taken offshore.

25 AGONIDAE

Body covered with bony plates, usually in eight rows. Two dorsal fins; finrays unbranched in all fins; no spines in the anal.

Agonus cataphractus (Linnaeus). Pogge FIG. 17.11
[WHITEHEAD et al. 1986*b*: 1265]

Head wide, depressed, wider than long, two pairs of massive spines on the snout; mouth inferior. Two barbels below the snout, and a divided barbel at each corner of the mouth; entire underside of head and branchiostegal membranes covered with barbels. A strong spine on the preoperculum; bony scutes spineless. Br. vi; 1D V–VI; 2D 6–8; A 6–7; C 12; P 16–17; V I/2; dorsal scutes 31–34; vert. 36. Dull brown above, white below. To 15 cm.

Inshore and in shallow water; all coasts; common. English Channel represents southern limit of range.

26 CYCLOPTERIDAE

Body short and stout, covered with thick skin armed with bony tubercles. 1D reduced, except in young fish. Pelvic fins transformed into a large sucking disk.

Cyclopterus lumpus Linnaeus. Lumpsucker FIG. 17.11
[WHITEHEAD et al. 1986*b*: 1271]

Body deep, thick, compressed. A crest of tubercles, with age incorporating 1D, passing from rear of the head to the (second) dorsal fin, there dividing to each side; three rows of tubercles along each flank. No lateral line. Br. vi; 1D VI–VIII; 2D I/10; A I/9–10; C 10–11; P 20–21; V I/5; vert. 26–29. Grey or bluish above, reddish (breeding males) or whitish below. To 60 cm.

Mainly bottom living, but also caught in pelagic nets; from the lower shore and shallow water; off all coasts; English Channel represents almost the southern limit of range.

27 LIPARIDAE

Head and trunk thick, tail compressed. Skin loose, with minute spinules; no scales or bony tubercles. Single long dorsal fin; anal long. Pelvic fins transformed into a sucking disk. Pectoral fins extending ventrally.

---

1. Anal and dorsal fins continuous with the base of the caudal fin; last anal finrays overlapping the base of the caudal fin; offshore **_Liparis liparis_**
   Anal and dorsal fins not continuous with the caudal fin; anal and dorsal finrays not overlapping the base of the caudal fin; on the shore and in shallow water **_L. montagui_**

Liparis liparis (Linnaeus). Sea Snail FIG. 17.11
[WHITEHEAD et al. 1986*b*: 1278]

Slimy and gelatinous, tadpole shaped. D low anteriorly, its membrane continuous with the caudal fin; A similarly continuous with caudal. Posterior nostrils well-developed. Br. vi; D 33–35; A 27–29; C 12–14; P 28; V I/5; vert. 38–40. Brownish, very variable. To 12 cm.

Found offshore; rare in the south and south-west. Apparently has been confused with the following species.

Liparis montagui (Donovan). FIG. 17.11
Montagu's Sea Snail [WHITEHEAD et al. 1986*b*: 1279]

Slimy and gelatinous, tadpole shaped. D low anteriorly, not reaching the caudal; A reaching the caudal, but its membrane not continuous with it. Posterior nostrils in adult covered by skin. Br. vi; D 28–30; A 22–25; C 14; P 27–30; V I/5; vert. 33–36. Brownish, very variable. To 6 cm.

Littoral and inshore, on rocky grounds; associated with algae, and may occur in *Saccorhiza* holdfasts. All coasts, except in the south-east.

Order **Perciformes**

28 MORONIDAE

Body fusiform, covered with small ctenoid scales which extend on to the head. Two dorsal fins, or the spinous fin continuous with the second; anal fin with three spines. Pelvic fins thoracic, with one spine and five rays. Mouth protrusible.

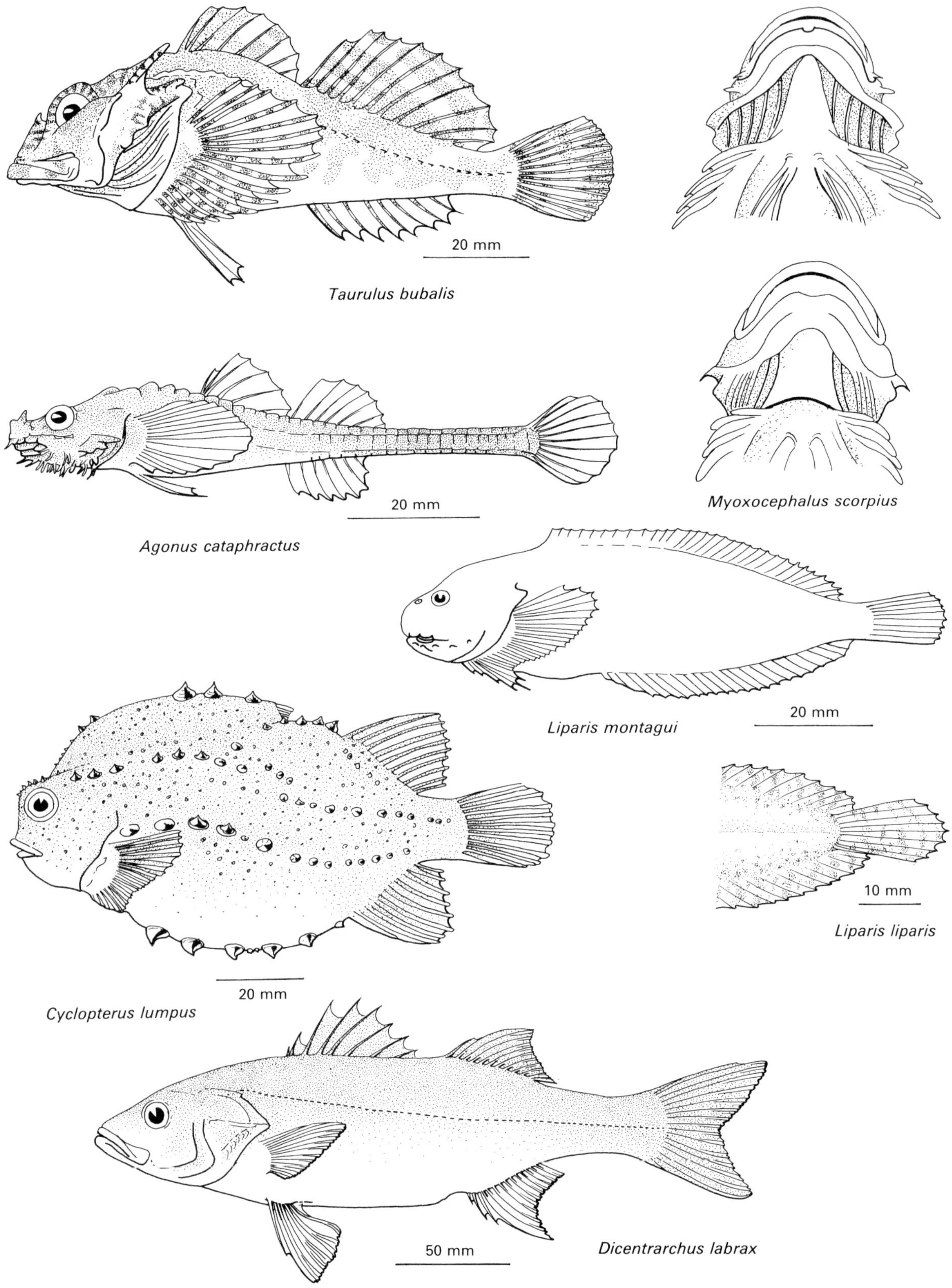
20 mm
Taurulus bubalis
20 mm
Agonus cataphractus
Myoxocephalus scorpius
20 mm
Liparis montagui
20 mm
Cyclopterus lumpus
10 mm
Liparis liparis
50 mm
Dicentrarchus labrax

Fig. 17.11

Dicentrarchus labrax (Linnaeus) FIG. 17.11
[Syn. *Morone labrax*]. Bass [WHITEHEAD et al. 1986*a*: 794]
Spiny- and soft-rayed dorsal fins separate; caudal emarginate. Teeth on the jaws. Posterior edge of preoperculum serrated, the teeth forward-pointing; two weak spines on the operculum. Br. vii; 1D VIII–IX; 2D I/12–13; A III/10–12; P 15–16; V I/5; lat. sc. 66–74; vert. 25–26. Grey or blue above, flanks silvery, belly white. To 1 m.

Inshore during the summer months, common off southern and western coasts, uncommon further north; migrating offshore during winter.

29 CARANGIDAE

Body fusiform, compressed. Two dorsal fins; the first spinous, often reduced to a series of short spines; anal fin preceded by two detached spines (neither dorsal nor anal fins succeeded by finlets); caudal fin forked. Pelvics placed below pectorals. Mouth protrusible. 24 vertebrae.

Trachurus trachurus (Linnaeus). FIG. 17.12
Scad, Horse Mackerel [WHITEHEAD et al. 1986*a*: 842]
Head large, about one-third of the total length. Two dorsal fins. Lateral line with a sharp curve below the origin of 2D, covered along its entire length with tall, slender scutes. A sensory canal passes from the back of the head, beside the fin base, to the rear of 2D. Br. vii; 1D VIII; 2D I/23–34; A II/25–34; P 19–21; V I/5; lat. sc. 70–80. Greenish blue above, flanks silvery; a black spot on the edge of the operculum. To 40 cm.

Pelagic; in large shoals, coming inshore during the summer; off all coasts.

30 MULLIDAE

Body fusiform with large caducous, faintly ctenoid scales on the body and head. Two well-separated dorsal fins, the first spinous. Pelvic fins below pectorals. Two long barbels on the lower jaw.

Mullus surmuletus. Linnaeus. Red Mullet FIG. 17.12
[WHITEHEAD et al. 1986*a*: 879]
Head about one-quarter of the total length; two very long barbels on the lower jaw, their length exceeding that of the pectoral fins. Preoperculum bearing a group of scales, the upper, suborbital row comprising two large scales only; operculum bearing two large scales. First spine of 1D very short. Br. iv; 1D VII–VIII; 2D 9; A I/6–7; P 16–18; V I/5; lat. sc. 35–40; vert. 24. Red, with yellowish stripes on the side; 1D crossed by a dark stripe. To 40 cm.

Coastal and offshore in the English Channel and to the south-west of the British Isles. Confusion has existed with *M. barbatus* Linnaeus [Whitehead et al. 1986*a*: 878], which is a more southern species. It has a very steep head profile, barbels not exceeding the length of the pectoral fins, and a third, small, anteriorly placed suborbital scale.

31 MUGILIDAE

Grey mullets. Body fusiform, covered with cycloid or faintly ctenoid scales which extend on to the head. Mouth terminal, broad; teeth minute or absent. Two widely separated dorsal fins, the first short, consisting of four strong spines. Pelvic fins abdominal (though the girdle is connected by a ligament to the pectoral girdle), but anterior to 1D. Lateral line absent. Stomach divided into two parts. The three British species are morphologically very alike. Trewavas and Ingham (1972) and Anderson (1982) have published keys. Small fish have been regarded as particularly difficult to identify but Reay and Cornell (1988: *J. Fish Biol.*, **32**, 95–99) have drawn attention to diagnostic melanophore patterns on the lower jaw of juveniles in the 25–50 mm length range. *Chelon labrosus* has large or dense melanophores anterior to the eye on the mandibular, ventro-opercular, and gular regions; in *Liza aurata* they are absent from the gular; in *L. ramada* they are present on the gular but absent from the ventro-opercular. Moreover, *C. labrosus* apparently spawns in spring and the *Liza* species in autumn.

---

1. Upper lip broad (its greatest depth equal to or exceeding half the eye diameter), bearing one to four rows of papillae in fish of standard length >90 mm. Six pyloric caeca ***Chelon labrosus***

   Upper lip thin (its greatest depth less than half the eye diameter), smooth. More than six pyloric caeca **2**

2. Posterior end of lachrymal bone tapered; pectoral fin 17–23% in standard length, reaching the orbit when turned forward; a well-defined golden spot on the operculum, no marked black spot at the pectoral axil; eight pyloric caeca, the dorsal ones longest ***Liza aurata***

   Posterior end of lachrymal bone truncate; pectoral fin less than 20% in standard length (not reaching the orbit when turned forward); a vague, diffuse golden spot on the operculum, and a black spot at the pectoral axil; seven pyloric caeca of uniform length ***Liza ramada***

Chelon labrosus (Risso). FIG. 17.12
Thick-lipped Grey Mullet
[WHITEHEAD et al. 1986*b*: 1198]
Upper lip very thick, its greatest depth equal to or exceeding half the eye diameter, bearing three to five rows of small papillae (which increase in number and size as the fish grows); a line of fine teeth also present. Lachrymal (preorbital) bone coming to an obtuse point posteriorly, its slightly convex lower margin finely serrated; end of maxilla hardly obscured by lachrymal; jugular space scarcely visible. Pectoral fin long, its length exceeding the distance from its base to the rear of the orbit; no long scale in the axil. Br. vi, 1D IV; 2D 9–10; A III/8–9; P 17–18; V I/5; lat. sc. 45–46; vert. 24; pyloric caeca 6. Dark grey above, flanks with longitudinal grey stripes, silvery below. To 66 cm.

The commonest grey mullet in the British Isles; all coasts;

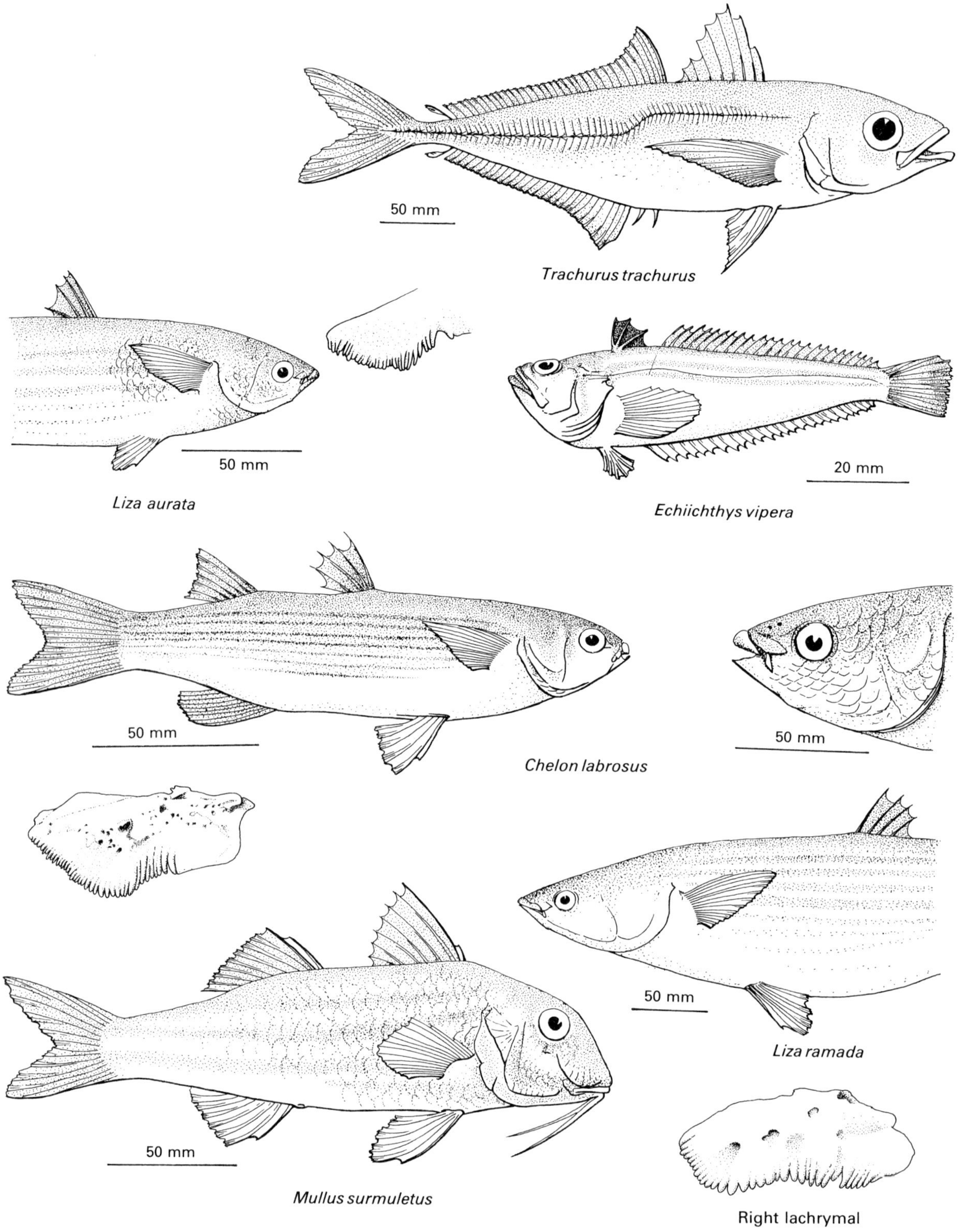
50 mm
Trachurus trachurus
50 mm
Liza aurata
20 mm
Echiichthys vipera
50 mm
Chelon labrosus
50 mm
50 mm
Liza ramada
50 mm
Mullus surmuletus
Right lachrymal

Fig. 17.12

probably offshore during winter, coming into coastal and estuarine waters during spring and summer; forms shoals.

Liza aurata (Risso). Golden Grey Mullet FIG. 17.12
[WHITEHEAD et al. 1986*b*: 1199]
Depth of upper lip not exceeding one-half of the eye diameter; without papillae. Labial teeth large. Lachrymal (preorbital) bone coming to an obtuse point posteriorly, serrated on its convex lower margin; maxilla covered by lachrymal; jugular space oval. Pectoral fin long, its length exceeding the distance from its base to the rear of the orbit; no long axillary scale. Br. vi; 1D IV; 2D III/7–9; A III/9; P 17; V I/5; lat. sc. 40–46; vert. 25; pyloric caeca 8. Blue-grey above, flanks somewhat yellowish with longitudinal grey stripes; golden spots on the preoperculum and operculum. To 44 cm.

Possibly on all coasts; uncommon or overlooked on account of similarity to other grey mullets; perhaps not entering brackish water.

Liza ramada (Risso). FIG. 17.12
Thin-lipped Grey Mullet [WHITEHEAD et al. 1986*b*: 1200]
Depth of upper lip not exceeding one-third of the eye diameter; without papillae; a line of fine teeth present. Lachrymal (preorbital) bone rounded posteriorly, with well-developed teeth around the angle and on the lower margin; only the end of the maxilla visible below the lachrymal; jugular space oval. Pectoral fin shorter than in the other species, its length less (or equalling in the young) than the distance from its base to the rear of the orbit; no long scale in the axil. Br. vi; 1D IV; 2D 8–9; A III/8–9; P 17; V I/5; lat. sc. 42–46; vert. 24; pyloric caeca 7. Dark grey above, flanks with longitudinal grey stripes, silvery below. To 50 cm.

May occasionally be found in the same habitats as *C. labrosus* on south and west coasts.

32 LABRIDAE

Wrasses. Body compressed, covered with cycloid scales which sometimes extend on to the head; the number of scale rows is important. Dorsal fin single, with a long, spinous anterior portion; anal with three or more spines. Lips thick, with powerful teeth; no palatal teeth. Hypopharyngeal bones fused, bearing large grinding teeth. No pyloric caeca. Closed swimbladder. Often brightly coloured.

---

1. Anal fin with five spiny rays — ***Centrolabrus exoletus***
   Anal fin with three to four spiny rays — **2**

2. 42–47 scales along the lateral line, and at least six rows of scales above it — **(*Labrus*) 3**
   32–38 scales along the lateral line, not more than four rows of scales above it — **4**

3. Dorsal fin with 19–20 spiny rays and nine to 11 soft rays; pectorals with 15 rays; mouth small — ***L. bergylta***
   Dorsal fin with 16–18 spiny rays and 11–14 soft rays; pectorals with 16–17 rays; mouth large — ***L. bimaculatus***

4. A dark spot at the base of the central rays of the caudal fin; 32–36 scales along the lateral line; teeth in a single row on each jaw — ***Crenilabrus melops***
   A dorsally situated dark spot on the caudal peduncle; 36–38 scales along the lateral line; teeth in several rows on each jaw — ***Ctenolabrus rupestris***

Centrolabrus exoletus (Linnaeus). Rock Cook FIG. 17.13
[WHITEHEAD et al. 1986*a*: 922]
Body oval-elongate, head profile rounded; mouth very small, the maxilla not reaching the level of the nostrils. Teeth in each jaw small, in a single row. Five spines in the anal fin. Preoperculum serrated along its lower margin and around the angle. Br. v; D XVIII–XX/5–7; A IV–VI/6–8; P 14; V I/5; lat. sc. 32–35; sc. rows 3–4/10–11; vert. 32–34. Greenish or brownish, paler below; caudal fin crossed by a broad, curved, dark band. To 15 cm.

Inshore over rocky grounds; south-west England and the west and north of the British Isles.

Crenilabrus melops (Linnaeus). Corkwing FIG. 17.13
[WHITEHEAD et al. 1986a: 936]
A small wrasse. Body oval; mouth small, the maxilla not reaching the level of the nostrils. Teeth in each jaw conical, in a single row. Hypopharyngeal bones with expanded outer angles; teeth merging into a single line on the forward process. Preoperculum serrated along its lower margin and around the angle. Br. v; D XV–XVIII/8–10; A III/8–10; P 14–16; V I/5; lat. sc. 32–36; sc. rows 4/12; vert. 32–34. Colour variable, bluish to yellowish-green, with an oval black spot below the lateral line on the caudal peduncle and a kidney-shaped mark behind the eye. To 15 cm.

Intertidal, in weedy shore pools, or inshore; all coasts, but commonest in the south and west.

A similar species, *Crenilabrus bailloni* Valenciennes [Whitehead et al. 1986*a*: 932], has been recorded from Jersey and the west of Ireland. *C. bailloni* differs from *C. melops* by lacking the crescentic dark mark behind the margin of the eye and, in life, by having a blue crescent at the base of the pectoral fin rays (A. Wheeler and P. Clark, 1984, *J. Mar. biol. Ass. U.K.*, **64,** 1–6).

Ctenolabrus rupestris (Linnaeus). Goldsinny FIG. 17.13
[WHITEHEAD et al. 1986*a*: 925]
Body elongate, head profile rounded; mouth small, the maxilla about reaching the level of the nostrils. A band of conical teeth in each jaw, those of the outer row being largest. Hypopharyngeal bones with rounded outer angles; teeth in three series on the forward process. Posterior margin of preoperculum evenly serrated. Br. v; D XVI–XVIII/8–10; A III–IV/7–8; P 14; V I/5; lat. sc. 36–39; sc. rows 3–4/14–15; vert. 32–35. Brownish red with a dark blotch on the commencement of the dorsal fin and another on the upper side of the caudal peduncle; paler below. To 18 cm.

Intertidal and inshore, amongst weed; all coasts, but

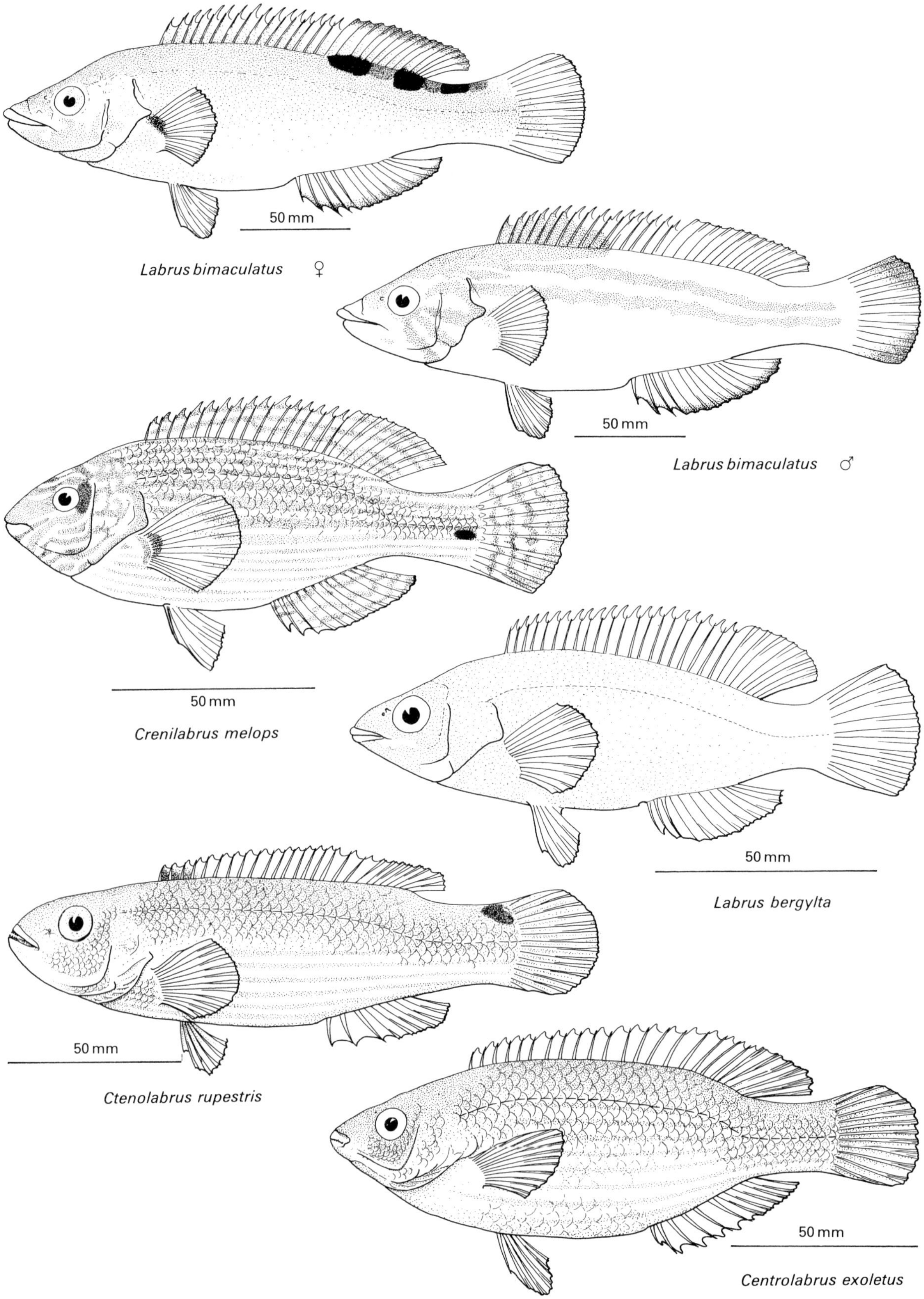
50 mm
Labrus bimaculatus ♀
50 mm
Labrus bimaculatus ♂
50 mm
Crenilabrus melops
50 mm
Labrus bergylta
50 mm
Ctenolabrus rupestris
50 mm
Centrolabrus exoletus

Fig. 17.13

uncommon on North Sea coasts south of the Moray Firth and in the eastern English channel.

Labrus bergylta Ascanius. Ballan Wrasse FIG. 17.13
[WHITEHEAD et al. 1986*a*: 926]
Body elongate, head profile rounded; mouth small, the maxilla scarcely reaching the level of the nostrils. Conspicuous teeth in a single row on each jaw, inside which is a second, scarcely visible row of minute teeth. Hypopharyngeal bones with rounded outer angles, teeth merging into one or two rows anteriorly. Margin of preoperculum entire (posterior edge serrated in young). Br. v; D XIX–XX/9–11; A III/8–10; P 14–15; V I/5; lat. sc. 41–47; sc. rows 6–7/12–15; vert. 37–38. Colour greenish or brownish, but very variable; each scale paler in the centre; flanks spotted or striped. To 50 cm.

Inshore, near rocks and over hard ground; common off southern and western coasts.

Labrus bimaculatus Linnaeus FIG. 17.13
[Syn. *L. mixtus* L.]. Cuckoo Wrasse [WHITEHEAD et al. 1986*a*: 927]
Body elongate, head profile tapering, snout elongated; mouth large, maxilla passing the level of the nostrils. Teeth in each jaw conical, in a single row. Hypopharyngeal bones with expanded outer angles; teeth merging into about three lines on the forward process. Posterior edge and angle of preoperculum usually smooth. Br. v; D XVI–XVIII/11–14; A III/9–11; P 17; V I/5; lat. sc. 50–60; sc. rows 5–7/17–21; vert. 38–39. Sexually dichromatic: males with blue heads and orange bodies boldly striped with blue; females orange with two dark brown spots at the base of the soft dorsal finrays and a third on the upper side of the caudal peduncle. To 35 cm.

Over rocks and hard ground, not on the shore or in very shallow water; mainly off southern and western coasts.

33 TRACHINIDAE

Benthic fishes with elongate, compressed bodies having dorsally placed eyes and an oblique mouth. The spinous dorsal fin is short, with a black membrane. Second dorsal and anal fins long; pelvics jugular. Operculum spined. Scales small, cycloid, arranged in distinctive oblique rows. No swimbladder. Lateral line high on the flank, downcurved on the caudal peduncle. Poison glands situated at the base of the fin spines and in the vicinity of the opercular spine. Wounds inflicted by the poison spine are painful and can be dangerous.

Echiichthys vipera (Cuvier). Lesser Weever FIG. 17.12
[WHITEHEAD et al. 1986*a*: 951]
Body rather short, compressed; mouth large, steeply sloping; eyes on top of the head. No preorbital spines. Margin of pectoral fin evenly rounded. Br. vi; 1D V–VI; 2D 21–24; A I/24–26; P 14; V I/5; lat. sc. 60; vert. 34–36. Yellowish brown above, paler below; first dorsal fin black. To 14 cm.

From about low water mark on sandy beaches and shallow water, living buried in sand with only the top of the head and the back uncovered; all coasts; common in suitable areas.

The much rarer, offshore Greater Weever *Trachinus draco* Linnaeus [Whitehead et al. 1986*a*: 953] is larger and has a less oblique mouth; a pair of small spines is present above and in front of each eye, and the pectoral fin is slightly emarginate.

34 BLENNIIDAE

Elongate fishes with a long, subdivided dorsal fin; pelvics jugular; caudal fin distinct. Skin scaleless. Each jaw with a single row of teeth, sometimes ending with a canine-like tooth. Swimbladder and pyloric caeca absent.

1.	Median cephalic tentacle and crest present	***Coryphoblennius galerita***
	Median cephalic tentacle and crest absent	**2**
2.	Orbital tentacles present	***Parablennius gattorugine***
	Orbital tentacles absent	***Lipophrys pholis***

Coryphoblennius galerita (Linnaeus). FIG. 17.14
Montagu's Blenny [WHITEHEAD et al. 1986*b*: 1098]
Body smooth, elongate, with D divided by a shallow notch into subequal parts. A crest along the top of the head commences as a tall, wide, fringed, erectile flap, and continues as a series of short filiform tentacles. No orbital tentacles. Canines present at the ends of the tooth row in the lower jaw only. Br. vi; D XII–XIII/15–18; A 17–18; P 12; V 2. Brownish with darker bands, and with bluish-white spots along the flank. To 8 cm.

Intertidal on rocky coasts, in mid-shore pools; very local, Cornwall, Pembrokeshire, and south-west Ireland north to Connemara.

Lipophrys pholis (Linnaeus) FIG. 17.14
[Syn. *Blennius pholis* L.]. Shanny
[WHITEHEAD et al. 1986*b*: 1104]
Body smooth, elongate, with D divided by a small notch into subequal parts. No orbital tentacles. Canines present at the ends of the tooth row in both jaws. Br. vi; D XI–XIII/18–20; A I/17–19; P 13; V 2. Dark brown and blotched, or blackish, with a dark spot behind the first dorsal finray. To 16 cm.

Intertidal; on any shore with rock and weed, or in pools; all coasts; common.

Parablennius gattorugine (Linnaeus). FIG. 17.14
Tompot Blenny [WHITEHEAD et al. 1986*b*: 1106]
Body smooth, elongate, with D divided by a dip into two parts of roughly equal height. Branched orbital tentacles present. No canine teeth. Br. vi; D XII–XIV/17–20; A I/20–21; P 14; V I/2–3; vert. 39. Brownish with dark bands. To 30 cm.

Mainly inshore, occasionally at extreme low water mark on rocky shores; south and west coasts.

The Butterfly Blenny *Blennius ocellaris* Linnaeus has the anterior part of the dorsal fin tall and bearing a white-edged black ocellus; it may be caught off south-western coasts.

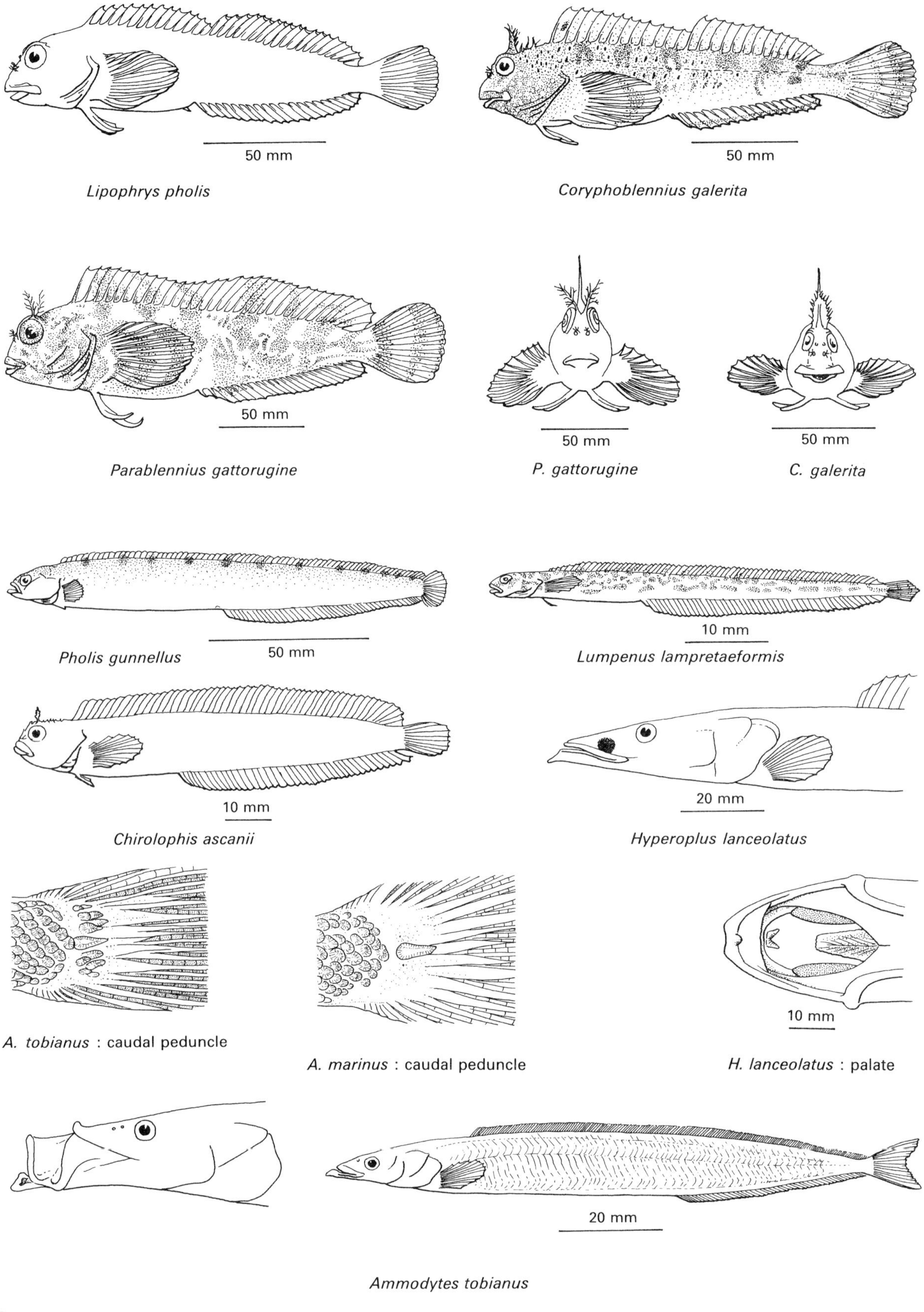
50 mm
Lipophrys pholis
50 mm
Coryphoblennius galerita
50 mm
Parablennius gattorugine
50 mm
P. gattorugine
50 mm
C. galerita
Pholis gunnellus
50 mm
10 mm
Lumpenus lampretaeformis
10 mm
Chirolophis ascanii
20 mm
Hyperoplus lanceolatus
A. tobianus : caudal peduncle
A. marinus : caudal peduncle
10 mm
H. lanceolatus : palate
20 mm
Ammodytes tobianus

Fig. 17.14

35 STICHAEIDAE

Elongate fishes with a long dorsal fin usually consisting of spiny rays only; pelvic fins jugular or absent. Scales small or absent. Swimbladder absent. Pyloric caeca present.

Chirolophis ascanii (Walbaum). FIG. 17.14
Yarrell's Blenny [WHITEHEAD et al. 1986*b*: 1122]

Body elongate, with minute scales on the body but not on the head. D long, of spiny rays only. Two pairs of supraorbital tentacles: the anterior pair short, the posterior pair longer and fringed. Branches or straight barbels on the first two or three dorsal finrays (best developed in males). Br. vi; D L–LIV; A I/35–40; P13–15; V I/3; vert. 55–57. Brownish, with dark transverse bands. To 25 cm.

Extreme low water mark on rocky shores and from shallow or moderately deep water over rough ground; Scotland, north-east England, Cornwall; Ireland; very local.

36 LUMPENIDAE

Elongate fishes with a long dorsal fin consisting of thin spiny rays only; pelvic fins under or slightly in advance of the pectorals, sometimes with a spiny ray. Scales small. Very similar to Stichaeidae and differing mainly in skeletal details.

Lumpenus lampretaeformis (Walbaum). FIG. 17.14
Snake Blenny [WHITEHEAD et al. 1986*b*: 1129]

Body very long and slender. D long, of spiny rays only; depressed anteriorly, with the first few rays not united by a membrane. D LXVIII–LXXVI; A I/47–53; P 14–15; V I/3; vert. 80–85. Brownish yellow, in places tinged blue; brown spots on the flank; usually eight to ten larger spots along the dorsal fin. To 40 cm.

Inshore and in shallow water, over muddy bottoms; off Scottish coasts.

37 PHOLIDAE

Elongate, laterally compressed fishes with a long dorsal fin consisting of short spiny rays only. Pelvic fins under the pectorals, reduced; or absent. No pyloric caeca. Very like Stichaeidae and differing mainly in skeletal details.

Pholis gunnellus (Linnaeus). Butterfish FIG. 17.14
[WHITEHEAD et al. 1986*b*: 1124]

Eel-like, compressed body; D of short, spiny rays only; pelvics small. Scales minute, embedded; skin slippery. Br. v; D LXXV–LXXXII; A II/39–45; P 10–12; V I/1; vert. 82–85. Yellowish brown, with 9–15 white-edged black spots along the base of the dorsal fin. To 25 cm.

Intertidal on rocky shores; also offshore; all coasts; common.

38 AMMODYTIDAE

Sand eels. Body elongate, covered with small cycloid scales; or scales absent. Head pointed, lower jaw projecting. Upper jaw bordered by premaxillary, extensible, and sometimes protrusible; teeth on jaws absent or reduced. Median fins without spiny rays; a single dorsal fin; caudal forked; pelvics usually absent. Lateral line high on the flank. Swimbladder absent. Pelagic or found buried in sand.

The offshore species *Gymnammodytes semisquamatus* (Jourdain), immediately distinguished by having the lateral line pores opening at the end of short transverse extensions from the main canal, is not further described.

---

1. Two 'teeth' on the vomer (in the roof of the mouth, Fig. 17.14). Dorsal fin commencing just behind the tip of the pectorals. Premaxillary not protractile **(*Hyperoplus*) 2**

   Vomer lacking 'teeth'. Dorsal fin commencing above pectorals. Premaxillary protractile **(*Ammodytes*) 3**

2. The upper jaw in adults bearing a dark spot; 52–61 finrays, 27–33 anal finrays ***H. lanceolatus***

   The upper jaw in adults uniformly dark; 59–62 dorsal and 31–34 anal finrays ***H. immaculatus***

3. Ventral body scales arranged in chevrons pointing towards the head (best seen when the fish is dry); dorsal and ventral muscle pads at the caudal finbase scaled; 50–56 dorsal finrays; typically inshore ***A. tobianus***

   Ventral body scales not in chevrons; dorsal and ventral muscle pads at the caudal finbase naked; 55–62 dorsal finrays; not inshore ***A. marinus***

Ammodytes marinus Raitt. FIG. 17.14
Raitt's Sand Eel [WHITEHEAD et al. 1986*a*: 946]

Eel-like body covered with small cycloid scales situated along numerous oblique dermal folds. Lateral line straight, unbranched, with pores along the lower side of the canal. Premaxilla protractile; anterior end of maxillae not fused, joined by a tendon. Vomer toothless. D commencing in front of the free end of the pectorals. Distinguished from *A. tobianus* by: back in advance of the dorsal fin partly scaled, the scales embedded in a band of skin on each side above the lateral line, leaving a naked median strip; belly between the ventro-lateral dermal folds covered by small, irregularly arranged scales, divided by a low mid-line fold; no scales, apart from the central one, on the muscle pads at the base of the caudal finlobes. Br. vii; D 55–62; A 26–35; P 11–13; vert. 66–72. Back greeny blue, belly and sides silvery. To 24 cm.

Offshore only; abundant except off the south-east.

Ammodytes tobianus Linnaeus. FIG. 17.14
Lesser Sand Eel [WHITEHEAD et al. 1986*a*: 947]

As preceding species except: back in advance of the dorsal fin completely covered with well-developed scales; scales on the belly arranged in chevrons, no dermal fold in the mid-line; dorsal and ventral muscle pads (which are separated by a single elongated scale) bearing scales. Br. vii; D 50–56; A 25–31; P 10–13; vert. 60–66. Back greenish, belly and sides silvery. To 20 cm.

Inshore and offshore; may be dug out of sandy beaches below MTL; all coasts; very common.

Hyperoplus immaculatus (Corbin). Corbin's Sand Eel [WHITEHEAD et al. 1986*a*: 949]

Eel-like body covered with minute cycloid scales not lying along oblique dermal folds. Lateral line straight, unbranched, with pores along the lower side of the canal. Premaxillae not protractile; anterior ends of maxillae broadened and united at the tip of the snout. Vomer bearing two strong tooth-like processes. Dorsal fin commencing behind the pectorals. D 59–62; A 31–34; vert. 70–74. Back greenish, belly and sides silvery; snout dark greenish-blue. To 30 cm.

Offshore in the west and south-west.

Hyperoplus lanceolatus (Lesauvage). FIG. 17.14
Greater Sand Eel [WHITEHEAD et al. 1986*a*: 950]

As preceding species except for meristic characters and snout colour. D 52–61; A 27–33; P 12–14; vert. 65–69. Back greenish, belly and sides silvery; a distinct dark spot on the side of the snout. To 32 cm.

Inshore and offshore; may sometimes be dug out of sandy beaches; off all coasts; common.

39 CALLIONYMIDAE

Dragonets. Body elongate, somewhat depressed. Two dorsal fins, the first spinous; pelvic fins in front of pectorals. Skin scaleless. Jaws toothed. Eyes situated dorsally, interorbital space very narrow. Angle of preoperculum bearing a variously armoured spine. No swimbladder. Sexually dimorphic.

1. Second dorsal fin (2D) with ten rays (the last one branched); the preopercular bone with three recurved and dorsally directed spines only. (2D of adult males marked with oblique rows of blackish spots) ***Callionymus reticulatus***

   2D with nine rays (the last one branched); the preopercular bone with three recurved and dorsally directed spines, plus one anteriorly directed spine **2**

2. Base of anal fin (A) at least 70% of the distance between its commencement and the tip of the snout; caudal fin at least 80% of the distance between the commencement of 1D and the tip of the snout. Two rows of dark spots laterally on the body, and four parallel rows of dark spots on 2D of the adult male. Lateral line displaced dorsally at the caudal peduncle (seen in preserved specimens) ***C. maculatus***

   Base of A not above 60% of the distance between its commencement and the tip of the snout; caudal fin less than 80% of the distance between the commencement of 1D and the tip of the snout. Body variously patterned but without dark spots laterally or on 2D. Lateral line not displaced dorsally at the caudal peduncle (seen in preserved specimens) ***C. lyra***

Callionymus lyra Linnaeus. FIG. 17.15
Common Dragonet [WHITEHEAD et al. 1986*b*: 1089]

Body elongate, head broad, somewhat depressed. Angle of the preoperculum tapering into a strong spine bearing three terminal processes, two directed upwards and one backwards; shortly in front of this a fourth, forwards-directed process. 1D short, the rays very long in males; triangular in females and immatures; rays of 2D long in males. Br. vi; 1D IV; 2D 9 (the last ray divided from its base); A 9; P 20; V I/5; vert. 21. Body of male brownish, with blue spots dorsally and a bluish band along the flank; the second dorsal fin yellow with bright blue longitudinal stripes. Body of female and immature brown, paler below, with a series of dark spots along the flank and, from 1D, three ill-defined brown bands across the back; posterior two-thirds of first dorsal fin bluish-black. Males to 30 cm, females to 20 cm.

In shallow and coastal waters; small specimens even in sandy shore pools; all coasts; common.

Callionymus maculatus Rafinesque. FIG. 17.15
Spotted Dragonet [WHITEHEAD et al. 1986*b*: 1089]

Body form as in preceding species. Br. vi; 1D IV; 2D 9, rarely 10 (the last ray divided from its base); A 9; C 9–10; P 16; V I/5. Body greyish-brown, paler below, with two rows of dark spots and some blue spots on the flank and, from 1D, four irregular, softly outlined brown patches across the back. Males with 2D light coloured, marked with four parallel rows of pearly-edged dark spots; females and immatures with indistinct spots on fin. Males to 40 cm, females to 11 cm.

In shallow and moderately deep water, less coastal and less common than *C. lyra*; apparently absent from the east coast.

Callionymus reticulatus Valenciennes. FIG 17.15
Reticulate Dragonet [CHANG 1951]

Body form as in other dragonets, but angle of the preoperculum lacks the fourth, forwards-directed process. 1D IV; 2D 10 (the last ray divided from its base); A 9–10; V I/5. Body pale orange-brown above, with bluish spots and, from 1D, four dark sharply outlined orange-red patches on the back. Males with oblique mottled blue bands across both dorsal fins, and lines of dark spots between them; anal fin with longitudinal blue and yellow lines; females with 1D dusky, 2D pale. Males to 11 cm; females to 8 cm.

Recorded in shallow water from extreme low water mark; evidently widespread around southern coasts of British Isles, reaching the west of Ireland, Isle of Man, and the southern North Sea, but overlooked or confused with *C. maculatus*. (Whitehead et al. 1986*b*, 1091 have transposed the figures of *C. recticulatus* and *C. risso*.)

40 GOBIIDAE

Small fishes, generally of benthic habit, with elongate bodies of rounded section; head large, blunt, eyes dorsolateral and close together. Two dorsal fins, the first short and with spiny rays.

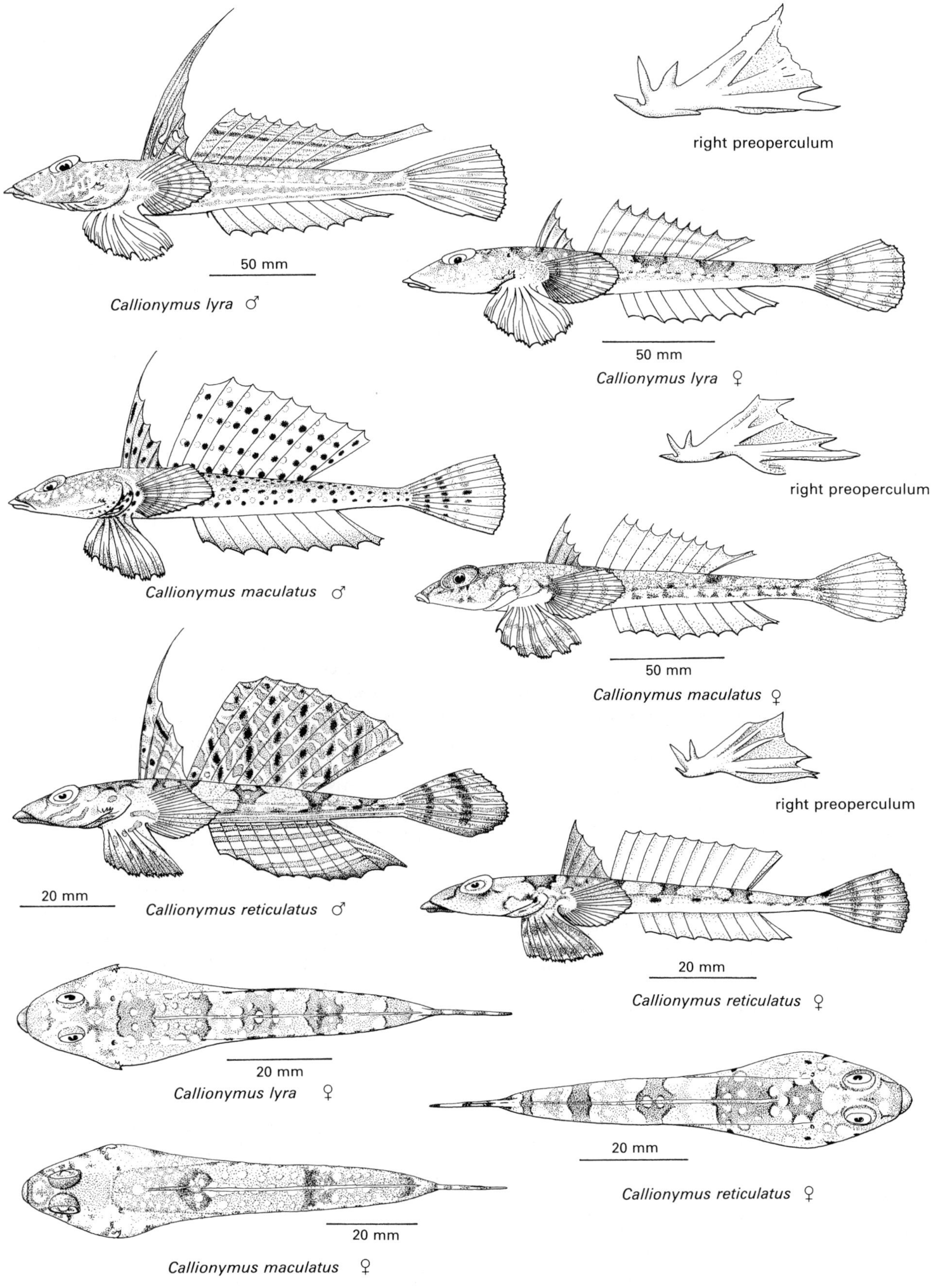
right preoperculum
50 mm
Callionymus lyra ♂
50 mm
Callionymus lyra ♀
right preoperculum
Callionymus maculatus ♂
50 mm
Callionymus maculatus ♀
right preoperculum
20 mm
Callionymus reticulatus ♂
20 mm
Callionymus reticulatus ♀
20 mm
Callionymus lyra ♀
20 mm
Callionymus reticulatus ♀
20 mm
Callionymus maculatus ♀

Fig. 17.15

Pelvic fins joined, forming a disk (Fig. 17.16); the fin membrane (except in *Lebetus*) also extending across the disk as an anterior fold; V I/5. Frequently sexually dimorphic or dichromatic.

Lines of minute sensory papillae ramify over the head of gobies, and their arrangement is important in taxonomy (Fig. 17.15). To reveal these papillae the fish should first be fixed in formalin (4 per cent formaldehyde in seawater) for 45 min, well washed and then transferred to chromic acid (2 per cent aqueous solution of chromium trioxide) for 24 h. Treated fish should be thoroughly washed and stored in 70 per cent ethanol.

The rarer British gobies are still poorly known, and are not described here; but one species, *Lesueurigobius friesii* (Malm), found on muddy grounds in deepish water, deserves mention because it lives in association with the crustacean *Nephrops norvegicus*; it is grey-brown with a row of large yellow spots on the flank, and has 24–29 scales in the lateral series.

Miller has provided a useful summary of European gobies in Lythgoe and Lythgoe's *Fishes of the sea* (1971).

---

1. Pelagic fishes with translucent, unpigmented bodies **Aphia minuta**

   Bottom-living fishes with opaque and generally pigmented bodies **2**

2. First dorsal fin (1D) with seven(to eight) spiny rays; a conspicuous black spot at the base of the tail **Gobiusculus flavescens**

   1D with six(to seven) spiny rays; base of tail lacking a conspicuous spot **3**

3. Gobies often exceeding 8 cm in length, with superior rays of pectoral fins separate and filamentous distally or with flanks prominently spotted with bright orange; four or more vertical rows of papillae on the side of the head below the orbit may be fairly conspicuous **4**

   Small gobies generally less than 8 cm in length, without filamentous rays to the pectoral fins and not prominently spotted with orange; suborbital papillae inconspicuous **(*Pomatoschistus*) 6**

4. Ground colour pale, bearing orange spots; hinder part of 1D black (34–38 scales in a lateral series) **Thorogobius ephippiatus**

   Colour pale or dark, but without orange spots **(*Gobius*) 5**

5. Fewer than 50 scales in a lateral series; usually a dark marking on the upper anterior corner of each dorsal fin **G. niger**

   Over 50 scales in a lateral series; 1D with a pale superior band **G. paganellus**

6. 55–70 scales in a lateral series, without a marked increase in size below 2D **P. minutus**

   Not more than 50 scales in a lateral series, showing a marked increase in size below 2D **7**

7. 35–40 scales in a lateral series; dorsal fins with one to two rows of black spots; three small double dark marks along the lateral mid-line below three pale dorsal saddles. **P. pictus**

   40–50 scales in a lateral series; dorsal fins without rows of black spots; often in brackish water **P. microps**

Aphia minuta (Risso). Transparent Goby FIG. 17.16
[WHITEHEAD et al. 1986*b*: 1023]
Body compressed, transparent, with vertebral column, swim-bladder, and the viscera visible in life. 1D quite high; pelvics united. Br. v; 1D V; 2D I/10–12; A I/11–13; P 16. To 6 cm.

Pelagic in shoals; inshore waters; all coasts; common.

A rarer transparent goby, *Crystallogobius linearis* (Von Düben), differs in having 1D small (2-rayed) or rudimentary, and 2D and A long and low.

Gobius niger Linnaeus. Black Goby FIG. 17.16
[WHITEHEAD et al. 1986*b*: 1042]
Thick-set goby; head about one-quarter of the total length; caudal peduncle about half of the head length. 1D with central rays the longest, especially in the male; fin membrane of 1D confluent with the start of 2D; upper two to three finrays of pectoral free at their tips (for one-quarter of their length at most); margin of pelvic disc rounded, the anterior membrane without lateral lobes. Rim of anterior nostril with a simple flap. 1D V–VII; 2D I/11–13; A I/10–13; P 15–20; lat. sc. 32–42; vert. 27–28. Dark grey or brown with lighter markings; each dorsal fin with a small black mark on the anterior upper corner. To 17 cm.

In shallow coastal or estuarine waters, rarely on the shore, amongst *Zostera* or seaweed on soft substrata; all coasts; frequent or common.

Two other rather similar gobies are present in south-western waters. *G. couchi* Miller and El-Tawil (1974) and *G. gasteveni* Miller (1974) lack the elongated rays in 1D and the dark anterior spots on both dorsals. *G. couchi*, 1D V–VII; 2D I/12–14; A I/11–13; P 15–18; lat. sc. 35–45, is grey-brown with several lateral blotches, a higher-than-wide spot on the pectorals, and three gular spots under the head. *G. gasteveni*, 1D V–VI; 2D I/13–15; A I/12–14; P 18–22; lat. sc. 37–45, has similar though fewer lateral markings, a wider-than-high blotch on the pectorals, and no gular spots.

Gobius paganellus Linnaeus. Rock Goby FIG. 17.16
[WHITEHEAD et al. 1986*b*: 1043]
Thick-set goby; head one-quarter to one-fifth of total length; caudal peduncle about two-thirds of the head length. 1D with the anterior ray longest; fin membrane of 1D extending to base of 2D; upper four to five finrays of pectoral free for much of their length. Margin of pelvic disk rounded, the anterior membrane with feebly developed lobes (Fig. 17.16). Anterior nostril with a digitate tentacle. Br. v; 1D VI; 2D I/11–13; A I/10–12; P 20; lat. sc. 50–58; vert. 28–29. Uniform dark brown

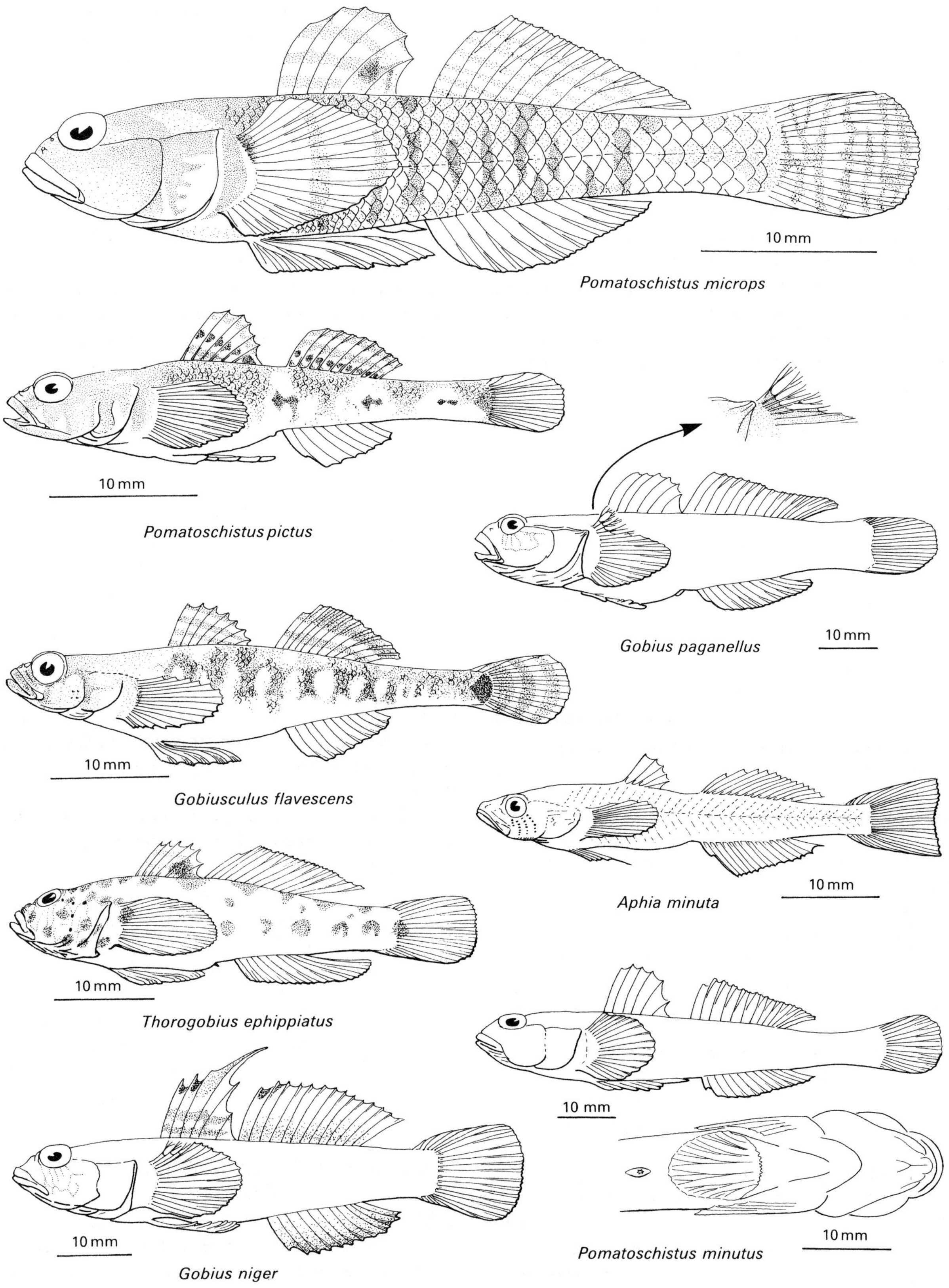
10 mm
Pomatoschistus microps
10 mm
Pomatoschistus pictus
Gobius paganellus
10 mm
10 mm
Gobiusculus flavescens
Aphia minuta
10 mm
10 mm
Thorogobius ephippiatus
10 mm
10 mm
Pomatoschistus minutus
10 mm
Gobius niger

Fig. 17.16

or nearly black; first dorsal fin with a pale yellow or orange marginal band. To 12 cm.

Intertidal on rocky shores, particularly in tide pools, and on hard grounds in shallow water; common on south-west coasts, absent from the east.

In Cornwall the Giant Goby *G. cobitis* Pallas, with lat. sc. 60–68, the membrane of the pelvic disk lobed, and the first dorsal fin uniformly coloured, occurs in high-level shore pools (Wheeler 1960). *Gobius cruentatus* Gmelin, with the lips and cheeks streaked with red, the rows of sensory papillae black, and lat. sc. 52–58, has been reported from Southern Ireland (Wheeler 1970; Miller, in Lythgoe and Lythgoe 1971).

Gobiusculus flavescens (Fabricius) [Syns. *Chapparudo; Gobius ruthensparri*]. Two-spot Goby FIG. 17.16
[WHITEHEAD et al. 1986*b*: 1046]
Slender body; interorbital space wide, at least half the eye diameter. Head about one-fifth total length; caudal peduncle long, about equal to the head length. 1D usually with one ray more than in other gobies. Margin of pelvic anterior membrane entire. 1D VII (rarely VI or VIII); 2D I/9–11; A I/10–11; P 19; lat. sc. 34–40; vert. 31. Reddish brown, darker above with a series of lighter bands across the back; a distinct and conspicuous black spot at the base of the caudal fin, and a smaller dark spot just behind the pectoral origin; dorsal fins banded with red and off-white, and caudal fin transversely banded.

Inshore and in intertidal pools, amongst weed; does not rest on the bottom as much as other gobies; all coasts; common in the south and west.

Pomatoschistus microps (Krøyer). FIG. 17.16
Common Goby [WHITEHEAD et al. 1986*b*: 1073]
Small goby; head one-quarter to one-fifth of the total length; caudal peduncle about equal to the head length. 1D and 2D separated. Margin of pelvic anterior membrane entire (Fig. 17.16). Size of the scales increasing quite markedly at about the commencement of 2D. Top of head, nape, and throat scaleless. 1D V–VI; 2D I/8–10; A I/8–10; P 18; lat. sc. 43–50. Grey or sandy coloured, often with darker spots or bars; dorsal fins with spots or bands, males with a black spot on the posterior edge of 1D; upper part of pectoral fin base dark. To 7 cm.

In tide pools, estuaries, salt-marshes, and brackish water; all coasts; abundant.

Pomatoschistus minutus (Pallas) *sensu lato*. FIG. 17.16
Sand Goby [WHITEHEAD et al. 1986*b*: 1074]
A small elongate goby; head about one-quarter of the total length; caudal peduncle almost equal to the head length. 1D and 2D slightly separated. Margin of pelvic anterior membrane fringed with villi (Fig. 17.16). Scales small and numerous, reaching forward on to the nape and on to the throat in front of the pelvic disk. Cleft between branchiostegal membranes reaching forward one-quarter to one-third of head length. 1D VI-VII; 2D I/9–12; A I/9–12; P 18–21; lat. sc. 55–75; vert. 30–34. Sandy-brown above, with darker speckles. First dorsal fin often with a dark spot on its posterior margin. To 11 cm.

Inshore on soft bottoms and near MLW on sandy shores; all coasts; common.

P. minutus s. lato comprises two similar but biologically recognizable species (Webb 1980). *P. minutus s. stricto* tends to inhabit estuarine waters. It has a mean vertebral count of 33 and reaches 11 cm. There is always a blue-black spot, which does not reach the edge of the membrane, on 1D. *P. lozanoi* (de Buen) is coastal, has a mean vertebral count of 32 and reaches only 8 cm. The dark spot on 1D is present on males but only in spring; it reaches the edge of the membrane. The two species can best be discriminated by the arrangement of sensory papillae on the head (Webb 1980; Whitehead *et al.* 1986*b*).

Further offshore, usually in depths exceeding 35 m, both species are replaced by *P. norvegicus* (Collett), in which the nape and throat are scaleless; the spot on 1D is present in maturing fish of both sexes and reaches to the edge of the membrane; and the cleft between the branchiostegal membranes reaches forward for about one-half of the head length. 2D I/8–10; A I/8–10; P 16–18; lat. sc. 55–60; vert. 32 (Webb and Miller 1975; Webb 1980; Whitehead et al. 1986*b*: 1075]). To 6.5 cm.

Pomatoschistus pictus (Malm). Painted Goby FIG. 17.16
[WHITEHEAD et al. 1986*b*: 1075]
Small goby; head one-quarter to one-fifth of the total length; caudal peduncle about three-quarters of head length. Dorsal fins slightly separated. Margin of pelvic anterior membrane entire. The size of the scales increases under 2D. No scales on top of the head or on the throat. 1D VI; 2D I/8–9; A I/8–9; P 18; lat. sc. 35–41; vert. 28. Brown; a line of four to five diffuse spots along the flank; four light bands across the back; dorsal fin with one to two rows of small black spots along the finbase, one (or two) between each ray, and longitudinal reddish bands (especially developed in breeding males). To 5.5 cm.

Inshore on rough, shelly, or gravelly grounds, occasionally found intertidally; south and west coasts; frequent.

Thorogobius ephippiatus (Lowe) [Syn. *Gobius forsteri* Corbin]. Leopard-spotted Goby FIG. 17.16
[WHITEHEAD et al. 1986*b*; 1082]
Head about one-fifth of total length; caudal peduncle about four-fifths of head length. Scales on body not extending forward on to the top of the head. 1D and 2D separated by a small gap; upper finrays of pectoral not separate. 1D V-VII; 2D I/10–12; A I/10; P 17–20; lat. sc. 33–42; vert. 28. Pale brown with orange spots and blotches on the head, flanks, and dorsal fins; posterior part of first dorsal fin black. To 14 cm.

Inshore, on rocky grounds; shelters in crevices but seen frequently by skin-divers; western coasts.

41 SCOMBRIDAE

Mackerels and tunnies. Body fusiform, rounded in section, tapering to a thin caudal peduncle; caudal fin strongly forked. First dorsal fin short, with spiny rays, widely separated from the second dorsal; both the second dorsal and the anal succeeded by a series of finlets extending on to the caudal peduncle. Pectoral fins high on the flank; pelvics below pectorals. Upper jaw not protractile. Scales reduced. Swimbladder absent (in mackerels) or present.

Scomber scombrus Linnaeus. Mackerel FIG. 17.17
[WHITEHEAD et al. 1986*a*: 991]
Br. vii; 1D XI-XIII; 2D 11–13+5(–6); A I/11–13+5; P 17; A I/5; vert. 30–31. Deep greeny-blue above, crossed by narrow, black curving bands; silvery on the flanks, white below. To 50 cm.

Pelagic, forming shoals; off all coasts, except in the south-east, especially during the summer months.

Order **Pleuronectiformes**

42 BOTHIDAE

Flatfish with eyes on the left side of the head. In addition to the species described below, immature Topknot *Zeugopterus punctatus* (Bloch) may occasionally be found near MLW of rocky shores and on hard inshore grounds. The Topknot (Fig. 17.17) is immediately distinguished by its almost oblong shape and large black 'eyebrows'.

---

1. Fish oval in outline (width about half of the length) **2**
   Fish rounded in outline (width about three-quarters of the length) **3**

2. Mouth small (half the length of the pectoral fin); pelvic fin on the blind side shorter-based than that on the eyed side. (Scales very readily detached) ***Arnoglossus laterna***
   Mouth large (as long as the pectoral fin); pelvic fin bases equal ***Lepidorhombus whiffiagonis***

3. Skin smooth; distal half of first dorsal finrays divided into filaments ***Scophthalmus rhombus***
   Skin on the eyed side bearing tubercles (except in small fish); only the tips of the first dorsal finrays divided ***Psetta maxima***

Arnoglossus laterna (Walbaum). Scaldfish
[WHITEHEAD et al. 1986*b*: 1296]
Elongate-oval outline; head about two-ninths of total length; mouth fairly small, the maxilla scarcely passing the front of the left eye. Finbase of pelvic shorter on the blind side. First dorsal finrays free for part of their length. Scales large but caducous. Lateral line arched above the pectoral fin. Br. vii; D 87–93; A 65–74; P 10–11; V 6; lat. sc. 48–56; vert. 37–42. Brownish above, white below. To 19 cm.

Inshore waters on sandy grounds around southern Britain; quite common; rarer northwards.

Lepidorhombus whiffiagonis (Walbaum). FIG. 17.17
Megrim [WHITEHEAD et al. 1986*b*: 1288]
Elongate-oval outline; head long, about one-quarter of total length; mouth large, maxilla reaching to the middle of the left eye. Pelvic finbases equal. Scales small. Lateral line strongly arched above the pectoral fin. Br. vii; D 85–94; A 64–74; P 13; V 6; lat. sc. 95–110; vert. 40–42. Brownish above, with darker spots; white below. To 61 cm.

In deep water off western coasts; rather rarely inshore.

Psetta maxima (Linnaeus) [Syn. *Scophthalmus maximus*]. Turbot FIG. 17.17 [WHITEHEAD et al. 1986*b*: 1290]
Rhomboid outline, width about two-thirds of length; head large, two-sevenths of total length. Dorsal and anal fins not reaching the caudal; tips of the first dorsal finrays free and branched. No scales but prominent tubercles scattered irregularly over the body on the eyed side. Lateral line strongly arched above the pectoral fin. Br. vii; D 57–71; A 43–52; P 11–12; V 6; vert. 30–31. Colour variable, often greyish-brown with darker speckles; white below. To 80 cm.

Mainly inshore on sandy or gravelly grounds; off all coasts but uncommon around Scotland.

Scophthalmus rhombus (Linnaeus). Brill FIG. 17.17
[WHITEHEAD et al. 1986*b*: 1291]
Oval outline, width about three-fifths of length; head one-quarter to two-ninths of total length. Dorsal fin reaching caudal, but not extending underneath it. First dorsal finrays branched and free for about half their length. Scales present; tubercles absent. Lateral line strongly arched above the pectoral fin. Br. vii; D 73–83; A 56–62; P 11–12; A 6; lat. sc. 150; vert. 34–37. Greyish brown with darker speckles, white below. To 61 cm.

Mainly inshore on sandy grounds; off all coasts but uncommon around Scotland.

43 PLEURONECTIDAE

Flatfish with eyes on the right side of the head (reversals may occasionally be found); snout pointed and the mouth terminal.

---

1. Mouth large, reaching or passing the middle of the eye ***Hippoglossoides platessoides***
   Mouth small, extending only to the front of the eye **2**

2. Eyed surface covered with ctenoid scales, rough to the touch ***Limanda limanda***
   Eyed surface smooth to the touch (except where prickles are present) **3**

3. Head small (about one-sixth of the total length); body outline oval; dorsal and anal fins almost reaching the caudal fin **4**
   Head large (about one-quarter of the total length); body outline rhombic; dorsal and anal fins distant from the caudal fin **5**

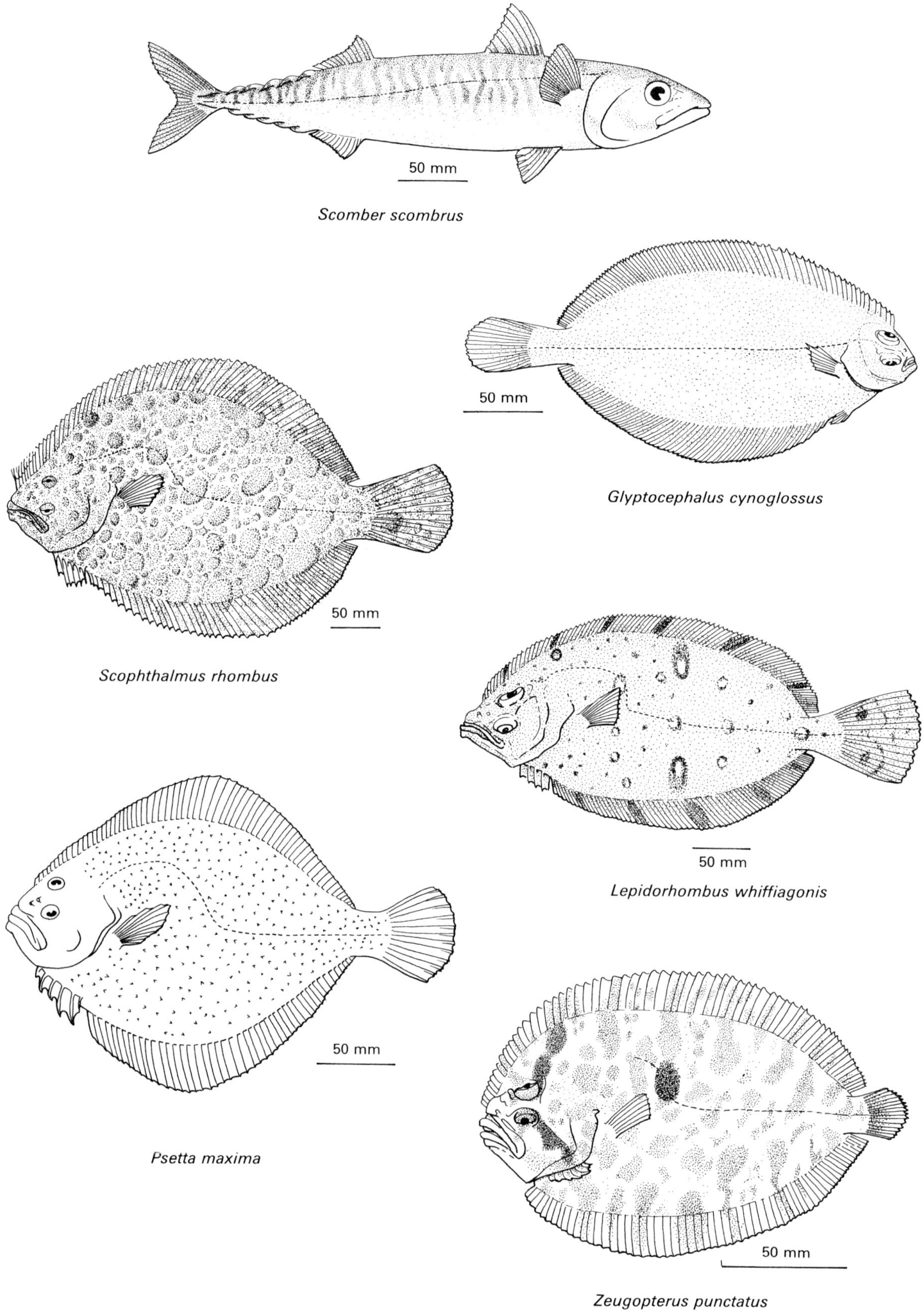
50 mm
Scomber scombrus
50 mm
Glyptocephalus cynoglossus
50 mm
Scophthalmus rhombus
50 mm
Lepidorhombus whiffiagonis
50 mm
Psetta maxima
50 mm
Zeugopterus punctatus

Fig. 17.17

4. Lateral line straight; body slender, width less than half of length **_Glyptocephalus cynoglossus_**

   Lateral line curved over pectoral fin; body oval, width more than half of length **_Microstomus kitt_**

5. Short spines present along the base of the dorsal and anal fins, and in the region of the lateral line; eyed side typically very dark, blind side opaque white **_Platichthys flesus_**

   No spines along the base of the dorsal and anal fins or in the region of the lateral line; eyed side typically brown with orange spots, blind side pearly white **_Pleuronectes platessa_**

Hippoglossoides platessoides (Fabricius). FIG. 17.18
Long Rough Dab [WHITEHEAD et al. 1986*b*: 1300]
Outline slender-oval; head two-ninths to one-fifth of the total length. Mouth large, the maxilla reaching the middle of the right eye. Skin on the eyed side rough from the ctenoid scales. Lateral line slightly curved above pectoral fin. Caudal fin obtusely wedge-shaped. Br. viii; D 78–98; A 60–79; P 10; V 6; lat. sc. 85–95; vert. 42–46. Brown above, white below. To 30 cm.

Offshore on fine grounds; common around Scotland, absent from south and south-east England.

Small Halibut *Hippoglossus hippoglossus* (Linnaeus), which might be found in the same geographical area, have the lateral line strongly arched above the pectoral fin and the caudal fin margin gently concave.

Limanda limanda (Linnaeus). Dab FIG. 17.18
[WHITEHEAD et al. 1986*b*: 1302]
Outline oval; head two-ninths to one-fifth of the total length. Mouth small, the maxilla just reaching the right eye. Skin on the eyed side rough from the ctenoid scales. Lateral line strongly arched above the pectoral fin. Br. viii; D 65–81; A 50–64; P 10–11; V 6; lat. sc. 73–90; vert. 40–41. Sandy brown above, white below. To 25 cm.

Inshore and in shallow offshore waters, on sandy grounds; off all coasts; abundant.

Microstomus kitt (Walbaum). Lemon Sole FIG. 17.18
[WHITEHEAD et al. 1986*b*: 1304]
Outline oval, the greatest width about one-half of the length; head small, less than one-sixth of the total length. Mouth small, the maxilla extending to the front of the right eye. Right eye in advance of the left. Scales cycloid, the skin smooth and slimy. Lateral line with a shallow arch above the pectoral fin. Caudal peduncle very short. Br. vii; D 85–97; A 69–76; P 10; V 5–6; lat. sc. 130; vert. 48. Light brown with darker and lighter irregular blotches. To 66 cm.

Offshore on sandy or shelly grounds; off all coasts but commoner in the north.

The Witch *Glyptocephalus cynoglossus* (Linnaeus) (Fig. 17.17), from deepish, northern waters, has a slender oval outline, the greatest width only one-third of the length, and a straight lateral line.

Platichthys flesus (Linnaeus). Flounder, Fluke FIG. 17.18
[WHITEHEAD et al. 1986*b*: 1304]
Outline oval; head large, one-quarter of the total length. Mouth small; maxilla not extending beneath the right eye. Two rows of conical, blunt teeth in the upper jaw, one on the lower. Scales small, cycloid; the skin smooth except for a band of prickles along the lateral line and a row of tubercles along the base of the dorsal and anal fins. Lateral line slightly curved above the pectoral fin. Anal fin preceded by a spine. Caudal peduncle long, half the length of the fin. Br. v; D 52–67; A 35–46; P 10; V 6; vert. 36. Colour variable above, but often dark greeny-brown, occasionally with pale orange blotches; dense, opaque white below. To 50 cm.

Inshore, particularly in estuaries; all coasts; common.

Pleuronectes platessa Linnaeus. Plaice FIG. 17.18
[WHITEHEAD et al. 1986*b*: 1305]
Outline oval; head slightly less than one-quarter of the total length. Mouth fairly small, the maxilla reaching to just below the right eye. Teeth compressed, similar on the two jaws. Scales cycloid; skin smooth, without prickles. Lateral line slightly curved above the pectoral fin; a line of tubercles extending between the lateral line and the interorbital ridge. Anal fin preceded by a spine. Caudal peduncle about one-third the length of the fin. Br. vii; D 65–79; A 48–59; P 10–11; V 6; vert. 42–43. Dark brown above, brightly blotched with orange; pearly (not opaque) white below. To 50 cm.

Shelf waters on sandy grounds, the young fish in particular coming right inshore; off all coasts; common.

44 SOLEIDAE

Flatfishes with eyes on the right side; head margin rounded, the mouth not terminal. Margin of preoperculum covered by skin. Scales small, ctenoid; the skin rough.

---

1. Pectoral fin on the eyed side unspotted, that on the blind side vestigial; every 5th or 6th dorsal and anal finray black. (A small fish not exceeding 13 cm in length) **_Buglossidium luteum_**

   Pectoral fin on the eyed side with a black spot; the blind side fin hardly smaller than that on the eyed side; dorsal and anal finrays uniformly coloured **2**

2. Anterior nostril on the blind side tubular, but not enlarged; black spot extending to the free end of the pectoral fin **_Solea solea_**

   Anterior nostril on the blind side enlarged, rosette-like; black spot not extending to the free end of the pectoral fin. (South and west) **_Solea lascaris_**

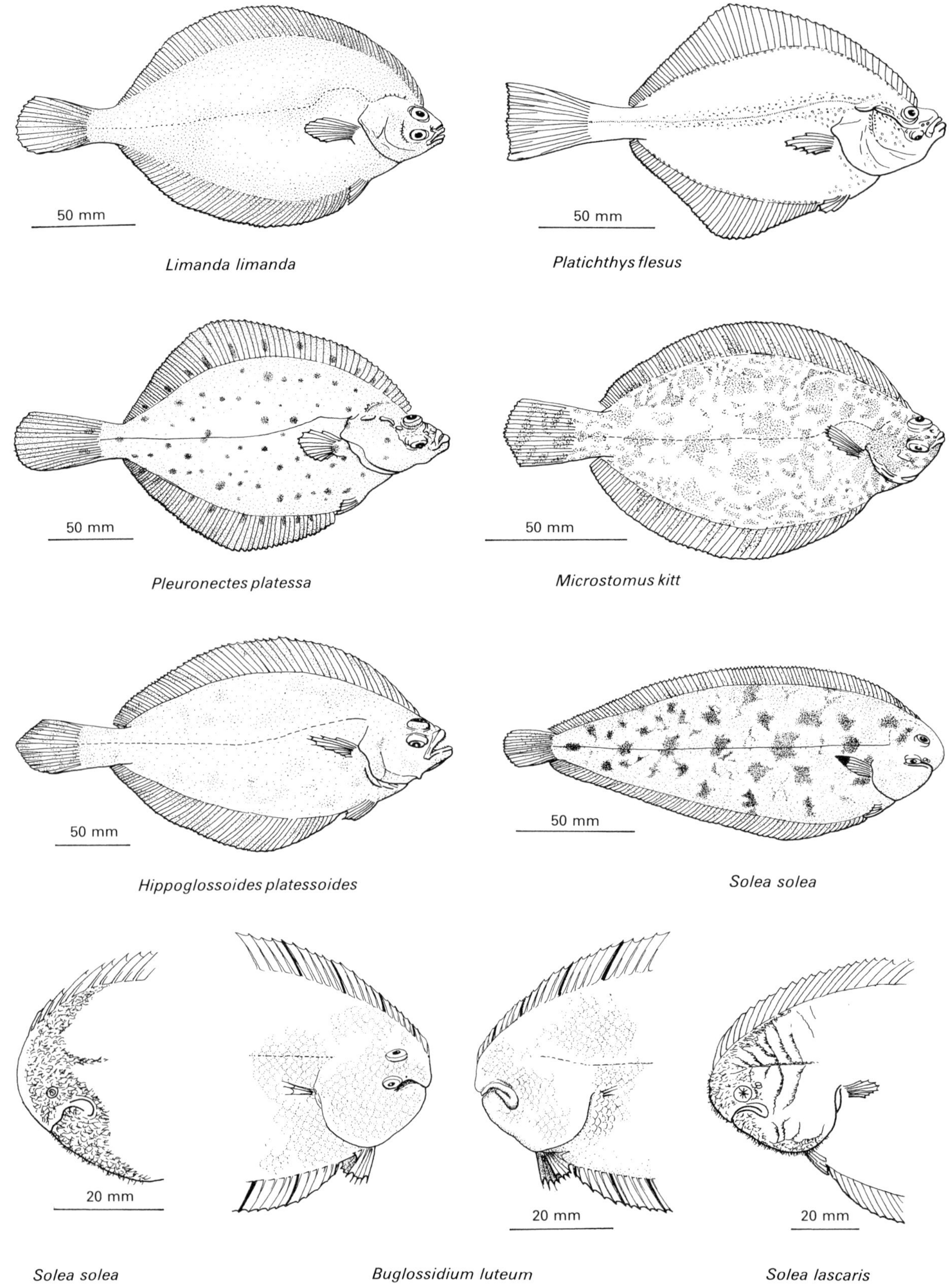
50 mm
Limanda limanda
50 mm
Platichthys flesus
50 mm
Pleuronectes platessa
50 mm
Microstomus kitt
50 mm
Hippoglossoides platessoides
50 mm
Solea solea
20 mm
20 mm
20 mm
Solea solea
Buglossidium luteum
Solea lascaris

Fig. 17.18

Buglossidium luteum (Risso) FIG. 17.18
[Syn. *Microchirus boscanion* of authors]. Solenette
[WHITEHEAD et al. 1986*b*: 1311]

Small sole of elongate-oval outline. Pectoral fins small, unequal; that on the eyed side shorter than the pelvics, its uppermost ray much longer than the others; blind side pectoral vestigial. Dorsal and anal fins terminating just short of the commencement of the caudal. Br. viii; D 71–78; A 55–56; RP 5, LP 3; V 5; lat. sc. 70–75; vert. 36. Light sandy brown, speckled darker; pectoral fin with a dark central spot; dorsal and anal fins pale with every fifth or sixth (occasionally fourth or seventh) ray black. To 13 cm.

On sandy grounds in shallow water, offshore and inshore; off all coasts; common.

Solea lascaris (Risso) [Syn. *Pegusa lascaris*]. FIG. 17.18
Sand Sole [WHITEHEAD et al. 1986*b*: 1320]

Outline elongate-oval. Pectoral fins approximately equal, much larger than pelvics. Dorsal and anal fin membranes continuous with the caudal. Anterior blind side nostril dilated, encircled by short branched filaments. Br. vii; D 71–90; A 55–75; P 10; V 5; lat. sc. 96–140; vert 46-67. Light brown above, speckled and spotted with dark brown; a black spot on the right pectoral does not reach the fin margin. To 35 cm.

On sandy grounds in coastal waters, the young coming right inshore; south-west Britain only; quite common.

Solea solea (Linnaeus) [Syn. *S. vulgaris* FIG. 17.8
Quensel]. Sole [WHITEHEAD et al. 1986*b*: 1322]

Outline elongate-oval. Pectoral fins approximately equal, longer than the pelvics. Dorsal and anal fin membranes continuous with the caudal. Anterior blind side nostril small and tubular, not fringed. Br. vii; D 75–93; A 59–79; P 7; V 5–6; lat. sc. 140–165; vert. 49–52. Brown above, with darker blotches, a black spot on the right pectoral extends to the fin margin. To 50 cm.

On sandy and fine grounds in coastal waters, including estuaries. Off all coasts, but rare towards the north of Scotland.

INDEX OF TECHNICAL TERMS

D

E

F

G

N

O

P

TAXONOMIC INDEX

Correct names of species and genera are given in *italic*.
Synonyms and the names of families and higher taxa are given in roman.

B

D

F

G

M

T

W

X

Y

Z